Progress in Atomic Spectroscopy

Part C

PHYSICS OF ATOMS AND MOLECULES

ELECTRON AND PHOTON INTERACTIONS WITH ATOMS
Edited by H. Kleinpoppen and M. R. C. McDowell

PROGRESS IN ATOMIC SPECTROSCOPY, Parts A and B
Edited by W. Hanle and H. Kleinpoppen

ATOM – MOLECULE COLLISION THEORY: A Guide for the Experimentalist
Edited by Richard B. Bernstein

COHERENCE AND CORRELATION IN ATOMIC COLLISIONS
Edited by H. Kleinpoppen and J. F. Williams

VARIATIONAL METHODS IN ELECTRON – ATOM SCATTERING THEORY
R. K. Nesbet

DENSITY MATRIX THEORY AND APPLICATIONS
Karl Blum

INNER-SHELL AND X-RAY PHYSICS OF ATOMS AND SOLIDS
Edited by Derek J. Fabian, Hans Kleinpoppen, and Lewis M. Watson

INTRODUCTION TO THE THEORY OF LASER – ATOM INTERACTIONS
Marvin H. Mittleman

ATOMS IN ASTROPHYSICS
Edited by P. G. Burke, W. B. Eissner, D. G. Hummer, and I. C. Percival

ELECTRON – ATOM AND ELECTRON – MOLECULE COLLISIONS
Edited by Juergen Hinze

PROGRESS IN ATOMIC SPECTROSCOPY, Part C
Edited by H. J. Beyer and Hans Kleinpoppen

A Continuation Order Plan is available for this series. A continuation order will bring delivery of each new volume immediately upon publication. Volumes are billed only upon actual shipment. For further information please contact the publisher.

Progress in Atomic Spectroscopy

Part C

Edited by

H. J. Beyer

and

Hans Kleinpoppen

University of Stirling
Stirling, United Kingdom

Plenum Press · New York and London

PHYSICS

Library of Congress Cataloging in Publication Data

Main entry under title:

Progress in atomic spectroscopy.

(Physics of atoms and molecules.)
Part C- edited by H. J. Beyer and Hans Kleinpoppen.
Includes bibliographical references and indexes.
1. Atomic spectra. I. Hanle, Wilhelm, 1901–   . II. Kleinpoppen, Hans. III.
Beyer, H. J.
QC454.A8P76 1983   539.7   78-18230

ISBN 0-306-41300-0

© 1984 Plenum Press, New York
A Division of Plenum Publishing Corporation
233 Spring Street, New York, N.Y. 10013

Printed in the United States of America

To Wilhelm Hanle

Professor Wilhelm Hanle

Contents of Part C

Chapter 1

The *k* Ordering of Atomic Structure
R. M. Sternheimer

Chapter 2

Multiconfiguration Hartree–Fock Calculations for Complex Atoms
Charlotte Froese Fischer

Chapter 3

New Methods in High-Resolution Laser Spectroscopy

B. Couillaud and A. Ducasse

Chapter 4

Resonance Ionization Spectroscopy: Inert Atom Detection

C. H. Chen, G. S. Hurst, and M. G. Payne

Chapter 5

Trapped Ion Spectroscopy
Günther Werth

Chapter 6

High-Magnetic-Field Atomic Physics
J. C. Gay

Chapter 7

Effects of Magnetic and Electric Fields on Highly Excited Atoms
Charles W. Clark, K. T. Lu, and Anthony F. Starace

Chapter 11
Impact Ionization by Fast Projectiles
Rainer Hippler

Chapter 12
Amplitudes and State Parameters from Ion– and Atom–Atom Excitation Processes
T. Andersen and E. Horsdal-Pedersen

Contents of Part A

Chapter 3
Perturbation of Atoms
Stig Stenholm

Chapter 4
Quantum Electrodynamical Effects in Atomic Spectra
A. M. Ermolaev

Chapter 5

Inner Shells

B. Fricke

Chapter 6

Interatomic Potentials for Collisions of Excited Atoms

W. E. Baylis

II. METHODS AND APPLICATIONS OF ATOMIC SPECTROSCOPY

Chapter 7

New Developments of Classical Optical Spectroscopy
Klaus Heilig and Andreas Steudel

Chapter 8

Excitation of Atoms by Impact Processes
H. Kleinpoppen and A. Scharmann

Chapter 9

Perturbed Fluorescence Spectroscopy
W. Happer and R. Gupta

Chapter 10

Recent Developments and Results of the Atomic-Beam Magnetic-Resonance Method
Siegfried Penselin

Chapter 11

The Microwave–Optical Resonance Method
William H. Wing and Keith B. MacAdam

Chapter 12

Lamb-Shift and Fine-Structure Measurements on One-Electron Systems
H.-J. Beyer

Chapter 13

Anticrossing Spectroscopy
H.-J. Beyer and H. Kleinpoppen

Chapter 14

Time-Resolved Fluorescence Spectroscopy
J. N. Dodd and G. W. Series

Chapter 15
Laser High-Resolution Spectroscopy
W. Demtröder

Contents of Part B

Chapter 16

The Spectroscopy of Highly Excited Atoms
Daniel Kleppner

Chapter 17

Optical Spectroscopy of Short-Lived Isotopes
H.-Jürgen Kluge

Chapter 18

Spin and Coherence Transfer in Penning Ionization

L. D. Schearer and W. F. Parks

Chapter 19

Multiphoton Spectroscopy

Peter Bräunlich

Chapter 20

Fast-Beam (Beam-Foil) Spectroscopy

H. J. Andrä

Chapter 21

Stark Effect

K. J. Kollath and M. C. Standage

Chapter 22

Stored Ion Spectroscopy

Hans A. Schuessler

Chapter 23

The Spectroscopy of Atomic Compound States
J. F. Williams

Chapter 24

Optical Oscillator Strengths by Electron Impact Spectroscopy
W. R. Newell

Chapter 25

Atomic Transition Probabilities and Lifetimes
W. L. Wiese

Chapter 26
Lifetime Measurement by Temporal Transients
Richard G. Fowler

Chapter 27
Line Shapes
W. Behmenburg

Chapter 28

Collisional Depolarization in the Excited State

W. E. Baylis

Chapter 29

Energy and Polarization Transfer

M. Elbel

Chapter 30

X-Ray Spectroscopy

K.-H. Schartner

Chapter 31

Exotic Atoms

G. Backenstoss

Introduction

H. J. BEYER AND H. KLEINPOPPEN

During the preparation of Parts A and B of *Progress in Atomic Spectroscopy* a few years ago, it soon became obvious that a comprehensive review and description of this field of modern atomic physics could not be achieved within the limitations of a two-volume book. While it was possible to include a large variety of spectroscopic methods, inevitably some fields had to be cut short or left out altogether. Other fields have developed so rapidly that they demand full cover in an additional volume.

One of the major problems, already encountered during the preparation of the first volumes, was to keep track of new developments and approaches which result in spectroscopic data. We have to look far beyond the area of traditional atomic spectroscopy since methods of atomic and ion collision physics, nuclear physics, and even particle physics all make important contributions to our knowledge of the static and dynamical state of atoms and ions, and thereby greatly add to the continuing fascination of a field of research which has given us so much fundamental knowledge since the middle of the last century.

In this volume, we have tried to strike a balance between contributions belonging to the more established fields of atomic structure and spectroscopy and those fields where atomic spectroscopy overlaps with other areas.

Present day atomic structure theory is successfully dealing with more and more complex atoms and making considerable progress in improving and extending approximation methods, as is borne out by the contribution on multiconfiguration calculations. On the other hand, there is still scope for surprisingly basic problems like that of ordering atomic levels in simple terms. The method of k-ordering of atomic structure is remarkably successful and reminds us of the classical period of unraveling spectroscopic data in terms of spectral line series.

Efforts are continually being made to improve established methods and to develop new ones with the aim of advancing the accuracy of atomic structure measurements and of extending the range of application to more atomic systems. Recently this was to some extent linked to the steady development of the laser both in terms of output characteristics and of spectral range. It was, therefore, felt that the latest developments in the field of lasers and laser spectroscopy should be reviewed again. Lasers are proving to be particularly powerful tools when used in conjunction with other techniques. In this way resonant laser absorption combined with ion detection techniques allows single atom detection and counting. Similarly, the already high resolution obtained in trapped ion spectroscopy can be increased even further by laser absorption cooling which results in a relative accuracy of the order of 10^{-15} and may ultimately lead to the time standard of the future.

Atoms in high fields have always represented a fascinating area of research. Landau spectroscopy, atomic diamagnetism, neutral vacuum decay in ultrastrong fields are not only interesting from the point of view of fundamental knowledge but have far-reaching consequences in astrophysics. Until recently experimental studies were severely restricted by the available magnetic field strength, but progress in the development of superconducting magnets and in the spectroscopy of highly excited states is rapidly closing this gap.

The introduction of heavy ion accelerators and the resulting possibility of studying fast heavy ion collisions is leading to completely new types of spectroscopy. Such collisions may cause the transient formation of super-heavy quasi atoms or quasi molecules having an effective nuclear charge given by the sum of the charges of the colliding partners. As a result of the short time scale of these collisions the systems can be studied in ultra-high electromagnetic fields. Tools for such investigations are x-ray spectroscopy, ion recoil spectroscopy, and spectroscopy of δ rays and positrons. Also, highly charged ions can be produced in single-ion collisions and open new avenues of research. New methods have been developed for the detection of these collision products and the extraction of spectroscopic data. The methods and their results are described in three chapters.

Low-energy ion–atom collisions are also providing typical spectroscopic data for collision products. Alignment and orientation parameters of states excited by ion bombardment can be extracted from angular correlation measurements. Coherent impact excitation of atoms manifests itself in angular correlations between emitted photons and scattered particles or in quantum beats of the related photon intensities. Population amplitudes and their phase differences can also be obtained and provide insight into the excitation mechanism of ions and atoms. The interplay between atomic potential curves and the quasimolecular structure of colliding atomic particles is important to the understanding of atomic excitation processes. There are significant links between such low-energy spectroscopic collision data and fields of applied research such as plasma and fusion physics.

Wilhelm Hanle's name will always be connected with his pioneering work on quantum mechanical interference of atomic states, now generally known as the Hanle effect, which he carried out shortly after the advent of modern quantum mechanics. Of course his research activity, which has been summarized and praised on many occasions, was not restricted to atomic spectroscopy. In fact, it reaches far beyond even the field of atomic physics. Nevertheless, his primary interest has always been in atomic spectroscopy and atomic collision physics, two fields from which much of the experimental spectroscopic information reported in this volume is drawn. It is, therefore, fitting to dedicate this volume and the following Part D to Professor Hanle. This dedication is even more appropriate as Professor Hanle was an enthusiastic editor of Parts A and B of *Progress in Atomic Spectroscopy*, but did not wish to take the burden of continued editorship. Nevertheless, he took great interest in this new project, and the present editors are very grateful for his advice and for suggestions of topics to be included. The authors and editors are most happy to honor Wilhelm Hanle by their contributions.

1

The k Ordering of
Atomic Structure

R. M. STERNHEIMER

1. Introduction

The discovery by the author[1] of a new ordering principle or extended symmetry in atomic physics (characterized by the quantum number k) in 1976 pertains to the excited-state energy spectra of atoms and ions consisting of a single valence electron outside a core of closed shells. The k ordering applies specifically to medium and heavy atoms and ions, i.e., those with atomic number $Z \geqq 11$ (Na), so that there is at least one filled p shell in the electron core of the atom or ion. The k-ordering symmetry thus discovered represents in a basic sense the opposite limiting case to the level ordering according to n (n degeneracy) of the hydrogen atom, which will be referred to as the H ordering.

The existence of the k ordering can certainly be used as a tool in the description and the systematization of the level structure of the excited states of pseudo-one-electron atoms and ions, such as the alkali-metal atoms and the alkaline-earth ions. We have thus introduced the concepts of the k gross structure, the l semifine structure, and the associated l patterns (see Figure 10). It is also hoped that the eventual understanding of the origin of the k-ordering symmetry will give rise to a deeper insight into the behavior of atomic many-body systems, i.e., specifically the effect on the energy spectrum of the many-body interactions between a core of

R. M. STERNHEIMER • Department of Physics, Brookhaven National Laboratory, Upton, New York 11973. The work of this paper was supported by the U.S. Department of Energy under Contract No. DE-AC02-76CH00016.

closed shells and an external valence electron, all in the field of a nucleus of charge Z. I have examined further correlations of the k ordering[1-5] with other properties of the atomic and ionic energy levels in the three-year period from 1976 to 1979.

In five recent papers,[1-5] I have introduced the concept of the quantum number k

$$k \equiv n + l \tag{1}$$

as an energy-ordering quantum number for the excited-state energy levels of the neutral alkali-metal atoms (i.e., Na, K, Rb, and Cs) and the singly ionized alkaline-earth atoms (i.e., Mg^+, Ca^+, Sr^+, Ba^+, and Ra^+)[1] and, in addition, states with one electron outside closed shells in the Group IB, IIA, IIB, and IIIA elements of the Periodic Table, and their isoelectronic ions.[2] In Eq. (1), n is the principal quantum number and l is the azimuthal quantum number of the valence electron outside the core of closed shells in the atom or ion considered.

For the spectra of Ref. 1, we have considered a total[1,3] of 416 excited states, while for the spectra of Ref. 2, we have analyzed a total[2,3] of 858 additional energy levels, giving a combined total of 1274 levels, which provide overwhelming evidence for the existence of a phenomenon which we have called "k ordering," namely, the grouping together of levels having the same value of k and having nearly the same energy (term value in the spectrum). Thus, the excited states of each spectrum can be divided into successive k groups, and within each k group (or "k band"), the levels increase (slightly) in energy according to a fixed sequence of l values, which we have called the "l pattern." Except in a few cases, the l pattern does not change with increasing k, and as an outstanding example, the l pattern is *pdsf* for a total of 158 excited states of rubidium extending from $k = 6$ to $k = 55$, i.e., over a range of 50 k values.

The k ordering and the associated l sequences of the levels nl have been exhibited specifically in nine j-averaged spectra in Ref. 1 and in ten j-averaged spectra in Ref. 2. By j-averaged spectra, we mean that we have averaged the energy values E_{nlj} listed in the tables of Moore[6] using the weighting factors $(2j + 1)$ for the two levels with $j = l + 1/2$ and $j = l - 1/2$, so as to average over the effects of the fine structure. Altogether a total of 42 spectra have been analyzed in this fashion, and the l patterns of these spectra have been tabulated in Table XIV of Ref. 2.

Two examples of k-ordered spectra are shown in Tables 1 and 2 of the present paper, which pertain to the spectra of neutral rubidium (Rb I) and of the singly ionized lead ion, Pb^+ (spectrum Pb II), respectively. In Table 1, we have listed only 45 of the 158 term values (i.e., energy values above the ground state $5s$), in order to keep the table to a manageable size. The value of k is listed in the second column, the energy value E_{nl}

Table 1. Spectrum of the Neutral Rubidium Atom, Rb I[a]

nl	k	E_{nl} (cm^{-1})	δ_{nl}	nl	k	E_{nl} (cm^{-1})	δ_{nl}
5s	5	0	3.195	9d	11	31,832	1.317
				11s	11	31,917	3.135
5p	6	12,737	2.712	8f	11	31,969	0.017
4d	6	19,355	1.233				
6s	6	20,134	3,155	11p	12	32,117	2.650
				10d	12	32,228	1.339
6p	7	23,767	2,675	12s	12	32,295	3.134
5d	7	25,702	1,294				
7s	7	26,311	3,144	12p	13	32,436	2.649
4f	7	26,792	0.012	11d	13	32,515	1.340
7p	8	27,858	2,663	13p	14	32,667	2.648
6d	8	28,689	1.316	12d	14	32,725	1.342
8s	8	29,047	3,139				
5f	8	29,278	0.013	31p	32	33,554.5	2.657
				30d	32	33,557.0	1.394
5g	9	29,298	≈ 0	32s	32	33,559.2	3.156
8p	9	29,848	2.656	29f	32	33,559.9	0.079
7d	9	30,281	1.327				
9s	9	30,499	3,137	32p	33	33,563.6	2.663
6f	9	30,628	0.014	31d	33	33,566.0	1.383
				33s	33	33,567.7	3.179
6g	10	30,637	≈ 0	30f	33	33,568.5	0.082
9p	10	30,966	2.654				
8d	10	31,222	1.333	49p	50	33,639.9	2.704
10s	10	31,362	3,136	48d	50	33,640.5	1.431
7f	10	31,442	0.015	50s	50	33,641.0	3.199
				47f	50	33,641.3	~0.08
6h	11	30,644	≈ 0				
10p	11	31,659	2.651	Limit		33,691.1	

[a] The excitation energies E_{nl} (in units cm^{-1}) are measured from the ground state (5s). The corresponding spectroscopic quantum defects δ_{nl} as obtained from Eq. (3) are listed in the last column of the table. The series limit L is 33,691.1 cm^{-1}. The values of E_{nl} are the j-averaged excitation energies, as derived from the tables of Moore[6] (see Table IV of Ref. 1).

(in units cm^{-1}) above the ground state is listed in the third column, and the spectroscopic quantum defect δ_{nl} (which will be discussed below) is given in the last column. In connection with the energy scale, we note that 8068 cm^{-1} = 1 eV. The series limit L is listed at the bottom of the table. Thus $L - E_{nl}$ is the binding energy of the electron in the state nl. In Table 2, we have listed the 50 lowest levels of Pb II, which show perfect k ordering.

If we denote the ionicity by δ (=$Z - N$, where N is the number of electrons), the equation for the quantum defect δ_{nl} can be obtained by means of the following expression for the binding energy $L - E_{nl}$:

$$L - E_{nl} = \frac{(1+\delta)^2 \text{Ry}}{(n - \delta_{nl})^2} = \frac{(1+\delta)^2 \text{Ry}}{n_{nl}^{*2}} \tag{2}$$

Table 2. Spectrum of the Singly Ionized Lead Ion, Pb$^+$ (Pb II)[a]

nl	k	E_{nl} (cm^{-1})	δ_{nl}	nl	k	E_{nl} (cm^{-1})	δ_{nl}
6p	7	9,387	4.019	11d	13	114,491	2.937
7s	7	59.448	4.335	12p	13	114,685	3.819
				13s	13	115,496	4.261
6d	8	69,274	3.094	10f	13	115,655	1.137
7p	8	76.334	3.874	9g	13	115,797	0.022
8s	8	89,180	4.300				
5f	8	92,520	1.091	12d	14	115,901	2.935
				13p	14	116,037	3.818
7d	9	94,896	2.918	14s	14	116,615	4.261
8p	9	95,851	3.842	11f	14	116,729	1.139
9s	9	101,346	4.303	10g	14	116,833	0.023
6f	9	102,874	1.112				
5g	9	103,559	0.018	13d	15	116,912	2.933
				14p	15	117,008	3.819
8d	10	103,872	2.973	12f	15	117,521	1.140
9p	10	104,821	3.830	11g	15	117,600	0.023
10s	10	107,930	4.258				
7f	10	108,533	1.123	14d	16	117,660	2.932
6g	10	108,968	0.020	13f	16	118,121	1.143
				12g	16	118,183	0.023
9d	11	109,304	2.937				
10p	11	109,734	3.824	15d	17	118,230	2.930
11s	11	111,574	4.262	14f	17	118,588	1.142
8f	11	111,942	1.130	13g	17	118,637	0.022
7g	11	112,230	0.021				
				16d	18	118,675	2.926
10d	12	112,444	2.937	14g	18	118,996	0.023
11p	12	112,726	3.806				
12s	12	113,912	4.262	17d	19	119,027	2.926
9f	12	114,147	1.135	15g	19	119,286	0.023
8g	12	114,346	0.022	Limit		121,243	

[a] The excitation energies E_{nl} (in units cm^{-1}) are measured from the $6p_{1/2}$ ground state. The corresponding spectroscopic quantum defects δ_{nl} as obtained from Eq. (3) (with $\delta = +1$) are listed in the last column of the table. The series limit L is 121,243 cm^{-1}. The values of E_{nl} are the j-averaged excitation energies, as derived from the tables of Moore[6] (see Table VII of Ref. 2).

where Ry = Rydberg unit = 109,737.3 cm^{-1} (for infinite nuclear mass), n_{nl}^* is usually called the "effective principal quantum number" for the state nl. Upon solving Eq. (2) for δ_{nl}, we obtain

$$\delta_{nl} = n - (1 + \delta)\left(\frac{\text{Ry}}{L - E_{nl}}\right)^{1/2} = n - n_{nl}^* \tag{3}$$

The values of δ_{nl} as obtained from Eq. (3) are listed in the last column of Tables 1 and 2. We note that the l pattern is obviously *pdsf* for the Rb I spectrum and *dpsfg* for the Pb II spectrum. It may be noted that the reason for the nonzero entry for E_{6p} of Pb II ($E_{6p} = 9387$ cm^{-1}) is the fact that E_{6p} represents the weighted average

$$E_{6p} = \tfrac{1}{3}E_{6p_{1/2}} + \tfrac{2}{3}E_{6p_{3/2}} \qquad (4)$$

measured from the ground state $E_{6p_{1/2}}$ taken as zero.

In connection with Eq. (1) for $k = n + l$, it should be pointed out that the combination $n + l$ was first considered by Madelung[7] for the order of filling of the orbitals nl of the ground states of the atoms of the Periodic Table. This combination is involved in Madelung's Rules A and B for the progressive filling up of shells in the *ground state*. Although these rules were discovered by Madelung as early as 1926, according to a statement published by Goudsmit,[8] they were published only in 1936 as an appendix in a textbook.[7] These rules were subsequently rediscovered by several authors; the appropriate references are listed in an unpublished report of Mann[9] and also in Refs. 4–9 of our paper in Ref. 1. In connection with the possibility of the existence of superheavy elements with Z in the region from $Z = 120$ to $Z = 130$, we rediscovered the Madelung rules independently in 1976. In this connection, it occurred to us that the quantum number $k = n + l$ may also be relevant to the ordering of the energy levels of *excited states* of atoms and ions with one electron outside closed shells, and this idea led us directly to the discovery of the *k* ordering of these levels, as described above.

In the present chapter, we will describe a number of properties of atomic spectra associated with the *k* ordering, in particular the existence of a limiting ionicity δ_1 and a limiting angular momentum l_1 and the related phase transitions from *k* ordering to hydrogenic ordering (according to n) (see Section 2), the connection with the phenomenon of inverted fine structure (see Section 3), the identification of the spectroscopic quantum defects δ_{nl} as the appropriate "order parameter" for the *k* ordering (see Section 4), and finally the possibility of a symmetry breaking in atomic spectra associated with the *k* ordering (see Section 5). General considerations concerning the possible significance of the existence of the *k* ordering, which prevails for such a large number of spectra and energy levels, will be presented in Section 6.

2. The Limiting Ionicity δ_1 and the Limiting Angular Momentum l_1

In the work of Refs. 1 and 2, it was noticed that if the ionicity $\delta(=Z - N)$ exceeds a certain limiting value δ_1, there is an abrupt transition from k ordering to hydrogenic ordering, i.e., ordering according to the quantum number n, which will be denoted by "H ordering." We first noticed this sharp transition in going from the spectrum of Mg^+ to that of the isoelectronic ion Al^{2+}. In Figure 1, we have shown the levels nl in the order of increasing energy, i.e., decreasing binding energy $E_L - E_{nl}$ on a logarithmic scale, going from right to left, for both Mg^+ (lower scale) and Al^{2+} (upper scale). For the case of Mg^+, the group of levels with the same value of k ("k bands") are shown bracketed with the k value indicated below the abscissa axis. On the other hand, for the case of Al^{2+}, it is quite obvious that the levels are arranged according to n value, with the energy E_{nl} increasing with increasing l for a given n, exactly as for hydrogenic levels. Levels with the same values of n and l in the two spectra are joined by approximately vertical lines. The frequent crossings of these "vertical" lines indicates the rearrangement of the spectrum in going from k ordering for Mg^+ to H ordering for Al^{2+}. In Ref. 2, we also noted that the heavier

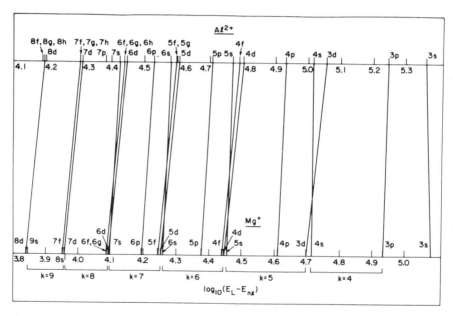

Figure 1. Energies of the excited-state levels of Mg^+ (lower abscissa) and Al^{2+} (upper abscissa) on a logarithmic scale, on which are plotted the values of $\log_{10}(E_L - E_{nl})$, where E_L is the series limit, $E_L = 121,267 \text{ cm}^{-1}$ for Mg^+ and $E_L = 229,454 \text{ cm}^{-1}$ for Al^{2+}. Levels with the same nl values in the two spectra are connected by (nearly vertical) lines. The frequent crossing of these lines indicates the rearrangement in the spectrum in going from Mg^+ to Al^{2+}, i.e., from k ordering to hydrogenic (H) ordering. At the bottom of the lower abscissa, we have shown the successive k regions or bands, from $k = 4$ to $k = 9$.

isoelectronic ions (Si^{3+}, P^{4+}, S^{5+}, Cl^{6+}, Ar^{7+}, etc.) also exhibit pure hydrogenic ordering.

The abrupt change in going from Mg^+ to Al^{2+} made us think of a phase transition between k ordering and hydrogenic ordering, at the time we wrote the first paper[1] (in 1976). After investigating the case of Mg^+ and Al^{2+}, we found many additional examples of the "phase transition" between k ordering and hydrogenic ordering as the degree of ionization $Z - N$ (or ionicity) is increased. As an additional example of the transition to H ordering, we may refer to the excited-state spectrum of Ge III (Ge^{2+}), which is isoelectronic with Zn I and Ga II. An inspection of the appropriate table in Moore's compilation[6] (Vol. II) shows that the spectrum of Ge III is almost purely hydrogenic, and the same is true for the heavier isoelectronic ions As IV and Se V. These results are in contrast to those for the neutral Zn I spectrum and the (singly ionized) Ga II spectrum, which are purely k-ordered. We note that for the elements in Group IIB of the Periodic Table such as Zn I, we consider only those levels with the configuration $3d^{10}4snl$, so that there is a single nl electron outside a spherical core consisting of the filled core shells including $3d^{10}$ plus a spherically distributed $4s$ electron.

Thus we are led to the concept of a *limiting ionicity* δ_1 which separates the two phases of k ordering and hydrogenic ordering. For $Z = 12$, we may assume that $\delta_1 \cong 1.5$, midway between the δ values ($\delta = Z - N$) for Mg^+ ($\delta = 1$) and Al^{2+} ($\delta = 2$). Also for $Z = 30$, we may assume that δ_1 is close to 1.5, in view of the phase transition occurring between Ga^+ ($\delta = 1$) and Ge^{2+} ($\delta = 2$). However, for $Z = 57$, we expect that δ_1 is at least 2.5 to take into account the fact that the spectrum of La^{2+}, which is isoelectronic with Ba^+, is still almost completely k-ordered, as discussed in Ref. 1 at the end of Section III. Thus there is a tendency for a gradual increase of δ_1 with Z for $Z > 30$. This tendency is even more pronounced at very low Z ($Z < 10$) for the following reason. It was already noticed in Ref. 1 that for Li ($Z = 3$) and its isoelectronic ions, namely, Be^+, B^{2+}, C^{3+}, etc., the ordering of the excited-state levels is completely hydrogenic (according to n) (see Table XVII of Ref. 1). This observation coupled with the fact that the k ordering appears for the first time at $Z = 11$ (Na) and persists up to the largest Z values investigated (Ra^+, $Z = 88$) implies that we need at least a filled p shell in the electron core (e.g., the $2p$ shell of Na) and preferably a large number of filled p and d shells in order for the k ordering of the excited-state valence electron levels to occur. The wider implications of this observation will be discussed in subsequent sections of this chapter, but for the present, we use this fact in order to derive the curve of δ_1 vs. Z at low Z. We will assume that δ_1 abruptly decreases to zero for $Z < 11$. The value of Z for which $\delta_1 = 0$ is somewhat arbitrary. It seems most logical to assume that $\delta_1(Z_0) = 0$ for a value of Z, Z_0, in the middle of the

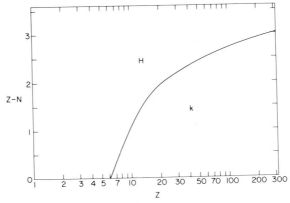

Figure 2. Regions of validity of k ordering and hydrogenic (H) ordering of the excited-state levels of neutral atoms and ions with a small degree of ionization $Z - N$. This figure represents the phase diagram in the plane of $Z - N$ vs. Z.

region of filling the $2p$ shell. We choose $Z_0 = 6$. The resulting curve of δ_1 vs. Z (on a logarithmic scale of Z) is shown in Figure 2, which is taken from Ref. 2. This curve separates the k and H phases of the excited-state spectra as a function of the ionicity $\delta = Z - N$.

A similar observation concerning the dependence of the k ordering on the angular momentum l was made in Ref. 3. It had been observed in the spectra of Refs. 1 and 2 that there were a number of so-called $(k + \lambda)$ exceptions, in which a level nl with $k_0 = n + l$ fell into a region of levels having the k values $k_0 - 1$ or $k_0 + 1$. It was soon realized (in Ref. 3) that the $(k + \lambda)$ exceptions generally occur in the region of large angular momentum l, where l exceeds a certain value l_1 which we have called the *limiting angular momentum*. It was found that l_1 increases slowly with increasing Z, in the same general manner as δ_1. Thus it was found that $l_1 \cong 2$ for Na $(Z = 11)$, $l_1 \cong 3$ for K $(Z = 19)$, $l_1 \cong 4$ for Rb $(Z = 37)$ and $l_1 \cong 4$–5 for Cs $(Z = 55)$. The resulting curve of l_1 vs. Z (with Z on a logarithmic scale) is shown in Figure 3. This figure is taken from Ref. 3. Again we have assumed that $l_1 = 0$ at $Z = Z_0 = 6$ to take into account the fact that at least one filled p shell is required to produce the observed k ordering. It may be noted that later work (carried out in Ref. 5) indicates that an alternative derivation of l_1 from the observed quantum defects δ_{nl} indicates that for $Z > 30$, l_1 increases somewhat less rapidly with increasing Z than is shown in Figure 3, which would imply that the curve of l_1 vs. Z bends over in the same manner as the curve of δ_1 vs. Z in Figure 2 (with Z plotted on a logarithmic scale in both cases).

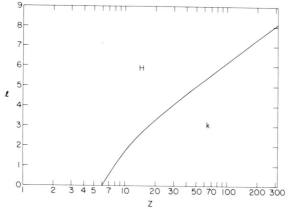

Figure 3. Regions of validity of *k* ordering and hydrogenic (H) ordering of the excited-state levels of neutral atoms and ions with a small degree of ionization $Z - N (\leqslant 2)$. This figure represents the phase diagram in the plane of *l* vs. *Z*. The curve separating the two phases represents the limiting angular momentum l_1.

The question as to the mechanism which leads to both the limiting ionicity δ_1 and the limiting angular momentum l_1 has been discussed in some detail in Ref. 3 (see pp. 1755, 1756). To restate the argument briefly, we believe that the *k* ordering is related in some basic way to the overlap of the valence wave function $v_0(nl)$ with the core wave functions $u_0(n_c l_c)$ and their perturbations $u_1(n_c l_c \rightarrow l'_c)$ by the exchange electrostatic interaction with the valence electron. Now if *l* is too large, the radial wave function (times *r*) of the valence electron in the region of the core, which is given by

$$v_0(nl) = ar^{l+1} + \cdots \tag{5}$$

becomes vanishingly small, thus reducing the overlap with $u_1(n_c l_c \rightarrow l'_c)$ to negligible values, which on the present hypothesis would destroy the *k* ordering, as is indeed observed.

In order to give a more general discussion, which also involves the ionicity δ, we write down the radial part of the Hamiltonian for the valence electron *nl*, including the centrifugal term $l(l + 1)/r^2$ (in Rydberg units):

$$H_l = \frac{l(l+1)}{r^2} - \frac{2(Z - \delta - 1)\chi(r)}{r} - \frac{2(1 + \delta)}{r} \tag{6}$$

where $\chi(r)$ is the Thomas–Fermi function,[10] with the following properties:

$\chi(0) = 1$, $\chi(r \to \infty) = 0$. It can be easily verified from Eq. (6) that the potential near the nucleus ($\chi \to 1$) becomes $-2Z/r$ which is, of course, the correct value independent of the degree of ionization δ. [The factors of 2 in Eq. (6) arise from the use of Rydberg units.] On the other hand, for large $r (\chi \to 0)$, the potential becomes $-2(1 + \delta)/r = -2(1 + Z - N)/r$, which is also correct.

We now note that by far the major part of the valence electron density lies in the region outside the core, and in that region the potential is hydrogenic, namely $-2(1 + \delta)/r$. We believe that there is a delicate balance between the k-ordering properties caused by the exchange-core polarization, and therefore the magnitude of the valence wave function inside the core and, on the other hand, the large region of r where the potential is essentially hydrogenic. If δ is made too large, the hydrogenic potential tends to predominate at a smaller radius over the nuclear potential, which is essentially $-2Z\chi(r)/r$. This mechanism would explain the existence of a limiting ionicity δ_1, such that when $\delta > \delta_1$, the ordering of the levels becomes hydrogenic. The increase of δ_1 with increasing Z (Figure 2) can also be understood in terms of the competition between the core region with potential $-2Z\chi/r$ and the hydrogenic region with potential $-2(1 + \delta)/r$. In a similar manner, with increasing Z, the coefficient a in Eq. (5) for $v_0(nl)$ will increase, leading to larger values of $v_0(nl)$ in the core region, thus explaining the increase of the limiting anguilar momentum l_1 with increasing Z (Figure 3).

The fact that k is a good quantum number both in the nonrelativistic region (small Z, $\alpha Z \ll 1$) and in the relativistic region (large Z, $\alpha Z \sim 1$, e.g., for Ra^+, $\alpha Z = 0.64$) has been emphasized in the papers of Refs. 1–3. Moreover, the validity of k is found to extend all the way from the ground state up to highly excited states, e.g., for Rb, from $k = 5$ (ground state) up to $k = 56$, i.e., for the states $55p$ and $54d$ which lie only $40 \, \text{cm}^{-1}$ (~ 5 meV) below the ionization limit. These two observations lead us to believe that k is a fundamental quantum number for all one-electron states of atoms and ions, provided that $l \leq l_1$, $\delta \leq \delta_1$, and that $Z \geq 11$, so that there is at least one filled p shell of the atomic core. We will return to a discussion of the probable significance of k in Section 5 of the present chapter.

3. The k Ordering of Atomic Energy Levels and its Relation to the Fine-Structure Inversion in Atomic and Ionic Spectra

In this section, we wish to discuss an approximate correlation between the angular momentum l_{inv} of the states of the alkali-metal atoms and ions for which the fine structure is inverted and the limiting angular momentum l_1 for k ordering. By inverted fine structure, we mean that the state with $j = l + 1/2$ lies *below* the state with $j = l - 1/2$, in contrast to the level ordering predicted by the usual one-electron approximation to the fine-structure Hamiltonian. Moreover, the inverted fine structure has been shown to occur only for neutral atoms and for those isoelectronic ions for which the ionicity is less than the limiting ionicity δ_1 for k ordering, i.e., the ionicity δ_{inv} for which the inversion occurs satisfies the following equations:

$$\delta_{inv} \lesssim \delta_{inv,max} \tag{7}$$

where

$$\delta_{inv,max} \simeq \delta_1 \tag{8}$$

The experimental situation for the inverted fine structure has been described extensively in Ref. 4. Thus for Na, the nd states are inverted ($nd_{5/2}$ below $nd_{3/2}$), but the nf states are noninverted, with a fine-structure splitting $\Delta\nu_{nf}$ which is within a few percent of the hydrogenic value. The deviation of the fine-structure splitting $\Delta\nu_{nf}$ of Na from the hydrogenic value has been calculated in the paper of Sternheimer, Rodgers, and Das,[11] and the calculations are in essential agreement with the experimental values of Gallagher *et al.*[12] (deviations ~5% for $11f$–$17f$). Since only the nd states of Na are inverted, we find $l_{inv} = 2$ for Na, which is identical with the value of l_1 for Na, as discussed above (in Section 2). Concerning the inverted fine structure of Na nd, we may note the calculations of Foley and Sternheimer[13] for Na $3d$, and also related calculations of Lee *et al.*,[14] Holmgren *et al.*,[15] and Luc-Koenig.[16] In connection with the situation for Na $3d$, we note that for Li, the fine structure of the d states is normal, i.e., essentially hydrogenic, and simultaneously, and most probably not by accident, the k ordering does not occur (see Table VIII of Ref. 1).

Turning now to the higher members of the Na isoelectronic sequence, we have the following results. For Mg II, $3d$–$8d$, $\Delta\nu_{nd}$ is negative (inverted); for $4f$–$6f$, $\Delta\nu_{nf}$ is positive. For Al III, $3d$–$6d$, $\Delta\nu_{nd} < 0$. However, $\Delta\nu_{nd}$ for Si IV becomes positive for $n \geqslant 5$. Similarly for P V and the higher members of the Na isoelectronic sequence, $\Delta\nu_{nd}$ is positive for all n. The appropriate experimental references are listed in Ref. 4 and also in the very useful paper of Persson and Pira.[17] The fact that the fine structure changes from inverted to normal as the ionicity δ is increased is well illustrated by the preceding results. In the present case, we find $\delta_{inv,max} = 2$, corresponding to Al III, as compared to $\delta_1 = 1.5$ for $Z = 12$, as derived above (in Section 2). Thus $\delta_{inv,max} \cong \delta_1$, as we have anticipated in Eq. (8).

We will now discuss the K isoelectronic sequence. For K I, the nd states are inverted, but for the isoelectronic Ca II, the nd level fine structure is large and positive, i.e., noninverted, as already shown by the tables of Moore.[6] Similarly, for Sc III and Ti IV, the data of Moore[6] show that the nd fine structure intervals $\Delta\nu_{nd}$ are positive in all cases. These earlier data are supplemented by several more recent works, for which the references are given in our paper of Ref. 4. In particular, Gallagher and Cooke[18] at Stanford Research Institute found that $\Delta\nu_{nd} < 0$ for K I in the region of $15d$–$20d$. We can conclude from these results that $l_{inv} = 2$ for K, and moreover $\delta_{inv,max} \sim 1$.

We will now discuss briefly the Rb I isoelectronic sequence. The tables of Moore[6] show that the $4d$ interval is negative, but the higher nd intervals $\Delta\nu_{nd}$ are positive, i.e., noninverted. On the other hand, Farley and Gupta[19] have shown that the $6f$ and $7f$ fine structures are clearly inverted, with the values of the intervals being $\Delta\nu_{6f} = -0.0162\ \text{cm}^{-1}$ and $\Delta\nu_{7f} = -0.0116\ \text{cm}^{-1}$. For the case of the isoelectronic Sr II spectrum (Sr$^+$ ion), the extensive measurements of Persson and Pira[17] show that the intervals $\Delta\nu_{nf}$ are negative for the states $4f$–$8f$, in complete similarity to the results for Rb I. However, we may note that the negative values of $\Delta\nu_{nf}$ for Sr II are considerably larger than those for the Rb I states. As an example, Persson and Pira[17] obtained $\Delta\nu_{6f}(\text{Sr}^+) = -0.569 \pm 0.005\ \text{cm}^{-1}$, which exceeds the value of $\Delta\nu_{6f}(\text{Rb}) = -0.0162\ \text{cm}^{-1}$ by a factor of 35.1.

For Y III (Y^{2+}), $\Delta\nu_{nf}$ passes through zero between $n = 4$ and $n = 5$ (i.e., $\Delta\nu_{4f} < 0$, $\Delta\nu_{nf} > 0$ for $n \geqq 5$), as found by Epstein and Reader.[20] For Zr IV (Zr^{3+}), $\Delta\nu_{nf}$ is positive for all n values, according to measurements of Kiess.[21] Similarly for Mo VI (Mo^{5+}), Romanov and Striganov[22] obtained no negative nf fine-structure intervals.

As pointed out by Persson and Pira,[17] the behavior of the $\Delta\nu_{nf}$ interval for the Rb I isoelectronic sequence is very similar to that for the $\Delta\nu_{nd}$ interval for the Na I sequence. In both cases, the fine-structure splitting is negative for a large number of states of the neutral atom and the few-times ionized ions, but becomes positive when the ionicity becomes larger than a certain critical value. This behavior is exactly that which is described by the Eqs. (7) and (8) above, and in fact, it led us to write down these equations.

To be more specific, for the Na I isoelectronic sequence, we find $\delta_{inv,max} = 2$ since all of the $n\,^2D$ states of Si IV (Si^{3+}) except $3\,^2D$ and $4\,^2D$ have $\Delta\nu_{nd} > 0$. The value $\delta_{inv,max} = 2$ can be compared with the limiting ionicity for k ordering in the Na–Mg region, which is $\delta_1 = 1.5$ to take into account the phase transition which occurs in going from Mg$^+$ to Al^{2+} (see Figure 1).

Similarly for the Rb I isoelectronic sequence, for Y III (Y^{2+}), $\Delta\nu_{nf}$ passes through zero between[20] $n = 4$ and $n = 5$, and for Zr IV (Zr^{3+}) $\Delta\nu_{nf}$ is positive for all n values.[21] According to our criterion for $\delta_{inv, max}$, we may write $\delta_{inv, max} \cong 2$ for the nf states of the Rb I sequence. On the other hand, the limiting ionicity for k ordering at $Z = 37$ is also $\delta_1 = 2$ (see Figure 2), so that Eq. (8) is also satisfied in this case.

Next we wish to discuss the correlation between the angular momentum l_{inv} of the states of the atoms and ions of the Na, K, and Rb isoelectronic sequences and the limiting angular momentum l_1 for k ordering, as discussed in Ref. 3 and in Section 2 of the present chapter. Thus $l_{inv} = 2$ for the Na I sequence, which is identical with the value of $l_1 = 2$ for Na, as discussed above and in Ref. 3. For the K I sequence, the nd states are inverted, so that $l_{inv} = 2$, the same as for Na. The K I spectrum exhibits k ordering up to and including the nf states (see Table III of Ref. 1), so that the derived value of l_1 is 3, and we find

$$l_{inv} = l_1 - 1 \qquad (Z \geqslant 19) \qquad (9)$$

Considering now the Rb I sequence, the only d state which is inverted is $4d$ ($\Delta\nu_{4d} = -0.44$ cm^{-1}).[6] However, the nf states are clearly inverted, as is shown by the results of the tables of Moore[6] ($4f, 5f, 6f, 7f$ inverted) and the more recent results of Farley and Gupta.[19] Moreover, as discussed above, the results of Persson and Pira[17] demonstrate that the $4f$–$8f$ states of Sr II have the inverted fine structure, and the experiment of Epstein and Reader[20] shows that $\Delta\nu_{4f} < 0$ for Y III. Thus $l_{inv} = 3$ for the Rb I sequence, and the value of l_1 as deduced from the spectrum of Rb I (Table 1) is $l_1 = 4$. Thus Eq. (9) is also satisfied for the Rb I sequence.

We will now discuss the Cs I isoelectronic sequence. The tables of Moore[6] list seven inverted nf levels of Cs I ($4f$–$10f$), and the more recent experimental data of Eriksson and Wenåker[23] give more accurate values for nine inverted nf levels ($4f$–$12f$). An additional paper of Popescu et al.[24] confirms that the nd levels $5d$–$34d$ have $\Delta\nu_{nd}$ positive. Thus we find $l_{inv} = 3$ for the Cs I spectrum, which is comparable to the value of l_1 for Cs ($l_1 = 3$–5 depending upon the precise definition of l_1). Thus the results for Cs I are essentially compatible with Eq. (9).

For the Ba II spectrum, the levels $4f$–$8f$ are well resolved and noninverted.[6] A similar result has been observed for the isoelectronic La III spectrum (La^{2+}), according to measurements of Sugar and Kaufman.[25] Thus we have $\delta_{inv,max} \cong 1$, which is somewhat smaller than the value of $\delta_1 = 2.5$, as deduced above (in Section 2). This isolated deviation from the relation of Eq. (8) may perhaps be due to the proximity of these elements (Cs, Ba, La) to the onset of the rare-earth region with its progressive filling of the $4f$ shell.

Summarizing this discussion of the value of l_{inv} for the Na, K, Rb, and Cs isoelectronic sequences, we can conclude that Eq. (9) is generally at least approximately satisfied, at least for $Z \geq 19$, although for the lighter elements in the Na region ($Z = 11$), the following relation holds:

$$l_{inv} = l_1 \qquad (Z < 19) \qquad (10)$$

The restrictions on the validity of Eqs. (9) and (10) as a function of Z have been indicated in parentheses.

In Ref. 4, we have also investigated the case of inverted fine structure in the spectra of the Group IB isoelectronic sequences (Cu I, Ag I, and Au I) and of the isoelectronic sequences of the Group IIIA elements, namely Al, Ga, In, and Tl. For the noble metals (Cu, Ag, and Au), we consider only those levels with the configuration $nd^{10}n'l$, i.e., having a single electron ($n'l$) outside the closed nd shell. Similarly, for the Group IIIA elements, only the levels with the configuration $ns^2n'l$ are considered. Both types of isoelectronic sequences provide important additional support for the relation of Eq. (9) for l_{inv} in terms of l_1. It may be noted that for the spectra of Pb II and Bi III (isoelectronic with Tl I), both the nf and ng levels exhibit the inverted fine structure, so that l_{inv} has both the values $l_{inv} = 3$ and $l_{inv} = 4$. Since l_1 for $Z \sim 82$ lies in the range $l_1 = 4$ to 5, Eq. (9) is again satisfied.

The result of Eqs. (9) and (10) is similar to that observed above for the ionicity values δ_{inv} for which the inverted fine structure occurs [Eqs. (7) and (8)], with the important difference that *all* of the atomic or ionic species with $\delta \leq \delta_1$ in the phase diagram of δ vs. Z (Figure 2) are expected to exhibit inverted fine structure, whereas only one or at most two l values are involved, if we consider the phase diagram of l vs. Z, and the affected l values lie close to but somewhat below the phase boundary given by the limiting angular momentum l [see Eq. (9)].

It is apparent from the preceding discussion that the phenomenon of inverted fine structure is directly connected with the existence of the phase diagrams for the k ordering. A possible explanation of these correlations is that a necessary condition for the existence of k ordering is an adequate amount of penetration of the valence orbital inside the core, which can occur only if l is sufficiently small, i.e., $l \leq l_1$. This point has been extensively discussed in Ref. 3 (pp. 1755, 1756). Such a penetration ensures an adequate overlap of the valence wave function $v(nl)$ with the core wave functions $u_0(n_cl_c)$ and their perturbations $u_1(n_cl_c \rightarrow l_c')$, and such an overlap is believed to be responsible for the k ordering via exchange core polarization effects, similar to those which lead to both the inverted fine structure[13, 14] and the quadrupole shielding and antishielding effects first introduced and calculated by the author.[26] What is of crucial importance in making this correlation is the observation[1–5,11,13] that all three effects (k ordering, inverted fine structure, and Sternheimer antishielding) depend directly on the presence of *closed np* (*and nd*) *shells* in the core of the atom or ion considered.

4. The k Ordering of Atomic Energy Levels and its Relation to the Spectroscopic Quantum Defects

In this section, we will demonstrate that the spectroscopic quantum defects δ_{nl} [as given by Eq. (3)] play the role of the "order parameter" for the k-ordering phase of the excited state spectra of atoms and ions with a single electron outside a core of closed shells.[5] Thus, only penetrating orbitals, i.e., those having a large $\delta_{nl}(\delta_{nl} \geq 0.2)$ exhibit the phenomenon of k ordering, and in particular, the curves of δ_{nl} vs. l (for fixed k) are generally curved downwards, with an abrupt decrease to values of δ_{nl} close to zero at the limiting angular momentum l_1 which has been previously introduced. Thus the curves of δ_{nl} vs. l are basically similar to the curves of the magnetic field **H** as a function of temperature T in a ferromagnet, with an abrupt decrease to $\delta_{nl} \sim 0$ at $l = l_1$, which can therefore be regarded as the analogue of the Curie temperature T_C.

In order to illustrate the preceding statements, we show in Figures 4–6 the curves of δ_{nl} vs. l for the spectra of Sn II ($k = 8$), Cs I ($k = 9$), and Pb II ($k = 9$), respectively. The l pattern is listed in the upper right-hand corner for each spectrum, e.g., $\{l_i\} = dpsfg$ for Pb II, as can also be seen from the spectrum of Pb II which is listed in Table 2. If we define l_1 for the present purpose as that value of l for which δ_{nl} becomes approximately zero, we find $l_1 = 3$ for Sn II and Cs I, and $l_1 = 4$ for Pb II. (We note that the values of δ_{nl} for the Pb II spectrum are listed in Table 2.)

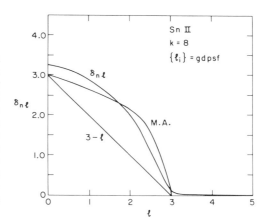

Figure 4. The spectroscopic quantum defect δ_{nl} for the Sn II spectrum (Sn$^+$ ion) as a function of l, for fixed $k = 8$. We note the abrupt decrease of δ_{nl} as l approaches the limiting angular momentum $l_1 = 3$. For $l > l_1$, the δ_{nl} are very small ($\delta_{nl} \leq 0.1$). The curve marked "M.A." refers to the magnetic analogy discussed in the text. The straight line $3 - l = l_1 - l$ represents a less accurate approximation to δ_{nl}.

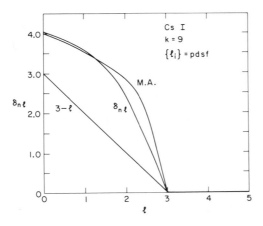

Figure 5. The spectroscopic quantum defect δ_{nl} for the Cs spectrum (Cs I) as a function of l, for fixed $k = 9$. The curve M.A. (magnetic analogy) approximates the behavior of δ_{nl} very closely. The straight line $l_1 - l = 3 - l$ is shown for comparison.

In addition to the curves of δ_{nl} vs. l, Figures 4–6 also show two additional features: (1) the straight line $\delta_{nl} = l_1 - l$ which is a very approximate representation of δ_{nl}, which has also been considered in a recent paper of Foley[27]; (2) a closer approximation to δ_{nl} indicated as M.A. for "magnetic analogy." As already indicated, the magnetic analogy curves refer to the dependence of the field **H** in a ferromagnet on the temperature T, where δ_{nl} takes the place of **H** and l takes the place of T (with l_1 the analog of the transition temperature T_C). The M.A. curves were obtained from the Brillouin function[28] B_J for $J \to \infty$. The maximum of the M.A. curve was taken as l_1 or $l_1 + 1$, and the corresponding value of T_C was taken as l_1. It is seen that the M.A. curve is quite a good approximation to δ_{nl} for the cases of Sn II and Cs I. For the case of Pb II, the approximation is less close, presumably because of the larger value of l_1 assumed ($l_1 = 4$) in Figure 6.

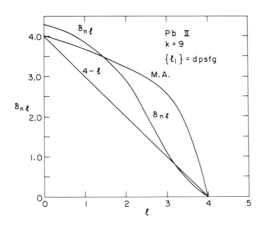

Figure 6. The spectroscopic quantum defect δ_{nl} for the Pb$^+$ spectrum (Pb II) as a function of l, for fixed $k = 9$. The curve M.A. and the straight line $l_1 - l = 4 - l$ are shown for comparison.

As mentioned above, the relation of the k ordering to the quantum defects δ_{nl} has been independently investigated by Foley.[27] In this paper, the approximate relation $\delta_{nl} \sim g - l$ has been derived by utilizing a relation to the phase shifts for elastic scattering at zero energy previously discovered by Seaton.[29]

Besides the preceding discussion and the Figures 4–6 which are taken from our paper of Ref. 5, we have also introduced a quantity called the "reduced quantum defect" η_{nl}, which is defined as follows[5]:

$$\eta_{nl} \equiv \delta_{nl} - (l_1 - l) = \delta_{nl} + l - l_1 \qquad (l \leqslant l_1) \tag{11}$$

$$\eta_{nl} \equiv \delta_{nl} \qquad (l \geqslant l_1) \tag{12}$$

where we have used two different definitions for the regions $l \leqslant l_1$ and $l \geqslant l_1$. Of course, for $l \geqslant l_1$, δ_{nl} is always small (generally $\delta_{nl} < 0.1$), and the definition of Eq. (12) ensures that η_{nl} is always positive for $l \geqslant l_1$, since $\delta_{nl} > 0$ in all cases.

From Eq. (11), we can derive the relation

$$\delta_{nl} = \eta_{nl} - l + l_1 \qquad (l \leqq l_1) \tag{13}$$

so that the denominator $(n - \delta_{nl})^2$ of Eq. (2) can be written as follows:

$$(n - \delta_{nl})^2 = (n + l - l_1 - \eta_{nl})^2 = (k - l_1 - \eta_{nl})^2 \qquad (l \leqq l_1) \tag{14}$$

For $l > l_1$, hydrogenic ordering prevails, and the energy levels are practically degenerate with the level $n = n_1$, where n_1 is defined as

$$n_1 \equiv k - l_1 \tag{15}$$

Thus we can write, both for $l \leqq l_1$ and for $l \geqq l_1$, the following general equation for the binding energy $L - E_{nl}$ of all of the levels in a given k band:

$$L - E_{nl} = \frac{(1 + \delta)^2 \mathrm{Ry}}{(k - l_1 - \eta_{nl})^2} = \frac{(1 + \delta)^2 \mathrm{Ry}}{(n_1 - \eta_{nl})^2} \tag{16}$$

As an example, for the case of Rb I, the k band with $k = 50$ contains besides the levels $49p$, $48d$, $50s$, and $47f$, which are listed in Table 1, also the following 43 levels which are essentially degenerate with $47f$, due to the hydrogenic ordering for $l > l_1$ $(l_1 = 3)$: $47g$, $47h$, $47i, \ldots, n = 47$, $l = 45$, and $n = 47$, $l = 46$. Thus the $k = 50$ band contains a total of

$$k - l_1 = 50 - 3 = 47 \text{ levels} \tag{17}$$

The approximate location of a given k band can be obtained from the Rydberg formula:

$$L - E_k \simeq \frac{(1 + \delta)^2 Ry}{(k - l_1)^2} \tag{18}$$

Of course, for Rb I, we have $\delta = 0$ (zero ionicity). Thus Eq. (18) with $k - l_1 = 47$ gives for the $k = 50$ band:

$$L - E_{50} \cong Ry/47^2 = 49.7 \text{ cm}^{-1} \tag{19}$$

so that $E_{50} = 33,691.1 - 49.7 = 33,641.4 \text{ cm}^{-1}$, which is very near the upper limit of the $k = 50$ band at $E_{47f} = 33,641.3 \text{ cm}^{-1}$.

The separation of levels with the same l value in neighboring k bands can be obtained from the derivative $\partial E_k/\partial k$, which is obtained from Eq. (18), where L, δ, and l_1 are constants:

$$\frac{\partial E_k}{\partial k} = \frac{2(1 + \delta)^2 Ry}{(k - l_1)^3} \tag{20}$$

Thus the spacing between the $nl = 46f$ and $47f$ levels is given approximately by using the average $k = 49.5$:

$$\Delta E_k(49, 50) = \frac{2 \text{ Ry}}{(49.5 - 3)^3} = 2.18 \text{ cm}^{-1} \tag{21}$$

From Eq. (16) it is readily seen that the levels within a given k band are energy-ordered as follows: $E_{n_a l_a} < E_{n_b l_b} < E_{n_c l_c} < E_{n_d l_d}$, if we have $\eta_{n_a l_a} > \eta_{n_b l_b} > \eta_{n_c l_c} > \eta_{n_d l_d}$. Thus the level with the highest algebraic η_{nl} value lies lowest, and the other levels with the same k value are arranged in the order of decreasing η_{nl} values, corresponding to the l pattern $\{l_i\} = l_a l_b l_c l_d$.

In Figures 7–9 we have plotted the values of η_{nl} vs. l for nine representative spectra—namely, Ca I, Ga I, and Sn II in Figure 7; K I, Rb I, and Ba II in Figure 8; and Cs I, Tl I, and Pb II in Figure 9. In the upper right-hand corner of each figure, we have indicated the values of k and l_1 and the l pattern $\{l_a l_b l_c l_d\}$ for each of the three spectra to which the figure refers.

The most striking feature of Figures 7–9 is that with the exception of two spectra (Ca I and K I), the two highest η_{nl} values involve the np and nd levels, while the two lowest η_{nl} values involve the ns and nf levels. Thus we expect that four l patterns will be predominant, namely $dpsf$, $pdsf$,

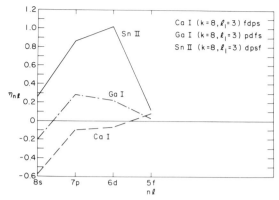

Figure 7. The reduced quantum defects η_{nl}, as defined by Eqs. (11) and (12), as a function of l for the $k = 8$ energy levels of the spectra Ca I, Ga I, and Sn II. The corresponding nl levels are listed on the abscissa. The values of k and l_1, and the $\{l_i\}$ patterns are listed in the upper right-hand corner of the figure.

dpfs, and *pdfs*, and this expectation is generally borne out by the frequencies of the actual l patterns, as presented in Table XIV of Ref. 2. Thus the l patterns *dpsf*, *dpfs*, and *pdsf* account for 32 cases out of a total of 48 which are listed in this table of Ref. 2. We believe that the maxima of η_{nl} at np and nd are due to the finer details of the valence-core overlap, which is believed to be responsible for the k ordering phenomenon,[3-5] as well as for the inverted fine structure of excited d, f, and g levels,[4,11,13,14] as discussed above in Sections 2 and 3.

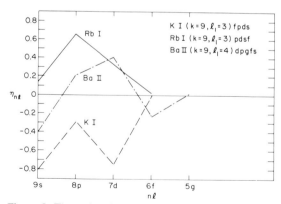

Figure 8. The reduced quantum defects η_{nl} as a function of l for the $k = 9$ energy levels of the spectra K I, Rb I, and Ba II. The corresponding nl levels are listed on the abscissa.

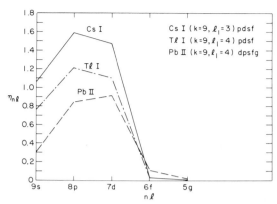

Figure 9. The reduced quantum defects η_{nl} as a function of l for the $k = 9$ energy levels of the spectra Cs I, Tl I, and Pb II. The corresponding nl levels are listed on the abscissa.

5. k-Symmetry Breaking in Atomic Spectra

In Refs. 3 and 4, we have extensively discussed the possibility that the arrangements of atomic levels in nonoverlapping k bands, and the regular l-dependent splittings within each k band, may be the result of a symmetry-breaking mechanism, similar to the SU(3) scheme of Gell-Mann and Ne'eman[30] for the mass spectrum of the elementary particles and their excited states (see Ref. 3, p. 1756; Ref. 4, Section IV). Thus we assume that there are primitive (unobserved) levels having only the quantum number k, and we refer to this structure of levels as the k *gross structure*. The primitive k levels are split into several levels with different l values, and with $n = k - l$ in each case. This structure which is caused by an l-dependent interaction, will be called the *semifine structure* of the atomic levels.

An example of the primitive $k = 9$ and $k = 10$ levels and the associated semifine structure is shown in Figure 10 for the case of the rubidium atom Rb (see Table 1 for the j-averaged energy values E_{nl} of the individual nl levels). The energy values of the primitive (unobserved) k levels are obtained by taking the average (centroid) of the E_{nl} values for the s, p, d, f, and g levels in each k band. Thus for $k = 9$, we find

$$E(k = 9) = \tfrac{1}{5}(29{,}298 + 29{,}848 + 30{,}281 + 30{,}499 + 30{,}628)$$

$$= 30{,}111 \text{ cm}^{-1} \tag{22}$$

and similarly, $E(k = 10) = 31{,}126 \text{ cm}^{-1}$.

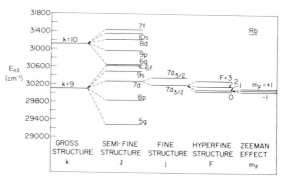

Figure 10. Levels with $k = 9$ and $k = 10$ of the spectrum of neutral rubidium, Rb I. The primitive $k = 9$ and $k = 10$ levels which are unobserved were obtained by taking an average of the observed $k = 9$ and $k = 10$ levels, respectively [see Eq. (22)]. These observed levels constitute the semifine structure. The energy scale E_{nl} is shown on the left-hand ordinate. This scale does not apply to the fine structure of $7d$, the hyperfine structure of the $7d_{3/2}$ level for Rb^{87} (I = 3/2), and the Zeeman splitting of the $F = 1$ level, which are shown schematically enlarged in the right-hand part of the figure.

In Figure 10, we note that the two k bands do not overlap, but the distance between $E_{6f} = 30{,}628 \text{ cm}^{-1}$ (top of $k = 9$ band) and $E_{6g} = 30{,}637 \text{ cm}^{-1}$ (bottom of $k = 10$ band) is only 9 cm^{-1}. This small separation of neighboring k bands has been extensively discussed in Ref. 3 (see pp. 1757–1759 and Table II). Besides the gross structure (quantum number k) and the semifine structure (l), we have shown schematically and on a progressively enlarged scale, the fine structure of the $7d$ state ($7d_{3/2}$ and $7d_{5/2}$), the hyperfine structure of $7d_{3/2}$ for the case of ^{87}Rb with nuclear spin $I = 3/2$, and finally the Zeeman effect of the $F = 1$ hyperfine level. The quantum numbers appropriate to each level structure are shown at the bottom of Figure 10, i.e., k, l, j, F, and m_F.

Thus there exists a hierarchy of five level structures for the rubidium atom, and, of course, also for other pseudo-one-electron atoms and ions with $Z > 11$, as explained in Refs. 1–5. If we consider now again the k gross structure and the l-dependent semifine structure, we wish to point out an important analogy to the SU(3) scheme[30] of the strongly interacting particles. In the latter case, for both the mesons and the baryons, the primitive J^P level is also unobserved, and is split into a number of observed particle levels having different strangeness S and isotopic spin I. Thus as mentioned in Refs. 3 and 4, k is the analog of J^P (spin-parity) and l is the analog of the doublet (S, I) in the particle case. The lack of overlap of neighboring k bands, i.e., what we have called the perfect (or near-perfect) k ordering, is similar to the approximate nonoverlap of different J^P multiplets, e.g., π, K, η, η' ($J^P = 0^-$) and ρ, ω, K^*, φ ($J^P = 1^-$) for the low-lying meson states.

In view of these similarities, we believe that a symmetry breaking, similar to that for the hadronic particle spectrum operates also in the present case. However, the relevant group for the pseudo-one-electron atomic and ionic spectra may well be different from the SU(3) group for the strong interactions (hadrons). Nevertheless, it would indeed be interesting if the same type of physics would apply in both situations, in spite of the fact that the energy spacings involved in the two cases are vastly different. Thus for the hadronic case, considering the baryon octet (N, Λ, Σ, Ξ), the average level spacing is

$$\langle \Delta \rangle_{\text{baryon}} = \tfrac{1}{3}(1320 - 939) = 127 \text{ MeV} \tag{23}$$

as compared to the following average spacing $\langle \Delta \rangle_{k=9}$ for the $k = 9$ group of levels in the Rb spectrum:

$$\langle \Delta \rangle_{k=9} = \frac{E_{6f} - E_{5g}}{4} = \frac{30,628 - 29,298}{4} = 332.5 \text{ cm}^{-1} \tag{24}$$

Here we have divided by 4, since there are four successive intervals involving the five levels shown in Figure 10. Now $332.5 \text{ cm}^{-1} = 332.5/8.07 = 41.2 \text{ meV}$, and the ratio of the level spacings in the two cases is

$$\frac{\langle \Delta \rangle_{\text{baryon}}}{\langle \Delta \rangle_{k=9}} = \frac{127 \times 10^6}{41.2 \times 10^{-3}} = 3.08 \times 10^9 \tag{25}$$

In the preceding context, it is not clear what is the nature of the l-dependent interaction which leads to the breaking of the primitive (global) k symmetry and to the precise sequence of levels with different l values (and $n = k - l$), which constitutes the l patterns which have been extensively discussed by the author in Refs. 1 and 2. Whatever the precise origin of the l-dependent interaction, it must explain the following three features: (a) In general, the l pattern does not change with increasing k, e.g., for Rb, it is $\{l_i\} = gpdsf$ from $k = 5$ to $k \sim 55$, i.e., from the lowest-lying level (ground state) up to highly excited levels very close to the ionization limit. (b) The k bands are separated from each other, i.e., sets of levels with neighboring k values do not overlap in general, provided that $l \leqslant l_1$ and the ionicity $\delta \leqslant \delta_1$. This is what is meant by "perfect k ordering" (see, e.g., Tables 1 and 2 of the present chapter). (c) k is a "good quantum number" for the total energy E_{nl} throughout the region of Z from $Z = 11$ (Na) ($\alpha Z = 0.08 \ll 1$, nonrelativistic region) up to $Z \sim 120$ ($\alpha Z = 0.88$, relativistic region). As discussed in Refs. 2 and 3, the presence of a particular *l* pattern (e.g., *fdps, dpsf, pdsf, dpfs*) depends mainly on the atomic number Z, i.e., on the number and type of closed shells of the core, in particular the *np* and *nd* shells which appear to be mainly responsible for the *k* ordering, as well as the inverted fine structure of excited *d*, *f*, and *g* levels of pseudo-one-electron atoms and ions.

6. Conclusions

In this section, we wish to discuss the general implications and the possible significance of the existence of the *k* ordering of the energy levels of pseudo-one-electron atoms and ions. We have already discussed the very wide prevalence of the *k* ordering for neutral atoms and few-times ionized ions. Incidentally, the number of levels quoted here, namely, 1274 levels (in 42 spectra), does not represent the entire set of levels in Moore's tables[6] which exhibit *k* ordering. In counting these levels, we have generally included only those cases for which at least three levels with the same *k* value (i.e., in the same "*k* band") have been observed. There are also many examples at high excitation energies where only two levels with the same *k* have been listed in Ref. 6. These levels provide additional evidence for the *k* ordering in two ways: (1) the levels with the same value of *k* lie in the order prescribed by the *l* pattern established for the lower-lying levels, e.g., the order is $55p$, $54d$ for Rb, where the *l* pattern is *pdsf*; (2) the doublets of levels always increase in energy with increasing *k*; e.g., for Rb, the doublet $55p$, $54d$ ($k = 56$) lies above the doublet $54p$, $53d$ ($k = 55$). Thus if in the 42 spectra enumerated in Refs. 1 and 2, we had included *all* of the levels which contribute evidence for the *k* ordering, the total would probably have been in the range of 1400–1500 levels. Of course, more recent spectroscopic data than those included in the tables of Moore[6] (1949–1958) would also provide additional evidence for the *k* ordering.

Concerning now the semifine structure of the levels with the same value of *k* as shown in Figure 10 for the $k = 9$ and $k = 10$ levels of Rb, we note that the *relative size* of the intervals $E(k, l_b) - E(k, l_a)$ remains almost constant in going from $k = 9$ to $k = 10$. Thus in Table 3, we have listed the relevant energy levels $E_{nl} = E(k, l)$ as given in Table 1, and in the next column the energy differences $\Delta_{l_b l_a}(k) \equiv E(k, l_b) - E(k, l_a)$, where l_b is the value of *l* pertaining to the line on which $E_{nl} = E_{n_b l_b}$ is listed and l_a is the value of *l* pertaining the next lower energy level (on the line which is immediately above the line for $n_b l_b$). Finally, in the last column of the

Table 3. The $k = 9$ and $k = 10$ Energy Levels E_{nl} of the Neutral Rubidium Atom (in units cm^{-1}), the Energy Differences $\Delta_{l_b l_a} \equiv E(k, l_b) - E(k, l_a)$ and the Ratios $\xi_{l_b l_a} \equiv \Delta_{l_b l_a}/W(k)$, Where $W(k)$ is the Total Width of the k Band

nl	E_{nl}	$\Delta_{l_b l_a}$	$\xi_{l_b l_a}$
$5g$	29,298		
$8p$	29,848	550	0.413
$7d$	30,281	433	0.326
$9s$	30,499	218	0.164
$6f$	30,628	129	0.097
$W(k = 9)$		1330	
$6g$	30,637		
$9p$	30,966	329	0.409
$8d$	31,222	256	0.318
$10s$	31,362	140	0.174
$7f$	31,442	80	0.099
$W(k = 10)$		805	

table, we have listed the values of the ratio $\xi_{l_b l_a} \equiv \Delta_{l_b l_a}/W$, where W is the total width of the k band, e.g., for $k = 9$, $W = 30,628-29,298 = 1330 \text{ cm}^{-1}$.

We note that the ratios $\xi_{l_b l_a}$ are almost independent of k. This result is a direct consequence of the fact that the quantum defects δ_{nl} are approximately independent of n for sufficiently large n, as was noted in our paper in Ref. 5. We may further note that for the particular case of Rb, the intervals $\Delta_{l_b l_a}$ decrease with increasing energy E_{nl} i.e., $\Delta_{14} > \Delta_{21} > \Delta_{02} > \Delta_{30}$. Similar regularities have been pointed out in Ref. 1 for the spectra of Na, K, and Cs (see Tables IX and X and the discussion on p. 1824). These regularities are again a consequence of the behavior of the reduced quantum defects[5] η_{nl} as a function of l (see Figures 7, 8, and 9).

We will now discuss the more general question of the probable significance and the underlying explanation of the existence of the k ordering of atomic and ionic energy levels. First of all, it should be noted that in view of the wide prevalence of the phenomenon of k ordering (1400–1500 levels, as discussed above), it is likely that the underlying effect is a very fundamental one and is the expression of a very basic aspect of atomic structure which has not been incorporated in the present descriptions of atomic energy levels. These descriptions belong to two types: (1) the calculation of the energy levels by means of a suitable "realistic" effective one-electron potential which approaches $-2Z/r$ (Ry) near the nucleus and

$-2(1 + \delta)/r$ (Ry) at large radii r; (2) the modification of the Rydberg formula for a hydrogenic atom by the inclusion of a suitable quantum defect δ_{nl}, i.e., Eq. (2) of the present article.

Concerning (1), I do not believe that a suitable effective potential $V(r)$ can be constructed which (with the aid of the Schrödinger equation) will reproduce all of the fine details of the energy spectra which constitute both the k ordering and the regularities of the associated l patterns. As an example, even for the $k = 9$ band of Rb, the energy intervals $\Delta_{l_b l_a}$ expressed in millivolts are only $\Delta_{14} = 68.2$, $\Delta_{21} = 53.7$, $\Delta_{02} = 27.0$, and $\Delta_{30} = 16.0$ meV. In the course of my calculations of the Sternheimer antishielding factors γ_∞ and R,[26] and the electronic polarizabilities α_d and α_q,[31] and in earlier calculations,[32] I have had extensive experience with the solution of the radial Schrödinger equation using a wide variety of potentials $V(r)$, and in my opinion, it would be nearly miraculous to reproduce accurately such small energy differences between states with different values of both l and n. More importantly, I believe that the k ordering and the associated l patterns represent basically a many-electron effect, and thus are not amenable to a description by a one-electron potential.

Concerning (2), it is noteworthy that the quantum defects δ_{nl}, which are supposed to describe the departure from a hydrogenic level, are quite large, up to $\delta_{nl} \sim 3$–4 for low l values (see Figures 4–6). Hence one might have expected that the use of a pseudohydrogenic formula, namely, Eq. (2), will be inadequate for the description of the excited-state spectrum, both in its general features and in its finer details. This is exactly what I have found in the papers of Refs. 1–5. In fact, the k ordering could be discovered only by examining directly the energy level values E_{nl}, rather than the indirectly derived quantum defects δ_{nl}. Nevertheless, as discussed in Ref. 5, the quantum defects δ_{nl} still play a role in the k ordering, namely, as the order parameter for the k-ordering phase of the spectrum as a function of l.

Thus the k ordering is basically a many-body effect involving both the core electrons (in particular the np and nd shells) and the valence electron. Specifically, the overlap of the core electron wave functions $u_0(n_c l_c)$ and their perturbations $u_1(n_c l_c \to l'_c)$ with the valence wave function $v(nl)$ is probably directly responsible for the k ordering, as discussed in Ref. 3 (pp. 1755, 1756) and in Sections 2 and 3 above. Nevertheless, it is not clear why the complicated and numerous exchange integrals involving the products $u_0 v$ and $u_1 v$ (with their associated angular coefficients) should conspire to give rise to such a beautiful and simple structure of the excited-state level spectra, i.e., the perfect k ordering of the different k bands and the constancy of the l pattern and the regular behavior of the intervals $\Delta_{l_b l_a}$ within each k band. This situation has led me to hypothesize the existence of a basic extended symmetry, which I have called the "k symmetry" and which would be broken by an "l-dependent interaction" to give rise to the sequence of observed nl levels, with $n = k - l$, i.e., the l semifine structure (see Section 5 and Figure 10).

Finally, it may be hoped that the introduction of the concept of k symmetry will ultimately explain another feature of the excited-state spectra of pseudo-one-electron atoms and ions, namely, the fact that neighboring k bands generally do not overlap ("perfect k ordering") and that the separation $S_{k,k+1}$ between neighboring bands is usually much smaller than the widths W_k and W_{k+1} of the bands.[3] Thus the energy region between the ground state and the ionization limit is occupied mostly by the successive k bands consisting of levels with $l \leqslant l_1$ (l_1 is the limiting angular momentum), with relatively little space between adjacent k bands.

Note Added in Proof

Two papers pertaining to the k ordering of atomic and ionic energy levels have been recently published by V. N. Ostrovsky [*J. Phys. B* **14**, 4425 (1981)] and by L. Armstrong Jr. [*Phys. Rev. A* **25**, 1794 (1982)].

Acknowledgments

In connection with the work of Refs. 1–5, I wish to thank Professor H. M. Foley, Drs. V. J. Emery, T. F. Gallagher, J. F. Herbst, R. F. Peierls, J. E. Rodgers, and G. Scharff-Goldhaber, and Professor G. zu Putlitz for helpful discussions. I am also indebted to Professors B. Edlén, C. K. Jørgensen, and W. Persson for valuable correspondence. I wish to thank Miss Anna Kissel for her excellent typing of the manuscript.

References and Notes

1. R. M. Sternheimer, *Phys. Rev. A* **15**, 1817 (1977).
2. R. M. Sternheimer, *Phys. Rev. A* **16**, 459 (1977).
3. R. M. Sternheimer, *Phys. Rev. A* **16**, 1752 (1977).
4. R. M. Sternheimer, *Phys. Rev. A* **19**, 474 (1979).
5. R. M. Sternheimer, *Phys. Rev. A* **20**, 18 (1979).
6. C. E. Moore, *Atomic Energy Levels*, Vols. I–III, Natl. Bur. Stand. (U.S.) Circular No. 467, Washington, D.C., U.S. Government Printing Office (1949–1958).
7. E. Madelung, *Die Mathematischen Hilfsmittel des Physikers*, p. 359; Springer-Verlag, Berlin, Appendix 15 ("Atombau") of 3rd edition (1936).
8. S. A. Goudsmit and P. I. Richards, *Proc. Natl. Acad. Sci. USA* **51**, 664 (1964).
9. J. B. Mann, Los Alamos Scientific Laboratory Report LA-UR-73-1442 (1973).

10. See, for example, E. U. Condon and G. H. Shortley, *The Theory of Atomic Spectra*, p. 337; Cambridge University Press, London (1935).
11. R. M. Sternheimer, J. E. Rodgers, and T. P. Das, *Phys. Rev. A* **17**, 505 (1978).
12. T. F. Gallagher, W. E. Cooke, S. A. Edelstein, and R. M. Hill, *Phys. Rev. A* **16**, 273 (1977).
13. H. M. Foley and R. M. Sternheimer, *Phys. Lett.* **55A**, 276 (1976).
14. T. Lee, J. E. Rodgers, T. P. Das, and R. M. Sternheimer, *Phys. Rev. A* **14**, 51 (1976).
15. L. Holmgren, I. Lindgren, J. Morrison, and A.-M. Mårtensson, *Z. Phys.* **A276**, 179 (1976).
16. E. Luc-Koenig, *Phys. Rev. A* **13**, 2114 (1976).
17. W. Persson and K. Pira, *Phys. Lett.* **66A**, 22 (1978).
18. T. F. Gallagher and W. E. Cooke, *Phys. Rev. A* **18**, 2510 (1978).
19. J. Farley and R. Gupta, *Phys. Rev. A* **15**, 1952 (1977).
20. G. L. Epstein and J. Reader, *J. Opt. Soc. Am.* **65**, 310 (1975).
21. C. C. Kiess, *J. Res. Natl. Bur. Stand.* **56**, 167 (1956).
22. N. P. Romanov and A. R. Striganov, *Opt. Spectrosc.* **27**, 8 (1969).
23. K. B. S. Eriksson and I. Wenåker, *Phys. Scr.* **1**, 21 (1970).
24. D. Popescu, M. L. Pascu, C. B. Collins, B. W. Johnson, and I. Popescu, *Phys. Rev. A* **8**, 1666 (1973).
25. J. Sugar and V. Kaufman, *J. Opt. Soc. Am.* **55**, 1283 (1965).
26. R. M. Sternheimer, *Phys. Rev.* **80**, 102 (1950); **84**, 244 (1951); **86**, 316 (1952); **95**, 736 (1954); **105**, 158 (1957); H. M. Foley, R. M. Sternheimer, and D. Tycko, *Phys. Rev.* **93**, 734 (1954); R. M. Sternheimer and H. M. Foley, *Phys. Rev.* **92**, 1460 (1953); **102**, 731 (1956); T. P. Das and R. Bersohn, *Phys. Rev.* **102**, 733 (1956). For the work carried out after 1957, we note that a bibliography of the articles on the Sternheimer shielding and antishielding factors R, γ_∞, η_∞ and R_H, which is essentially complete up to 1966 is given in the author's paper: R. M. Sternheimer, *Phys. Rev.* **146**, 140 (1966). A more extensive bibliography is given in the list of references (up to 1976) in the Ph.D. thesis of K. D. Sen, entitled "Hartree–Fock–Slater Wave Functions and Sternheimer Shielding–Anti-shielding Factors in Atoms and Ions" (pp. 158–170), submitted to the Department of Chemistry of the Indian Institute of Technology, Kanpur, India, in August 1976. We note that relativistic calculations of γ_∞ have been carried out by F. D. Feiock and W. R. Johnson, *Phys. Rev.* **187**, 39 (1969).
27. H. M. Foley, *Phys. Rev. A* **19**, 2134 (1979).
28. H. Eyring, D. Henderson, B. J. Stover, and E. M. Eyring, *Statistical Mechanics and Dynamics*, p. 295; Wiley, New York (1964).
29. M. J. Seaton, *C. R. Acad. Sci.* **240**, 1317 (1955).
30. M. Gell-Mann, *Phys. Rev.* **125**, 1067 (1962); Y. Ne'eman, *Nucl. Phys.* **26**, 222 (1961).
31. R. M. Sternheimer, *Phys. Rev.* **96**, 951 (1954); **107**, 1565 (1957); **115**, 1198 (1959); **127**, 1220 (1962); **183**, 112 (1969); *A* **1**, 321 (1970).
32. R. M. Sternheimer, *Phys. Rev.* **78**, 235 (1950).
33. It should be noted that the present chapter is a revised and enlarged version of an earlier review article on the k ordering, published by the New York Academy of Sciences as part of "A Festschrift for Maurice Goldhaber," Transactions of the New York Academy of Sciences, Series II, **40**, 190–210 (1980).

2

Multiconfiguration Hartree–Fock Calculations for Complex Atoms

CHARLOTTE FROESE FISCHER

1. Introduction

The Hartree–Fock method has become a standard in atomic structure theory. Simpler methods are often compared with it when accessing their reliability or worth[1] and the notion of correlation, which intuitively may be thought of as the correction needed to account for the fact that electrons do not move independently in a central field, is defined with respect to the Hartree–Fock method rather than some other independent-particle model.[2] In fact, in an earlier article in this series, Fricke[3] states, "The so-called HF method is the basis of all good atomic calculations." In some sense, the Hartree–Fock method is the "best" method. Let us briefly review its properties. (A more detailed discussion may be found in Ref. 4.)

1.1. The Hartree–Fock Approximation

In nonrelativistic quantum mechanics, the total wave function Ψ of a many-electron system is a solution of Schrödinger's equation, $H\Psi = E\Psi$, where H is the nonrelativistic Hamiltonian and E the total energy of the system. The solutions of this equation describe states of the system, and since they also are eigenfunctions of the total angular momentum L and the total spin S they are designated by a label such as γLS where γ usually specifies a configuration.

CHARLOTTE FROESE FISCHER • Department of Computer Science, Vanderbilt University, Nashville, Tennessee 37235.

Wave functions for a single electron in a central potential are called *orbitals* or *spin orbitals* if the spin factor is also included. The radial *factors* of the orbitals are functions $R(nl; r) = (1/r)P(nl; r)$, where $P(nl; r)$ is referred to as the radial *function*. For a given potential the bound-state orbitals form an orthonormal set with $\langle nl|n'l \rangle = 0$ for $n' \neq n$. Let $(nl) = (n_i l_i, i = 1, \ldots, N)$ define a configuration for an N-electron atom. A *configuration state function* (csf) is an antisymmetrized sum of products of orthonormal spin orbitals, one for each electron in the configuration, constructed so as to be an eigenfunction of LS. In this regard, Racah's vector coupling algebra[5] for the coupling of the angular and spin momenta provides a powerful tool for the manipulation of the angular and spin factors. Let (l) denote the set of one-electron angular quantum numbers and α any other information such as the seniority or the coupling scheme required to uniquely specify the spin–angular coupling. Then the configuration state function $\Phi((nl)\alpha LS)$ may be defined as

$$\Phi((nl)\alpha LS) = \mathscr{A}\left\{ \left[\prod_{i=1}^{N} R(n_i l_i; r) \right] |(l)\alpha LS\rangle \right\} \tag{1}$$

In the Hartree–Fock approximation, $\Psi(\gamma LS) \approx \Phi(\gamma LS)$ but with the radial functions selected in a very special manner, namely, the energy functional, $\langle \Phi|H|\Phi \rangle / \langle \Phi|\Phi \rangle$, must be stationary with respect to all allowed variations in the radial functions. It is this stationary condition that is special to the Hartree–Fock method. It leads to a system of coupled integrodifferential equations referred to by Hartree as the "equations with exchange."

1.2. Brillouin's Theorem

The allowed perturbations of the Hartree–Fock solutions that leave the energy functional stationary are of two types:

(i) $P(nl; r) \rightarrow P(nl; r) + \lambda P(n''l; r)$, $\langle n'l|n''l \rangle = 0$ for all occupied orbitals $n'l$

(ii) $\begin{pmatrix} P(nl; r) \\ P(n'l; r) \end{pmatrix} \rightarrow \dfrac{1}{(1 + \lambda^2)^{1/2}} \begin{pmatrix} 1 & \lambda \\ -\lambda & 1 \end{pmatrix} \begin{pmatrix} P(nl; r) \\ P(n'l; r) \end{pmatrix}$

In the first the $n''l$ orbital is an unoccupied orbital whereas in the second both nl and $n'l$ are occupied and the perturbation is, in effect, a rotation of the radial basis. For these perturbations a generalized Brillouin theorem holds, namely,

$$\langle \Phi|H|\Phi_{nl \rightarrow n''l} \rangle = 0 \quad \text{and} \quad \langle \Phi|H|\Phi_{nl \rightarrow n'l, n'l \rightarrow nl} \rangle = 0 \tag{2}$$

The first condition of Eq. (2) states that the interaction with a configuration obtained by replacing an occupied orbital by an unoccupied orbital is zero, provided there is no change in the spin–angular coupling. Another, more positive, way of interpreting this theorem is to say that the configuration obtained by this replacement has been included in the wave function to first order since a 2×2 configuration interaction calculation in which $\Psi \approx c_1 \Phi + c_2 \Phi_{nl \to n'l}$ would not change the Hartree–Fock result. The second condition of Eq. (2) states that the perturbation resulting from a rotation of the orbital basis has been included to first order.

Brillouin's theorem is usually stated with respect to configuration state functions whereas the generalized theorem stated above refers to functions that result from a perturbation. Often the two are the same, but exceptions do occur and then Brillouin's theorem is said not to hold. In such cases the Hartree–Fock results may not be quite as good an approximation. An example is a $2p \to 3p$ replacement in $2p^5$ which, at first sight, would yield the configuration $2p^4 3p \, {}^2P$. The latter, however, does not correspond to a configuration state function as defined by Eq. (1) since there are several ways of coupling $2p^4 (LS) 3p \, {}^2P$. The perturbation of Φ in this case is a linear combination of configuration state functions over the various couplings, with the coefficients proportional to the coefficients of fractional parentage for the $2p^4 (LS)$ group. These cases were first pointed out by Bauche and Klapisch[6] and the similar rotation perturbation by Labarthe.[7] An example of the latter is the $1s2s \, {}^1S$ case. Here the rotation of the radial basis produces a perturbation proportional to $\{\Phi(1s^2 \, {}^1S) - \Phi(2s^2 \, {}^1S)\}$ with the consequence that, with Hartree–Fock orbitals,

$$\langle 1s2s \, {}^1S | H | 1s^2 \, {}^1S \rangle = \langle 1s2s \, {}^1S | H | 2s^2 \, {}^1S \rangle$$

A similar situation occurs with all $ss'' \, {}^1S$ couplings and accounts for the fact that Hartree–Fock results for such configuration states are not upper bounds to the energy. This will be discussed further in the section on excited states.

1.3. LS Dependence

The Hartree–Fock calculation may be performed in a variety of different ways. In a fixed-core calculation certain orbitals, usually inner core orbitals, are not allowed to vary but are kept fixed at some predetermined value. Sometimes LS dependence is suppressed. Since the energy expression for a configuration state depends on the coupling as well as the configuration, fully variational Hartree–Fock calculations yield different radial functions for the different couplings of a configuration. At times it is convenient to constrain the radial functions to be the same for all couplings of a given configuration. The variational calculations are then performed

Table 1. ^1P to ^3P Separations[a] (cm^{-1}) in Some p^5d Configurations

		Average	LS	Obs.
K II	$3p^53d$	67,670	37,000	37,229[b]
Ca III	$3p^53d$	107,180	84,000	74,500[c]
Rb II	$4p^54d$	45,890	27,600	25,600[d]
Sr III	$4p^54d$	74,320	56,400	48,400[e]

[a] Hansen (Ref. 8).
[b] Minnhagen (Ref. 10).
[c] Borgström (Ref. 11).
[d] Reader and Epstein (Ref. 12).
[e] Persson and Valind (Ref. 13).

for the average energy of the configuration, with the energies of the various LS terms computed as deviations from the average energy, expressed in terms of a few Slater integrals that can easily be evaluated. This may result in a considerable saving in computer time. For example, for the $3s3p3d$ configuration there are *nine* configuration states. The fully variational calculation that is the standard Hartree–Fock calculation would require nine different sets of calculations whereas there would only be one for the average energy. The question that arises immediately is whether the additional calculations result in better approximations.

Hansen[8] has studied the LS dependence of radial functions in a large number of configurations. Among the earliest were the p^5d configurations of spectra in the Ar I and Kr I isoelectronic sequence. A remarkable feature of these configurations in ionized atoms is the very high-lying ^1P term. In Ca III, for example, the $3p^53d\ ^1P$ level is more than 50,000 cm^{-1} above any other level of the configuration and Cowan[9] predicted the term to be a further 20,000 cm^{-1} above the observed. Table 1 shows the improvement that can be achieved in predicted level separation when LS dependence is taken into account.

In the configuration model of an atom one tends to think of the state as being described by the nl quantum numbers of the electrons. This is a very simplified concept and the above example shows that allowing the orbitals for an electron to vary from one configuration state to another may already improve the quality of a predicted energy.

2. Correlation and the MCHF Approximation

The effect of correlation in the motion of the electrons in a many-electron system is an important factor in the theoretical determination of atomic properties. It may be included in the description of the wave function by expanding the wave function over a complete, preferably orthonormal, set of configuration states.

2.1. The MCHF Method

In the MCHF method, the wave function Ψ is approximated by a function ψ that is a linear combination of configuration state functions such that

$$\Psi \approx \psi = \sum_{i}^{m} c_i \Phi(\gamma_i LS)$$

but again the radial functions that enter into the definition of the configuration state functions and now also the mixing coefficients, c_i, are such that the energy functional for the approximation, $\langle \psi|H|\psi \rangle / \langle \psi|\psi \rangle$, is stationary with respect to all allowed variations of these quantities. The stationary condition with respect to variations in the radial functions again leads to a system of coupled integrodifferential equations like the Hartree–Fock equations except that mixing coefficients enter into the definition of the potential and exchange, and an interaction term is present as well. The stationary condition with respect to the mixing coefficients leads to the well-known secular problem. This is simplified if the configuration states are orthogonal, a condition always assumed in MCHF calculations. Then the MCHF energy is an eigenvalue of the interaction matrix and the mixing coefficients are components of the corresponding eigenvector.

Many configurations could be included in an MCHF calculation, but often a Hartree–Fock result can be improved significantly through the inclusion of only a few configurations. In order to select configurations wisely it is helpful to consider the question of correlation in greater detail. A more complete review has been presented by Hibbert in an earlier article.[14]

2.2. Zero- and First-Order Sets

In an accurate correlation study, many configurations must be considered and a systematic procedure needs to be adopted. In many-body perturbation theory the notion of "order" is convenient and calculations are often performed to second, third, or higher order. When the MCHF method is used it is convenient instead to partition the configuration states in such a way that the first-order set contains all configuration states interacting with one or more of the zero-order configuration states. Consequently these configuration states may differ by no more than two electrons from at least one of the zero-order states. Unlike perturbation theory, however, in the MCHF method the radial functions for the zero-order configuration states do not remain fixed as more and more configuration states are added to the description of the wavefunction. Thus the notion of zero-order, first-order, or higher-order configuration states is a convenient classification, but with dynamic rather than static properties. This is similar, in some respects, to the difference between the term dependent Hartree–Fock and the configuration averaged Hartree–Fock results for a single configuration. Of course, the MCHF calculation can be performed in a manner whereby certain effects remain fixed, though the wave function then is not quite as accurate. On the other hand, some variations are exceedingly small, and extensive calculations sometimes can be avoided in a "fixed" approximation.

By considering each electron pair in turn and making replacements from each pair, the first-order set of configuration states may readily be enumerated. The set arising from replacements from a given LS-coupled pair defines a *symmetry adapted pair-correlation function*, or more simply a *pair-correlation function*. Configurations that involve replacements to occupied orbitals may contribute to more than one pair-correlation function. In the MCHF method it is convenient to include such configuration states simply in one set.

The above discussion has not defined the zero-order set. In most many-body theories, the zero-order configuration state is often just the Hartree–Fock configuration state, but with this definition it may be necessary to go to higher-order terms (three- and four-electron replacements) to obtain highly accurate results. In the MCHF approach, which in some sense is a pragmatic approach, the zero-order set should contain all the configuration states with an appreciable component in the wave function. In many instances, the zero-order set may already define a sufficiently accurate, approximate wave function.

Through experience the strongly interacting configurations are becoming fairly well known. For highly ionized systems, the *complex*[15] of configuration states with the same parity and the same principal quantum numbers defines the potentially important interactions. Thus for the ground state of beryllium[16] accurate energies could be obtained from a first-order set provided the zero-order set included both $1s^2 2s^2$ and $1s^2 2p^2$.

2.3. Brillouin's Theorem and Interaction with Series

Like the Hartree–Fock approximation, when radial functions are determined variationally the resulting wave function has already included certain configurations differing by one electron, in the sense discussed earlier. This can be an important advantage when certain configurations interact with a whole series of states.

The $sp^6\,^2S$ term of the neutral halogens is an interesting example. Here there is strong interaction with s^2p^4d (all principal quantum numbers the same). Reader[17] noted that the unperturbed Hartree–Fock energy for sp^6 was above that of $s^2p^4d\,^2S$ and, in fact, above the ionization limit. He concluded that configuration interaction would tend to push the $sp^6\,^2S$ levels even higher into the continuum. Cowan, Radziemski, and Kaufman[18] performed a configuration interaction calculation in order to determine the position of sp^6 in which this configuration was allowed to interact with the bound members of the $s^2p^4d\,^2S$ series as well as the $s^2p^4\varepsilon d\,^2S$ continuum. Twenty three basis states were used to represent the d series, both bound and continuum. Their calculation showed that the interaction with the high-lying continuum was sufficiently strong to push the $sp^6\,^2S$ state below the continuum in Cl I. In Br I, the effect of the continuum was smaller and severe mixing with the series was observed with no level being primarily sp^6 in character. Hansen[19] later showed that essentially the same results could be obtained from a two-configuration MCHF calculation in which the whole $s^2p^4d\,^2S$ series, including the continuum, is represented by a single term.

The wave-function compositions for this example, tabulated in Table 2, show some of the severe degeneracy effects that may be present in complex atoms. Clearly, when such situations arise, both configurations must be included in the zero-order set.

Table 2. Eigenvector Compositions[a] (in Percent) for the Lowest ^2S State in the Cl I, Br I, and I I Isoelectronic Sequence Obtained from MCHF Calculations Including sp^6 and s^2p^4d

	Cl I	Ar II	K III	Hydrogenic limit
$3s3p^6\,^2S$	53.3	62.7	66.4	76.2
$3s^23p^4(^1D)\,^2S$	46.7	37.3	33.6	23.8
	Br I	Kr II	Rb III	Hydrogenic limit
$4s4p^6\,^2S$	45.0	61.0	67.0	88.0
$4s^24p^4(^1D)4d\,^2S$	55.0	39.0	33.0	11.9
	I I	Xe II	Cs III	Hydrogenic limit
$5s5p^6\,^2S$	50.7	61.1	56.0	89.7
$5s^25p^4(^1D)5d\,^2S$	49.3	38.9	34.0	10.3

[a] Hansen (Ref. 19).

2.4. Unconstrained Orbitals

When several configurations are present in the zero-order set, it often is advantageous to "unconstrain" the radial functions and not require that they be the same in different configuration states. In order that the secular problem be an eigenvalue problem rather than a generalized eigenvalue problem the different configuration state functions in the expansion of the wave function should be orthogonal. Orthogonality is often guaranteed by the different spin–angular coupling so that there really is no need to constrain the radial functions to be the same. The only additional factors that must be considered are the overlap integrals, if any, that enter into the expression for the interaction when orbitals are unconstrained.

Consider the case of $2s^2 2p \, ^2P$ in boron. Here the most important interacting configuration is $2p^3 \, ^2P$. In Table 3 the results of both a constrained and unconstrained MCHF calculation are reported. Note the lower energy for the unconstrained calculation and also the variation in the mean radii of the different $2p$ electrons. In the $2s^2 2p$ configuration the $2p$ is screened by the $2s^2$ subshell so that its mean radius is fairly large, whereas in the $2p^3$ configuration the self-screening is less and the mean radius smaller. The single, constrained MCHF radial function represents a compromise solution between the two.

The better accuracy of the unconstrained approximation can be explained by Brillouin's theorem. Let the MCHF approximation be

$$\psi = c_1 \Phi(2s^2 2p_1 \, ^2P) + c_2 \Phi(2p_2^3 \, ^2P)$$

where $2p_1 \equiv 2p_2 \equiv 2p$ in the constrained calculation. Then in the constrained calculation the $2p \rightarrow 3p$ replacement leads to the condition

$$\left\langle \psi \middle| H \middle| c_1 \Phi(2s^2 3p \, ^2P) + c_2 \left\{ \sum_{LS} \langle 2p^2 \, LS \| 2p^3 \, ^2P \rangle \Phi(2p^2 (LS) 3p \, ^2P) \right\} \right\rangle = 0 \quad (3)$$

where $\langle 2p^2 \, LS \| 2p^3 \, ^2P \rangle$ represents a coefficient of fractional parentage. In

Table 3. A Comparison of Results for Constrained and Unconstrained MCHF Calculations for Boron over the Set of Configurations $2s^2 2p_1$, $2p_2^3 \, ^2P$

	Constrained	Unconstrained
$\langle r_{2p_1} \rangle$	2.119	2.187
$\langle r_{2p_2} \rangle$	2.119	1.973
$-E^{\text{MCHF}}$ (a.u.)	24.56034	24.56201

the unconstrained calculation both parts of Eq. (3) individually would have a zero interaction and the replacement, $2p_1 \to 3p_1$, for example, would not require the $3p_1$ orbital to be orthogonal to the $2p_2$ orbital, though it would have to be orthogonal to that for $2p_1$. In a sense, the constrained MCHF calculation has included an average effect of the one-electron replacements, whereas the unconstrained calculation has included $2p_1 \to np_1$ replacements exactly to within the present approximation.

2.5. Reduced Forms

Many of the above ideas are related to the notion of a "reduced form" for a configuration interaction expansion.[20] For example, the ability to represent the interaction with a whole series by a single term follows from the transformation

$$c\Phi(\gamma(L'S')l^*LS) = \sum_n c_n \Phi(\gamma(L'S')nlLS) \qquad (4)$$

where

$$P(l^*; r) = \sum_n c_n P(nl; r)/c, \qquad c^2 = \sum_n c_n^2 \qquad (5)$$

The orbital associated with the transformed radial function $P(l^*; r)$ is very different from what is sometimes called a "spectroscopic" orbital. Often such transformed orbitals are labeled in the same way as spectroscopic orbitals, but are then referred to as "correlation" or "virtual" orbitals. The halogen example described earlier showed that the expansion defining the transformation included a summation over both bound and continuum states.

A second type of reduction can be used when double summations are present in the expansion but with no change in spin–angular coupling. For example, an orthogonal transformation of an orbital basis exists such that

$$\sum_{n \geq 1} \sum_{m \geq 2} c_{n,m} \Phi(nsmp\ {}^3P) = \sum_{n \geq 1} c_n^* \Phi(ns^*(n+1)p^*\ {}^1P) \qquad (6)$$

The transformation of the basis now depends on the matrix of coefficients $(c_{n,m})$. The reduced form makes it possible to represent pair-correlation functions with a minimum number of configurations. Furthermore, the configurations may all differ by two electrons. Reducing the number of configurations clearly has the potential for increasing the efficiency of a method, but the ability to deal with configurations that differ by two electrons also increases computational efficiency because the interaction between configurations may then be expressed in terms of a relatively small number of Slater integrals. (When configurations differ by one electron the whole core may enter into the interaction, greatly increasing the number of Slater integrals.) On the other hand, since the transformations, in general, are different for different sums, more radial functions are needed to represent the wave functions and more differential equations need to be solved in an MCHF calculation. But for a large problem, the amount of computation depends primarily on the number of Slater integrals and the number of configurations, and to a lesser extent on the number of equations.

The exact solution of a two-electron problem is equivalent to a pair-correlation problem, though somewhat simpler since orthogonality to a core need not be considered. As an example, let us consider the $2\,^3S$ and $2\,^3P$ states in Li II. In Li II we have a two-electron system for which the wave function can be represented in a reduced form with a minimum number of configurations. For ^3S the MCHF expansion was taken to be over the set of configurations

$$\{1s2s,\ 3s4s,\ 2p3p,\ 4p5p,\ 3d4d,\ 4f5f\}$$

whereas for ^3P the expansion was over the set

$$\{1s2p_1,\ 2s3p_1,\ 3s4p_1,\ 2p_23d_1,\ 3p_24d_1,\ 3d_24f\}$$

In the above, orbitals with the same subscript (or none) are orthogonal, but orthogonality is not required when the subscripts are different since the different angular coupling assures the orthogonality of the configuration states without any *radial* orthogonality constraint.

The total energies achieved with this approximation are reported in Table 4, in atomic units, and compared with similar values obtained by Accad, Pekeris, and Schiff[22] and also experiment.[23] The MCHF results include relativistic shift corrections to be described in a later section. Accad

Table 4. Total Energies for the $2\,^3S$ and $2\,^3P$ States
of Li II in Atomic Units

	$2\,^3S$	$2\,^3P$
MCHF[a]	−5.111325	−5.028211
Expansion[b]	−5.111345	−5.028286
Experiment[c]	−5.111201	−5.028146

[a] Froese Fischer (Ref. 21).
[b] Accad *et al.* values (Ref. 22), revised to include −0.000539 a.u.
as the relativistic shift correction for the 1s electron.
[c] Moore (Ref. 23).

et al. only report the ionization energy but include similar relativistic corrections to this quantity. In converting their results to total energies, a relativistic shift of $-\alpha^2 Z^4/8$ a.u. was added to the nonrelativistic energy of the 1s electron and the Rydberg constant $R_{Li} = 109728.7$ was used to convert their ionization energy, given in cm^{-1}, to atomic units. Neither of the theoretical calculations include the effect of the Lamb shift. Table 3 shows that the discrepancy between the MCHF and the Hylleraas expansion method is smaller than the Lamb shift.

The above examples have shown that the MCHF method can be used most effectively when a configuration interacts with a whole series, in which case the effect of the series can be represented by a single term, and when a pair-correlation function needs to be computed, in which case a rapidly converging, reduced form for the expansion can be obtained, frequently requiring only a few terms. Now let us focus on some of the problems that can be encountered.

3. Excited States

Like the Hartree–Fock method, the MCHF method determines stationary states of an energy functional. In many cases the solutions obtained are also upper bounds to the exact energy, but this is not always guaranteed. The Hylleraas–Undheim–MacDonald (HUM) theorem together with Brillouin's theorem can sometimes be used to prove that an MCHF energy is indeed an upper bound. According to the HUM theorem, the kth eigenvalue of an interaction matrix is an upper bound to the kth eigenvalue of the exact Hamiltonian.

Consider the ^3F series of Al II where the fourth member is designated $3p3d\ ^3F$ and the configuration $3p3d$ is a perturber for the whole $3snf\ ^3F$ series. In the MCHF method the wave function for the second ^3F, for example, through transformations such as those of Eqs. (4) and (5), could be represented as

$$\psi(3s5f\ ^3F) = c_1 \Phi(3s5f\ ^3F) + c_2 \Phi(3p3d\ ^3F)$$

where $3s$, $5f$, $3p$, $3d$ are all variational or, more briefly,

$$\psi(3s5f\ ^3F) = c_1 \Phi_1 + c_2 \Phi_2$$

Then the MCHF total energy is an eigenvalue of a 2×2 interaction matrix. Now suppose we extend the nf orbital basis to include a $4f$-like orbital and define $\Phi_3 \equiv \Phi(3s4f\ ^3F)$. Consider now the 3×3 interaction matrix where the new configuration state has been added to the MCHF expansion.

By Brillouin's theorem

$$\langle \psi(3s5f\,^3F)|H|\Phi(3s4f\,^3F)\rangle = 0$$

or, equivalently,

$$c_1H_{13} + c_2H_{23} = 0$$

The latter condition is sufficient to assure that the MCHF energy is also an eigenvalue of this extended matrix. If the diagonal energy of the new basis state, H_{33}, is such that $H_{33} < E^{\text{MCHF}}$, then the 3×3 interaction matrix has an eigenvalue lower than the MCHF energy, and the latter must be an upper bound to the second exact ^3F energy (neglecting relativistic effects). The proof depends on finding an appropriate $4f$ function orthogonal to the MCHF $5f$. In the present case, it is not difficult to find such a function. The same idea can be extended to higher members of the series, at least in principle. In each case, to prove that the MCHF energy is an upper bound to the kth eigenvalue, one must show that the MCHF energy is the kth eigenvalue of a possibly extended interaction matrix.

Although, in principle, the two configuration expansion could be used for all members of the ^3F series, numerical instabilities arise. For example, the Hartree–Fock energy of $3p3d\,^3F$ changed very little when the interaction with the $3snf\,^3F$ series was included,[24] but the wave function changed drastically to the point where the $3p3d\,^3F$ configuration state accounted for only about 33%–36% of the composition of the final wave function. Thus a small change in the energy produced a large change in the wave function, a typical example of instability. In a study of this series, the instabilities were circumvented to a large degree by introducing a fixed basis for the lower-lying configuration states and using variational functions to "mop up" the rest of the series.

4. Transition Metals

The transition metals are examples of complex atoms that have not been studied extensively by *ab initio* methods. Lines in the spectra of these atoms, particularly the neutral species, were identified largely by least-squares fitting of the levels assuming a fairly simple interaction model.[25] It was from such studies that the near degeneracies of the $3d$ and $4s$ electrons in the neutral species was discovered. It was somewhat surprising, therefore, in a study of the $3d^n4s \rightarrow 3d^n4p$ transition using the MCHF method, to find that in most instances the interaction was relatively small.[26] On closer inspection it was found that the Hartree–Fock multiplets were not always predicted in the correct order.[27] Figure 1 shows some of the energy levels of Mn I.[28] On the left the energy scale is adjusted so that the energy of the $3d^6$ ^5D configuration coincides with the ionization limit. Two configurations have energies above the observed levels, namely, $3d^6(^5D)4p$ ^6P and $3d^6(^5D)4s$ ^6D whereas all the configurations $3d^54s(^7S)nl$ ^6L have energies that are too low. Since correlation of outer electrons with the core tends to *increase* the ionization energy, this factor cannot explain energy levels that are too low. On the right the HF energy levels are shifted so that now the energy of $3d^54s$ ^7S agrees with this ionization limit and we see that now the energy levels are too high. Those for $3d^54s(^7S)nl$ are in good agreement with observation implying that correlation of the outer nl electron with the $3d^54s$ ^7S core is small. On the other hand, those for $3d^6(^5D)4l$ are much too high and produce some inversions in the order of the levels. The two ^6D states are too close together whereas the two ^6P states are too far apart. In the case of the latter, configuration interaction between these configurations will only shift them further apart. Thus, merely including the $3d, 4s$ degeneracy in a calculation for these states is not likely to improve the quality of the wave function.

The main cause of the problem in this case, as well as others in the iron series, lies with the fact that there is far more correlation in the $3d^n$ core than the $3d^{n-1}4s$ core. In the pair correlation theory as analyzed by

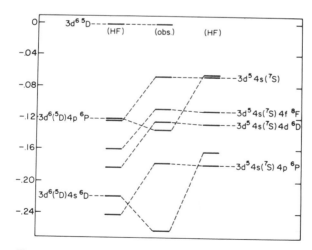

Figure 1. A comparison of the observed and Hartree–Fock term structure in Mn I. On the left the Hartree–Fock levels are relative to the $3d^6$ ^5D ionization limit. On the right the levels are relative to the $3d^54s$ ^7S ionization limit.

Layzer et al.,[29] the contribution to correlation energy from a particular electron pair is approximately proportional to the number of pairs, and subsequent calculations have shown that the per pair correlation energy depends largely on the proximity of the two electrons. Thus, the $(3d, 4s)$ per pair correlation energy is less than the $(3d, 3d)$ per pair correlation energy and there are $(n - 1)$ more $(3d, 3d)$ pairs in $3d^n$ than in $3d^{n-1}$. This difference in correlation must be accounted for before the $3d$, $4s$ degeneracy can be introduced in a meaningful way in an *ab initio* calculation.

Actually, Figure 1 suggests that possibly calculations should be performed relative to the $3d^5$ core, or $3d^{n-1}$ in general. This also suggests that a partially extended model could be used in which the last $3d$ electron is not constrained to be the same as all the other electrons. This model had been used in the study of the $4s\,^2S$–$4p\,^2P$ transition of the Cu I sequence.[30] In effect the $3d^{10}$ subshell was replaced by $3d^9 3d'\,^1S$ and since the $3d'$ orbital can be decomposed into a $3d$ component and its orthogonal projection, say a $4d$ component, we can represent the configuration state function for $3d^9 3d'\,^1S$ as a multiconfiguration wave function, namely,

$$\Phi(3d^9 3d'\,^1S) = c_{3d}\Phi(3d^{10}\,^1S) + c_{4d}\Phi(3d^9 4d\,^1S)$$

For the neutral atom, the improvement in the theoretical ionization energy was appreciable, particularly for $4s$. Table 5 shows the steady improvement in the ionization energy as first the core is replaced by the partially extended model described above and then the correlation between the outer electron and the core (called core polarization) is taken into account.

The above model has also been applied to a study of the $3d^5 4s\,^7S$–$3d^5 4p\,^7P$ transition in Cr I where there is significant interaction between $3d^5 4p\,^7P$ and $3d^4(^6D)4s4p\,^7P$ in the upper state. Earlier MCHF studies,[27] including correlation of the outer electron with the $3d^5$ core, predicted the wrong dominant component for the lowest ^7P, with a composition of 81%

Table 5. Some Theoretical Ionization Energies for $4s$ and $4p$ in Cu I Compared with Observation[a,b]

	$4s$	$4p$
Hartree–Fock	0.23550	0.12319
$3d^9 3d'$ core	0.25200	0.12975
+ core polarization	0.28416	0.14291
Observation[c]	0.28394	0.14406

[a] All units are atomic units.
[b] Froese Fischer (Ref. 30).
[c] Moore (Ref. 23).

$3d^4 4s(^6D)4p \; ^7P$ for the state, whereas Roth's analysis,[25] based on the fitting of 296 experimental levels, yielded a composition of about 70% $3d^5(^6S)4p \; ^7P$ for this same state. In the partially extended model the energy of the latter configuration state was lowered sufficiently that it became the lowest and when the two configurations were allowed to interact, the results were in reasonable agreement with observation.[28]

The Cr I study illustrated the importance of the upper bound property of a wave function. Even though the MCHF wave function for the lowest state had the wrong dominant component, it was still an approximation for the lowest state, and this was evident from the f values. The f value for a transition from the $3d^5 4s \; ^7S$ ground state to the lowest ^7P is smaller than for the next ^7P. Thus in an MCHF calculation the phases of the mixing coefficients for the two interacting ^7P configuration states must be such that there is cancellation in the f value for a transition to the lowest state and enhancement for the next. The relative phases of the largest two components is actually a more important characteristic of a configuration state expansion of a wave function than the dominant component, although the latter has been accepted, in practice, as a means of identifying the state.

The partially extended model has also been used to investigate the energy separation between $3d^3 4s \; ^5F$ and $3d^2 4s^2 \; ^3F$ in titanium.[31] The latter is the ground state and here the $3d^2 4p^2 \; ^3F$ configuration accounts for the dominant outer correlation effect. The partially extended model was used for $3d^2 3d' 4s \; ^5F$ and produced a $3d'$ orbital very different from the $3d$ orbital. In Table 6 some energy separation results are presented. The Hartree–Fock value is too small and including only the dominant correlation effect in the ground state produces a result that is much too large. When the partially extended model is used for the ^5F state in the form of an MCHF wavefunction, theoretical energies in reasonable agreement with observation are obtained. Because of the diffuse nature of the $3d'$ orbital, it has a large $4d$ component so that there also will be a large $3d^2 4s 4d \; ^5F$ component in the wave function. It is known that $3d^2 4s 4d \; ^5F$

Table 6. The Energy Separation^a (cm^{-1}) of the $3d^2 4s^2 \; ^3F$ and $3d^3 4s \; ^5F$ Levels in Ti I

	$^5F - ^3F$
Hartree–Fock	4,460
MCHF (ground state)	10,700
MCHF (both states)	7,020
Observation	6,500

^{a} Dunning, Botch, and Harrison (Ref. 31).

interacts fairly strongly with $3d^2 4p4f\ ^5F$. The inclusion of the latter configuration may bring the theoretical results in even better agreement.

Correlation effects play a decisive role in determining properties of the neutral atoms and ions in the transition metals, but they have not been studied extensively. Recently Jankowski *et al.*[32,33] have begun to investigate correlation in some of the closed shell systems, using a variational-perturbation theory method for obtaining second-order, pair-correlation energies. For Zn III, a correlation energy of -0.50249 a.u. was obtained for the $3d^{10}$ subshell of which -0.4116 a.u. was attributed to $3d^2 \rightarrow \{dd, ff\}$ replacements. This is a large correlation effect and may have important implications in a variety of situations.

5. Relativistic Corrections

Further experience needs to be acquired, but it appears that the relativistic effects on the energy are often larger than on the wave function, and that simple corrections suffice, at least for the smaller atoms. For a closed shell system the Dirac–Hartree–Fock (DHF) calculation is not much lengthier than a Hartree–Fock calculation and if all cases were of this type one should simply forget about nonrelativistic calculations. However, as soon as correlation becomes important the number of configuration states that need to be included in the jj coupling scheme increases rapidly, more rapidly than in the nonrelativistic approximation. This point has been made by several authors. For example, in their study of Au and Hg, Desclaux and Kim[34] point out that the *nine* configuration states for the $6\ ^1S$ ground state are equivalent to *three* in the nonrelativistic equivalent. Thus for neutral atoms or ions of low degree, where correlation is of primary importance, a simple relativistic correction may be advantageous.

In the Breit–Pauli approximation,[35] the Hamiltonian is expressed as a sum of the nonrelativistic Hamiltonian and the relativistic correction

$$H_{BP} = H_{NR} + H_R$$

where H_R is a sum of the following contributions: H_{SO}, the one-electron spin–orbit interaction; H_M, the relativistic mass correction; H_{D_1}, the one-body Darwin term; H_{SOO}, the spin–other-orbit interaction; H_{SS}, the dipolar spin–spin interaction; H_{OO}, the orbit–orbit interaction; H_{D_2}, the two-body Darwin term; and H_{SSC}, the spin–spin contact term. For our purposes it is more convenient to regroup these operators so that

$$H_{BP} = H_{NF} + H_F$$

where H_{NF} is the non-fine-structure Hamiltonian, diagonal in the LS coupling scheme, namely,

$$H_{NF} = H_{NR} + H_M + H_{D_1} + H_{OO} + H_{D_2} + H_{SSC}$$

and H_F is the fine-structure operator

$$H_F = H_{SO} + H_{SOO} + H_{SS}$$

The latter was approximated by Blume and Watson[36] so that

$$H_F \approx \zeta' \sum_i' l_i \cdot s_i$$

where the sum on i is over the indices of nonclosed shells.

Among the first to consider extensive configuration interaction (six or more configurations) along with spin–orbit interaction were Dankwort and Trefftz,[37] who computed oscillator strengths for boron-like Si X. As off-diagonal spin–orbit parameters they used a Hermitian generalization of the Blume and Watson parameters ζ'. Most of their predicted energy levels were in good agreement with observation, but for $2s2p^2\,^4P$ and $2p^3\,^4S$ the energy levels were too low. The neglected relativistic components of the non-fine-structure term tend to shift the energy of a configuration with very little LS dependence. A configuration with two $2s$ electrons will be shifted more than one with only one $2s$ electron. With these corrections the energies of the configurations mentioned above would shift upwards relative to the ground state.

Intercombination lines are a sensitive test of a relativistic correction, since they have their origin in the fine-structure interaction. In his study of the $3s3p\,^3P_1$–$3s^2\,^1S_0$ transition in Mg I, Dankwort[38] found the best agreement with observation for the fine-structure splitting of $3s3p\,^3P$ from an approximation that included correlation with the core, relativistic shift corrections, as well as spin–orbit interaction. This is shown in Table 7 where the $^3P_0$–$^3P_1$ separation is tabulated. Note that including all Breit–Pauli interactions (which we define to include the spin–orbit interaction) has no effect when outer correlation alone is included. Only when the complete $2p^6$ subshell is broken to include correlation with the core, do these terms contribute. The splitting improves appreciably when correlation with the core is introduced. The final result is in excellent agreement with experiment.

The gf value obtained from this calculation for the intercombination line was 2.67×10^{-6} which agrees very well with the relativistic-potential

Table 7. The Dependence of the Fine-Structure Splitting of $3s3p\ ^3P$ in Mg I on Various Interactions[a] (i) Outer Correlation; (ii) Spin–Orbit Interaction; (iii) Breit–Pauli Interactions; (iv) Correlation with the Core

Interactions	E_{0-1} (cm^{-1})
(i) + (ii)	15.4
(i) + (iii)	15.4
(i) + (ii) + (iv)	21.3
(i) + (iii) + (iv)	20.0
Observation[b]	20.06

[a] Dankwort (Ref. 38).
[b] Moore (Ref. 23).

method value of 2.6×10^{-6} reported by Aymar and Luc-Koenig.[39] Unfortunately there is considerable scatter in the experimental values but one value of 2.5×10^{-6}, obtained by Mitchell[40] using the hook method, agrees well with these theoretical values.

There are some difficulties associated with the use of the Blume and Watson spin–orbit parameter in an intermediate coupling type of calculation. First, this parameter was defined by Blume and Watson only for configurations with at most one open shell. When several open shells are present, the MCHF program[41] used by Dankwort and Trefftz in their computation introduced some simplifying assumptions. Second, the special Hermitian approximation mentioned earlier had to be introduced when off-diagonal spin–orbit parameters entered into the interaction matrix. The configuration interaction approach adopted by Hibbert and Glass[42] is similar in many respects but relies only upon a well-defined nuclear spin-orbit term and other terms of the Breit–Pauli operator.

In their calculations, Hibbert and Glass expand the wave function in terms of a set of configuration states for which an orthogonal set of orbitals has already been determined. Usually the orbitals in this set are close to Hartree–Fock orbitals for selected configurations. The energy interaction matrix is then computed, including various contributions from the Breit–Pauli Hamiltonian, and diagonalized. However, usually H_{OO}, H_{D_2}, and H_{SSC} are omitted.

The berylliumlike sequence has been studied extensively by Glass[43] and it is interesting to compare some of the fine-structure splitting obtained in this way with the multiconfiguration Dirac–Hartree–Fock data where the relativistic effects also affect the wave function. Cheng, Kim, and Desclaux[44] have published extensive data on energy levels for isoelectronic sequences of the atoms lithium to neon. Their calculations take into account all the degeneracy effects within the $n = 2$ shell, the Breit correction, and the Lamb shift. In Table 8 some of their results are compared with those obtained by Glass.

The fine-structure separation $E(^3P_2 - {}^3P_1)$ depends primarily on the nuclear spin–orbit interaction and other two-body effects whereas the transition energy $E(^3P_1 - {}^1S_0)$ is affected to a greater extent by correlation, the mass correction, one-body Darwin term, and the Lamb shift, at least for higher Z values. Glass's calculations neglect the orbit–orbit, two-body Darwin, and spin–spin contact terms of the Breit–Pauli interaction, all of which are two-body effects that would tend to affect the fine-structure separation more than the transition energy. His wave-function expansion appears to include considerably more correlation than the MCDHF calculation but there is a basic difference in the two approaches compared here. In the MCDHF calculations, separate variational calculations were performed for each J value. Glass, on the other hand, uses the same orbitals for a variety of states (not all of them mentioned here) and so a larger basis is needed to represent the allowed perturbations of a variational calculation. However, there is no counterpart in the MCDHF calculation of the pd type of configurations. Glass's fine-structure splitting and energy

Table 8. Multiconfiguration Breit–Pauli Results[a] Compared with Multiconfiguration Dirac–Hartree–Fock[b] Results and Observation[c] for Highly Ionized Atoms of the Beryllium Sequence. BP: $\{2s2p, 2s3p, 3s2p, 3s3p, 2p3d, 3p3d\}$ ${}^3P^0$; $\{$same as above plus $2s4p, 3s4p, 2s5p, 3s5p\}$ ${}^1P^0$; MCDHF: $\{2s2p_{1/2}, 2s2p_{3/2}\}$ ${}^3P_J, J = 0, 1, 2$

Z		$E(^3P_2 - {}^3P_1)$ (cm^{-1})	$E(^3P_1 - {}^1S_0)$ (a.u.)	f
12	BP	2447	0.6470	2.02(−3)
	MCDHF	2443	0.6534	1.93(−3)
	Obs.	2462	0.6453	
14	BP	5159	0.7865	5.23(−3)
	MCDHF	5137	0.7922	4.99(−3)
	Obs.	5181	0.7843	
18	BP	16,889	1.0793	2.29(−2)
	MCDHF	16,724	1.0819	2.18(−2)
	Obs.	16,727	1.0746	
20	BP	27,631	1.2352	4.13(−2)
	MCDHF	27,264	1.2348	3.94(−2)
	Obs.	27,490	1.2279	
26	BP	94,616	1.7598	1.64(−1)
	MCDHF	92,101	1.7325	1.53(−1)
	Obs.	90,602	1.7275	

[a] Glass (Ref. 43).
[b] Cheng et al. (Ref. 44).
[c] Fawcett (Ref. 45) for the $E(^3P_2 - {}^3P_1)$ separation and Widing (Ref. 46) for the $E(^3P_1 - {}^1S_0)$ transition energy.

separation data is in slightly better agreement with observation up to at least Si XI, probably because of additional correlation effects. It also is interesting to note that, even for Ca XVII, the BP $E(^3P_1-{}^1S_0)$ energy separation, corrected for the Lamb shift by the same amount as the MCDHF value, yields a value of 1.2266 a.u., in better agreement with the observed value of 1.2279 a.u. than the MCDHF value of 1.2348 a.u. For Fe XXIII the BP value, corrected for the Lamb shift in a similar manner, is no longer in better agreement but differs from the MCDHF value by only 0.00596 a.u. or 0.34%. From this comparison we conclude that for the lighter atoms, at intermediate stages of ionization, where both correlation and relativistic effects are important, a nonrelativistic approach correcting for relativistic effects can compete favorably with a fully relativistic one.

The f value for an intercombination line is sensitive to the Breit interaction, yet in Table 8 the difference in the two f values remains constant at about 5% up to Ca XVII and increases to just over 6% for Fe XXIII.

6. Transition Probabilities

The discussion so far has concentrated primarily on the energy since that is the criterion used most often in assessing the quality of a wave function. Other properties such as oscillator strengths were introduced at times, but always in the same context. However, transition probabilities themselves are of interest, so some comments on the computation of these quantities seems appropriate.

6.1. A First-Order Theory for Oscillator Strengths

A first-order theory for oscillator strengths (FOTOS) was first proposed by Beck and Nicolaides[47] and derived independently[48] for a more general MCHF approach. The latter will be described here.

The zero-order and first-order configuration sets define the large and small contributions to the wave function. With this in mind, let us write the wave functions Ψ and Ψ' of the initial and final state, respectively, as

$$\Psi = |0\rangle + |1\rangle$$
$$\Psi' = |0'\rangle + |1'\rangle \tag{7}$$

The f value for the transition, in the length form, is given by the expression

$$f = \frac{2}{3}\Delta E \frac{\langle \Psi' \|\mathbf{r}\| \Psi \rangle^2}{(2L+1)}$$

with all quantities in atomic units. Substituting the expansions given in Eq. (7) into the definition of the transition matrix element we get

$$\langle \Psi' \| \mathbf{r} \| \Psi \rangle = \langle 0' \| \mathbf{r} \| 0 \rangle + \langle 1' \| \mathbf{r} \| 0 \rangle + \langle 1 \| \mathbf{r} \| 0' \rangle + \langle 1' \| \mathbf{r} \| 1 \rangle$$

As the notation implies, the four contributions to the transition matrix element are of zero-, first-, first-, and second-order, respectively. Because of the selection rules for the dipole moment operator, many of the configuration states in the first-order set do not contribute to the transition matrix element to first order. Thus a Hartree–Fock f value often can be improved significantly through the inclusion of only a few additional configurations in the description of the wave function, provided the observed value of ΔE is used in the computation of the f value. The theoretical ΔE depends on many more configuration states to first order in the wave function and in a series of calculations may vary from being too large to being too small, depending on the amount of correlation included in each state.

As an example, consider the $3\,^2S$–$3\,^2P$ transition in Na I where the zero-order sets of both the initial and final states contain only the single configurations, $1s^2 2s^2 2p^6 3s\,^2S$ and $1s^2 2s^2 2p^6 3p\,^2P$, respectively. The configuration sets defining a first-order transition matrix calculation can be obtained by allowing other electrons in the initial and final state to undergo the transition. Thus for the initial state we have

$$\{3s, 2p^5 ns\,(^1P)3p, 2p^5 nd\,(^1P)3p, 2s3p^2, 2smp\,(^1P)3p, 1s3p^2, 1smp\,(^1P)3p\}^2 S$$

where $n \geqslant 3$ and $m \geqslant 4$. In designating the configurations here we have omitted all references to complete subshells.

The MCHF method is particularly effective for such first-order calculations since each of the implied sums over m or n can be reduced to a single term by the transformations given by Eqs. (4) and (5). The effect of correlation on the f values for transitions in Na I sequence have been investigated[48] but it was found that for neutral atoms (or atoms of low degree of ionization) a first-order transition matrix element calculation does *not* yield an accurate f value. For outer electrons the Hartree–Fock orbital is often too diffuse, a property closely related to the ionization energy being too small and, in the present case, the f value being too large. The first-order calculation does not adequately correct for this effect.

6.2. Core Polarization and Ionization Energies

By definition, the ionization energy of a particular electron is the positive energy difference between the energy of a many-electron system before and after the electron has been removed. The presence of an electron introduces extra pair correlations and these contribute to its ionization energy.

Let us consider the $2s$ and $2p$ ionization energies of Li I that have been studied in detail.[21] There the pair correlations defining core polarization are $1s2s$ and $1s2p$. In many respects the problem of determining these pair correlations is similar to the expansion for the ^3S and ^3P states in Li II, where the most important contributers to correlation were configurations which in lithium could be labeled $1s2p^2$ or $1s2p(^{1,3}P)3p$ for the ^2S state and $1s2s(^{1,3}S)2p$ or $1s2p(^{1,3}P)3d$ for the ^2P state. In most cases the interacting configuration can be interpreted as an outer electron outside a core, usually a non-^1S core, and so this type of correlation is often referred to as *core polarization*. Configurations such as $1s2p^2 \, ^2S$ account for the orthogonality of the outer $3p$ electron to the $2p$ electron in the core.

As mentioned before, the Hartree–Fock ionization energy is always too low. It would be helpful if the problem of core polarization could be treated as one of maximizing the ionization energy relative to a fixed core. However, the results in Table 9 show that the observed ionization energy is not an upper bound. An accurate calculation including most of the correlation with the core may yield an energy that is too low with respect to the ionization limit. However, when correlation is then added to the core of both the atom and the ion there is interference between the correlation in the core and the core polarization for the system with the additional electron. The $(N-1)$-electron system consequently has slightly larger correlation in the core than the N-electron system, bringing the energy difference in better agreement with observation.

The effect of core polarization is sometimes introduced through a polarization potential, with parameters adjusted so that the ionization energies of selected electrons agree with the observed values. Typical results for Li I are included in Table 9. These methods predict the energies as well as (if not better than) the more elaborate MCHF results, but there is a question about the accuracy of this approach when predicting other

Table 9. Ionization Energies[a] for Li I in cm^{-1}

	$2s$	$2p$
Hartree–Fock	43,081	28,234
+ Core polarization	43,606	38,575
+ Core correlation	43,454	28,580
Polarization potential[b]	43,472	28,539
Observation[c]	43,487	28,583

[a] Froese Fischer (Ref. 21).
[b] Moore *et al.* (Ref. 48).
[c] Moore (Ref. 23).

Table 10. F Values for the $3s\ ^2S$–$3p\ ^2P$ Transition in Mg II

Method	f value
Hartree–Fock[a]	0.988
MCHF[a]	0.912
Statistical model potential[b]	0.961
plus core polarization[c]	0.956
Model potential[d]	0.897
Experiment	0.86^e, 0.88^f, 0.93^g

[a] Froese Fischer (Ref. 48).	[e] Liljeby et al. (Ref. 53).
[b] Saraph (Ref. 50).	[f] Sorenson (Ref. 54).
[c] Mendoza (Ref. 51).	[g] Smith and Liszt (Ref. 55).
[d] Black et al. (Ref. 52).	

atomic properties that may be more sensitive to the composition of the wave function, such as f values.

Table 10 compares f values obtained from a variety of methods for Mg II. Unfortunately, experimental values exhibit a great amount of scatter and only some of the more recent values in better agreement with the f value trend are included in this table. The model potential results also fluctuate. It is interesting to note, however, that the result obtained by Saraph[50] using a statistical model potential, but neglecting core polarization, is very similar to that obtained by Mendoza[51] including core polarization. Both used the same computer codes. In the MCHF computation the $2p^53d(^1P)3p$ and $2p^53d(^1P)3s$ core polarization corrections reduced the Hartree–Fock f value appreciably.

The close-coupling calculations for collision and scattering problems are closely related to MCHF calculations, including correlation with a core, for systems with an outer electron. However, in the scattering approach the "target" (core) wave functions are usually defined in terms of spectroscopic orbitals and not varied. Seaton and Wilson[56] refer to this as the "frozen cores" approximation.

6.3. f Value Trends

In an earlier article in this series, Wiese[57] discussed recent developments in the determination of atomic transitions and lifetimes. The study of systematic regularities in f value trends has proved valuable not only in assessing the reliability of data from a variety of sources, but also in increasing our understanding of the factors important in the accurate determination of f values. In theoretical calculations, correlation has been found important and also relativistic effects, at least for higher degrees of ionization. In beam-foil spectroscopy the cascading problem has been recognized.

Most of the isoelectronic sequences studied to date have been for the resonance transition of lighter atoms or larger atoms with only a few electrons outside a closed shell. Much less is known about f value trends in more complex systems. Even for relatively modest systems such as the Zn isoelectronic sequence, where there is strong interaction between $4s4d \ ^1D$ and $4p^2 \ ^1D^{[58]}$, the two lowest ^1D's in Ga II have only recently been identified.[59] This information is needed before experimental studies can be undertaken.

Many spectra for an isoelectronic sequence contain what Condon and Odabasi[60] and Weiss[61] referred to as "plunging" configurations. These are configurations that lie in the continuum for neutral atoms or atoms with a low degree of ionization, but as the nuclear charge increases (the number of electrons remaining unchanged) the levels become bound, crossing many other bound state levels with which they interact. In the Al I sequence, the configuration $3s3p^2 \ ^2S$ is such an example. In both Al I and Si II, it lies above $3s^24s$, but already in P III it is the lowest ^2S. In fact, the Hartree–Fock energies of $3s^24s$ and $3s3p^2$, considered as a function of the nuclear charge Z, cross in the interval (14, 15). Theoretical f values are usually plotted as a continuous curve but it should be remembered that in most instances values have only been computed for integral values of Z, and that these values are then joined by a continuous curve.

In a study of f value trends in the presence of level crossings,[61] it has been shown that strict continuity of the f value trend is maintained only when f values are plotted for transitions from the jth state of a given symmetry to the kth state of another symmetry. For example, in the Al I sequence the f value trend is continuous for a transition from the lowest ^2S to the lowest ^2P, the second lowest ^2S to the lowest ^2P, and so on. This is illustrated in Fig. 2, where we see a rapid change in the f value in the vicinity of the crossover. It is also clear that *in this case* the irregularity is not physically meaningful since a smooth curve could be joined between the values for integral Z values. However, some crossovers occur near integral values of Z,[62] and the cause of the irregularity must then be recognized.

Generally there are two types of interactions between crossing configurations—long-range and short-range interactions. For the former, the interaction persists at a significant level for a relatively large portion of the range of Z values of interest, whereas for the latter the interaction is appreciable only in the vicinity of the crossover. The $\{3s^23s, 3s3p^2\} \ ^2D$ interaction is a long-range interaction and it seems appropriate to plot the

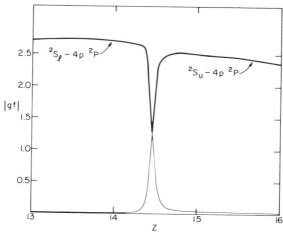

Figure 2. Theoretical $|gf|$ values for the ^2S_l (lower)-$4p$ ^2P and ^2S_u (upper)-$4p$ ^2P transitions as a function of Z. The dark curve depicts the gf values for the $4s$ ^2S–$4p$ ^2P transition and the light curve the $|gf|$ values for the $3s3p^2$ ^2S–$4p$ ^2P transition. A two-configuration model was used for the ^2S states and a single configuration for the ^2P state. (Reprint of figure appearing in Ref. 62.)

f value for transitions to the two lowest ^2D states in a manner consistent with mathematical continuity. On the other hand, the $\{3s^24s, 3s3p^2\}$ ^2S interaction is a short-range interaction. In studying a trend one may wish to ignore the irregularity, though it seems advisable to indicate such a smoothing over of an irregularity in some manner.

Crossovers do not always occur only at low stages of ionization. Consider the $6s^2$ ^1S–$6s6p$ ^1P transition in the barium sequence where the $n = 4$ and $n = 5$ shells consist of the electrons $4s^24p^64d^{10}5s^25p^6$. At some large Z value the $4s^24p^64d^{10}4f^8$ configuration will be the lowest ^1S state and along the way, many other configurations will have crossed $6s^2$.

One of the heaviest isoelectronic sequences to have been studied systematically[66] is the $4d^{10}$ ^1S–$4d^94f$ ^1P transition in the palladium sequence. This sequence still has a relatively simple excitation spectrum compared with most heavy ions, but the results show some anomalous behavior. Typically, the f value trend for a $\Delta n = 0$ transition shows a slight increase for the first few stages of ionization and then decreases to zero as Z increases, at least when nonrelativistic transition energies are used. In the palladium sequence, the $4f$-orbital is so diffuse for low stages of ionization that the f value starts exceptionally small in Ag II, increasing about an order of magnitude before peaking at Ba XII. This increase of the f value is due to the collapse of the $4f$ orbital with increasing degree of ionization. An important factor as well was the $4d^{10}$, $4d^8 4f^2$ interaction in the ^1S initial state which reduced the f value by about 25% at the peak value. This interaction represents the first-order correction to the transition matrix element, the other corrections having a much smaller effect.

6.4. Relativistic Effects

For high degrees of ionization the relativistic effects cannot be ignored, but for $\Delta n = 0$ transitions the nonrelativistic f values appear to be in good agreement with relativistic values for many stages of ionization, provided a relativistic transition energy is used and the wave function is corrected for spin–orbit interaction.

Consider the f values for the resonance transitions of Mo XIII and W XLV in the Zn isoelectronic sequence reported in Table 11. In this table, outer correlation refers to correlation between the outer two electrons, and core polarization actually only includes the correlation between the outer electrons and the $3d^{10}$ core. The difference in the values reported in columns (i) and (ii) represents the effect of this core polarization on the f value. In Mo XIII, core polarization is a more important effect than the relativistic effect on the transition matrix element. In W XLV ($Z = 74$), relativistic effects are clearly more important, but the MCHF results corrected for spin–orbit interaction are in better agreement with the RRPA values than one might expect. In a study of the ratio of relativistic and nonrelativistic line strengths of hydrogenic transitions, Younger and Weiss[67] found the ratio to be within 1 ± 0.1 for some fairly large degrees of ionization, though the precise degree depended on the transition. For $2s–2p$ and $3s–3p$

Table 11. A Comparison of Relativistic RRPA[a] f Values and MCHF f Values with Relativistic Corrections for Mo XIII and W XLV. (i) Outer Correlation; (ii) Outer Correlation and Correlation with $3d^{10}$

	Mo XIII		W XLV	
	(i)	(ii)	(i)	(ii)
MCHF:				
Nonrelativistic	1.57	1.43	1.52	1.46
With spin–orbit interaction	1.54	1.41	1.19	1.15
RRPA	1.54	1.39	1.15	1.09

[a] Shorer (Ref. 63).
[b] Froese Fischer and Hansen (Ref. 64).

transitions it included more than 40 stages of ionization. The comparison in Table 11 of relativistic and MCHF f values together with the comparison of energy level data presented in Table 8 suggests that nonrelativistic wave functions can be used to predict reliable data in many cases.

Acknowledgment

The work was supported, in part, by a contract with the U.S. Department of Energy.

References

1. F. Herman and S. Skillman, *Atomic Structure Calculations*, Prentice Hall, Inc., Englewood Cliffs, New Jersey (1963).
2. P.-O. Löwdin, *Phys. Rev.* **97**, 1509 (1955).
3. B. Fricke, *Progress in Atomic Spectroscopy, Part A* (Edited by W. Hanle and H. Kleinpoppen), Plenum Press, New York (1978).
4. C. Froese Fischer, *The Hartree–Fock Method for Atoms*, John Wiley & Sons, New York (1977).
5. G. Racah, *Phys. Rev.* **62**, 438 (1942).
6. J. Bauche and M. Klapisch, *J. Phys. B* **5**, 29 (1972).
7. J. Labarthe, *J. Phys. B* **5**, L181 (1972).
8. J. E. Hansen, *J. Phys. B* **5**, 1083 (1972).
9. R. D. Cowan, *J. Opt. Soc. Am.* **58**, 924 (1968).
10. Minnhagen (private communication).
11. A. Borgström, *Ark. Fys.* **38**, 243 (1968).
12. J. Reader and G. I. Epstein, *J. Opt. Soc. Am.* **58**, 924 (1968).
13. W. Persson and S. Valind, *Phys. Lett.* **35A**, 71 (1971).
14. A. Hibbert, *Progress in Atomic Spectroscopy, Part A* (Edited by W. Hanle and H. Kleinpoppen), Plenum Press, New York (1978).
15. D. Layzer, *Ann. Phys. (N.Y.)* **8**, 271 (1959).
16. C. Froese Fischer and K. M. S. Saxena, *Phys. Rev. A* **9**, 1498 (1974).
17. J. Reader, *J. Opt. Soc. Am.* **64**, 1017 (1974).
18. R. D. Cowan, L. J. Radziemski, Jr., and V. Kaufman, *J. Opt. Soc. Am.* **64**, 1474 (1974).
19. J. E. Hansen, *J. Opt. Soc. Am.* **67**, 754 (1977).
20. C. Froese Fischer, *J. Comput. Phys.* **13**, 502 (1973).
21. C. Froese Fischer, in preparation.
22. Y. Accad, C. L. Pekeris, and B. Schiff, *Phys. Rev. A* **4**, 516 (1971).
23. C. E. Moore, *Atomic Energy Levels*, NBS Circ. No. 467, U.S. Government Printing Office, Washington D.C. (1952).
24. C. Froese Fischer, *Phys. Scr.* **21**, 466 (1980).
25. C. Roth, *J. Res. Natl. Bur. Stand. Sect. A* **74**, 157 (1970).
26. C. Froese Fischer, *J. Quant. Spectrosc. Radiat. Transfer* **13**, 201 (1973).
27. C. Froese Fischer, J. E. Hansen, and M. Barwell, *J. Phys. B* **9**, 1841 (1976).
28. C. Froese Fischer, *Proceedings of the Sixth International Conference on Atomic Physics* (Edited by R. Damburg and O. Kukaine), Plenum Press, New York (1978).

29. D. Layzer, Z. Horak, M. N. Lewis, and D. P. Thompson, *Ann. Phys.* (*N.Y.*) **29**, 101 (1964).
30. C. Froese Fischer, *J. Phys. B* **10**, 1241 (1977).
31. T. H. Dunning, B. H. Botch, and J. F. Harrison, *J. Chem. Phys.* **72**, 3419 (1980).
32. K. Jankowski, P. Malinowski, and M. Polasik, *J. Phys. B* **12**, 345 (1979).
33. K. Jankowski, P. Malinowski, and M. Polasik, *J. Chem. Phys.* **76**, 448 (1982).
34. J. P. Desclaux and Y.-K. Kim, *J. Phys. B* **8**, 1177 (1975).
35. H. A. Bethe and E. E. Salpeter, *Quantum Mechanics of One- and Two-Electron Systems*, Springer-Verlag, Berlin (1957).
36. M. Blum and R. E. Watson, *Proc. R. Soc.* (*London*) **270**, 127 (1962).
37. W. Dankwort and E. Trefftz, *Astron. Astrophys.* **47**, 365 (1976).
38. W. Dankwort, *J. Phys. B.* **10**, 3617 (1977).
39. A. Aymar and E. Luc-Koenig, *Phys. Rev. A* **15**, 821 (1977).
40. C. J. Mitchell, *J. Phys. B.* **8**, 25 (1975).
41. C. Froese Fischer, *Comput. Phys. Commun.* **14**, 145 (1978).
42. A. Hibbert and R. Glass, *Comput. Phys. Commun.* **16**, 19 (1978).
43. R. Glass, *J. Phys. B* **12**, 697 (1979).
44. K. T. Cheng, Y.-K. Kim, and J. P. Desclaux, *At. Data Nucl. Data Tables* **24**, 111 (1979).
45. B. C. Fawcett, *At. Data Nucl. Data Tables* **16**, 135 (1975).
46. K. G. Widing, *Astrophys. J.* **197**, L33 (1975).
47. D. R. Beck and C. A. Nicolaides, *Chem. Phys. Lett.* **36**, 79 (1975).
48. C. Froese Fischer, *Can. J. Phys.* **54**, 1465 (1976).
49. R. A. Moore, J. A. Reed, W. T. Hyde, and C. F. Liu, *J. Phys. B* **12**, 1103 (1979).
50. H. Saraph, *J. Phys. B* **9**, 2379 (1976).
51. C. Mendoza, *J. Phys. B* **14**, 397 (1981).
52. J. H. Black, J. C. Weisheit, and E. Laviana, *Astrophys. J.* **177**, 567 (1972).
53. L. Liljeby, A. Lindgard, S. Mannervik, E. Veje, and B. Jenlenkovic, *Phys. Scr.* **21**, 805 (1980).
54. G. Sorenson, *Phys. Rev. A* **7**, 85 (1973).
55. W. H. Smith and H. S. Liszt, *J. Opt. Soc. Am.* **61**, 938 (1971).
56. M. J. Seaton and P. M. H. Wilson, *J. Phys. B* **5**, L175 (1978).
57. W. L. Wise, *Progress in Atomic Spectroscopy, Part B* (Edited by W. Hanle and H. Kleinpoppen), Plenum Press, New York (1979).
58. C. Froese Fischer and J. E. Hansen, *Phys. Rev. A* **19**, 1819 (1979).
59. B. Denne, U. Litzen, and L. J. Curtis, *Phys. Lett. A* **71**, 35 (1979).
60. E. U. Condon and H. Odabasi, Jila Report No. 95 (1968).
61. A. W. Weiss, *Beam-Foil Spectroscopy 1* (Edited by I. A. Sellin and D. J. Pegg), Plenum Press, New York (1976).
62. C. Froese Fischer, *Phys. Rev. A* **22**, 551 (1980); *see also* C. A. Nicolaides and D. R. Beck, *Chem. Phys. Lett.* **53**, 87 (1978).
63. C. Froese Fischer, *Can. J. Phys.* **54**, 740 (1978).
64. P. Shorer, *Phys. Rev. A* **18**, 1060 (1978).
65. C. Froese Fischer and J. E. Hansen, *Phys. Rev. A* **17**, 1856 (1978).
66. S. M. Younger, *Phys. Rev. A* **22**, 2682 (1980).
67. S. M. Younger and A. W. Weiss, *J. Res. Natl. Bur. Stand. Sect. A*

3

New Methods in High-Resolution Laser Spectroscopy

B. COUILLAUD AND A. DUCASSE

1. Introduction

It is not necessary any more to point out that optical spectroscopy has been revolutionized by the advent of highly monochromatic laser sources. The laser light is far superior to the light of conventional sources in brightness, spectral purity, and spatial coherence. As a result, almost all classical spectroscopic techniques become very much more sensitive and convenient through the use of lasers. In addition to the improvements of the already existing techniques, a large number of new methods, only conceivable on the basis of the laser characteristics, have been developed. It is not the aim of this chapter to give a review of all that has been done in the last 15 years. A very good review article has already been published in Part A of this monograph by Demtröder.[1] As two years have already passed since the publication of that volume, and as the developments during this period have been quite fast, it seems worthwhile to complement that work. The field of laser spectroscopy has become so wide and diverse that it is impossible to give a meaningful review of all the interesting recent advances within the limits of this presentation. We will restrict ourselves to what is now known as "high-resolution spectroscopy," that is, the spectroscopy of the gaseous medium with a resolution better than the Doppler width of the transitions involved.

B. COUILLAUD AND A. DUCASSE • Centre de Physique Moléculaire, Université de Bordeaux I, 33405 Talence, France.

Among the various methods to suppress the Doppler broadening of a transition, one can distinguish two different classes: the methods where the direction of the velocity of the molecules (and sometimes modulus of their velocity) is imposed by the apparatus, and the methods using gases in a cell. The methods using an atomic or molecular beam apparatus were known long before the advent of lasers but have been regenerated by these new light sources. They have the advantage of suppressing the interaction of the analyzed species with the wall and of reducing the collisions. The disadvantage of the beams is the cost and the complexity of the machines involved. Among the techniques using a cell, one can separate the resonant methods from the nonresonant. The resonant methods use the high intensity of the laser light to saturate the transition studied. Saturated absorption spectroscopy or Lamb dip spectroscopy is the oldest and perhaps the most widely used of these methods. Here the spread of atomic velocities along the direction of observation is effectively reduced by velocity-selective bleaching and probing with two counterpropagating monochromatic laser beams. Saturated fluorescence[2] spectroscopy and in particular the method of intermodulated fluorescence[3] extend the potential of this method to optically very thin fluorescent samples. The nonlinear interaction of two counterpropagating beams can also be detected via changes in the refractive index rather than in absorption, as demonstrated by saturated dispersion spectroscopy,[4,5] or via the induced dichroism, as demonstrated in laser polarization spectroscopy.[6] The nonresonant methods consist mainly of the Doppler-free multiphoton spectroscopies extensively used in the case of two-photon transitions.[7,8] They are nonresonant in the sense that none of the frequencies of the lasers used is resonant with an electric dipole transition. Among the other nonresonant methods are the use of intense static electric or magnetic fields to compensate the Doppler shift and the use of light shifts to produce a similar effect.[9]

Each of these methods has its own limitations. The residual Doppler broadening in a beam due to the finite collimation can be overcome by the simultaneous use of a saturated absorption technique; in saturation spectroscopy the residual Doppler broadening can be canceled by the use of perfectly colinear beams. More difficult to overcome, the transit time broadening due to the finite time for an atom or molecule to cross the light beams, can nevertheless be suppressed by the use of the Ramsey fringes technique,[10] a technique which seems really promising for the ultimate resolution. At the present time the second-order Doppler effect and the recoil effect are still present in the experiments unless one uses a trap to cool down the atoms[11,12] to cancel the first of these effects.

The generalization and improvement of the Doppler-free methods are directly linked to the development of the laser technology. The great breakthrough came with the development of tunable lasers and particularly with the advent of dye lasers. The extremely small linewidth required in most of the experiments requires the use of cw systems working in single mode. For most experiments the performance of the free-running laser turns out to be not good enough and an active frequency stabilization is required if the laser linewidth is not to limit the resolution.

In Section 2 of this chapter a survey will be given of the sources now available for high-resolution spectroscopy in the visible, near infrared, and near uv. The advantages of "ring lasers" will be pointed out and an introduction to the mode-locked lasers presented. The characteristics and possibilities of color center lasers will then be depicted, and, finally, the different methods presently available to precisely calibrate the wavelength of the tunable lasers will be investigated.

In Section 3, the different techniques to produce uv radiation with convenient characteristics for high-resolution spectroscopy will be presented. The use of nonlinear crystals inside and outside the cavity of a single-mode dye laser, and the possibilities of frequency mixing will mainly be discussed and the recent results presented.

Section 4 will give a review of the new methods of high-resolution spectroscopy which have been developed during the last few years. It is interesting to note that since the publication of Part A of this series, no real breakthrough was made in the methods used in Doppler-free spectroscopy. Following the revolution which characterizes the early 1970s, we have worked on refinements of the different techniques already discovered, and on first applications of these methods to obtain spectroscopic data.

Finally in Section 5 the techniques of high-resolution spectroscopy with short light pulses will be presented. It will be shown that with pulsed laser sources of relatively broad bandwith, it is possible to increase drastically the resolution of Doppler-free two-photon spectroscopy, as well as saturated absorption or polarization spectroscopy by excitation with two or more phase-coherent light pulses.

2. New Sources in High-Resolution Spectroscopy and Their Frequency Calibration

Since the development of Doppler-free spectroscopy is strongly related to the development of new laser sources, we will briefly discuss in this section the different systems which have appeared in the last few years.

2.1. The Ring Laser

Surprisingly it is only recently that the superiority of the traveling-wave laser as opposed to standing-wave dye lasers has been recognized for single-frequency operation, although ring dye lasers have been under investigation for over seven years now. As a matter of fact, the ring cavities are even older; they were used in the early development of gas lasers.

The first cw dye lasers were standing wave lasers.[13-15] Very rapidly, the design of a cavity permitting a very efficient use of the small amplifier medium was given by Shank, Ippen, and Dienes.[16] The dye laser cavity must sustain a stationary mode with a focusing point or beam waist whose dimensions are comparable to the size of the amplifying medium. In fact, a two-mirror cavity can be used, as demonstrated by several experiments. But these cavities, if they are working in the center of their stability range, are then too short to allow intracavity elements; or if the mirror separation is large enough, they must be adjusted at the limit of their stability domain, and are then very sensitive to any perturbation. The use of resonators with intracavity focusing elements is the answer to the problem. The cavity can work at the center of the stability domain, while tuning elements like prisms or Lyot filters and etalons can be introduced in the long arm of the cavity. At the very beginning, the intracavity focusing element was a lens;[13,14] the losses produced were then quite high, and the etalon effects were often present in the system in such a way that the single-mode operation was difficult to obtain. The idea of using a spherical mirror in reflection instead of a lens in transmission led to the design of the well-known three-mirror cavity.[16] The reflection losses on the focusing element as well as the etalon effect are then overcome. The overlap of the stability domains in sagittal and tangential planes not assured in the case of an empty three-mirror cavity with short focal folded mirror, is realized by the use of a brewster angle amplifier medium with the correct thickness. The astigmatism introduced by focusing in the dye jet is then compensated by the astigmatism introduced by the off-axis mirror.

Except for very specific uses, the three-mirror cavity has been the synonym of cw dye lasers during the last seven years. It has been superseded by the ring laser cavities which have been recognized as more efficient for single-mode operation only recently.[17] Nevertheless, the fundamental concepts of the three-mirror cavities survive in most of the ring cavity designs.

It has been known for a long time that spatial hole burning is responsible for the low power obtained in single-mode operation from standing-wave dye lasers.[18] The spatial hole burning effect is especially pronounced if the active medium only fills a small fraction of the resonator length as in the cw dye lasers. The saturation of the gain for a given longitudinal mode is most pronounced at the maxima of the electric field of the standing wave inside the resonator. The inversion of the active medium shows a local variation with a modulation period of $\lambda_1/2$, where λ_1 stands for the wavelength of the oscillating mode. This saturation pattern leaves unsaturated zones in the amplifier medium, namely, those places where the electric field is zero. These zones, where the gain is high, even in the presence of the mode λ_1, can then generate a second longitudinal mode with convenient frequency λ_2. The condition on the wavelength λ_2 of the so-called "hole burning mode" is that its maxima of the electric field are located at the minima of the electric field of the main mode in the thin amplifier medium. The two limitations in high-power single-mode operation for standing-wave cw dye lasers are then obvious. Since the oscillating mode uses only a fraction of the inverted molecules, the conversion efficiency from multimode operation to single mode is low; when the gain of the amplifier medium is increased, the hole burning mode oscillation is more and more difficult to suppress. Typical output power of 200 mW can be obtained with a standing-wave laser.

In a ring laser the optical cavity does not send the light going out of the amplifier medium back on itself, but instead by-passes the amplifier medium in such a way that the new amplification process starts at the same input as the previous one. The oscillator scheme is, in this case, quite similar to the oscillator in electronics where part of the output of the amplifier is fed back to the input with the right phase and amplitude. The only difference is that the amplifier medium is bidirectional in optics and thus a ring laser can be viewed as a superposition of two oscillators with counterpropagating beams in the feedback loop. As a result, without nonreciprocal losses or suitable mode inhibiting effects, the ring laser supports running waves in both directions, clockwise and counterclockwise. In this configuration, the electric field forms a standing-wave pattern leading to a nonuniform distribution of the inverted population in the active medium. By changing over from standing-wave (SW) operation to traveling-wave (TW) operation, the multimode output power is then roughly the sum of the powers on the two outputs in SW operation, and single-mode operation is quite easily obtained with low-selectivity intracavity elements. The unidirectional ring laser, because of the running wave involved instead of a standing-wave pattern, has then two advantages. Firstly, the spatial hole burning modes effect does not exist in these systems and therefore high-power single-mode operation can easily be achieved; secondly, since the whole active medium participates to the amplification, the efficiency of the system is increased. Typical output powers in the R6G range are of the order of 1.5 W in single-mode operation, and output powers of 2.5 W have recently been reported.[19]

It is worthwhile to note that if the single-mode operation is not required, a standing-wave system is certainly to be preferred to a traveling-wave one, since the output powers in multimode operation are comparable for the two systems, while the ring laser, containing much more elements, is more expensive and more troublesome to use. In the particular case of single-mode operation, the ring system, with its efficiency of about 80% between multi- and single-mode operation, is at present the only answer to stable single-mode operation at high power over the whole visible range. The ring lasers, today, are all constructed around the same basic ideas (Figure 1). The optical cavity is of the type "astigmatically compensated cavity," previously developed for SW lasers; a third beam waist is often provided by use of convenient spherical mirrors to allow intracavity experiments (mainly intracavity doubling). The device which forces the laser to operate stably in a preferred direction consists in a nonreciprocal element: a Faraday rotator, associated with a reciprocal element: an optically active plate of quartz crystal. The combined action of this device along with the polarization-dependent losses on the Brewster faces in the optical cavity, gives unequal losses for the two waves traveling in opposite directions, leading to unidirectional operation of the system.[20] In order to compensate the Faraday rotation, a device where the path of the beam is not contained in a plane to produce the rotation of the polarization has also been proposed.[21] A broad-band birefringent filter is used to tune the laser over the bandwidth of the amplifier medium.[22,23] Single-mode operation is assured by low-finesse etalons, while a large scanning range can be obtained from tilting Brewster plates to change the cavity length. The linewidth for these systems is typically of the order of several megahertz in free-running operation, and longitudinal mode hops are unfortunately frequent. Stabilization techniques locking the laser frequency to a reference cavity are commonly used to reduce the jitter and cancel the mode hops as well. A short-term frequency stability of about 1 MHz is obtained without too much effort. Special design using reference cavities in reflection and then allowing phase-lock techniques have recently been reported which have provided stabilities of better than 10 kHz.[24,25]

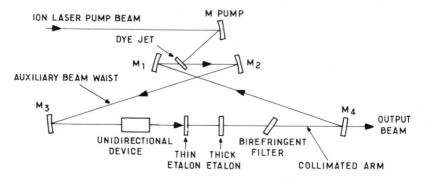

Figure 1. Optical scheme of a ring dye laser. (From Ref. 17.)

2.2. Mode-Locked Lasers

The mode-locked operation of a laser, and particularly of a dye laser, gives the shortest optical pulses that can be produced at present. These sources are the basic tools for time-resolved spectroscopy, and their use in high-resolution spectroscopy does not seem obvious when one remembers the basic principle: the shortest in time, the broadest in frequency. In order to demonstrate the interest of mode-locked lasers in high-resolution spectroscopy, we will briefly discuss below some basic properties of these systems.

A mode-locked laser is essentially a multimode laser where the frequency spacing, relative phases, and amplitudes of the oscillating longitudinal modes have been fixed. The modes which, in a simple multimode operation, are oscillating independently, one from the other, are now interfering, giving rise to a laser output which is a well-defined function of time. Consider a multimode laser with N longitudinal modes. The nth mode has the amplitude E_n, angular frequency ω_n, and phase ϕ_n.

The total laser output field E_T can be written

$$E_T = \sum_{n=-N/2}^{N/2} E_n \exp i |\omega_n[(t-z)/c] + \phi_n| + \text{c.c.}$$

where c.c. stands for the complex conjugate and we have assumed the radiation is traveling in the $+z$ direction. If we have equal mode frequency spacings $\omega_n = \omega_0 + n\Delta$ where $\Delta = 2\pi(c/2L)$ and ω_0 is the optical frequency of the central mode, we can write

$$E_T = \exp i\omega_0[(t-z)/c]\left\{ \sum_{n=-N/2}^{N/2} E_n \exp i |n\Delta[(t-z)/c] + \Phi_n| + \text{c.c.}\right\}$$

This corresponds to a carrier wave of frequency ω_0 whose envelope depends on the values of E_n and Φ_n. We note, however, that the envelope travels with the velocity of light and is periodic with period $T = 2\pi/\Delta = 2L/c$. Moreover, for $\Phi_n = $ a constant independent of n, this envelope consists of a single pulse, whose width is approximately the reciprocal of the frequency range over which the E_n's have an appreciable value. This situation corresponds to an optical pulse, traveling back and forth between the laser mirrors. The laser output in time is then a continuous train of pulses equally spaced by the cavity round trip time $2L/c$ and generated at the output mirror by the successive partial transmissions of the intracavity pulse. In the frequency domain, however, the laser output consists of a regular set of modes, each separated by the pulse repetition rate. The mode-locked laser can then be viewed in many ways as an ensemble of single-mode lasers, where the forced oscillations due to the mode-locking operation impose an equal frequency spacing between successive modes all over the spectrum. As a result, the frequency jitters of the oscillating modes are similar to the jitter in single-mode operation but are now not independent. Although the frequency spectra of mode-locked lasers are quite broad (typically 10 Å for pulses of 1 psec duration), since these spectra consist of a particular set of very stable fringes whose individual width is comparable to the linewidth of a single-mode laser, it will be demonstrated in Section 5 that they can introduce some interesting features when used in Doppler-free spectroscopy.

Mode-locked operation of a multimode laser can be achieved using different techniques. Among these techniques, one can make a distinction between the so-called "active" and "passive" ones, the difference arising from the fact that external energy has, or has not, to be provided in order to obtain the mode-locked operation. Since we are only interested in the application of mode-locked lasers to high-resolution spectroscopy, we will only consider the mode-locked dye laser systems, which are the only ones so far to have been used in this domain. At the beginning of cw dye lasers, the mode-locked operation was obtained by inserting a saturable absorber in the optical cavity.[26] This passive mode locking was very efficient at that time to provide short pulses. Nevertheless the use of a saturable absorber was troublesome in many ways. The laser had to operate close to the threshold in order to obtain a single pulse in the optical cavity. The characteristics of the dye used as a saturable absorber had to match carefully the characteristics of the dye amplifier medium. Only a good couple DODCI-R6G was used which covered only part of the complete range of the R6G. Despite some attempts to extend the frequency range of passive mode-locked dye lasers, no real breakthrough has been found. The advent of the so-called "synchronous pumping" of dye lasers provides attractive means of producing tunable picosecond pulses over the whole visible range.[27] In a synchronously pumped dye laser system, as shown in Figure 2, an acoustooptically mode-locked ion laser is used to pump a dye laser whose cavity is extended so that the dye laser intermode spacing is an integral multiple of the argon laser mode-locked frequency. Under this condition, a circulating dye laser pulse is generated inside the extended cavity and the dye amplifier medium is pumped synchronously by an argon pulse each time the dye pulse arrives at the amplifying dye sheet. The output is then a continuous train of pulses spaced by the cavity round trip time. Synchronous pumping has been shown to produce picosecond pulses with many lasing dyes providing tunable pulses from 420 to 1000 nm.[28-30] Till recently, only standing-wave dye lasers were used for mode-locked operation; the first experiments using ring lasers have just been reported,[31] the requirement of a low-Q cavity in standing-wave lasers under intense

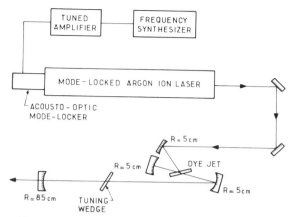

Figure 2. Scheme of a synchronous pumped dye laser.

pumping (in order to get rid of satellite pulses) being unnecessary with unidirectional ring lasers where a good and stable locking at high pumping levels with very high-Q cavities has been demonstrated. A significant improvement in the peak power of mode-locked dye lasers can then reasonably be expected from these systems.

2.3. Color Center Lasers

Development of tunable lasers with emission in the near infrared between 1 and 20 μ has provided spectroscopists with much higher resolution than had been thought possible in this infrared "fingerprint" region, rich in characteristic molecular vibration rotation lines. The major problem with lasers has been their limited continuous tuning range; as recently as 1970, continuous tuning in the infrared was limited to within 0.1 cm^{-1} of discrete gas laser lines. Since then, lasers with continuous tuning ranges broader than 1 cm^{-1} have been developed and spectral coverage of certain types of tunable lasers has exceeded 300 cm^{-1}. A good review article describing many of these devices like diode lasers, spin–flip Raman lasers, nonlinear devices, Zeeman tuned gas lasers... has been published by Hinkley, Nill, and Blum.[32] The reader is referred to this article and to the references cited for further details. We will, in the following, restrict ourselves to the wavelength range between 0.8 and 3.3 μ where the advent of cw color center lasers seems quite promising in terms of sub-Doppler spectroscopy. In fact, cw color center lasers have been known since 1974,[33] but because of the interaction of laser cavity, cryogenic and high-vacuum technologies, their development has been difficult, and it was only five years later that the first spectroscopic applications were reported.

A cw color center laser is essentially a cw dye laser where the dye sheet is replaced by a thin slab of colored crystal mounted at the Brewster angle. The pump intensity required in the region of the amplifier medium to get an inversion, is of the same order of magnitude as for the dye amplifier, and so the three-mirror cavity already mentioned is quite suitable for use in a color center laser. The color centers suitable for laser action are those F-like centers based on a simple anion (halide ion) vacancy in an alkali halide crystal having the rock salt structure.[34] The ordinary F center consists of a single electron trapped at such a vacancy. It is something of an archetype to the other varieties of center, and as such, should be discussed and understood.

The optical pumping cycle of the F center is shown in Figure 3. It consists of four steps: excitation, relaxation, luminescence, and relaxation back to the normal configuration. The relaxation process consists of a simple expansion of the vacancy, and of a corresponding adjustment in the electronic wave function. In the case of the F center, when the system reaches its relaxed excited state, the associated electronic wave function becomes spatially very diffuse, in such a way that the overlap between this wave function and that of the ground state of the relaxed system is very poor. As a result, the oscillator strength of the luminescence band is quite small. These properties associated with the possibility of self-absorption of the emitted photons into the conduction band make rather unlikely the attainment of a net optical gain.

The F-like centers that have been found eminently suited for laser action are, respectively, the $F_A(II)F_B(II)$, F_2^+, and $(F_2^+)_A$ centers. The $F_A(II)$ center is an F center where one of the six metal ions which surround the vacancy is foreign and which relaxes to a double-well configuration following optical excitation. This last property makes it highly suitable for laser action since the self-absorption of the emitted photon cannot occur. $F_A(II)$

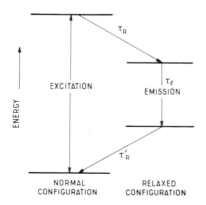

Figure 3. The optical pumping cycle of F centers. (From Ref. 34.)

centers can be formed for example in lithium-doped KCl crystals.[35] $F_B(II)$ centers are quite similar to the $F_A(II)$ except that two metal ions surrounding the vacancy are now foreign. An example of $F_B(II)$ centers can be found in sodium-doped KCl.[35] Structurally, the F_2^+ center consists of an electron trapped at two adjacent anion vacancies along a $\langle 110 \rangle$ lattice direction, while the $(F_2^+)_A$ center is an F_2^+ center trapped next to an impurity alkali ion.

It is certainly not possible, within our limited space, to discuss all the interesting results on F-like centers suitable for laser action that have been reported. We will restrict ourselves instead to giving a rough idea of the presently available frequency range that can be obtained by cw laser action, since these domains are typically of interest for high-resolution spectroscopy.

$F_A(II)$ and $F_B(II)$ centers generally emit between 2 and 3 μm, depending on the host lattice. They are relatively inefficient owing to the large Stokes shifts, but can be used for lasing even after long periods of storage at room temperature. F_2^+ centers generally emit roughly between 0.8 to 2 μm and they are very efficient, yielding over 1 W of cw power with a 60% light-to-light conversion efficiency. A serious barrier, however, to the widespread use of lasers employing the ordinary F_2^+ center is the need for essentially constant refrigeration, since warming up at room temperature results in an irreversible destruction of the centers. Recently, several new centers have been proposed that overcome this problem. Nevertheless, they still have to be used at low temperature, even if it has been demonstrated that laser action could occur at room temperature for a short period of time. Among these new centers one can note the $(F_2^+)_A$ center in lithium-doped KCl which is highly efficient, has a low threshold power (few mW), is continuously tunable from about 2.00 to 2.50 μ, and retains its laser capability after long storage at room temperature.[36] Another interesting center is the F_2^+-like center in NaF which fills an important gap (1.0–1.2 μ) in the tuning range possible with F_2^+ and F_2^+-like centers. Thus, the entire tuning range from 0.82 to 1.91 μm is now covered with no gaps.[37] This new center, whose configuration was not completely determined at the time this chapter was written, can be stored at room temperature for several months with no significant destruction of its properties.

Although color center lasers are known for the ease of single-mode operation, and their remarkably small emission linewidth, it is only several years after their discovery that the first spectroscopic applications have been reported, and it is only very recently that high-resolution spectroscopy studies have been undertaken with these systems. Among the results that have been published one can note the observation of infrared spectra of molecules by crossing a molecular beam with the output of a color center laser[38] and the Doppler-free optogalvanic spectroscopy of excited states of helium and neon in a hollow cathode discharge tube.[39] In the first

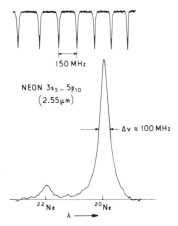

150 MHz

NEON $3s_5 - 5p_{10}$
$(2.55\mu m)$

$\Delta\nu \simeq 100$ MHz

^{22}Ne $\qquad$ ^{20}Ne
$\lambda \longrightarrow$

Figure 4. Doppler-free scan of the neon $3s_5-5p_{10}$ transition obtained with a color center laser. (From Ref. 39.)

experiments, molecules like HF and NO have been investigated in the range 2.2 to 3.2 μ and the reported linewidth is of the order of 1.5 MHz. The second experiment, using a technique described in Section 4 of this chapter, has been performed with a color center laser near 2.6 μm. For helium $n = 2$ to 6 transitions, the resolution was limited to about 320 MHz (FWHM) by Holtzmark broadening due to the presence of charged particles in the discharge. But lines as narrow as 60 MHz have been observed for the neon $3s_5-5p_{10}$ transition (Figure 4).

2.4. Precise Wavelength Calibration for Tunable Lasers

The high resolution that is almost routinely achieved in spectroscopic studies with single-mode lasers is creating a growing need for methods to measure laser wavelengths conveniently and rapidly, with commensurate precision. Quite a few different schemes for laser wavemeters have been proposed. These devices compare the unknown wavelength λ_x to a reference wavelength provided by an etalon source. Among the successful techniques that have been developed, the fringe-counting wavemeter, first introduced by Kowalski et al.[40] and by Hall and Lee,[41] is certainly the most popular. In the Kowalski design (Figure 5), a laser beam of unknown frequency and a reference beam of known frequency travel along exactly the same path, but in opposite directions, inside a two-arm interferometer, consisting of two beam splitters and a moving corner cube reflector. Two photodetectors register the number of interference fringes for each beam, as the corner cube travels along its path. The ratio of the fringe counts immediately gives the ratio of the respective wave numbers. A travel path of 1 m, for instance, gives counts on the order of several millions at visible wavelengths, and

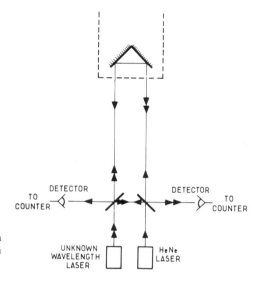

Figure 5. Schematic diagram of a wavemeter for cw lasers. (From Ref. 40.)

permits very simple measurements, good to a few parts in 10^7. More sophisticated electronic data processing makes it possible to measure fractions of a fringe, and accuracies of a few parts in 10^9 have been reported. The identical path of the two laser beams in the Kowalski wavemeter, makes the instrument highly insensitive to perturbations caused by gas flow turbulence or acoustical vibrations.

The choice of a system is directly related to the required accuracy. If, as is usual for most of the tunable laser users, one only needs a help to set the laser frequency, an accuracy of one part in 10^6 is very often sufficient in the visible range. This accuracy corresponds roughly to the Doppler width of a transition and can be attained with very simple devices based on the Kowalski design where no fancy electronics have to be used. The measurement gives the ratio of wave numbers in air. However, the use of identical paths ensures that this ratio is equal, within one part in 10^6, to the ratio of wave numbers in the vacuum. It is also the ratio of wavelengths under standard atmospheric conditions (15° C, 760 Torr) to the same accuracy, even though the measurement conditions may be quite different. A commercial HeNe laser, where a single longitudinal mode is selected at the output with a polarizer, is quite convenient at this level of accuracy as frequency standard. An accuracy of a few parts in 10^6 requires only to count 10^6 fringes which corresponds to a path length of about 30 cm at 6000 Å for the corner cube. The value of the unknown wavelength is then nearly instantaneously known (fraction of a second), which makes this device quite convenient for measurement and control.

Of course the number of relevant applications in physics increases with the available accuracy. As long as spectroscopists were interested in the relative positions of lines distributed in a very restricted frequency domain, the Fabry–Perot markers were sufficient and have been extensively used. The new step, in which several laboratories are already engaged, requires the determination of the absolute frequency of the lines investigated. As a result, very refined devices have been developed. It seems, nevertheless, that an accuracy of one part in 10^9 is the best that can be achieved at the moment. This accuracy is mainly limited by alignment, diffraction effects, and the knowledge of the air refractive index. The frequency standard common to these apparatus is the HeNe laser locked to one hyperfine component of I_2^{127}.

Among these apparatus, the so-called "sigma meter" and "lambda meter" have made it possible to measure the dye laser's wavelengths with sub-Doppler absolute accuracy.[42]

The lambda meter[43] is basically an automatic scanning Michelson interferometer, utilizing corner cube reflectors, with phase multiplication for extending the resolution. Motion of the carriage, holding the corner cubes, lengthens one arm and shortens the other arm of this interferometer. For a given distance traveled by the carriage, a different number of fringes will be counted for the unknown laser and the reference laser. The fixed fringe rate generated by a uniform motion of the carriage allows one to use the technique of frequency metrology rather than distance metrology in the fringe interpolation. The resolution of each optical fringe into 100 distinct levels is obtained by phase locking the $\nu/100$ digital output of an oscillator to the optical fringe rate. Digitally counting the phase-locked oscillator output ν provides the fringe interpolation. This concept enables one to obtain wavelength information of a given resolution level in $1/100$ of the time required by direct fringe counting.

The "sigma meter" (Figure 6) is essentially a Michelson interferometer with a fixed path difference δ, where a totally reflecting prism has been included in one arm. The laser beam enters the interferometer polarized at $45°$ to the prism axis. Thus, one has two rectangular polarizations interfering independently and the two corresponding fringe systems present a phase difference of $\pi/2$ for a proper choice of the incidence angle on the totally reflecting prism. The two components of the beam are separated at the output of the interferometer using a polarization beam splitter. Two signals are then available, expressed by $I_0(1 + \cos 2\pi\sigma\delta)$ and $I_0(1 + \sin 2\pi\sigma\delta)$, where σ is the wave number to be measured. From these two signals, one deduces, after an electronic treatment, the value of σ modulo $1/\delta$. In order to overcome this inaccuracy, several interferometers having a common mirror, with path differences in geometrical ratios, are used. The wave number of the radiation can then be obtained with an accuracy determined by the interferometer highest path difference.

It is impossible in a chapter of this nature to give an exhaustive review of all the systems recently developed for wavelength measurements. Nevertheless, before closing this section, we would mention at least two other devices. The first one, developed by Snyder and called a "Fizeau wavelength meter,"[44] presents many advantages, among which are the compatibility with cw or pulsed sources, the completely static operation, and the real optical and mechanical simplicity. It is one of the few systems directly usable for a pulse laser. The instrument is based on a Fizeau or "optical wedge" interferometer. The fringe pattern produced by the interferometer is digitized and stored in a small computer which converts the fringe pattern into the wavelength of the interfering light. A resolution of one part in 10^8 has been demonstrated.

The second apparatus using a high-finesse spherical Fabry–Pérot interferometer has been developed in order to get a compact device where the motion-related problems, usually encountered in other systems, are reduced.[45] By combining a moderately long mirror motion of a few centimeters with a Vernier-type coincidence detection, a measurement accuracy of 1 part in 10^7 has been achieved. The two radiations, known

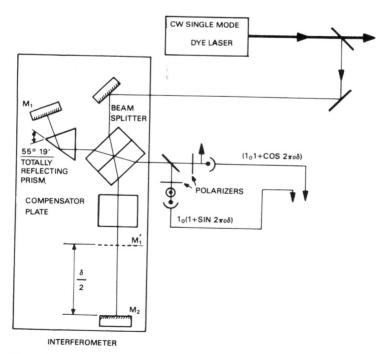

Figure 6. Schematic diagram of the interferometer used in the sigma meter.
(From Ref. 42.)

and unknown, are sent into a spherical Fabry–Pérot interferometer, along the same direction. The two transmitted beams are then separated with appropriate filters, and the fringes counted on each channel, when the mirror spacing of the interferometer is increased. The multiple beam interference gives narrow fringes on each channel, in such a way that a coincidence method can be used very efficiently. The fringe counting is started and ended when two fringes from the two channels coincide. This leads to a measurement precision that is proportional to the fringe counts and the finesse of the interferometer. Namely, the precision is enhanced by a factor equal to the finesse F of the interferometer over that obtained by simple fringe counting. In other words, the mirror motion is then F times smaller than in the case of the simple fringe counting for the same precision in the measurements. This coincidence method, developed by Salimbeni and Pole,[45] has already been extended to the Michelson-type wavemeter,[46] where electronic circuits have to be used in order to artificially increase the finesse of the devices to take full advantage of the method.

3. Frequency Doubling and Sum Frequency Mixing for High-Resolution Spectroscopy in the Near uv

Spectroscopists are generally more familiar with the restricted semi-classical calculation of the macroscopic polarization induced by an electromagnetic field in a medium, developed for laser purposes by Lamb,[47] than with the general formulation of this polarization, introduced to summarize all the effects discovered in the field of nonlinear optics. For a good review of the Lamb theory, one can refer to the book of Sargent, Sculley, and Lamb Jr.,[48] while the Bible for nonlinear optics remains the Bloembergen work of 1965.[49] It is not the subject of this chapter to give a review on the fundamentals of nonlinear optics, but since we are concerned with effects which typically belong to this domain, we would like to emphasize some points in order to introduce the spectroscopist to this field.

The restricted semiclassical theory as developed by W. E. Lamb is concerned with resonant effects at the same frequency as the inducing field. Other effects are not taken in account as soon as the rotating wave approximation has been used. It is not surprising then that, for example, the second harmonic generation (SHG) discovered in 1961 by Franken et al.,[50] does not show up in this calculation. An integration of the component equations of motion for the population matrix given by Lamb, can of course be performed in which all the different terms are kept. The system is generally solved in a perturbative method, by performing iterations from a knowledge of the values of the unsaturated populations. The well-known result is a development of the polarization in terms of the powers of the electric field which only contains the odd power terms. This development, which is no longer restricted to the resonant effects at ω, nevertheless does not include second-order terms, responsible among other effects for the second harmonic generation and sum frequency mixing. This result originates in the choice of a scalar model, validated by the study of a gaseous medium. In the more general case of noncentrosymmetric systems, the polarization contains terms at all the powers of the field. For a polarization induced by two monochromatic fields at frequencies ω_1 and ω_2 propagating in the same direction, the high-frequency second-order term of the polarization involving the two fields is usually written in the electric dipole approximation:

$$\mathbf{P}^{(2)}(\omega_1 + \omega_2) = \overleftrightarrow{\chi}^{(2)}(\omega_1, \omega_2)\mathbf{E}_1(\omega_1)\mathbf{E}_2(\omega_2)$$

where the second-order susceptibility $\chi^{(2)}(\omega_1, \omega_2)$ is a third-rank tensor accounting for the anisotropy of the medium. If $\omega_1 = \omega_2 = \omega$, a polarization at 2ω (as well as a static polarization: $\omega_1 - \omega_2 = 0$) is induced in the medium. The polarization at 2ω, acting as a source in the Maxwell equations, induces the generation of a second harmonic field (while the static polarization leads to the optical rectification effect, which will not be considered here). If ω_1 is different from ω_2, waves at the sum $\omega_1 + \omega_2$ (and difference frequencies $\omega_1 - \omega_2$) are then generated. In the case of the sum frequency mixing, the wave vector of the induced polarization is $(k_1 + k_2)\mathbf{u}$ where k_1 and k_2 are, respectively, equal to $n_1\omega_1/c$ and $n_2\omega_2/c$, with n_1 and n_2 refractive indexes of the medium at the frequencies ω_1 and ω_2. The field generated at $\omega_1 + \omega_2$ is characterized by a wave vector $k_3\mathbf{u}$ equal to $n_3(\omega_1 + \omega_2)/c$, where n_3 is the index of refraction of the medium at $\omega_1 + \omega_2$. In order to get the maximum energy conversion from ω_1 and ω_2 into the sum $\omega_1 + \omega_2$, the polarization at the sum frequency must behave spatially as the generated field. This condition, known as "the phase-matching" condition, can be expressed as

$$\mathbf{k}_1 + \mathbf{k}_2 = \mathbf{k}_3 \quad \text{or} \quad n_1\omega_1 + n_2\omega_2 = n_3(\omega_1 + \omega_2)$$

a condition which reduces to $n(\omega) = n(2\omega)$ in the second harmonic generation case. All optical materials have sufficient spectral dispersion to require that special techniques be used to satisfy the conditions of the preceding equation. Two techniques using birefringent materials are commonly used. The first technique, critical or angle phase matching, utilizes uniaxial crystals and propagation of the fundamental waves at some angle to the optical axis; the polarization of the different beams is chosen in such a way that the birefringence overcomes the spectral dispersion. The second, noncritical or temperature phase matching, allows propagation of the conveniently polarized beams, perpendicular to the optical axis of a uniaxial crystal and utilizes the temperature dispersion of the birefringence to overcome the spectral dispersion. Phase matching schemes are utilized in which the two interacting fundamental waves propagating in the birefringent medium have either parallel or orthogonal polarizations. These two schemes are, respectively, referred to as type 1 and 2 phase matching.

3.1. Second-Harmonic Generation (SHG)

The production of tunable uv radiation for high-resolution spectroscopy imposes a high monochromaticity on the fundamental radiations used in the nonlinear generation process. The single-mode cw lasers are the only sources fulfilling this requirement. Their output power is generally low (few watts at most) in such a way that great care has to be taken in the design of the doubling or summing system, in order to get the maximum efficiency. Among the two possible techniques of phase matching described above, the angle phase matching has to be avoided whenever possible, since double refraction, which occurs for directions of propagation other than parallel or perpendicular to the optical axis, greatly reduces the overlap of the interacting beams and thus the efficiency of the system. Only some of the nonlinear crystals presently available present a large 90° phase-matched temperature tuning range, over a significant range of fundamental wavelengths. In terms of highly monochromatic uv generation, the situation can be summarized as follows. Among the existing nonlinear crystals, ADP (ammonium dihydrogen phosphate) and ADA (ammonium dihydrogen arsenate) are of particular interest because of their large 90° phase-matched temperature tuning range. These two crystals will normally cover in SHG the whole range between 245 and 310 nm with the exception of a small domain around 280 nm. The fundamental radiation can then be provided either by gas lasers (krypton or argon) for some particular frequencies where high uv power is required, or by single-mode cw dye lasers operating between 490 and 620 nm. Other crystals like KDP, RDP, RDA can also be used but their narrow temperature tuning range greatly restricts their interest.

This brief summary would not be complete if we did not include Urea[51] and the potassium pentaborate KB5.[52] Although they cannot be used in 90° temperature phase matching, because of the very low sensitivity of their refractive index with temperature, they nevertheless present some interesting characteristics which make them attractive. Although not commercially available at present, recent studies have shown that Urea has properties which point to its future widespread application as a nonlinear medium. Urea has a nonlinear coefficient that is larger than those of the KDP isomorphs and provides efficient SHG for fundamental wavelengths down to 480 nm. The KB5 crystal, in angle matched SHG, is the only crystal available to generate uv under 240 nm, and so has to be considered, despite its small nonlinear coefficient, which, combined with the double refraction effect, provides low conversion efficiencies.

The output power to be expected in SHG, when using Gaussian input beams, can be calculated by use of the theory developed by Boyd and Kleinman.[53] Under small signal conditions, the output power $P_{2\omega}$ is proportional to the square of the input power P_ω. The conversion efficiency $P_{2\omega}/P_\omega^2$ depends on many parameters, including the direction of propagation, the focusing and linewidth of the input beam, the nonlinear coefficient of the medium, and the length of the nonlinear crystal. For a 90° phase matching, under the optimal focusing conditions established by Boyd and Kleinman, the conversion efficiency is about $10^{-3}\,\mathrm{W}^{-1}$ for a 25-mm-long ADP or ADA crystal. This value falls rapidly when the incidence angle is decreased, reaching for example $5.10^{-5}\,\mathrm{W}^{-1}$ at $\theta = 60°$.

It is obvious from the preceding considerations on the conversion efficiency, and from the value of the available output power of the cw lasers, that the uv power generated by SHG is expected to be low, typically in the mW range. Tunable cw uv radiations have been produced by frequency-doubling the output of a gas laser or of a dye laser, using nonlinear crystals located external to the laser cavity,[51,54,55] in configuration of angle or temperature tuning. This simple approach provides powers under a mW and a tunable uv source of sufficient intensity for a certain ' number of spectroscopic studies. However, in order to obtain higher powers, an elegant solution consists of locating the nonlinear crystal within a laser cavity where the losses have been reduced as much as possible, in order to raise the effective intracavity power of the fundamental wave. This approach, first tested with argon ion lasers and linear cw dye lasers,[56–61] has found its full power with the advent of ring lasers.

The first intracavity doubling experiment, using a cw ring dye laser, was performed in 1976.[62] From that time till 1980, several studies of the generation of radiation at wavelengths in the vicinity of 300 nm, by intracavity doubling a rhodamine 6G ring dye laser, have been reported.[63–68] However, the problems encountered in the different setups were such that practically no spectroscopic studies were completed. The difficulties in intracavity doubling are mainly caused by the insertion losses afforded by the nonlinear crystal. These crystals are soft, and an optical polish is difficult to obtain; moreover, they are hygroscopic to the extent that they have to be used in a water-vapor-free environment. The solution commonly used to get rid of these problems is to protect the crystal against environmental conditions by sealing it in a cell filled with an index matching fluid, which also serves as a good heat conductor and minimizes the reflection losses at the crystal surfaces. Typical output powers of a few tens of milliwatts have been achieved with this method, but only during a few minutes, the uv degrading with time. The nominal uv power could be restored in these experiments by slightly translating the nonlinear crystal in order to change the beam path in the active medium. The matching index fluid used during the last few years in the different intracavity doubling attempts has recently been proved to be responsible for power instability in the uv generation. Whether these instabilities are related to the photochemistry of the matching index fluid under uv excitation, or to more complex reactions, is not presently known, but the recent experiments performed with no liquid coatings on the optical surfaces of the crystal have clearly demonstrated that the uv degradation was directly attributable to the presence of this fluid.

An example of an intracavity frequency-doubled rhodamine 6G ring dye laser system is the system recently developed in Stanford[68] and shown schematically in Figure 7. In this experiment, a Coherent Radiation model 699-21 ring cavity cw dye laser has been employed to produce single-frequency ultraviolet powers of more than 10 mW near 296 nm, stable over many hours, and continuously scannable over 60 GHz.

A water-based rhodamine 6G solution dye permitted pumping by all visible lines (19 W) of an argon ion laser. The solvent was a mixture of 75% Ammonyx LO detergent and 25% ethylene glycol, cooled to near 10° C to obtain the proper viscosity for the dye jet. The pressure at the jet was increased from 40 to 70 psi. Up to 2.5 W of visible single-mode output power could be generated in this way with the standard output coupler of 15% transmission.

For ultraviolet operation, the light direction inside the dye laser was reversed by changing the polarity of the magnet for the Faraday rotator of the optical diode. The standard output coupler was replaced by a highly reflective mirror transmitting only about 50 mW of visible radiation, and a coated beam splitter was used to send most of this light into the external cavity and into the reference diode used for electronic frequency stabiliz-

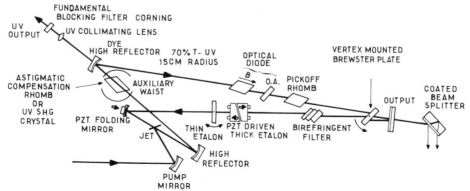

Figure 7. Arrangement of the ring laser cavity for SHG (from Coherent Radiation).

ation. In order to extract the second harmonic output, the standard upper folding mirror was replaced by a quartz substrate, coated to give 70% transmission in the uv while maintaining high reflectivity at the fundamental wavelength.

An 18-mm-long crystal of ADA with optical surfaces at the Brewster angle served as the frequency doubler. It was inserted into the dye laser cavity at the auxiliary beam waist between the upper folding mirror and the "high reflector," replacing the astigmatic compensation rhomb. The crystal, together with a small electric heater, was enclosed in a protective housing with openings for the beam, and a continuous flow of dry nitrogen gas helped to protect the delicate crystal surfaces.

Using this source, the technique of saturation spectroscopy has been extended into the ultraviolet. The cw dye laser with internal frequency doubler has been used to record Doppler-free spectra of the Hg I transitions $6^3P_0-6^3D_1$ at 296.728 nm and $6^3P_0-6^1D_2$ at 296.759 nm. The isotope shifts for naturally abundant mercury could be measured to within a few MHz.

A spectrum of the $6^3P_0-6^3D_1$ line is shown in Figure 8. The saturation intensity is low, since the metastable level can be accumulatively depleted by optical pumping. Very strong signals with obvious power broadening were recorded with less than 0.5 mW/mm^2. At lower intensities, the resolution approached the natural line width of 27 MHz (FWHM in the uv).

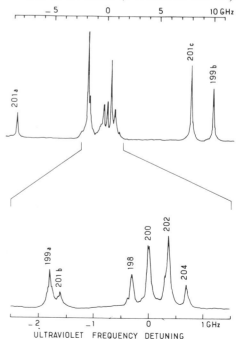

Figure 8. Ultraviolet Hg line $6^3P_0-6^3D_1$ at 296.72 nm, recorded by cw saturated absorption spectroscopy. The central portion with the closely spaced even isotopic components has been recorded below at reduced intensity. (From Ref. 68.)

The closely spaced lines of the even isotopes and the hyperfine components of the odd isotopes appear completely resolved. The intensities of the even components appear distorted in the upper trace because the absorption in the cell exceeded 90%. An expanded spectrum recorded at reduced discharge intensity is shown below.

Other intracavity doubling experiments, presenting interesting features, have been recently published. Among these, we would like to point out the generation of continuous wave, tunable uv radiation (250–260 nm) by intracavity doubling a coumarin 515 ring dye laser.[69] A cooled (200–280 K) ADP crystal with end faces cut at the Brewster angle, was placed inside the laser ring cavity. Ultraviolet powers at 254 nm of 120 and 60 μW were achieved with the laser operating multimode (bandwidth $\simeq 20$ GHz) and single mode (bandwidth $\leqslant 50$ MHz), respectively. The capabilities of the laser system are illustrated in Figure 9, where the sub-Doppler fluorescence excitation spectrum of natural mercury has been investigated in an atomic beam for the 253.7-nm $^3P_1$–$^1S_0$ transition.

In conclusion, a cw ring laser with intracavity frequency doubler can produce a stable, highly monochromatic tunable radiation at the second harmonic frequency. Such lasers extend the range of cw Doppler-free saturation spectroscopy into the important ultraviolet spectral region as has been demonstrated in the different studies of ultraviolet transitions in atomic mercury. The crystals now available however, do not permit, one to cover continuously the whole domain between 200 and 300 nm, because of the 90° phase-matching requirement for high efficiency. The complement to the range covered by SHG could nevertheless be obtained by sum frequency mixing of the dye laser and a fixed-frequency ion laser.

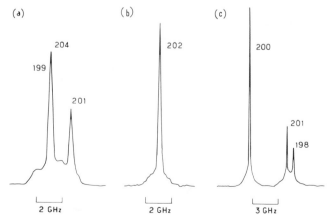

Figure 9. Single-mode continuous scans over selected hyperfine components of the Hg $^3P_1^0$–$^1S_0$ transition around 253 nm. (From Ref. 69.)

3.2. Sum Frequency Mixing (SFM)

Ultraviolet radiation can be generated by sum frequency mixing in nonlinear materials, but although straightforward, this technique has not found as wide application as harmonic generation. The requirement of two laser sources partly explains this situation, as well as the troublesome problems of focusing and aligning of the two fundamental beams. It turns out, however, that sum frequency mixing has certain advantages over second harmonic generation. Prominent among these advantages are a broader output tuning range and, in many cases, higher output power.

We have already given the condition which has to be fulfilled in SFM, in order to assure the maximum uv power. This condition, known as the "phase-matching condition," is given by

$$n_1\omega_1 + n_2\omega_2 = n_3\omega_3$$

It can be achieved in nonlinear crystals with a proper choice of input beam polarizations, direction of propagation, and crystal temperature. The two different arrangements of the respective polarizations of the fundamental beams already mentioned, and referred to as type 1 and type 2, are commonly used. The nonlinear crystals available for SFM are the same as those already mentioned for SHG; the low cw fundamental powers available impose, for the reasons already developed for SHG, the 90° temperature phase matching every time it is possible. ADA and ADP are therefore the more appropriate nonlinear crystals for generation of highly monochromatic uv by SFM. To illustrate some of the advantages of SFM over SHG, the set of sum wavelengths λ_3 that can be generated for different temperatures of an ADP crystal used in a 90° phase matching scheme has been plotted versus the two corresponding fundamental wavelengths λ_1 and λ_2 in Figure 10. The apex of the different curves (where λ_1 and λ_2 are degenerate) corresponds to the SHG. Since the Curie point of $-126°$ C in ADP limits the lowest temperature at which the crystal can be used, the generation of uv in SHG is limited to roughly 245 nm. Because of their concavity, turned toward the highest frequencies, it is readily seen from the curves that the use of SFM schemes extends the range of uv that can be produced in SHG toward the high frequencies. Another interesting feature of SFM is that the temperature tuning affords some flexibility in the input wavelength combinations λ_1 and λ_2 that may be used to generate a particular wavelength. Finally, in certain instances, radiations at wavelengths that can be generated by frequency doubling can be more efficiently generated by mixing, resulting in higher output powers. This is possible because the use of SFM may permit utilization of more efficient lasers, or the use of more efficient nonlinear mediums, or a more favorable phase-matching condition.

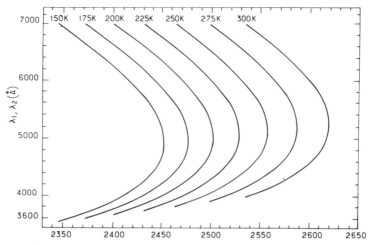

Figure 10. 90° phase-matched sum frequency generation in ADP.

In terms of sub-Doppler spectroscopy, and in particular with single-mode lasers, radiations tunable across a broad range of uv wavelengths can be generated simply by mixing the output of a commercial cw dye laser with various output lines from a krypton or argon ion laser. Figure 11 shows the sum frequency wavelength tuning range that can be obtained by mixing the output of a rhodamine 6G dye laser with selected output lines from a krypton or argon ion laser. The resultant output tuning range, extending from 211 to 406 nm, can be compared in this figure with the range that can be obtained by directly doubling the output of the dye laser.

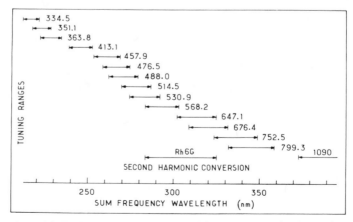

Figure 11. Tuning ranges obtained by mixing the output of a rhodamine 6G laser with selected ion laser lines. (From Ref. 70.)

Although frequency mixing seems quite promising to generate highly monochromatic uv in the range 200–300 nm, this technique has not really been used in high-resolution spectroscopy yet. Till now, people have been more concerned with the technology of the systems than with applications. Among the different results that have been published in the past, we must mention the generation of continuously tunable uv between 257 and 357 nm by S. Blit and co-workers.[71,72] This radiation was obtained by mixing the output of a rhodamine 6G laser with selected output lines from an ion laser. Several nonlinear crystals have been used in an angle-matching configuration to cover the whole range: ADP, ADA, KDP, and RDP. The output power of the uv generated radiation was typically of the order of a few tens of microwatts. We have already mentioned that even for the wavelength in the SHG range, sum frequency mixing, because of the high output power available in several of the lines of ion lasers, is expected to yield higher uv power. This has been demonstrated[73] recently in studies of the production of cw radiation in the vicinity of 247.5 nm, which can be generated, in a 90° phase matched temperature-tuned ADP crystal, both by mixing the 413.1 nm output of a krypton ion laser with radiation from a rhodamine 6G laser and by frequency doubling the output of a coumarin 480 laser.[74] The use of SFM yielded output powers of $\simeq 1$ mW, an order of magnitude greater than those obtained by SHG. Moreover, the phase-matching temperature required for SFM was close to $-60°$ C, about 40° C above the temperature for SHG, making the system more easy to handle.

A certain number of attempts have been made to generate efficiently uv radiation by 90° temperature matching SFM in ADP, at wavelengths below those attainable directly by SHG in this material. For example, radiation at 243.5 nm has been generated by mixing the 406.7-nm line of a krypton ion laser with 606.6 nm radiation from a rhodamine 6G laser.[73] Another interesting combination has been used to generate highly monochromatic uv radiation at the same wavelength (twice Lyman α) in order to perform sub-Doppler two-photon spectroscopy on the $1S$–$2S$ transition in hydrogen.[75] This radiation was produced by mixing the 413.1-nm line of a single-mode frequency-locked krypton ion laser, with 590.6-nm radiation from a rhodamine 6G single-mode, frequency-stabilized ring laser. With 600 mW output power for the ion laser and 1 W from the dye laser, as much as 0.7 mW was generated with a linewidth under 2 MHz and a scannable range of 30 GHz. In order to obtain a reasonable intensity for the two-photon experiment, an enhancement cavity was used to provide intracavity power higher than 5 mW in each direction. This experiment, however, has not yet been completed, because of a degradation with time of the uv power, whose mechanism is not completely understood now.

To close this—certainly very incomplete—list of results, let us mention that radiation tunable from 211 to 215 nm, a wavelength region not directly accessible by SHG, has been generated by mixing, in an angle-tuned KB5 crystal, the output of a rhodamine 6G laser with the 334.5-nm line of an argon ion laser.[76]

In conclusion, as is evident from the foregoing discussion, cw radiation can now be generated across an extended range of uv wavelengths, by use of SHG and SFM techniques. It remains, nevertheless, that the output powers generated in the uv, even after the advent of the ring dye lasers which have brought an improvement of nearly an order of magnitude, are low (typically under a mW for SFM, and few tens of mW in the more favorable case in SHG). These low powers make nonlinear spectroscopy experiments difficult and restrict the high-resolution studies in the domain between 200 and 300 nm to those transitions presenting a high probability. The use of systems to enhance the uv power, as in the hydrogen experiment, are only partly a solution to the problem since the possibilities of such systems are limited, and because the energy confinement imposes a large transit time broadening effect which generally imposes a strong limitation on the resolution.

4. New High-Resolution Spectroscopy Techniques Using Single-Mode Lasers

The advent of Doppler-free spectroscopy has been a real breakthrough in producing a better understanding of gaseous medium. In the early developments, numerous methods have been developed and demonstrated by the study of what could be called "easy elements" like molecular iodine or sodium. It appeared at the same time that the potential of these methods was limited by a certain number of factors. Some of them, fundamental, are quite general and it seems, at least till now, that the spectroscopists have to live with them. This is the case, for example, for the second-order Doppler shift and the recoil effect (in one-photon experiments) which limit the ultimate resolution. The other factors are more directly related to the experimental environment.

In terms of resolution, the laser linewidth, residual Doppler broadening, pressure broadening, power broadening, as well as transit time broadening are typical limiting experimental factors whose effects have to be overcome in order to take full advantage of the elimination of the Doppler width. Improvements on the basic methods had then to be made in order to increase the resolution, and also the sensitivity, since the Doppler-free spectroscopy was to be applied to weak absorbers as well.

During the last few years we have assisted in the development and refinement of the already existing methods. We will report in this section on a certain number of these improvements, limiting ourselves to the more significant. It is indeed impossible within our limited space to discuss all the interesting applications that have been reported or suggested. The section will be divided into two parts. The first part will be devoted to the different attempts proposed in order to improve the resolution. The second part will deal with the different techniques that have been developed to increase the sensitivity of the already existing methods. We will include in this last part the description of new techniques of detection like optoacoustic or optogalvanic detection.

4.1. New Techniques Improving the Resolution

We have already listed in the introduction to this section the different effects which limit the resolution in Doppler-free spectroscopy and noted that the second-order Doppler shift as well as the recoil effect in one-photon spectroscopy cannot be overcome in experiments where atomic particles are moving. One method to get rid of the second-order Doppler shift would be to study the particle at rest.[77,78] This very promising approach has already been demonstrated with the cooling and trapping of ions.[79,80] The projection of the techniques developed for the ions are nevertheless not directly applicable to neutral particles and further developments have to be made in order to generalize the method.

Among the other factors limiting the resolution, the performances of the existing lasers have already been discussed in Section 2. The residual Doppler broadening has been greatly reduced very early in saturated absorption spectroscopy by using perfectly collinear beams. In fact, the more significant improvements in the last few years in terms of resolution come from the reductions of the contribution from the transit time, the velocity redistributing collisions, and the power broadening. We will describe below the respective techniques that have been developed in each case, starting with the Ramsey fringes for the transit time. The new method of polarization intermodulated excitation supressing the Doppler background due to velocity redistributing collisions will then be described; and finally, a review of the laser saturation spectroscopy with optical pumping techniques will be given.

4.1.1. Optical Ramsey Fringes

One of the essential contributions to line broadening in spectroscopy with collimated laser beams is the finite interaction time of atoms with the laser field, which is limited by the transit time of atoms with thermal velocities through the laser beam. It has been proved possible to resolve the recoil-induced doublets in the three main hyperfine components of methane at 3.39 μm, but this high resolution (2 parts in 10^{11}), derived from an external absorption cell with a 30 cm aperture, nevertheless remained two orders of magnitude removed from the natural linewidth.[81] Since the enlargement of the laser beam is made difficult by the restrictions imposed by the decreasing power density and the curvature of the wavefronts of a "plane wave" caused by diffraction, one is induced to find alternative schemes to reduce transit time broadening and so approach natural linewidth resolution.

The methods which have been proposed by Y. V. Baklanov and his co-workers[82] are directly related to the method of successive oscillatory fields developed for the microwave range by N. F. Ramsey.[83]

Figure 12 represents schematically the principal idea of the method in the microwave range. A beam of particles interacts with two fields considered as plane traveling waves. The interaction time with each field is τ and the distance between the two beams is L. u is the projection of the velocity of a particle along the beam axis, and $L/u = \mathcal{T}$ is defined as the transit time of a particle.

The two fields oscillating at the same frequency ω interact with particles considered as two-level systems. The transition frequency is ω_{21} between the two levels and p is the matrix element of the electric dipole moment of the transition.

After interaction with the first field $E_1(t) = E_1 \cos(\omega t - \phi_1)$, the dipole moment carried by a particle, calculated (from the amplitude probabilities) in a first-order approximation is found to be

$$d = p\frac{G}{2}\tau e^{-j\phi_1} e^{j\omega_{21}t} + \text{c.c.} \tag{1}$$

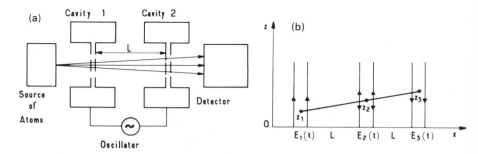

Figure 12. (a) Schematic diagram of the method of separated fields in the microwave range. (b) Interaction of particles with three separated optical fields. (From Ref. 82.)

where $G = jpE_1/\hbar$. This dipole moment is oscillating at the Bohr frequency since the particle is in a region of space free of electromagnetic radiation. The energy W absorbed in the second field $E_2(t) = E_2 \cos(\omega t - \phi_2)$ at $t = \mathcal{T}$ is then found to be

$$W = -\frac{\hbar\omega}{2}\frac{p^2}{\hbar^2} E_1 E_2 \tau \sin(\Omega\mathcal{T} + \phi_2 - \phi_1) \qquad (2)$$

where $\Omega = \omega_{21} - \omega$ is the laser detuning. The important result is that this energy depends on the difference of the dipole moment phase and of the field phase, in such a way that when the laser frequency is swept over the transition, the power absorbed on the second beam is sinusoidaly modulated with a period $\mathcal{T}^{-1}$ in the frequency range. If the transit time $\mathcal{T}$ is larger than the interaction time τ with one beam, the resonance profile resulting from the interaction with the first beam is modulated by the interaction with the second beam and exhibits the so-called "Ramsey fringes." This very simple approach demonstrates that the resolution in a given experiment can be greatly improved by the use of the method of successive oscillatory fields. A certain number of factors have obviously not been taken in account; in particular, the dipole moment damping rate will, in the actual experiment, present a finite value, so that the fringe contrast will decrease with increasing values of L. Another factor limiting the resolution in this method comes from the dispersion of the time-of-flight of the different particles between the two fields. In the interaction of an ensemble of particles, averaging is related to this time-of-flight and the dispersion results in a rapid damping of the absorption oscillations, with an increase of the laser detuning.

The preceding method, using two fields and developed for the microwave range, cannot be used without modifications in the optical band. In optics, field dimensions are usually larger than a wavelength, in such a way that the dipole moment associated with each particle of the beam will acquire a phase ϕ_1 in the field E_1 depending on the position of the particle with respect to the z axis. A similar effect will occur for the second interaction with the field E_2. Because of the beam divergence and velocity spread, the particles coming from the same point z_1 of the first interaction zone having their dipole moments in phase will experience fields with different phases in the second interaction zone. In spite of the fact that individual particles transfer coherence, absorption of the ensemble of particles associated with the polarization transfer is zero. Thus, observation of the effects connected with coherence transfer requires methods eliminating the phase spread which prevents the direct use of the Ramsey experiment in the optical range.

Over the last few years, various methods have been proposed to obtain resonances in separated fields based on the nonlinear interaction of particles with time and spatially limited optical fields: a three-beam system for two-level atoms[82]; a two-photon absorption in the field of two spatially separated standing waves[84]; a three-level system based on the use of processes like stimulated Raman scattering in three-level systems.[85] We will briefly describe below the first of these methods for which several experimental results are available, the reader being referred to the original papers for the two others.

Baklanov and his colleagues have introduced the idea of a third equally spaced interaction zone as a method to recover the Ramsey fringes. They have shown[86] that in a scheme similar to the one described previously in the microwave range, the dipole moment phase of a particle fully coincides, at a distance L from the second beam, with the phase of a third plane traveling wave propagating in the same direction as the two other waves and oscillating at the same laser frequency ω. In order to obtain this result, the relative phases of the first two fields were chosen in such a way that at all times, the two fields were in phase at the crossing points between their direction of propagation and the particle beam axis. It can then be shown that the third field fulfils the same condition. Since all particles will lock in spatial and time synchronism with the third wave, resonance phenomena (optical Ramsey fringes) are not observed in the process of absorption. Nevertheless, this method has the advantage of demonstrating that polarization transfer can be obtained in the optical range despite the "Doppler phase effect" which tends to average the macroscopic polarization to zero. The mechanism which locks all the particles in synchronism with the third wave whatever their velocity or direction of propagation is closely related in nature to a photon echo.

The preceding method, which does not afford improvement in terms of resolution, can be modified in order to provide an optical polarization transfer at the distance L from the second beam, with the peculiarity that the phase of the third field and of the particle dipole moment is shifted by a value of $2\Omega\mathcal{T}$, a situation which leads to the appearance of resonance phenomena.

The scheme proposed by Baklanov and colleagues is depicted in Figure 12b. It is essentially the same three separated oscillating fields scheme already mentioned, except that the second field is now a standing wave field instead of a traveling wave one. Since we are interested in the processes near the line center, we will assume that $\Omega\tau \ll 1$. From the values of a_1 and b_1, the probability amplitudes for finding a particle on the upper and lower levels after interaction with the first field $E_1(t) = E_1 \cos(\omega t - \phi_1)$, one can calculate the probability amplitudes a_2, b_2 after the interaction with the second standing field $E_2(t) = E_2 \cos \omega t \cos \phi_2$, where $\phi_1 = kz_1$, $\phi_2 = kz_2$, and $k = 2\pi/\lambda$. In a second-order perturbation theory, these amplitudes are found to be

$$b_2 = b_1 + a_1 G_2 \tau e^{j\Omega \mathcal{T}} \cos k z_2 - b_1 |G_2|^2 \frac{\tau^2}{2} \cos^2 kz$$

$$a_2 = a_1 + b_1 G_2 \tau e^{-j\Omega \mathcal{T}} \cos k z_2 - a_1 |G_2|^2 \frac{\tau^2}{2} \cos^2 kz$$

$$(3)$$

The dipole moment d_2 after the second interaction can easily be calculated from

$$d_2 = p a_2^* b_2 e^{-j\omega_{21} t} + \text{c.c.}$$ \qquad (4)

The contribution to d_2 from the terms nonlinear in G_2 is then found to be

$$a_1 b_1^* |G_2|^2 \frac{\tau^2}{2} e^{j2\Omega \mathcal{T}} e^{-j\omega_{21} t}$$

$$-a_1 b_1^* |G_2|^2 \frac{\tau^2}{4} e^{2j\Omega \mathcal{T}} [e^{-j2kz_2} + e^{2jkz_2}] e^{-j\omega_{21} t}$$

$$-2 a_1^* b_1 |G_2|^2 \frac{\tau^2}{2} \cos^2 kz_2 e^{-j\omega_{21} t}$$

$$(5)$$

The first two terms result from the product of two processes involving one quantum. The first term corresponds to the interaction with the two oppositely traveling waves of the standing wave field E_2; in the second term, one recognizes the product of two one-quantum interactions with the traveling wave of E_2 propagating in the same direction as E_1 for the term in e^{-2jkz_2}, and the same process but with the oppositely traveling wave from the term in e^{2jkz_2}. These different terms do not lead to resonance phenomena. The interesting term to consider is the third term of the nonlinear contribution to d_2. This term, connected with two quantum processes, can be written

$$-G_1 \tau |G_2|^2 \frac{\tau^2}{4} e^{j\phi_1} e^{-j\omega_{21} t} - \frac{G_1}{2} \tau |G_2|^2 \frac{\tau^2}{4} e^{j\phi_1} e^{2j\phi_2} e^{-j\omega_{21} t}$$

$$-\frac{G_1}{2} \tau |G_2|^2 \frac{\tau^2}{4} e^{j\phi_1} e^{-2j\phi_2} e^{-j\omega_{21} t} \qquad (6)$$

where the first order developments of the probability amplitudes $a_1 = 1$ and $b_1 = (G_1/2)\tau e^{j\phi_1}$, with $G_1 = jpE_1/\hbar$, have been used.

The only term leading to cancelation of the "Doppler phase effect" is the term involving the phase difference between ϕ_1 and $2\phi_2$. At the distance $2L$ from the first field, the value of the phase of this term is given by

$$\phi = -k(2z_2 - z_1) - 2\omega_{21}\mathcal{T} = -kz_3 - 2\omega_{21}\mathcal{T} \tag{7}$$

This expression demonstrates that a transfer of polarization occurs at the distance L from the second beam, since as long as the transit time $\mathcal{T}$ of the different particles is the same, the phase at a given point z_3 is independent of z_1 and the dipole moment does not average to zero. It can be readily seen that the spatial and temporal dependence of the phase $\phi(z_3, 2\mathcal{T})$ is characteristic of a traveling wave oscillating at ω_{21} and propagating oppositely to E_1. The macroscopic polarization transferred after interaction of the particles with E_1 and E_2 respectively, can be observed in absorption of a probe wave E_3 propagating parallel to the direction of the first two fields, at distance L from E_2, and oscillating at the laser frequency ω. The occurrence of resonance phenomena is directly related to the presence of a phase angle $2\Omega T$ in the difference $\Delta\psi$, between the probe field phase and the dipole moment phase. This situation is obtained when the traveling wave $E_3(t) = E_3 \cos(\omega t + kz_3)$ propagates in a direction which is opposite to the direction of the first traveling wave E_1. Under these conditions the value of $\Delta\psi$ is given by

$$\Delta\psi = 2\omega\mathcal{T} + kz_3 - kz_3 - 2\omega_{21}\mathcal{T} = -2\Omega\mathcal{T}$$

at the distance $2L$ so the phase of the field and of the particle dipole moment is shifted by a value $2\Omega\mathcal{T}$. It is this result which is responsible for the appearance of optical Ramsey fringes in the observation of the probe beam absorption.

The model above has been simplified in order to illustrate the principle of the method. It is evident that in order to precisely describe the actual experiments, a certain number of factors, omitted above, have to be taken into account. The reader is referred to the articles of C. J. Bordé for a more realistic calculation, taking into account the relaxations and the Gaussian structure of the laser beam.[87,88]

The first experimental observation of three-zone optical "Ramsey fringes" has been presented by J. C. Berquist et al.[89] The experiment, shown in Figure 13, used a monovelocity beam of metastable neon atoms interacting with separated light beams from a frequency-stabilized single-mode dye laser. The metastable atoms were produced by charge exchange in a sodium oven between accelerated neon ions and sodium atoms. The transition investigated was the $1s_5$–$2p_2$ at 5882 Å, and the interaction was

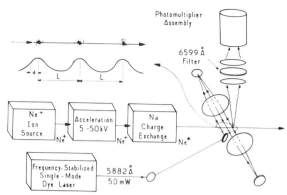

Figure 13. Schematic diagram of the experimental setup used in the first experimental observation of three zone optical Ramsey fringes. (From Ref. 89.)

detected via the strong $2p_2-1s_2$ fluorescence at 6599 Å with an appropriately filtered photomultiplier. In the experiment, the spatial separation of the adjacent radiation zones was set at $\simeq 0.5$ cm, this value being imposed by a population decay length of 0.8 cm and a corresponding phase coherence length of 1.6 cm. The position of the fringes in a three separated field experiment depends on the relative spatial phases of the different fields involved. The spatial phases in the three zones have been set in such a way that they appear to be determined from a large planar wave front. This situation corresponds to symmetric fringes with a fringe maximum at the central frequency of the transition. It is obvious that, for metrological purposes, this symmetry of the fringes must be essential. A simple yet elegant optical system that intrinsically provides this condition was formed in the Berquist *et al.* experiment by the use of two cat's-eye retroflectors set opposite each other. Fermat's principle applied to the optical system depicted in Figure 13 shows that the necessary constant phase condition was indeed satisfied. The experimental observations are shown in Figure 14. Figure 14(a) shows most of the Doppler profile, the saturated absorption dip, and the fringes due to the interaction with three equally spaced standing-wave radiation zones. The upper trace, in the inset, is the fluorescence profile produced when the atomic beam intercepts only two standing waves. No fringes due to the interaction with two separated beams are expected. The lower trace shows the profile in the three-field case where one can easily observe the cosine modulation of the saturation dip which constitutes the Ramsey fringes.

Further examples and more detailed information about optical Ramsey fringes can be found in the Proceedings of the Fourth Laser Spectroscopy Conference.[90]

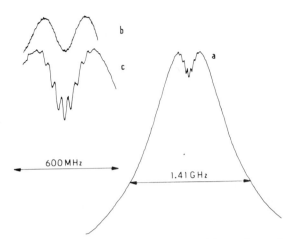

Figure 14. Saturated absorption Ramsey fringes observed
in neon. (From Ref. 89.)

4.1.2. Polarization Intermodulated Excitation Spectroscopy

Polarization intermodulated excitation (POLINEX)[91-93] is a technique of nonlinear laser spectroscopy developed in order to obtain sub-Doppler spectra free of Doppler-broadened background, despite velocity changing collisions. This technique is related to intermodulated fluorescence.[3] Two beams of equal intensities are sent in opposite directions through the cell containing the gas of atoms under study, but the polarization of one or both beams is modulated instead of the intensity. When the two beams interact with the same velocity class of atoms, the combined absorption then depends on the relative polarization of the two beams, and an intermodulation in the total rate of excitation is observed. If the cross section for disorienting collisions is at least of the same magnitude as the cross section for velocity-changing collisions, the direction of the dipole moment is randomized before the collisions redistribute the velocities of the pumped atoms over the Doppler profile of the transition. Consequently, as pointed out by Delsart and Keller,[92] the broad background which can appear in intermodulated fluorescence experiments is suppressed. This type of experiment has also been discussed by Colomb and Dumont[91] while studying the Zeeman coherences in saturated absorption.

In order to demonstrate the advantages of the technique over the previous methods, Doppler-free Polinex spectroscopy with fluorescence detection has been used to record the spectrum of the $1s_5-2p_2$ transition in neon at 5882 Å, which was observed previously by intermodulated fluorescence spectroscopy. The output of the dye laser was sent through a linear polarizer and $\lambda/4$ wave plate to produce circularly polarized light. The laser beam was then split into two parts of approximately the same intensity. A polarizing filter, spinning at about 150 revolutions per second, served as polarization modulator for one beam. It produced linearly polarized light with the polarization axis rotating around the beam direction. The second laser beam passed through a linear polarizer, which remained at a selected fixed angle during the experiment. The two beams entered a neon cell in opposite directions where a mild rf discharge was sustained. Fluroescence light emitted perpendicularly to the cell axis was observed with a photomultiplier. The signal, modulated at twice the polarizer frequency, was recorded with a lock-in amplifier. The Doppler-free Polinex spectrum of Ne $1s_5-2p_2$ transition is shown in the upper part of Figure 15, while the lower part gives, in comparison, the same spectrum recorded by the intermodulated fluorescence technique. The suppression of the Doppler background is obvious in these experimental results but does not constitute the only advantage of the technique. Since Polinex is related to intermodulated fluorescence as well as polarization spectroscopy, it combines the advantages of the two methods. In particular, it permits the studies even of very dilute and weakly absorbing samples and one can take advantage of the selection rules for the absorption of polarized light to yield information on the angular momenta of the participating energy levels. Another interesting feature is that the intermodulated signal does not have to be detected on a strongly modulated background if the total excitation rate is observed (optogalvanic detection for example), because neither beam alone can produce a modulated signal in an isotropic medium.

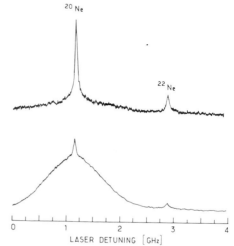

Figure. 15. (a) Doppler-free Polinex spectrum of the Ne $1s_5-2p_2$ transition. (b) Same spectrum recorded by the intermodulated fluorescence method of M. S. Sorem and A. L. Schawlow. The Doppler-broadened background in (b) is caused by velocity changing elastic collisions. (From Ref. 93.)

This technique has already found an application in the Doppler-free spectroscopy of refractive elements, where the spectra obtained from hollow cathodes were strongly limited in resolution by the Doppler background due to velocity changing collisions.

4.1.3. Laser Saturation Spectroscopy with Optical Pumping

In the Polinex method previously described, experimental situations where optical pumping plays a role have not been investigated. It is only relatively recently that the special features, which can occur in laser saturation experiments where optical pumping is present, have been investigated, by Pinard[94] and Feld.[96]

If the lower state of an atomic transition is a stable ground state or a metastable state, and if the excited-state relaxations occur predominantly through radiative decay to this lower state, the necessary conditions for optical pumping are met. In an experiment using two counterpropagating beams to perturb and probe the gas of atoms, the modifications of the absorption of the probe beam induced by the saturating beam depend on the relative polarizations of the two beams. In addition to the total population modifications, the optical pumping produces an alignment or an orientation in both levels involved. This unbalance of the Zeeman populations generally greatly reduces the intensity required in order to obtain a significant signal in nonlinear spectroscopy.

The narrow bandwidth of the laser introduces a strong correlation between atom velocities and internal variables (orientation, alignment), unlike conventional optical pumping. The method is also distinct from saturated absorption since the atomic observables are created by a pumping beam which is not strong enough to saturate the optical transition. As a consequence, the induced optical coherence is always completely negligible, and the atomic observables evolve accordingly to rate equations. Therefore, the evolution of these observables can be exactly described by application of the theory of optical pumping to the case of velocity selective interaction with the light beam. This last feature is of particular interest in the study of collisions, since the extraction of information from the experimental curves will be substantially less complicated than in conventional saturated absorption spectroscopy. Another consequence of the absence of optical saturation is that this method provides a Doppler-free resonance absorption line essentially without power broadening of the excited state. Finally, the greatly reduced saturation thresholds make possible the saturation studied using relatively weak sources.

Experimental verifications of the special features mentioned above have been performed in neon[94] where the accent has been put mainly on the study of collisional processes. In particular, the effects of velocity-changing depolarization and metastability exchange collisions have been investigated and discussed. Other features including the demonstrating of greatly reduced saturation thresholds have been demonstrated on the resonance lines of atomic sodium and barium.[95]

In conclusion, the method of saturated absorption can be improved by a convenient choice of the relative polarizations of the light beams, in order to fully take into account the optical pumping possibilities. The advantages that can be afforded are twofold: an increase in the resolution, and an increase in the sensitivity. These improvements are nevertheless linked to the occurrence of optical pumping, whose main requirement is that the lifetime of one of the atomic levels involved in the optical transitions has to be long compared to the lifetime of the other level.

4.2. New Techniques Improving the Sensitivity

Most of the techniques improving the resolution that have been described in the preceding subsection could have been included in this part since they improve the sensitivity of the detection as well. For example, the laser saturation spectroscopy with optical pumping, by decreasing the saturation intensities, permits the study of molecular transitions with low oscillator strengths.

It has also been pointed out that Polinex improves the sensitivity of intermodulated fluorescence spectroscopy, since the intermodulated signal does not have to be observed on a strongly modulated background. We will describe in this subsection, several new experiments, as well as detection methods which have been developed or recently applied to sub-Doppler spectroscopy in order to improve the signal-to-noise ratio in already known techniques.

4.2.1. High-Frequency Optically Heterodyned Saturation Spectroscopy

In practice, the actual sensitivity for Doppler-free experiments external to the laser cavity is often degraded by the presence of an unstable coherent background due to the interference on the detector of a weak parasite portion of the saturating beam, with the probe beam. The amplitude of this background depends on both the amplitude and the phase of the parasite field in such a way that the accidental interference is directly dependent on the accidental changes produced in the different optical paths by air turbulences, mechanical vibrations, etc. Several techniques have already been used for reducing this background. The most common consists simply of allowing a small angle between saturating and probe beam,[96,97] at the expense of some residual Doppler broadening. Another method consists of inducing one mirror to vibrate in order to eliminate the interferences.[98] Of course, whenever it is possible, the direct modulation of the level population will also remove the background. The method of high-frequency optically heterodyned saturation spectroscopy via resonant degerate four-wave mixing recently proposed by Raj *et al.*[99,100] presents, in addition to the background suppression, the advantage of increasing the sensitivity of heterodyne detection by going out of the laser noise spectrum. The experimental arrangement is depicted in Figure 16. The laser beam is split into two orthogonally polarized beams in order to minimize, with a judicious use of polarizer, the optical feedbacks. The saturating beam is frequency shifted and modulated by a traveling-wave acoustooptic modulator. If ω is the laser frequency, Δ the radiofrequency carrier of the acoustooptic modulator, and δ the low-frequency amplitude modulation of the pump beam, the two fields can be written

$$E_{pr} = E_1 \exp\left(j\omega t\right) + \text{c.c.}$$

$$E_{pump} = E_2\{\exp\left|j(\omega + \Delta + \delta)t\right| + \exp\left|j(\omega + \Delta - \delta)t\right|\} + \text{c.c.}$$

The resonant interaction of the probe field and the pump fields in the nonlinear medium generates signal field at ω_4 through the third-order susceptibility $\chi^{(3)}(\omega_1, -\omega_2, \omega_3)$. The frequency of the signal wave is found by simple energy conservation to be

$$\omega_4 = \omega_1 - \omega_2 + \omega_3$$

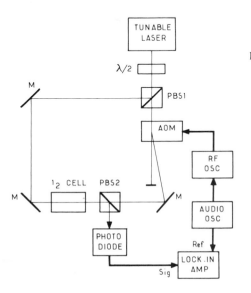

Figure 16. Schematic of the experimental configuration. The polarizing beam splitter (PBS1) splits the light from the laser into two beams. The pump beam transmitted by PBS1, is frequency shifted and modulated by the acoustooptic modulator (AOM), and passes through the I_2 cell. The probe beam is reflected at PBS1, passes through the I_2 cell, and then is reflected by PBS2 onto the photodiode. The relative power in the pump and probe beams is adjusted by the $\lambda/2$ plate. The photodiode output at the AOM modulation frequency is amplified and demodulated by the lock-in amplifier. (From Ref. 99.)

where $\omega_1 = \omega + \Delta \pm \delta$, $\omega_2 = \omega + \Delta \mp \delta$, and $\omega_3 = \omega$. That is,

$$\omega_4 = \omega \pm 2\delta$$

From a perturbative calculation, it can be shown that, in a Doppler-broadened gas, the maximum intensity is obtained when one-photon absorption and three-photon transition to the excited state are simultaneously resonant for the same axial velocity group, v. This condition, which can be written

$$\omega_1 + kv = \omega_0$$

$$\omega_1 + kv - (\omega_2 + kv) + (\omega_3 - kv) = \omega_0$$

gives the condition $2\omega_1 - \omega_2 + \omega_3 = 2\omega_0$ for the occurence of Doppler-free resonance. From the two possible choices for ω_1 and ω_2, Doppler-free resonances occur whenever the laser frequency is

$$\omega = \omega_0 + \frac{\Delta}{2} \pm \frac{3\delta}{2}$$

As might be expected, since probe and pump beams are not at the same frequency, the saturation resonance is shifted from peak of the Doppler profile by an amount $\simeq \Delta/2$. Moreover, if δ is larger than the homogeneous linewidth, each transition will yield Doppler-free doublets separated by $3\delta/2$.

The advantage of this method is twofold. First, since the pump-beam frequency differs from the probe-beam frequency by the rf carrier Δ, interference with the probe field regenerates the radiofrequency Δ which can be eliminated in the detection by appropriate filtering. The unstable background already mentioned at the begining of this subsection is thus easily eliminated. The second advantage is that the generated signal at $\omega_4 = \omega \pm 2\delta$ mixes with the probe beam at ω to generate a beat signal at frequency 2δ which can be detected using a lock-in detection. The sensitivity of heterodyne detection can then be increased by going out of the laser noise spectrum for large δ's.

Several experiments have been performed in order to demonstrate the effectiveness of the method. As an example, the left part of Figure 17 shows a spectrum of the $I_2[R(15)v = 0 \to v = 43]$ line at 5145 Å[100] recorded with $\Delta = 75$ MHz and $\delta = 20$ kHz. Since δ is smaller than the homogeneous linewidth, the doublet structure is not resolved. The signal-to-noise ratio in this recording can be shown to be close to the photon shot noise limit, while as expected, there is no detectable background. The

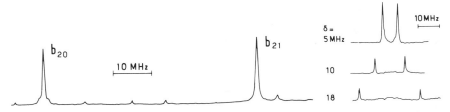

Figure 17. (a) Spectrum of a portion of the $I_2[R(15)v = 0 \rightarrow v = 43]$ line at 5145 Å for $\delta = 70$ kHz. (b) Power detection of the beat signal for various values of δ. (From Ref. 100.)

right part of the same figure gives an example of the doublets corresponding to a single line recorded for different δ's. Since δ can be accurately known, the doublet splitting can be used as a precise calibration scale.

Another interesting feature of the method pointed out by the authors is that if δ is of the order of the relaxation rates, the overall phase of the signal is shifted because the populations follow the modulation with some delay. The measurement technique is then the equivalent in high-resolution spectroscopy of the phase shift method in modulated fluorescence; This technique has yielded first direct measurement of the collisional quenching of the $2s_{1/2}$ metastable state of atomic hydrogen in a gas discharge.[101]

The method of high-frequency optically heterodyned saturation spectroscopy is in principle applicable to any nonlinear spectroscopy schemes where it allows an optimization of the signal-to-noise ratio by an adequate choice of the pump beam modulation frequency. This method seems promising in particular for two-photon spectroscopy where preliminary experiments are being performed at the present time.

4.2.2. Saturated Absorption Inside a Fabry–Pérot Resonator

The saturated absorption technique has been widely used in high-resolution spectroscopy as well as in metrology. This technique, using a cell external to the laser cavity, is directly derived from the Lamb dip spectroscopy where the nonlinear medium is placed inside the resonator of the laser. The advantages of the external cell will not be demonstrated here, the decoupling between laser and experiment justifying by itself the use of such a method. It turns out, nevertheless, that some of the interesting features of Lamb dip spectroscopy are lost in this method. It is indeed rather difficult to get a pure Gaussian beam in the spectrometer as well as matching the probe and saturating beams inside the absorption cell; the power available outside the laser is much lower than the intracavity power.

The saturated absorption technique can be greatly improved if the absorption cell is placed inside an enhancement cavity.[102] This cavity, consisting of two spherical mirrors, is locked to the laser frequency in order to ensure an enhancement factor of the empty cavity independent of the laser frequency. The laser beam is mode matched in the cavity in order to channel the maximum of energy in its fundamental mode.

It has been shown that, for weak absorptions, the relative variation of the intensity transmitted by an absorbing medium is multiplied by a factor $(1 + R)/(1 - R)$ if this medium is placed inside a cavity consisting of mirrors with identical reflection coefficients R. This property remains valid in saturated absorption in the case of low saturation parameters $(S \simeq 0.1)$.

The advantages of saturated absorption inside a Fabry–Pérot resonator, where R is close to 1 are then obvious; it allows high-resolution spectroscopy in the case of weak absorbers and provides a signal-to-noise enhancement. The signal can be multiplied by a factor of 2 by producing a difference between the transmitted and reflected beams in the interferometer, minimizing at the same time the laser intensity fluctuations. The beam structure inside the resonator is a pure Gaussian beam, and saturating and probe beam are naturally matched.

Using this method, the high-resolution spectroscopy of a weak iodine line [R(48)–15–5] lying in the scanning range of the He–Ne laser oscillating at 612 nm has been possible for the first time.[102] The gain in sensitivity over previous techniques has been estimated to be about 50 in this experiment. This increase in sensitivity has permitted the high-resolution spectroscopy of the $R(97)9-2$ line of iodine at low pressure, and low saturation $(S \simeq 0.1)$, giving the first direct evidence of the effects of hyperfine predissociation on the width of the different hyperfine lines.

4.2.3. Optogalvanic and Optoacoustic Detections Applied to High-Resolution Spectroscopy

Although optogalvanic detection and optoacoustic detection have been known for long time, it is only very recently that these detection schemes have been applied to high-resolution spectroscopy.

The optogalvanic effect, discovered over fifty years ago by Penning,[103] was not widely used as a spectroscopic detection method until the development of tunable dye lasers. The reader is referred to the excellent article by Goldsmith and Lawler[104] for a general review of the optogalvanic spectroscopy. We will restrict ourselves, in this section to the results obtained in the sub-Doppler spectroscopy domain.

Illumination of a gas discharge with a radiation at a wavelength corresponding to an atomic transition of a material in the discharge causes perturbations of the steady state of two or more levels. In general, the collisional ionization rates from different levels are unequal, so perturbations of the steady-state populations of bound levels produce a perturbation of the ionization balance of the discharge. This in turn causes a change in the electrical properties of the discharge. The optogalvanic effect relies on these electrical property changes. The optogalvanic spectroscopy is then based on the detection of an impedance change in a gaseous discharge, produced by convenient irradiation with a laser. This change in discharge impedance is generally detected as a change in voltage across a ballast resistor, so that no optical measurement is required in monitoring the absorption process. Consequently, problems associated with the measurement of weak fluorescence signals such as collection efficiency and scattered excitation light, do not limit the sensitivity and dynamic range.

To our knowledge, the first high-resolution experiment using optogalvanic detection was performed by Johnston[105] in a helium–neon discharge. The output from a single-mode cw dye laser was sent through a He–Ne laser plasma and retroreflected upon itself in order to obtain the two counterpropagating beams of the saturated absorption method. Despite the fact that the experimental arrangement did not make it possible to chop only one beam, the Doppler-free signal appeared very clearly as a small dip on top of a large Doppler profile. Several lines in neon were investigated in this experiment, and the behavior of the saturation dip with saturating intensity analyzed.

The Doppler-free intermodulated optogalvanic spectroscopy (IMOGS), introduced by Lawler et al.,[106] is closely related to the intermodulated fluorescence spectroscopy. The experimental setup is very similar. Two counterpropagating beams from a single-mode laser are sent through a discharge tube. One beam is chopped at frequency f_1 and the second beam at a different frequency f_2. When the laser is tuned within one homogeneous width of the line center, the two beams interact with the same group of atoms. A Doppler-free optogalvanic signal at sum $(f_1 + f_2)$ and difference $(f_1 - f_2)$ modulation frequencies is then obtained by monitoring the voltage across a ballast resistor. The sum or difference frequency is conveniently detected with a lock-in amplifier. As a demonstration of the method, the hyperfine splitting of the $2^3P-3\,^3D$ transition in He^3 has been investigated. A typical partial scan of this line is shown in Figure 18. The authors conclude that IMOGS, limited only by the shot noise of the discharge, compares favorably in sensitivity to other Doppler-free techniques, and that the method may be useful in regions of the spectrum where low noise detectors, interferometric-quality optics, or high-quality polarizers are unavailable. Another illustration of the capabilities of IMOGS is given by the isotope shift measurements performed in a hollow cathode discharge, where the spectra of nonvolatile materials have been observed using optogalvanic detection.[107]

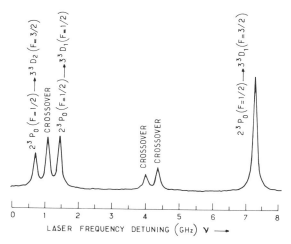

Figure 18. Intermodulated optogalvanic spectrum of part of the ^{3}He $2\,^3P$–$3\,^3D$ transition at 587.5 nm. (From Ref. 106.)

Optogalvanic detection has, more recently, been used in the high-resolution study of Doppler-free two-photon transitions.[108] A two-photon transition is typically observed by monitoring the fluorescence emitted when the atom or molecule radiatively cascades from the upper level to lower-lying levels. This detection scheme is quite suitable for studying transitions from the ground level of an atom or molecule, since the only fluorescence from the upper level is that induced by the two-photon excitation. If the lower level is instead a level populated by a discharge, the fluorescence induced by the two-photon excitation is generally masked by the discharge-induced fluorescence. The change in discharge impedance observed as a change in voltage across a ballast resistor provides a sensitive technique for detecting small change in level populations induced by two-photon excitation, thus avoiding the difficulty of discharge-induced fluorescence. This feature is even more interesting if the lower level of the transition is a nonmetastable level, since in this case the technique of excitation in the afterglow cannot be used. Two-photon optogalvanic spectroscopy (TOGS) has been demonstrated in ^{20}Ne, where several transitions have been observed, and measurements of several fine-structure intervals performed. The first Doppler-free observation of two-photon transitions originating from energy levels other than ground states of metastable levels has been reported in these experiments as well as the first Doppler-free measurement of the $4d[5/2]J = 2 - 4d[5/2]J = 3$ splitting in ^{20}Ne.

   The optoacoustic effect (OA) was known even before the optogalvanic effect, since it was discovered in 1881 by Bell. Spectroscopy using optoacoustic detection has become, in recent years, one of the most important techniques for the detection of very small absorptions in gases. However, the application of the OA methods to high-resolution spectroscopy is made difficult by the strong decrease in sensitivity when the pressure of the sample gas is diminished. If one trades off between sensitivity and pressure broadening, Doppler-free OA spectroscopy becomes then feasible as has been demonstrated in the infrared by A. di Lieto et al.[110] and in the visible by E. E. Marinero et al.[111] In the optoacoustic effect, the laser resonant with a transition of the gas sample acts as heat source that causes the pressure of the gas to change from equilibrium. This change of pressure is directly monitored by a microphone placed in a miniaturized nonresonant cell contaning the sample gas. In this type of experiment the cell has to be kept as small as possible since the optoacoustic signal is proportional to the reciprocal volume. The experimental setup that has been retained to demonstrate sub-Doppler optoacoustic spectroscopy is identical with the usual apparatus of intermodulated fluorescence detection, except that the photomultiplier is replaced by the microphone. The first experiment reported[110] used a CO_2 laser to investigate the 9–P(34) line of CH_3OH and the 10–P(30) line in pure CO_2. The feasibility of the method is clearly demonstrated, while an experimental analysis of the dependence of the OA signals on the gas pressure is given. Another experiment[111] has been

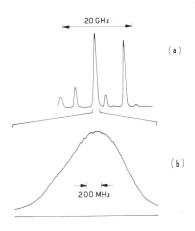

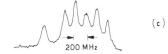

Figure 19. Optoacoustic spectra of $^{127}I_2$. (a) Doppler-limited 30 GHz scan around 5915 Å. (b) Expanded view of the largest peak in (a). (c) Doppler-free scan. (From Ref. 111.)

reported in the visible, with the study of the $P(93)$ line of 11-0 band of the $B-X$ transition of I_2 (Figure 19). The resolution obtained in these experiments, obviously limited by pressure broadening, is far beyond the resolution routinely obtained with conventional methods. The advantage of OA sub-Doppler spectroscopy with respect to the usual saturated absorption spectroscopy lies, above all, in the small sample size that can be analyzed and in the possibility of studying weakly or nonfluorescing molecules.

In this chapter, we have given a quick survey of some new techniques that have been recently introduced in the field of sub-Doppler spectroscopy; it was not possible within the limits of this presentation to give a complete review of all the interesting recent advances in this field. Subjects like compensation of Doppler broadening by light shift,[112] F.M. spectroscopy,[113] and spectroscopy beyond the natural linewidth[114,115] could have been included as well. The reader is referred to the original articles, for the methods mentioned above which are not considered in this article and to the proceedings of the fifth international conference on laser spectroscopy for a survey on very recent works.[116]

5. High-Resolution Spectroscopy with Short Light Pulses

High-resolution spectroscopy has been associated for a long time with highly monochromatic sources; the advantages of pulsed lasers (power, spectral range) were not considered since the resolution that could be achieved was limited by the spectral width $1/\tau$ of the pulse, (τ = pulse duration), thus making it impossible to resolve many closely spaced resonance lines. An important step was made when Salour and Cohen-Tannoudji[117] demonstrated in a two-photon experiment that, by exciting atoms with two time-delayed coherent pulses, resolution higher than the spectral width of one single pulse could be achieved. The technique, combining the advantages of pulsed lasers with the high resolution usually associated with a cw excitation, can be considered as an extension in the optical range of the well-known Ramsey method, where the two spacially separated fields are replaced by two time-delayed pulses. In view of these experiments, it was suggested that by using more than two coherent light pulses, important improvements would occur in resolution and signal strength. These interesting features were demonstrated shortly after, by using cw mode-locked dye lasers as an excitation source. In particular, it has been shown in different techniques (two-photon spectroscopy,[118] saturated absorption,[119] and polarization[120] spectroscopies) that the resolution that can be achieved in multicoherent pulse excitation is comparable to the resolution currently obtained in similar experiments using highly monochromatic sources. In order to introduce the reader to these new techniques and point our their respective advantages over continuous excitation, we will in the first part describe the experiment of Salour and Cohen-Tannoudji and the mehtod of high-resolution two-photon spectroscopy with multipulse excitation. The second part will describe the method of saturated absorption and polarization spectroscopies with coherent trains of short light pulses.

5.1. Quantum Interference Effects in Two-Photon Spectroscopy

5.1.1. Two-Pulse Excitation[117,121]

Interaction of atoms and molecules with intense monochromatic laser standing waves can give rise to very narrow Doppler-free two-photon resonances. The principle of this technique is well known. The moving atom absorbs one photon in each of the two counterpropagating waves constituting the standing wave field. The resulting energy is $h\nu(1 + v/c) + h\nu(1 - v/c) = 2h\nu$. Since this energy is velocity independent, all atoms, regardless of their velocity, participate in resonance, and the linewidth of the transition is limited in theory by the natural linewidths of the radiating atoms. In practice, however, the linewidth observed in two-photon spectroscopy has thus far been limited by the laser. The highly monochromatic cw lasers are then more suited to take full advantage of the elimination of the Doppler width, but, on the other hand, pulsed sources would be more suited to provide the appreciable intensity required to obtain two-photon resonances.

In an attempt to achieve the higher resolution associated with two-photon resonances when a pulsed laser is used, Salour and Cohen-Tannoudji have demonstrated that by exciting atoms with two delayed coherent laser pulses, one can obtain interference fringes in the Doppler-free profile of the transitions with a splitting $1/2T$ (T: delay between the two pulses) much smaller than the spectral width $1/\tau$ of one single pulse (τ: duration of each pulse). In a simplified picture, one can think of a two-level system excited by two pulses obtained by amplifying the output of a single-mode cw laser with two time-delayed pulsed amplifiers. The two generated pulses are then portions of the same sinusoid and have the same phase as the cw carrier wave. The first pulse creates a dipole moment with a phase related to the phase of the exciting field. During the dark period between the two pulses, the dipole processes freely at the Bohr frequency ω_0 of the transition and accumulates a phase $\omega_0 T$, while the field at frequency ω acquires in the same time a phase ωT. The interaction with the second pulse will either further excite the atoms or deexcite them by stimulated emission, depending on the relative phase $(\omega_0 - \omega)T$ between

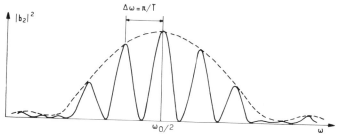

Figure 20. Sinusoidal modulation of the pulse envelope in a two-coherent-pulse experiment. (From Ref. 121.)

electric field and dipole oscillations. It can be shown in a very simple calculation that the Doppler-free probability $|b_2|^2$ for having the atom excited in the upper state after the second pulse is related to the probability $|b_1|^2$ of excitation after the first pulse by

$$|b_2|^2 = 4|b_1|^2[\cos(\omega_0 - 2\omega)T/2]^2$$

where the factor 2 comes from the fact that we deal with two-photon transitions. If now one varies the laser frequency ω with T fixed, to record the Doppler-free two-photon resonance profile, one gets (Figure 20), within the profile associated with $|b_1|^2$, interference fringes described by the last oscillatory term with a splitting in ω determined by π/T and a central fringe at $\omega = \omega_0/2$. This process is exactly the analog of the well-known Ramsey technique developed in the preceding section. An interesting feature is that the experiment can be performed in a gas cell, since the Doppler-free two-photon excitation with standing waves ensures that the fringe structure is not smeared out by dephasing due to the unavoidable spread of transverse atomic velocities.

In order to obtain interference fringes, two important requirements must be fulfilled. First, each pulse must be reflected against a mirror placed near the atomic cell in order to submit the atoms to a pulsed standing wave. Second, the phase difference between the two pulses must remain constant during the whole experiment, since any phase variation will produce a shift of the fringes and then completely smear out the fringe pattern.

Because of this last requirement, the experimental demonstration of the method has not been possible with two independent pulses, the variation from pulse to pulse of the refractive index in the amplifier medium giving rise to a random phase difference which, when averaged, washed out the interference fringes.

The experimental method that has been used consists of starting with a Fourier-limited pulse obtained by amplifying a cw wave, and to generate

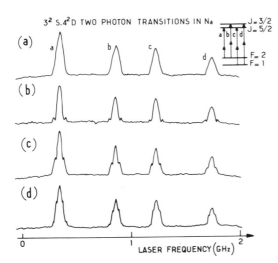

Figure 21. (a) Recording of the four Doppler-free $3\,^2S$–$4\,^2D$ two-photon resonances of Na observed in the reference cell excited with a single pulse. (b)–(d) Same resonances observed in the sample cell excited with two time-delayed coherent pulses. Width of each pulse, $\tau = 8$ nsec; effective delay between two pulses, $\tau_{\mathrm{eff}} = 2T = 17$, 25, and 33 nsec, respectively. The spacing between the fringes, $\Delta\nu \approx 60$, 40, and 30 MHz, respectively, is in good agreement with the theoretical prediction $\frac{1}{2}T$. (From Ref. 117.)

two pulses from it in an optical delay line. Since T is generally not an integer multiple of the period of the field, the two pulses do not belong to the same sinusoid; moreover, since T is fixed by the optical length of the delay line, the phase difference between the two pulses is ω dependent in such a way that no fringe will occur when the laser is scanned over the transition. The solution that has been used to lock the phase of the two pulses, consists of varying the length of the optical delay line (that is T) when ω is varied, so that $e^{i\omega T}$ remains equal to 1. In this situation the phases are locked, and since the two pulses belong to the same sinusoid, the central fringe of the interference pattern is centered exactly at $\omega = \omega_0/2$.

Figure 21 shows the four well-known two-photon resonances of the 3^2S–4^2D transition of Na observed for the upper trace with a single pulse, and for the other traces, with the two-pulse techniques. The delay times between two pulses are, respectively, from (b) to (d), 17, 25, and 33 nsec. An interference structure appears clearly on each resonance. The splitting between fringes decreases with π/T as expected, but the contrast becomes poor for the high T values. For a given value of T, only those atoms that remain in the excited state for a time at least equal to T would contribute to the interference effect. When T is increased, the number of contributing atoms is obviously decreased in such a way that one obtains a better resolution accompanied by a smaller contrast.

5.1.2. Multipulse Excitation

Shortly after the possibility of Doppler-free two-photon spectroscopy was first suggested, narrow spectral fringes were observed by exciting the sodium $3s$–$5s$ transition with a train of coherent pulses.[122] The idea was to increase the resolution obtained with the two-pulse method, where the sinusoidal fringe pattern makes it difficult, if not impossible, to resolve closely spaced line components, even if the number of fringes is small. From the analogy between the two-pulse excitation experiment and the well-known spatial interference pattern that is obtained in the diffraction of light from a double slit, one can expect that coherent excitation with a whole train of identical light pulses will produce a spectrum that resembles the interference pattern from an array of slits. The fringe spacing should remain the same for a given pulse separation, while the fringes should condense into narrow spectral lines. Another interpretation of these narrow fringes is to note that the Schrödinger equation provides the atoms simply with a means of taking the Fourier transform of the incoming pulse train, so that the Fourier transform, which gives sinusoidal modulation for two pulses becomes a Fourier series (fringes with width null) in the limit of an infinite train of pulses.

The first scheme used to demonstrate multipulse excitation in two-photon Doppler-free spectroscopy consisted of producing a pulse train by injecting a single short dye laser pulse into an optical cavity formed by two mirrors.[122] The gas sample was placed near one end mirror so that atoms saw a pulsed standing wave field once during each round trip, when the pulse was being reflected by the mirror. The sharp multipulse interference fringes observed when the resonator length was changed have been interpreted in terms of the modes of the optical resonator. This resonator filters narrow spectral lines out of the spectrum of the laser pulse. A fringe maximum is observed if the resonator is tuned so that its modes can excite the two-photon resonance. This can occur in two ways: either the frequency of a mode coincides with $\omega_0/2$ (ω_0 frequency of the transition), or the frequency $\omega_0/2$ falls exactly halfway between two cavity modes. In either case, all the modes contribute to the excitation since for a given mode one can find a second mode such that the sum frequency of the two modes is equal to ω_0. The interpretation in terms of cavity modes, that is to say in the frequency space, is of course complementary to the interpretation previously given in the time domain. If the fringe pattern can be understood easily by using the description in terms of passive spectral filtering, it turns out that the calculation of the magnitude of the signal is much easier if one uses the description in the time domain. From this calculation it has been demonstrated that the resonant signal, due to the cavity, is enhanced in the limit of negligible atomic relaxations by a factor equal to the square of the effective number of pulse round trips, while the Doppler-broadened background due to traveling wave excitation grows only linearly with N so that the contrast improves with increasing N. The Na $3s$–$5s$ transition has been studied with this method; the two-photon excitation was observed by monitoring the $4p$–$3s$ uv fluorescence light. Fringes of very high contrast

and of a few MHz width, much below the 300-MHz laser linewidth, have
been observed with a 2-m-long focal resonator and about 6 effective pulse
round trips.

Since the resolution in coherent excitation is directly related to the
number of participating pulses, it was rather important to find new ways
to produce a nearly infinite train of coherent pulses so that the laser does
not significantly contribute to the linewidth of the investigated transitions.
We have already mentioned in section 2 of this chapter that the pulse train
from a mode-locked laser can be described as a superposition of phase-
locked oscillating laser modes which are equally spaced in frequency.
Mode-locked lasers are then the equivalent in the frequency domain of a
single laser pulse filtered by an optical cavity. But, since the number of
pulses generated by the laser is nearly infinite, the modes are quite narrow,
a situation which would correspond in the optical filtering method to a
cavity with a nearly infinite finesse. Resonant two-photon Doppler-free
excitation with mode-locked lasers is then possible whenever two laser
mode frequencies coincide with an atomic transition frequency in a region
where two counter propagating pulses form a standing wave field. When
the laser frequency is scanned, each transition then appears as a comb of
resonances, separated, for the reason already mentioned, by half the laser
intermode frequency spacing.

A diagram of the apparatus used to demonstrate the feasibility of the
method[118] is shown in Figure 22. A mode-locked argon laser was used
to synchronously pump a dye laser whose cavity length had been extended
to equal half the length of the pump laser in order to increase the frequency
mode spacing. The dye laser bandwidth was restricted by Fabry–Pérot
etalons to limit to about 10 the number of oscillating modes. The output
of the laser was focused in a Na cell, collimated by a lens and reflected
back by a plane mirror. The distance of the plane mirror to the Na cell
was chosen in such a way that each pulse meets its counterpropagating

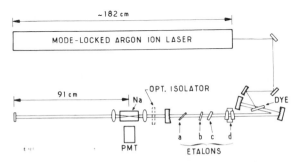

Figure 22. Experimental set up for multipulse two-photon
experiment. (From Ref. 118.)

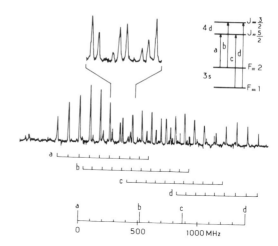

Figure 23. Multipulse two-photon spectrum of the sodium 3s–4d transition with expanded portion shown above, recorded with pulses of 500 psec duration. (From Ref. 118.)

predecessor in the cell, thus forming a pulse standing wave. Figure 23 shows a typical multipulse spectrum of the 3s–4d transition where each of the four components appears as a comb of resonances. The linewidth of each fringe is about 4 MHz for 500 psec pulse duration. As is apparent from the figure, the combs of fringes of the closely spaced transitions appear interleaved. This type of multipulse spectrum presents a certain number of advantages over the spectrum obtained by two-photon excitation with a single mode laser. Since the intermode spacing of mode-locked lasers is rigidly determined by the modulation frequency of the mode-locker system, the fringe pattern of each transition lays down an exact frequency calibration over the bandwidth of a single pulse. Moreover, in the case where several transitions are recorded on the same spectrum, the actual value of the atomic line splitting can be determined with high accuracy, if one has prior knowledge of the line separation to an accuracy better than the fringe spacing. This splitting value is determined modulo the fringe period, by measuring the small frequency interval between adjacent fringes, each of these two fringes belonging to one given transition. It can be noticed that by using ultrashort pulses, line splitting as large as 1000 GHz could be measured. Another interesting advantage is that since many two adjacent fringe intervals are available for one pair of transitions, a single scan permits many individual readings, in such a way that a statistical analysis of the measurements can be made from only one experiment. Finally, this technique holds promises for Doppler-free excitation of a two-photon transition, where the photon frequency lies in the uv domain. In that case the signal would be greatly enhanced over a similar experiment with a cw laser, since, for identical average power, the uv obtained by second harmonic generation with a mode-locked laser is larger by a factor equal to the inverse of the duty cycle, and the two-photon excitation rate depends on the square of the average ultraviolet power.

5.2. Saturated Absorption and Polarization Spectroscopy with a Coherent Train of Short Light Pulses

In Doppler-free saturation spectroscopy cw or pulsed quasimonochromatic sources have been extensively used. It has been demonstrated however, that a coherent train of short light pulses can provide high resolution as well, in saturated absorption[119] in which the highly monochromatic source is replaced by a synchronous mode-locked laser.

While two-photon excitation of a gas of atoms or molecules with a coherent train of standing wave light pulses can be viewed as a linear interaction between an effective field and a two-level system, and analyzed in the time domain, the fundamental nonlinear character of saturated absorption prevents a similar simple treatment. However, if the dipole relaxation time of the investigated transition is much larger than the delay time between two pulses, so that one molecule experiences many pulses before the decay of the coherences induced by the first pulse, the interaction can then be easily understood in the frequency domain, and a very clear analysis can be made. The two counterpropagating beams can be described as a superposition of phase-locked oscillating laser modes which are equally spaced in frequency. The saturating beam, propagating along the z axis, perturbs the thermal equilibrium of the velocity classes v_z of molecules resonant with each laser mode. The equally spaced holes burned in the velocity distribution are then analyzed by the frequency comb of the probe beam propagating in the opposite direction. If n denotes the number of laser modes, one then obtains for each transition ω_0, $2n - 1$ fringes corresponding to the enhancement of the probe beam transmission which occurs each time at least one of its modes interacts with the same molecules, as does one mode of the saturating beam. The fringe spacing is then equal to half the laser mode spacing. From this simple picture, one notices that in addition to the advantages already reported for two-photon spectroscopy (high resolution, precise frequency scale, etc.), and in contrast with saturation spectroscopy using cw lasers, many velocity classes can contribute to the signal. The result is an increase in the sensitivity.

The feasibility of saturated absorption with two counterpropagating trains of short light pulses has been demonstrated in I_2.[119] A mode-locked argon ion laser was synchronously pumping a R6G dye laser. In order to obtain the highest resolution, the jitter on each laser mode was reduced down to ± 250 kHz by frequency locking one of the laser modes to the transmission fringe of a high finesse optical cavity.[123] The dipole relaxation time of the investigated transitions belonging to the $B^3\Pi_0^+ \to X^1\Sigma_g^+$ system of molecular iodine, of the order of 1 μ sec was much larger than the delay time between two pulses ($\approx$5 nsec). In this way, the fundamental condition for obtaining fringelike spectra was fulfilled. The pulse duration was 500 psec. Figure 24 shows part of a multipulse saturated absorption spectrum of the P(93) 18–1 line of the B–X system near 5719 Å. In this spectrum, 14 of the 21 components a–n of the line contribute significantly to the fringe pattern. The measured linewidths are less than 3 MHz, comparable with the linewidth currently recorded with highly monochromatic sources.

In the saturated absorption experiment described above, only those modes within the Doppler width of the transition contribute to the saturation signal. Other modes merely contribute to the background intensity at the probe detector. The technique of polarization spectroscopy can be used to reduce this background.[120] In a typical polarization spectrometer a probe beam passes through two crossed polarizers between which is placed a cell with the gas sample. The strong saturating beam is circularly polarized and propagates in the opposite direction to the probe beam. This beam induces a frequency-dependent optical anisotropy, by causing a velocity-dependent angular momentum orientation of the atoms. A polarized light wave in resonance with the oriented atoms experiences a change in polarization due to dichroism and gyrotropic birefringence of the medium. The change in the probe polarization, which occurs when modes of both beams interact with the same atoms, manifests itself as an increased transmission through the crossed polarizers. The background is then greatly

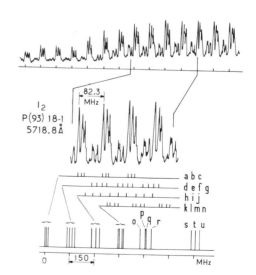

Figure 24. Part of the multipulse saturated absorption spectrum of the P(93) 18–1 Iodine line of the B–X system. (From Ref. 119.)

reduced, and the signal-to-noise ratio improved, since only the modes within the Doppler width contribute to the signal, the other modes being rejected by the crossed polarizers. This technique has been demonstrated with a train of picosecond pulses from a mode-locked laser. The transition studied was the $1s_5$ to $2p_2$ in ^{20}Ne at 588.2 nm. The corresponding line is an isolated line free of hyperfine structure, so its mode-locked polarization spectrum consists of a series of fringes, separated, as in saturated absorption spectroscopy, by one-half the intermode laser spacing.[120]

References

1. W. Demtröder, in *Progress in Atomic Spectroscopy*, Part A, p. 679, Eds. W. Hanle and H. Kleinpoppen, Plenum Press, New York 1978.
2. C. Freed and A. Javan, *Appl. Phys. Lett.* **17**, 53 (1970).
3. M. S. Sorem and A. L. Schawlow, *Opt. Commun.* **5**, 148 (1972).
4. C. Bordé, G. Camy, B. Decomps, and L. Pottier, *Colloq. Int.* CNRS **217**, 231 (1974).
5. B. Couillaud and A. Ducasse, *Phys. Rev. Lett.* **35**, 1276 (1975).
6. C. Wieman and T. W. Hänsch, *Phys. Rev. Lett.* **36**, 1170 (1976).
7. L. S. Vasilenko, V. P. Chebotaev, and A. V. Shishaev, *JETP Lett.* **12**, 113 (1970).
8. F. Biraben, G. Grynberg, and B. Gagnac, *Phys. Rev. Lett.* **32**, 643 (1974); M. D. Levenson and N. Bloembergen, *Phys. Rev. Lett.* **32**, 645 (1974); T. W. Hänsch, G. Meisel, and A. L. Schawlow, *Opt. Commun.* **11**, 50 (1974).
9. S. Reynaud, M. Himbert, J. Dupont-Roc, H. H. Stroke, and C. Cohen-Tannoudji, *Phys. Rev. Lett.* **42**, 756 (1979).
10. N. F. Ramsey, *Phys. Rev.* **78**, 695 (1950).
11. T. W. Hänsch and A. L. Schawlow, *Opt. Commun.* **13**, 68 (1975).
12. V. S. Letokhov, V. G. Minogin, and B. D. Pavlik, *Opt. Commun.* **19**, 92 (1976).
13. M. Hercher and H. A. Pike, *Opt. Commun.* **3**, 65 (1971).
14. S. A. Tuccio and F. C. Strome Jr., *Appl. Opt.* **11**, 64 (1972).
15. A. Dienes, E. P. Ippen, and C. V. Shank, *IEEE J. Quantum Electron.* **QE8**, 382 (1972).
16. H. W. Kogelnik, E. P. Ippen, A. Dienes, and C. V. Shank, *J. Quantum Electron.* **QE8**, 373 (1972).
17. H. W. Schröder, L. Stein, D. Fröhlich, and H. Welling, *Appl. Phys.* **14**, 377 (1977); S. M. Jarrett and J. F. Young, *Opt. Lett.* **4**, 176 (1979).
18. C. T. Pike, *Opt. Commun.* **10**, 14 (1974).
19. B. Couillaud, L. A. Bloomfield, J. E. Lawler, A. Siegel, and T. W. Hänsch, *Opt. Commun.* **35**, 359 (1980).
20. T. F. Johnston Jr, W. Proffitt, *IEEE J. Quantum. Electron.* **QE16**, 483 (1980).
21. F. Biraben, *Opt. Commun.* **29**, 353 (1979).
22. J. M. Yarborough and J. Hobart, IEEE/OSA, Conf. Laser Engineering and Applications, Washington D.C. (1973).
23. D. R. Preuss and J. L. Gole, *Appl. Opt.* **19**, 702 (1980).
24. J. L. Hall, 7th Conf. on Atomic Physics, Boston (1980).
25. T. W. Hänsch and B. Couillaud, *Opt. Commun.* **35**, 441 (1980).
26. E. P. Ippen, C. V. Shank, and A. Dienes, *Appl. Phys. Lett.* **21**, 348 (1972).
27. C. K. Chan and S. O. Sari, *Appl. Phys. Lett.* **25**, 403 (1974).
28. H. Mahr, *IEEE J. Quantum Electron.* **QE12**, 544 (1976).

29. J. Kuhl, R. Lambrich, and D. von der Linde, *Appl. Phys. Lett.* **31**, 657 (1977).
30. J. N. Eckstein, A. I. Ferguson, T. W. Hänsch, C. A. Minard, and C. K. Chan, *Opt. Commun.* **27**, 466 (1978).
31. S. Blit and C. L. Tang, *Appl. Phys. Lett.* **36**, 16 (1980).
32. E. D. Hinkley, K. W. Nill, and F. A. Blum, *Laser Focus* **12**, 47 (April 1976).
33. L. F. Mollenauer and D. H. Olson, *Appl. Phys. Lett.* **24**, 386 (1974).
34. L. F. Mollenauer and D. H. Olson, *J. Appl. Phys.* **46**, 3109 (1975).
35. H. Welling, G. Litfin, and R. Beigang, in *Laser Spectroscopy III*, p. 370, Eds. J. L. Hall and J. L. Carlsten, Springer-Verlag, Berlin (1977).
36. I. Schneider and C. I. Marquardt, *Opt. Lett.* **5**, 214 (1980).
37. L. F. Mollenauer, *Opt. Lett.* **5**, 188 (1980).
38. T. E. Gough, D. Gravel, R. E. Miller, and G. Scoles, in *Proceedings of the 11th International Quantum Electronic Conf.*, Boston (1980), p. 665.
39. D. J. Jackson, H. Gerhardt, and T. W. Hänsch, *Opt. Commun.* **37**, 23 (1981).
40. F. V. Kowalski, R. T. Hawkins, and A. L. Schawlow, *J. Opt. Soc. Am.* **66**, 965 (1976).
41. J. L. Hall, and S. A. Lee, *Appl. Phys. Lett.* **29**, 367 (1976).
42. P. Juncar and J. Pinard, *Opt. Commun.* **14**, 438 (1975).
43. J. Cachenaut, C. Man, P. Cerez, A. Brillet, F. Stoeckel, A. Jourdain, and F. Hartmann, *Rev. Phys. Appl.* **14**, 685 (1979).
44. J. J. Snyder, in *Laser Spectroscopy III*, p. 419 Eds. J. L. Hall and J. L. Carlsten, Springer-Verlag, Berlin (1977).
45. R. Salimbeni and R. V. Pole, *Opt. Lett.* **5**, 39 (1980).
46. X. H. Rong, S. V. Benson, and T. W. Hänsch, *Laser Focus* **17**, 54 (March 1981).
47. W. E. Lamb Jr., *Phys. Rev.* **134**A, 1429 (1964).
48. M. Sargent, M. O. Scully, and W. E. Lamb Jr., *Laser Physics*, Addison-Wesley, Reading, Massachusetts (1974).
49. N. Bloembergen, *Non-Linear Optics*, Benjamin, New York (1965).
50. P. A. Franken, A. E. Hill, C. W. Peters, and G. Weinreich, *Phys. Rev. Lett.* **7**, 118 (1961).
51. J. M. Halbout, S. Blit, W. Donaldson, and C. L. Tang, *IEEE J. Quantum Electron.* **QE15**, 1176, (1979).
52. R. E. Stickel Jr., *Appl. Opt.* **17**, 2270 (1978).
53. G. D. Boyd and D. A. Kleinman, *J. Appl. Phys.* **39**, 3597 (1968).
54. S. Blit, E. G. Weaver, F. B. Dunning, and F. K. Tittel, *Opt. Lett.* **1**, 58 (1977).
55. S. Blit, E. G. Weaver, F. B. Dunning, and F. K. Tittel, *Appl. Opt.* **17**, 721 (1978).
56. C. Gabel and M. Hercher, *IEEE J. Quantum Electron.* **QE8**, 850 (1972).
57. P. Huber, *Opt. Commun.* **15**, 196 (1975).
58. D. Frölich, L. Stein, H. W. Schröder, and H. Welling, *Appl. Phys.* **11**, 97 (1976).
59. A. I. Ferguson, M. H. Dunn, and A. Maitland, *Opt. Commun.* **19**, 10 (1976).
60. A. I. Ferguson and M. H. Dunn, *Opt. Commun.* **23**, 177 (1977).
61. B. R. Marx, K. P. Birch, R. C. Felton, B. W. Joliffe, W. R. C. Rowley, and P. T. Woods, *Opt. Commun.* **33**, 287 (1980).
62. H. W. Schröder, L. Stein, D. Frölich, B. Fugger, and H. Welling, *Appl. Phys.* **14**, 377 (1977).
63. A. I. Ferguson and M. H. Dunn, *IEEE J. Quantum Electron.* **QE13**, 751 (1977).
64. C. E. Wagstaff and M. H. Dunn, *J. Phys. D* **12**, 355 (1979).
65. C. M. Marshall, R. E. Stickel, F. B. Dunning, and F. K. Tittel, *Appl. Opt.* **19**, 1980 (1980).
66. S. J. Bastow and M. H. Dunn, *Opt. Commun.* **35**, 259 (1980).
67. C. E. Wagstaff and M. H. Dunn, *Opt. Commun.* **35**, 353 (1980).
68. B. Couillaud, L. A. Bloomfield, J. E. Lawler, A. Siegel, and T. W. Hänsch, *Opt. Commun.* **35**, 359 (1980).
69. C. R. Webster, L. Wöste, and R. N. Zare, *Opt. Commun.* **35**, 435 (1980).

70. F. B. Dunning, *Laser Focus* **14**, 72 (May 1978).
71. S. Blit, G. Weaver, F. B. Dunning, F. K. Tittel, *Opt. Lett.* **1**, 58 (1977).
72. S. Blit, E. G. Weaver, F. B. Dunning, and F. K. Tittel, *Appl. Opt.* **17**, 721 (1978).
73. R. P. Mariella Jr., *Opt. Commun.* **29**, 100 (1979).
74. R. P. Mariella Jr., *J. Chem. Phys.* **71**, 94 (1979).
75. A. I. Ferguson, J. E. M. Goldsmith, T. W. Hänsch, and E. W. Weber, in *Laser Spectroscopy IV*, p. 31, Eds. H. Walther and K. W. Rothe, Springer-Verlag, Berlin (1979).
76. R. E. Stickel Jr. and F. B. Dunning, *Appl. Opt.* **17**, 2270 (1978).
77. T. W. Hänsch and A. L. Schawlow, *Opt. Commun.* **13**, 68 (1975).
78. D. J. Wineland and H. Dehmelt, *Bull. Am. Phys. Soc.* **20**, 637 (1975).
79. D. J. Wineland, R. E. Drullinger, and F. J. Walls, *Phys. Rev. Lett.* **40**, 1639 (1978).
80. W. Neuhauser, M. Hohenstatt, P. Toschek, and H. Dehmelt, *Phys. Rev. Lett.* **41**, 233 (1978).
81. J. L. Hall, C. J. Bordé, and K. Vehara, *Phys. Rev. Lett.* **37**, 1339 (1976).
82. Y. V. Baklanov, B. Y. Dubetsky, and V. P. Chebotaev, *Appl. Phys.* **9**, 171 (1976).
83. N. F. Ramsey, *Phys. Rev.* **78**, 695 (1950).
84. V. P. Chebotaev, A. V. Shishaev, B. Y. Yurshin, and L. S. Vasilenko, *Appl. Phys.* **15**, 43 (1978).
85. V. P. Chebotaev and B. Y. Dubetsky, *Appl. Phys.* **18**, 217 (1979).
86. V. P. Chebotaev, *Appl. Phys.* **15**, 219 (1978).
87. C. J. Bordé, *C.R. Acad. Sci. Paris* **282**, 341 (26 April 1976).
88. C. J. Bordé, *C.R. Acad. Sci. Paris* **284**, 101 (7 February 1977).
89. J. C. Berquist, S. A. Lee, and J. L. Hall, *Phys. Rev. Lett.* **38**, 159 (1977).
90. *Laser Spectroscopy IV*, Eds. H. Walther and K. W. Rothe, Springer-Verlag, Berlin (1979).
91. I. Colomb and M. Dumont, *Opt. Commun.* **21**, 143 (1977).
92. C. Delsart and J. C. Keller in *Laser Spectroscopy III*, p. 154 Eds. J. L. Hall and J. L. Carlsten, Springer-Verlag, Berlin (1977).
93. T. W. Hänsch, D. R. Lyons, A. L. Schawlow, A. Siegel, Z. Y. Wang, and G. Y. Yan *Opt. Commun.* **37**, 87 (1981).
94. M. Pinard, C. G. Aminoff, and F. Laloë, *Phys. Rev. A* **19**, 2366 (1979).
95. M. S. Feld, M. M. Burns, T. U. Kühl, P. G. Pappas, and D. E. Murnick, *Opt. Lett.* **5**, 79 (1980).
96. T. W. Hänsch, M. D. Levenson, and A. L. Schawlow, *Phys. Rev. Lett.* **26**, 946 (1971).
97. B. Couillaud and A. Ducasse, *Opt. Commun.* **13**, 398 (1975).
98. C. N. Man, P. Cerez, A. Brillet, and F. Hartmann, *J. Phys. (Paris) Lett.* **38**, 287 (1977).
99. J. J. Snyder, R. K. Raj, D. Bloch, and M. Ducloy, *Opt. Lett.* **5**, 163 (1980).
100. R. K. Raj, D. Bloch, J. J. Snyder, G. Camy, and M. Ducloy, *Phys. Rev. Lett.* **44**, 1251 (1980).
101. D. Bloch, R. K. Raj, and M. Ducloy, *Opt. Commun.* **37**, 183 (1981).
102. P. Cerez, A. Brillet, C. Man-Pichot, and J. Umezu, *C.R. Acad. Sci. Paris*, **290**, 515 (23 Juin 1980).
103. F. M. Penning, *Physica* **8**, 137 (1928).
104. J. E. M. Goldsmith and J. E. Lawler, to be published in *Contemp. Phys.*
105. T. F. Johnston Jr., *Laser Focus* **14**, 58 (1978).
106. J. E. Lawler, A. I. Ferguson, J. E. M. Goldsmith, D. J. Jackson, and A. L. Schawlow, *Phys. Rev. Lett.* **42**, 1046 (1979).
107. A. Siegel, J. E. Lawler, B. Couillaud, and T. W. Hänsch, *Phys. Rev. A* **23**, 2457 (1981).
108. J. E. M. Goldsmith, A. I. Ferguson, J. E. Lawler, and A. L. Schawlow, *Opt. Lett.* **4**, 230 (1979).
109. A. G. Bell, *Philos. Mag.* **11**, 510 (1881).

110. A. Di Lieto, P. Minguzzi, and M. Tonelli, *Opt. Commun.* **31**, 25 (1979).
111. E. E. Marinero and M. Stuke, *Opt. Commun.* **30**, 349 (1979).
112. S. Reynaud, M. Himbert, J. Dupont Roc, H. H. Stroke, and C. Cohen-Tannoudji, *Phys. Rev. Lett.* **42**, 756 (1979).
113. G. C. Bjorklund and M. D. Levenson, 11th I.Q.E.C., Boston (1980) Post-deadline paper.
114. H. M. Gibbs and T. N. C. Venkatesan, *Opt. Commun.* **17**, 87 (1976).
115. P. Meystre, M. O. Scully, and H. Walther, *Opt. Commun.* **33**, 153 (1980).
116. *Fifth International Conference on Laser Spectroscopy*, Jasper (1981), Eds. A. R. W. McKellar, T. Oka, and B. Stoicheff, Springer-Verlag, Berlin.
117. M. Salour and C. Cohen-Tannoudji, *Phys. Rev. Lett.* **38**, 757 (1977).
118. J. N. Eckstein, A. I. Ferguson, and T. W. Hänsch, **40**, 847 (1978).
119. B. Couillaud, A. Ducasse, L. Sarger, and D. Boscher, *Appl. Phys. Lett.* **36**, 407 (1980).
120. A. I. Ferguson, J. N. Eckstein, and T. W. Hänsch, *Appl. Phys.* **18**, 257 (1979).
121. M. Salour, *Rev. Mod. Phys.* **50**, 667 (1978).
122. R. Teets, J. Eckstein, and T. W. Hänsch, *Phys. Rev. Lett.* **38**, 760 (1977).
123. B. Couillaud, A. Ducasse, L. Sarger, and D. Boscher, *Appl. Phys. Lett.* **36**, 1 (1980).

4

Resonance Ionization Spectroscopy: Inert Atom Detection

C. H. Chen, G. S. Hurst, and M. G. Payne

1. Introduction

The central role of the atom in our concepts of matter creates a great need to count individual atoms in most of the scientific disciplines. The recent appearance of experimental techniques for the counting of atoms can be directly tied to the development of the pulsed laser.

Rutherford[1] referred to "the counting of atoms" in an unpublished note. A form of "atom counting" or, more appropriately, "decay counting" has been an essential technique of nuclear physics (and its many applications) since 1908 when Rutherford and Geiger[2] developed the first electrical counter. Furthermore, the mass spectrometer of J. J. Thomson,[3] when refined with modern particle detectors, can be used to count individual ions of atoms.

Recently the tunable laser has made possible resonance ionization spectroscopy (RIS), which is the basis for the counting of atoms in a more general way. Resonance ionization spectroscopy is a laser technique in which a single electron can be removed from each atom of a population

C. H. Chen, G. S. Hurst, and M. G. Payne • Oak Ridge National Laboratory, Oak Ridge, Tennessee 37830. Research sponsored by the Office of Health and Environmental Research, U.S. Department of Energy under contract W-7405-eng-26 with the Union Carbide Corporation.

115

of a given type of atom—and it is spectroscopically selective. These characteristics are in sharp contrast with the ionization of gases by the traditional methods, such as X rays or alpha particles, which are neither selective nor sensitive.[4] Furthermore, with pulsed lasers, ionization can be created in a known volume at the time of choice of the observer. Combining RIS with the one ion-pair sensitivity of ionization detectors made it possible to count one atom with selectivity of the type of atom, with space resolution, and with time resolution.

One purpose of this article is to review RIS, especially recent developments which make possible the ionization of Ar, Kr, and Xe. Whereas RIS offers the potential for three fundamentally different ways of counting atoms, this article is limited to the "direct" method as shown in part (b) of Figure 1. A thorough discussion of this method is the second objective of this article. Finally, an analysis of applications of isotope selective noble gas atom counters in particle physics and weak interaction physics, and, more briefly, other areas of applications such as oceanography and ice cap dating is the third objective of this review.

2. Resonance Ionization Spectroscopy

It has been known since the discovery of X rays and of radioactivity that "ionization" can be measured in extremely sensitive as well as

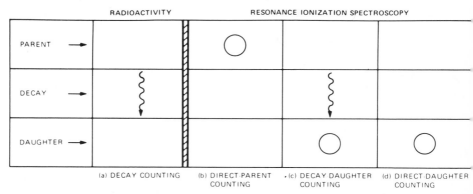

Figure 1. A perspective on methods of counting atoms. The traditional method (a) is based on the radioactive decay of a nucleus. Resonance ionization spectroscopy (RIS) provides three additional methods as illustrated. The direct counting method (b) will be discussed in this article with concentration on the noble gases Ar, Kr, and Xe. The decay-daughter atom counting (c) offers the potential for counting facilities with very low backgrounds, and has been discussed elsewhere (Refs. 5–8). Direct counting of daughter atoms (d) accumulated on a surface could provide an alternative to decay counting for long-lived parent atoms (Ref. 4).

extremely accurate ways. However, traditional ionization measurements lack the specificity (i.e., the spectroscopic features) of light emission or absorption. By using a tunable laser as a source of light to produce ionization in two or more absorption steps, spectroscopic features are attached to ionization. We refer to the process as *resonance ionization* to distinguish it from the traditional nonselective ionization associated with X rays and radioactivity.

A saturated resonance ionization process[9,10] was initially used to measure the number of excited states created in He gas by a pulse of protons from a Van de Graaff accelerator. Rather than waiting for the He metastable state to emit a photon as a result of complicated collision processes, we selected to fire a laser at a time of choice so that an electron could be removed and detected by photoionization of the excited state. If the wavelength of the laser is properly chosen, a long-lived or metastable state will be promoted to a higher excited state, i.e., an intermediate state. The population of the intermediate state grows at the expense of the excited state of lower energy. Usually, in a very short time the rate of stimulated emission from the intermediate state will just equal its rate of production. This equilibrium persists as long as the laser intensity is kept sufficiently high during a pulse. Such an equilibrium is actually a quasi steady state because photons of the same wavelength can slowly photoionize the intermediate state. If the *photon fluence* (photons per unit of beam area) is large enough, a necessary condition for saturation of the RIS process has been met. If, in addition, the rate of photoionization, which depends on the *photon flux* (photons per cm^2 per second), is larger than the rate at which the intermediate state is destroyed by possible competing processes (i.e., processes which convert the intermediate state to a species that can no longer be photoionized), then another necessary condition has been satisfied, in which case each selected state is converted to one electron plus one positive ion and the RIS process is said to be saturated. Furthermore, the process is so selective that only those excited states of the chosen type are ionized.

The RIS technique made use of a laser directed through a parallel plate ionization chamber to measure the number of long-lived excited states created by proton interactions with the He gas. Furthermore, by delaying the laser pulse with respect to the charged particle pulse that initially excited the gas and by observing the successively smaller signals, one can measure the lifetimes of metastable atoms in the system. These data on the abundance and the lifetimes of various kinds of excited states created in gases are useful to the understanding of radiation interaction with matter and to the development of new gas lasers.

The success of the first RIS experiment with He established a firm basis for the belief that single atoms could be detected. The plans for one-atom detection were, however, developed much earlier—within a few days after the initial RIS concept. Briefly, it was suggested in a patent[11] by the Oak Ridge group that an atom in its ground state could be ionized with a pulsed laser directed through a proportional counter. With the counter adjusted to detect one free electron, each atom of the selected type could be detected. Furthermore, a proportional counter gets its name from the property that the amplitude of its pulse is directly proportional to the number of the free electrons which initiate it. Therefore, if more than one atom is ionized, the height of the pulse measures the number of atoms which were in the laser beam.

To accomplish one-atom detection,[12,13] the saturated RIS process (scheme 1 in Figure 2) was used to produce free electrons, and a proportional counter was used as an electron detector. Recall that by one-atom detection we mean that one atom of a selected type can be detected at a given time and place without interference from other background atoms. In order to have a time-resolved method, use is made of a laser which provides enough photons in one pulse to remove an electron from each of the atoms of interest, and this electron is removed in a time which is less than the time required for an atom either to enter or to leave the beam. Thus, when the laser beam was pulsed through the proportional counter that contained about 10^{19} Ar atoms per cm^3 and about 10^{18} CH_4 molecules

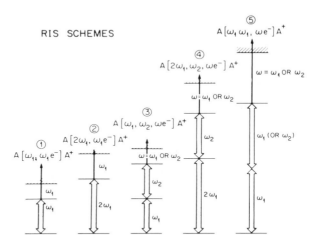

Figure 2. Classification of RIS schemes for selective ionization of the elements. Above each scheme is written a nuclear physics type notation. For example, for the noble gas (stressed in the present review), a two-photon process at angular frequency ω_1 excites an atom to a level where another photon at frequency $\omega = \omega_1$ or ω_2 photoionizes a state of even parity.

per cm^3, no ionization occurred due to the counting gas; however when just one atom of Cs diffused into the laser beam, it could be counted. In this way it was shown that even one atom of Cs could be detected without interference from 10^{19} or more atoms of some other type.

Fortunately, it turns out that with currently available lasers the RIS process can be used to selectively ionize almost every kind of atom in its normal ground state.[14] Various laser schemes (Figure 2) now make it feasible to detect each atom (except possibly He and Ne, see Figure 3).

It is shown[15,16] that one pulsed laser beam could be used to photodissociate each molecule of CsI into Cs and I and a second laser could be used to detect Cs. By dissociating each molecule in the beam and by detecting each dissociated atom, a basis for detecting just one molecule was established. As a consequence, a new method was developed for the absolute measurement of the cross section for the photodissociation of a molecule. Since the cross section for photodissociation of a molecule is a measure of the probability (expressed in geometrical units of cm^2) that a photon will interact with and dissociate one molecule of the alkali halide, one must know the number of photons, the number of dissociated molecules, and the initial number of molecules in order to obtain the cross section. But since each molecule in the source laser beam can be dissociated and since each molecule yields one atom which is detected, the latter two quantities come directly from the measurement; no independent knowledge of the density of the alkali halide is needed. Following the 1975 demonstration that the RIS process could be saturated,[9] several other groups[17-20] have used the process to obtain cross sections.

The demonstration[21] that a single neutral atom of Cs could be detected when a ^{252}Cf atom decays in a binary fission process has a dual significance. First, there is a technological importance for the following reason. When the fission daughter atom is born from the parent atom, it has a kinetic energy of about 80 MeV. When this energy is dissipated in the counting gas, nearly three million electrons are produced in the particle track, and other transient species such as excited states and negative ions may be formed. The one atom that we wish to detect is buried within this track near its end, yet we must detect it by removing from it just one electron! There is also a scientific significance to the observation of neutral atoms. When an atom is born with large kinetic energy, it is stripped of many of its electrons. As a positive ion slows down in a gas, it can capture electrons from the neutral gas; but it can lose electrons during other collisions. This capture and loss process has been studied in great detail as charged particles slow down, but techniques have not been available to see if the ions become neutral at the ends of their tracks. In the binary fission process there is a "heavy mass" and a complementary "light mass" traveling in exactly opposite directions, and from chemical studies it is known that about 15%

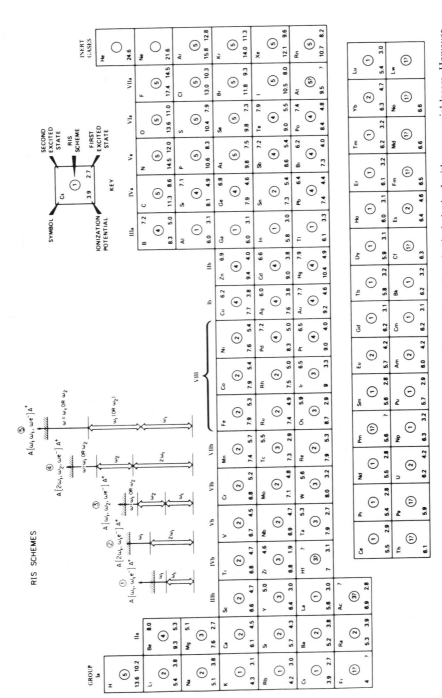

Figure 3. Using the five schemes of Figure 2, all of the elements except He and Ne can be selectively ionized with commercial lasers. However, the volumes which can be saturated vary greatly with the scheme used and the particular laser frequency required. Estimates of the small effective volumes for Ar, Kr, and Xe are given in this review.

of the heavy fragments issuing from Cf are Cs ions. Detection of neutral Cs atoms at the ends of approximately 14% of the heavy fission fragment tracks proved, therefore, that essentially all of the Cs ions were neutralized (even though this ionization potential was so low that they could not charge exchange with the counter gas at thermal energies) when they came into thermal equilibrium with the counting gas.

Having shown that a single neutral atom can be counted when it is created as a daughter product of a nuclear event, we are encouraged to speculate on a number of similar studies. Most of the elements have radioactive isotopes, and they can decay into just as many types of stable atoms. Previously we were able to detect only the radiation associated with the transmutation, but now we can detect the *stable atom and in time coincidence* with the radiation signal (method c in Figure 1). Discussion of this method is beyond the scope of this article and has been discussed elsewhere.[5-8]

Noble gas detection has recently become possible by using new commercial lasers for the RIS process. It is the purpose of this article to discuss in some detail the detection of Ar, Kr, and Xe at the one-atom level.

3. RIS on Rare Gas Atoms, Including Various Laser Schemes

3.1. Introduction

In this section we study theoretically what we believe to be the best RIS schemes for Ar, Kr, and Xe, using currently available lasers. In all cases the suggested scheme involves starting the ionization process with two-photon excitation from the np^6 ground state to a $np^5n'p$ level. In the case of Kr and Xe, two photons are absorbed from the same laser beam, while for Ar the resonance is achieved by absorbing photons of different frequency from each of two intense laser beams. The two-photon Rabi frequencies associated with each scheme are calculated, and the effect of using broad bandwidth lasers is discussed. Finally, we consider the yield of ions achieved by the use of focused laser beams. Expected ionization volumes are calculated for Ar, Kr, and Xe.

3.2. Two-Photon Excitation of Inert Gases with Broad Bandwidth Lasers

The topic of multiphoton excitation has been considered by several workers.[22] We present here a brief extension of our earlier work.[14] In Ref. 14 we have shown that if a laser is tuned very near a two-photon resonance, excitation is describable in terms of a set of Bloch equations. The off-diagonal elements of the density matrix (in the absence of relaxation processes) can be eliminated to obtain

$$Z_1 = -1 + \alpha^2 \int_{-\infty}^{t} Z_1(t')K(t, t') \, dt' \tag{1}$$

where

$$K(t, t') = -\int_{t'}^{t} dt'' \, E^2(t')E^2(t'') \cos\left[Q(t') - Q(t'')\right]$$

and

$$Q(t) = \delta t + 2\eta(t) - \beta \int_{0}^{t} E^2(t') \, dt'$$

In the present analysis we describe the problem through a two-state system coupled by two-photon processes with $Z_1 = \rho_{11} - \rho_{00}$ = population inversion.[14] We have taken $\hat{H} = \hat{H}_0 - \hat{P}_z E(t) \cos\left[\omega_L t + \eta(t)\right]$, and emphasis is on the two-photon excitation process in the absence of ionization, spontaneous emission, or any pressure-broadening effects. Correspondingly, $\rho_{11} + \rho_{00} = 1$. The quantity $\frac{1}{2}\alpha E^2(t)$ is the two-photon Rabi frequency and $\beta E^2(t)$ is the ac Stark shift in the position of the resonance. The detuning frequency δ is defined by $\hbar\delta = \hbar(2\omega_L - \omega_1 + \omega_0)$, where $\hbar\omega_1$ and $\hbar\omega_0$ are the energies of the excited and ground states, respectively. We wish to describe excitation for situations where the laser coherence time τ_c is very short compared with the pulse length τ but $\langle \alpha E^2(t)\tau_c \rangle \ll 1$. In the latter situation, the amplitude and phase of the laser change greatly in a few coherence times; but Z_1 hardly changes at all. Since β and α are typically of the same order of magnitude, we expect that dQ/dt is dominated by δ and $2d\eta(t)/dt$ ($|d\eta/dt| \sim \pi/\tau_c \gg |\beta E^2(t)|$) and ac Stark shifts can be neglected. Differentiating Eq. (1) and ensemble averaging over the phase space of the photon field, we can use $\langle Z_1(t')\partial K(t, t')/\partial t \rangle \approx \bar{Z}_1(t)\langle \partial K(t, t')/\partial t \rangle$, where we have used the fact that $\langle \partial K(t, t')/\partial t \rangle$ is only nonzero for $(t - t')$ smaller than a few τ_c and Z_1 does not change appreciably in this time. We find

$$\frac{d\bar{Z}_1}{dt} = -\alpha^2 M(\delta, t)\bar{Z}_1 \tag{2}$$

$$M = \text{Re} \int_{0}^{\infty} e^{i\delta\tau} \langle E^2(t - \tau)E^2(t) \exp\{2i[\eta(t) - \eta(t - \tau)]\}\rangle \, d\tau \tag{3}$$

Equation (2) can be shown to be equivalent to $d\rho_{00}/dt = R(\rho_{11} - \rho_{00})$ and $d\rho_{11}/dt = -R(\rho_{11} - \rho_{00})$, where $R = \alpha^2 M(\delta, t)/2$. In a collision-free situation, but in the presence of photoionization of the excited state, the validity of rate equations hinges on the presence or absence of coherent oscillations in $\bar{Z}_1$, i.e., on the validity of rate equations for the excitation process. Relaxation processes shorten the coherence and thereby enhance the validity of a rate equation description. When the ionization rate is small compared with the laser bandwidth, the rate equations for the diagonal components of the density matrix are

$$\frac{d\rho_{00}}{dt} = -R(\rho_{00} - \rho_{11})$$

$$\frac{d\rho_{11}}{dt} = R(\rho_{00} - \rho_{11}) - \Gamma_I \rho_{11}$$

(4)

with $\Gamma_I = \sigma_I \mathscr{F}$. We assume plane polarized light, and σ_I, the photoionization cross section of the excited state, must be calculated accordingly. The photon flux is represented by $\mathscr{F}$; and with our assumptions of $\langle \alpha E^2(t)\tau_c \rangle \ll 1$, $\Gamma_I \tau_c \ll 1$, Eqs. (4) should give a very accurate description with $M(\delta, t)$ given by Eq. (3). In the approximation of a square pulse envelope for the power density, the ionization probability of an atom described by Eqs. (4) is

$$P = 1 - \frac{\Gamma_I R}{2J} \left[\frac{\exp(-\beta_1 \tau)}{\beta_1} - \frac{\exp(-\beta_2 \tau)}{\beta_2} \right]$$

(5)

where $\beta_1 = R + \Gamma_I/2 - J$, $\beta_2 = R + \Gamma_I/2 + J$, and $J = (\sqrt{R^2 + \Gamma_I^2/4})^{1/2}$. We can think of $P = P(R\tau, \Gamma_I \tau)$. That is, P depends only on the values of the saturation parameters $R\tau$ and $\Gamma_I \tau$ at the location of the atom.

Higher laser field autocorrelation functions such as the one in M are usually not known for a laser. Many models have been developed, but nearly all of them yield Lorentzian line shapes.[23] Such line shapes are never observed for lasers operating far above threshold. We choose our model for a broad band laser to be the limit of a large number of field modes, all having independent phases. Thus, $E(t)\cos[\omega_L + \eta(t)] = (A + A^*)/2$, where

$$A(t) = e^{i\omega_L t} \int_{-\infty}^{\infty} B(\omega, t) \exp\{i[\omega t + \phi(\omega)]\} \, d\omega$$

(6)

and $B(\omega, t)$ has a very slow time dependence limited by the pulse envelope. The phase independence is introduced through the assumption $\langle \exp|i[\phi(\omega) - \phi(\omega')]|\rangle = \delta(\omega - \omega')$, $\langle \exp|i[\phi(\omega_1) + \phi(\omega_2) - \phi(\omega_3) -$

$\phi(\omega_4)]|\rangle = \delta(\omega_1 - \omega_3)\delta(\omega_2 - \omega_4) + \delta(\omega_1 - \omega_4)\delta(\omega_2 - \omega_3)$, etc. The power density $I(t)$ is given by

$$I(t) = C\langle A^*A\rangle/8\pi = (C/8\pi)\int_{-\infty}^{\infty}|B(\omega, t)|^2\, d\omega = \int_{-\infty}^{\infty}I(\omega, t)\, d\omega$$

Thus, $I(\omega, t) = (C/8\pi)|B(\omega, t)|^2$. This model yields $\langle E^{2n}(t)\rangle = n!|\langle E(t)\rangle|^n$ and thereby reproduces the $n!$ enhancement that has been observed when broad bandwidth lasers are used for off-resonant multiphoton ionization. Using the properties of the mode phases, we find

$$M = 2\,\mathrm{Re}\int_0^{\infty}d\tau\int_{-\infty}^{\infty}d\omega\,|B(\omega, t)|^2\int_{-\infty}^{\infty}d\omega'\,|B(\omega', t)|^2\exp\left[i(\omega + \omega' + \delta)\tau\right]$$

$$= \frac{128\pi^3}{C^2}\int_{-\infty}^{\infty}d\omega\,I(\omega, t)I(-\omega - \delta, t)\tag{7}$$

where

$$\mathrm{Re}\int_0^{\infty}\exp\left[i(\omega + \omega' + \delta)\tau\right]d\tau = \pi\delta(\omega + \omega' + \delta)$$

has been used. The form of Eq. (7) suggests a picture in which photons at $\omega_L + \omega$ and $\omega_L - \omega - \delta$ are absorbed to give a total energy absorbed of $\hbar(2\omega_L - \delta)$, which is exact resonance since ω_L is the central laser frequency. The rate involves a sum of such incremental rates over the laser line shape. A convenient form for $I(\omega, t)$, which is not unreasonable for many lasers, is $I(\omega, t) = I(t)(2\pi\sigma^2)^{1/2}\exp(-\omega^2/2\sigma^2)$, where σ is the laser bandwidth. We find for this line shape

$$M(\delta, t) = \frac{128\pi^3}{C^2}(4\pi\sigma^2)^{-1/2}e^{-\delta^2/4\sigma^2}I^2(t)\tag{8}$$

At lower power densities, observed line shapes for ionization would be determined by those of $M(\delta, t)$. From Eq. (8) we see that the linewidth for ionization based on this model is larger than that of the laser by a factor of $\sqrt{2}$. At higher power densities where $P(R\tau, \Gamma_I\tau) \to 1$, the ionization linewidth will be even broader. Two-photon excitation linewidths which are broader than the laser line width are common to nearly all stochastic theories. When excitation is carried out with two parallel laser beams of rather different frequencies and one photon must be absorbed from each to achieve two-photon excitation resonance, $M(\delta, t)$ must be modified to

$$M(\delta, t) = \frac{128\pi^3}{C^2}[2\pi(\sigma_1^2 + \sigma_2^2)]^{-1/2}\left\{\exp\left[-\frac{\delta^2}{2(\sigma_2^2 + \sigma_2^2)}\right]\right\}I_1(t)I_2(t) \quad (9)$$

where $I_1(\omega, t) = I_1(t)\exp(-\omega^2/2\sigma_1^2)$ and $I_2(\omega, t) = I_2(t)\exp(-\omega^2/2\sigma_2^2)$ are the two laser line shapes. In the latter case, α must be modified from Ref. 14:

$$\alpha = S_n \frac{\langle 1|\hat{P}_z|n\rangle\langle n|\hat{P}_z|0\rangle}{2\hbar^2(\omega_n - \omega_0 - \omega_L)} \quad (10a)$$

to

$$\alpha = S_n \frac{\langle 1|\hat{P}_z|n\rangle}{2\hbar^2}\langle n|\hat{P}_z|0\rangle\left(\frac{1}{\omega_n - \omega_0 - \omega_{L1}} + \frac{1}{\omega_n - \omega_0 - \omega_{L2}}\right) \quad (10b)$$

S_n indicates a sum over discrete states and an integral over the ionization continuum. For inert gases nearly all states for which $\langle n|\hat{P}_z|0\rangle$ is sizable are close to satisfying $\omega_n - \omega_0 \gtrsim 2\omega_L$. Also, such states lie in a narrow range of energy (~ 4 eV). A very reasonable approximation is achieved by replacing $\omega_n - \omega_0$ by the value $\bar{\omega}$, where $\hbar\bar{\omega}$ is the average of the ionization potential of $|0\rangle$ and E_{10} the lowest excitation energy for which both $\langle 1|\hat{P}_z|n\rangle$ and $\langle n|\hat{P}_z|0\rangle$ are nonzero. Thus, by closure, Eq. (10a) gives $\alpha \approx \langle 1|\hat{P}_z^2|0\rangle/[2\hbar^2(\bar{\omega} - \omega_L)]$, and Eq. (10b) yields a result which is a factor of 2 larger if $\bar{\omega} - \omega_{L1}$ and $\bar{\omega}_{L2}$ are not too different. The closure approximation has not been tested for inert gas two-photon excitation, but it stands an excellent chance of giving α within a factor of 2 or better, depending on the degree of cancellation involved in the infinite sum over all discrete and continuum states.

3.3. RIS Schemes for Ar, Kr, and Xe

In this section we use a single configuration approximation with jl coupling in order to estimate $\langle 1|\hat{P}_z^2|0\rangle$ matrix elements to go with the laser schemes that we will suggest for the ionization of Ar, Kr, and Xe. In this approximation one assumes that configuration mixing is unimportant and that an approximate set of angular constants of the motion are j, the total angular momentum quantum number of the ion core; l, the total orbital angular momentum quantum number of the single excited electron; k, the angular momentum quantum number obtained by diagonalizing the operator $(\hat{j} + \hat{l})^2$; J, the total angular momentum quantum number; and

M_J, the z projection quantum number of the total angular momentum. The approximation should be good when there is strong spin orbit coupling in the core, but the ionization energy of the excited electron is very small compared with ionization energy of the ground state, and there are no nearby states of the same configuration. All of our schemes for the inert gases involve excitation from the ground $np^6(J = 0)$ level to a member of the $np^5n'p$ multiplet with $j = 3/2$. With our approximation the relevant matrix elements break up into an angular part, depending on the angular momentum quantum numbers (i.e., the particular member of the multiplet) and a radial integral of r^2 which is independent of the level in the multiplet. The allowed transitions for plane polarized light are $\Delta M_J = 0$ and $\Delta J = 0$ or 2. With circularly polarized light, $\Delta M_J = \pm 2$ and $\Delta J = 2$. We consider only plane polarized light. The largest angular factor is for $|1\rangle = |n', np^5n'p[1/2]_0\rangle$. That is, $j = 3/2$, $k = 1/2$, $J = 0$, and $M_J = 0$. In the latter case,

$$|\langle 1|\hat{P}_z^2|0\rangle| = \tfrac{2}{3}R_2(n, l = 1; n', l' = 1) \tag{11}$$

where R_2 is the radial matrix element of r^2. In atomic units, R_2 will be of the form of a dimensionless constant times $e^2a_0^2$. If L_2 is the constant (i.e., $R_2 = e^2a_0^2L_2$) and we take $\bar{\omega} - \omega_L \simeq 9 \times 10^{15}/\text{sec}$ for all cases, we obtain at $\delta = 0$

$$R \simeq \frac{7L_2^2}{\Delta_{1/2}}I^2 \ (\text{W/cm}^2) \tag{12}$$

where $\Delta_{1/2} = 2(2\ln 2)^{1/2}\sigma$ is the full width at half-maximum of $I(\omega, t)$. When lasers of two different frequencies are used, Eq. (12) is replaced by $R = 14L_2^2I_1I_2/(\Delta_1^2 + \Delta_2^2)^{1/2}$, where Δ_1 and Δ_2 are full widths at half-maximum for the two-laser line. We now discuss the ionization schemes.

Argon. We will discuss two methods here. The first method would involve the use of an ArF excimer laser at 1933 Å and a parallel beam obtained by anti-Stokes shifting 1933 Å radiation in HD. Thus,

$$\text{Ar}(3p^6) + \hbar\omega_1(\lambda = 1933 \text{ Å}) + \hbar\omega_2(\lambda = 1807 \text{ Å}) \rightarrow \text{Ar}(3p^54p[1/2]_0)$$

One might expect[24] that ~2 mJ could be obtained at 1807 Å and 100 mJ at 1933 Å. It has also been demonstrated that ArF light can be achieved with very narrow bandwidth.[24] The second excitation method uses two beams, one from a He–F_2 laser at 1576.3 Å and a second is obtained at 2292.8 Å by frequency summing Nd–Yag light from a frequency doubled dye laser. Thus,

$$\text{Ar}(3p^6) + \hbar\omega_1(\lambda = 1576.3 \text{ Å}) + \hbar\omega_2(\lambda = 2292.8 \text{ Å}) \rightarrow \text{Ar}(3p^54p[1/2]_0)$$

The 1576.3 Å radiation is obtainable from commercial $He-F_2$ lasers with a linewidth $\Delta\lambda_{L1} = 0.01$ Å and an energy per pulse $\varepsilon = 10$ mJ. About 2 mJ with $\Delta\lambda_{L2} = 0.02$ Å can be obtained at 2292.8 Å. For Ar, as well as for Kr and Xe, we have used Hartree–Fock, Hartree–Fock–Slater, l-averaged Hartree–Fock,[25] and parametric potential in calculating the atomic wave functions employed in the calculation of R_2. For Ar the average of these calculations is $R = 1.7e^2a_0^2$. The spread of values suggests a factor of 2 in accuracy. We let $k'I_1I_2$ where $k' = 14L_2^2/(\Delta_1^2 + \Delta_2^2)^{1/2}$. For the last scheme, $k' \approx 1.0 \times 10^{-9} \sec^{-1} W^{-2} cm^4$. At 1576 Å one has $\sigma_I \simeq 1 \times 10^{-18} cm^2$. Details of the atomic structure calculations and some extensions will be published elsewhere.[26]

   Krypton. One scheme for Kr is that of Bokor *et al.*[27] In this scheme we propose to use a frequency-narrowed ArF laser to injection lock an ArF amplifier. However, this possibility has not been achieved. The excitation is

$$Kr(4p^6) + 2\hbar\omega_L(\lambda = 1927.48 \text{ Å}) \rightarrow Kr(4p^56p[1/2]_0)$$

Again, another photon can ionize with $\sigma_I \simeq 5 \times 10^{-19} cm^2$.[28] A cross section of this magnitude has been measured by Bokor *et al.*[27] Here the estimated value of R_2 is $R_2 = 0.9e^2a_0^2$. We find $R = kI^2$ (W/cm^2) with $\Delta\lambda_L = 0.8$ Å, $k = 7L_2^2/\Delta_{1/2} \simeq 1.2 \times 10^{-12} \sec^{-1} W^{-2} cm^4$. Another method combines a scheme by Hawkins *et al.*[29] to produce a laser beam at 248.3 nm with a line of the $He-F_2$ laser:

$$Kr[4p^6] + \hbar\omega_1(\lambda = 1575.3 \text{ Å}) + \hbar\omega_2(\lambda = 2482.6 \text{ Å}) \rightarrow Kr(4p^56p[1/2]_0)$$

At 2482.6 Å a line width of 10^{-4} Å has been achieved with $\varepsilon = 60$ mJ. The F_2 line has $\Delta\lambda_L = 0.01$ Å and $\varepsilon = 2$ mJ. Alternative (and less expensive) schemes are

$$Kr(4p^6) + \hbar\omega_1(\lambda = 1576.3 \text{ Å}) + \hbar\omega_2(\lambda = 3262 \text{ Å}) \rightarrow Kr(4p^55p[1/2]_0)$$

$$Kr(4p^6) + \hbar\omega_1(\lambda = 1576.3 \text{ Å}) + \hbar\omega_2(\lambda = 2824 \text{ Å}) \rightarrow Kr(4p^55p'[1/2]_0)$$

where at 1576.3 Å we have $\varepsilon = 10$ mJ, $\Delta\lambda_L = 0.01$ Å. At $\lambda = 3262$ Å and 2824 Å, one can have $\Delta\lambda_L \simeq 0.01$ Å and $\varepsilon = 5$ and 10 mJ, respectively, by frequency doubling dye lasers pumped by the second harmonic of a Nd–Yag. The value R_2 for $4p^55p[1/2]_0$ is $R_2 = 3e^2a_0^2$. Thus, it may be possible to ionize Kr very efficiently since $k' = 5 \times 10^{-9} \sec^{-1} W^{-2} cm^4$. At 3262 Å we have $\sigma_I = 10^{-18} cm^2$.

   Xenon. Excitation of $5p^56p[1/2]_0$ in Xe has been studied experimentally. In this case the scheme is

$$Xe(5p^6) + 2\hbar\omega_L(\lambda = 2496 \text{ Å}) \rightarrow Xe(5p^56p[1/2]_0)$$

Theory[28] gives $\sigma_I \approx 3 \times 10^{-18}$ cm^2. A cross section of this magnitude was confirmed experimentally. Chen et al.[30] also observed $k \approx 2 \times 10^{-9}$ sec^{-1} W^{-2} cm^4 with $0.05 < \Delta\lambda_L \leqslant 0.0$ Å. The width of the observed ionization resonance under strongly saturated conditions was 0.15 Å. The 2496 Å linewidth was inferred from the fact that 0.16 Å red light was frequency doubled, then frequency summed with 1.06 micron radiation to obtain the ~1 mJ of output. Calculations on Xe gave $R_2 = 7e^2a_0^2$. If we assume $\Delta\lambda_L = 0.08$ Å for Chen's case, we find $k = 1.3 \times 10^{-9}$ sec^{-1} W^{-3} cm^4. In the case of Xe, the variations in the four calculated values were large and there were signs of configuration mixing in the multiplet structure. Correspondingly, the apparent good agreement with experiment is probably accidental. There are two ways of increasing the rate of ionization of Xe. The simplest is to use

$$\text{Xe}(5p^6) + \hbar\omega_1(\lambda = 1576.3 \text{ Å}) + \hbar\omega_2(\lambda = 5995 \text{ Å}) \rightarrow \text{Xe}(5p^56p[1/2]_0)$$

The 1576.3 Å radiation has $\varepsilon = 10$ mJ and $\Delta\lambda_{L1} = 0.01$ Å. It is from a commercial excimer laser running with a He–F$_2$ mixture. Many commercial dye lasers can give $\lambda = 5995$ Å radiation with $\Delta\lambda_{L2} < 0.1$ Å and $\varepsilon > 30$ mJ. Dyes pumped with XeF radiation or the second harmonic of Nd–Yag are particularly effective. An alternative scheme at low repetition rates is that of Ref. 30 but with the light source of Hawkins et al.[29] The latter workers have obtained 60 mJ of transform-limited bandwidth light at 2496 Å. With this light, Doppler-free excitation can be obtained with counter-propagating beams. In this case it is appropriate to estimate a two-photon Rabi frequency. We get $\Omega_2 = 6.1I$ (W/cm^2). With 60 mJ in a 10-nsec pulse, a power density $I = 6 \times 10^6$ W/A is obtained where A is the laser beam size on the target. Thus, $\Omega_2 = 4 \times 10^7$ W/A, and a beam diameter of <0.3 cm would cause considerable power broadening. A study of power broadening with this light source would yield good estimates of Ω_2 from experiment.

3.4. Effective Volume for Ionization

We consider in this section the yield of ions in a focused Gaussian beam. Most of our schemes for rare gases will not use lasers with Gaussian beams, but many of the features do not depend on the details of the radial power density. Let I_0 be the power density at the focus and on the optical axis. If $\rho_0(z) = [(F\theta_{1/2})^2 + (z\rho_I/F)^2]^{1/2}$, where F is the lens focal length, $\theta_{1/2}$ is the beam divergence, z is the distance from focus and ρ_I is the beam radius at the lens, then a reasonable functional form for the power density when $\theta_{1/2}$ is much larger than the diffraction limit is

$$I(\rho, z) = I_0[F\theta_{1/2}/\rho_0(z)]^2 \exp[-\rho^2/\rho_0^2(z)] \qquad (13)$$

I_0 is determined through $\varepsilon = I_0 \tau \pi (F\theta_{1/2})^2$, where τ is the pulse length and ε is the energy per pulse. With our assumption of cylindrical symmetry, the number of ions produced per pulse is

$$n_I = 2\pi n \int_{-\infty}^{\infty} dz \int_{0}^{\infty} \rho \, d\rho \, P \qquad (14)$$

where the ionization probability P is that for an atom at z and ρ. Now, P depends on the local power density through R and Γ_I. The situation is simplified if we note that $R = kI^2$ and $\Gamma_I = \sigma_I F = \sigma_I I / \hbar \omega_L$. Let $R_0 = kI_0^2$, $\Gamma_0 = \sigma_I I_0 / \hbar \omega_L$, and $u = \exp[-\rho^2/\rho_0^2(z)]$. Also, defining w by letting $w = F^2\theta_{1/2}/\rho_I z$, we get

$$n_I = 2\pi n \frac{F}{\rho_I} (F\theta_{1/2})^3 \int_{0}^{\infty} d\omega \, (1 + w^2) \int_{0}^{1} \frac{du}{u} P\left(\frac{Su^2}{(1 + w^2)^2}, \frac{Tu}{1 + w^2}\right) \qquad (15)$$

where, as before, $P = P(R\tau, \Gamma_I\tau)$ and $S = R_0\tau$ with $T = \Gamma_0\tau$. The concept of effective volume of ionization is defined through

$$n_I = n \, \Delta V \qquad (16)$$

In the form of Eq. (15), the tabulation of ΔV involves only the determination of a function of S and T, which are the saturation parameters at $\rho = 0$ and $z = 0$.

When $S > 5$ and $T > 5$, as they will usually be with the schemes described in the last section, nearly all of the contribution comes from $w^2 \gg 1$ so that in Eq. (15) the $1 + w^2$ can be replaced by w^2. Thus, letting $v = Su^2/w^4$, we get

$$\Delta V = 2\pi \frac{F}{\rho_I} (F\theta_{1/2})^3 G(r) S^{3/4} \qquad (17)$$

where $r = T/\sqrt{S} = \sigma_I \tau^{1/2}/(\hbar \omega_L \sqrt{k})$. The parameter r depends only on atomic parameters, pulse length, wavelength, and bandwidth. Typically,

$$\sigma_I \approx 10^{-18} \text{ cm}^2, \tau = 10^{-8} \text{ sec}, \hbar \omega_L \approx 10^{-18} \text{ J}, k < 10^{-9} \text{ sec}^{-1} \text{ W}^{-2} \text{ cm}^4$$

Therefore, r is generally quite large except for very narrow bandwidth lasers and strong two-photon transitions. For $r > 5$, we obtain

$$G(r) = \tfrac{8}{9}\Gamma(5/4) - \tfrac{4}{3}\Gamma(3/2)r^{-1/2} + \tfrac{2}{3}\Gamma(7/4)r^{-1} + O(r^{-3/2})$$
$$\approx 0.8057 - 1.1816r^{-1/2} + 0.6127r^{-1} + O(r^{-3/2}) \qquad (18)$$

where $\Gamma(x)$ is the gamma function. When I_0 is written in terms of energy per pulse, $S = k\varepsilon^2/\pi^2\tau(F\theta_{1/2})^4$ and

$$\Delta V = \frac{2}{\sqrt{\pi}} \frac{F}{\rho_I} (k/\tau)^{3/4} \varepsilon^{3/2} G(r) \tag{19}$$

Table 1 gives a brief tabulation of ΔV based on various laser schemes discussed earlier. The numbers are based on Eq. (15), but Eq. (19) is very close for all entries. Equation (18) can be used for $G(r)$ in all but two cases.

The region of ionization for $S > 10$ and $T > 10$ can be visualized by imagining a surface on which $P(R\tau, \Gamma_2\tau) = 1/2$. Inside this surface, P is close to unity and outside it rapidly decreases to zero. The surface in question has a very small radius near $z = 0$; but as we move in either direction away from the focus, the radius grows. At some $|z_0|$ the radius will be a maximum, and for larger $|z|$ it will again decrease. In Figure 4 this surface is shown schematically. The beam radius is exaggerated for emphasis. For most lasers the volume will be very long compared with its radius. In some uses of inert gas, only a segment of length Δz of the long slender ionization volume can be utilized. The actual z at which the radial integration is largest can be shown for $S > 10$, $T > 10$, $r > 5$ to be given by $z_0 = F^2\theta_{1/2}S^{1/4}/\sqrt{2}\rho_I$. In this limit, if $z_0 - \Delta z/2 < z < z_0 + \Delta z/2$:

$$\Delta V = \frac{1}{e^2} \Delta z (k/\tau)^{1/2} \varepsilon \tag{20}$$

Since most of the schemes satisfy the conditions of validity for Eq. (20), it should be very useful when one desires to transmit an optimum number of ions through a quadrupole mass spectrometer and transmission is poor unless the ion is produced in a sphere of radius of 0.5 cm.

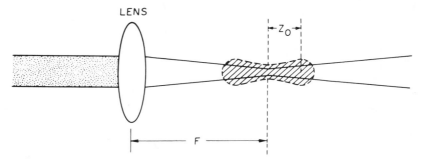

Figure 4. Schematic for effective volume of ionization of rare gases by resonance ionization spectroscopy.

Table 1. Estimates of Effective Volumes for Ionization of Heavy Rare Gases by Various Laser Schemes[a]

RIS Scheme	λ_L (Å)	$\Delta\lambda_L$ (Å)	ε (mJ)	ΔV (cm³)	S	T	r
$Kr + 2\hbar\omega_L \rightarrow Kr(4p^5 6p[1/2]_0)$	1927.5	4	100	3.0×10^{-4}	2.7	158	96
$Kr + \hbar\omega_1 + \hbar\omega_L \rightarrow Kr(4p^5 5p[1/2]_0)$	1576.3	0.01	10	2×10^{-13}	23	26.2	5.5
	3262	0.01	5				
$Ar + \hbar\omega_1 + \hbar\omega_L \rightarrow Ar(3p^5 4p[1/2]_0)$	1807	4	2	1×10^{-4}	1.0	315	315
	1934	4	100				
$Ar + \hbar\omega_1 + \hbar\omega_L \rightarrow Ar(3p^5 4p[1/2]_0)$	1576.3	0.01	10	3×10^{-4}	33.7	13.15	2.3
	2292.8	0.01	2				
$Xe + \hbar\omega_1 + \hbar\omega_L \rightarrow Xe(5p^5 6p[1/2]_0)$	1576.3	0.01	10	10^{-2}			
	5995	0.03	10				
$Xe + 2\hbar\omega_L \rightarrow Xe(5p^5 6p[1/2]_0)$	2496	0.08[b]	0.7[b]	1.2×10^{-4}	24.5	33.4	6.7

[a] All calculations were carried out assuming $F = 10$ cm, $\theta_{1/2} = 0.001$ rad, $\rho_l = 0.3$ cm, $R_2(Kr) = 0.9e^2 a_0^2$ for $4p^5 6p[1/2]_0$, $R_2(Kr) = 3e^2 a_0^2$ for $4p^5 5p[1/2]_0$, $R_2(Ar) = 1.7e^2 a_0^2$, and $R_2(Xe) = 7e^2 a_0^2$. Effective volumes were determined from Eq. (15), but in all cases Eq. (19) was close to the result of Eq. (15). Except for one Xe case, $\tau = 10^{-8}$ sec.

[b] All pulse lengths are 10^{-8} sec except here, where $\tau = 5 \times 10^{-9}$ sec. Also, here $\theta_{1/2} = 5 \times 10^{-4}$ rad instead of 10^{-3} rad.

4. RIS Experiments with Xe Atoms

Saturated resonance ionization spectroscopy and one-atom detection can be achieved, in principle, on every element of the Periodic Table except He and Ne. However, most of the actual work on single-atom detection has been done on alkali metal atoms by using proportional counters[12,14,21] as single-electron detectors. Here we are describing new work on the detection of Xe at the one-atom level and with isotopic selectivity. This can be achieved by using a quadrupole mass spectrometer to mass analyze ions produced by the RIS process[4,30] in a near vacuum and special counting technique to be described. A saturated two-photon transition process is required for rare gas atoms because of the high energy levels (>8 eV) of these atoms. Our ultimate goal is to use RIS to count each individual rare gas atom of a selected isotope even when the ratio of the selected isotope to a neighboring mass number is one in 10^{15}.

Following the availability of high-power uv lasers, a two-photon resonance excitation process has attracted much attention in the field of nonlinear optics,[31-33] spectroscopy,[34] and chemical physics.[35,36] Isotopically selective excitation of Xe atoms has been studied by Bushaw and Whitaker.[37] Bjorkholm and Liao[38] reported on the line shape and strength of two-photon absorption with a resonant and nearly resonant intermediate state. Wang and Davis[39] demonstrated the saturation of resonant two-photon transition of Tl vapor. Kligler et al.[36] studied collisional and radiative properties of excited H_2, and they measured ionization cross sections of $H_2(E, F^1\Sigma_u^+)$ prepared by two-photon excitation. Bokor et al.[27] and Bischsel et al.[40] estimated the two-photon transition rates for rare gases from a theory[23] of resonant multiphon ionization.

In Section 3 we described various methods for the RIS of Ar, Kr, and Xe. In each case, the RIS scheme utilizes high-power laser beams to saturate a two-photon transition and a third photon is absorbed to complete a RIS process. One of these schemes for Xe has been reported[30] and is shown in Figure 5. A Quanta-Ray Nd–Yag laser beam was frequency doubled and used to pump a red dye (Exciton DCM) to obtain tunable light between 645 and 675 nm with an output of about 40 mJ/pulse and a pulse duration of about 5 nsec. The dye laser output was doubled in a KDP crystal, and this light was mixed with the 1.06 μ radiation from the Nd–Yag to yield tunable light from 247 to 256 nm.

In the region of some two-photon resonance in Xe (e.g., 249.6 and 252.6 nm), the maximum energy per pulse was about 1 mJ and the pulse duration was about 4 nsec. In the experiment, the 249.6-nm radiation was focused through a 13.5-cm focal length lens into the center of the RIS cell which consists of a repeller plate and a channeltron detector. The channeltron detector was operated in a low-gain analog mode where its output is

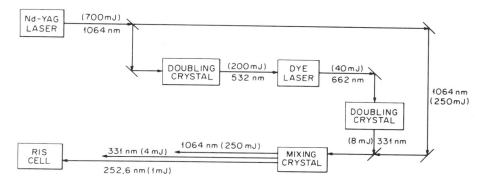

LASER SCHEME FOR RIS OF Xe ATOMS: $Xe(\omega_1\omega_1,\omega_1e^-)Xe^+$.

Figure 5. A laser method for producing 252.6-nm photons used for the RIS process $Xe(\omega_1\omega_1, \omega_1e^-)Xe^+$.

proportional to the number of ions detected. The pressure inside the RIS cell was approximately 10^{-6} Torr with the Xe partial pressure ranging from 10^{-8} to 10^{-6} Torr. Some ionization line shapes for the RIS process described above are shown in Figure 6. The observed peaks correspond to two-photon resonance with $Xe[1/2]_0(5p^56p)$ and $Xe[3/2]_2(5p^56p)$ and a final ionization of these respective states at the indicated wavelength. The FWHM of these

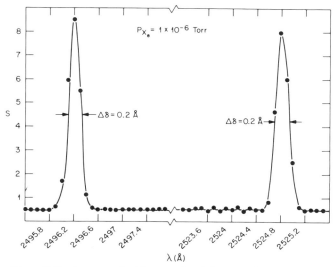

Figure 6. Tuning curve for Xe atoms involving two-photon excitation followed with one-photon ionization at the two indicated wavelengths.

resonances (about 0.2 Å) is strongly influenced by the laser width which is estimated to be about 0.1 Å. The ratio of background ions, which were produced from multiphoton ionization of residual gases, to the saturated resonance ionization signal is about 10^{-2}.

A quadrupole mass filter was introduced to further reduce the background ions. When the mass filter was tuned to pass amu = 131 and the resolution $(M/\Delta M)$ was approximately 300, the background signal was nonobservable, even when the gain of the detector was increased to permit the detection of single ions. Thus, for a heavy ion such as Xe^{+}, a relatively simple mass filter can be used to eliminate background ionization due to multiphoton ionization of the residual gases at power densities where the ionization probability for a rare gas atom at the focus is close to unity. A Xe mass spectrum produced from RIS is given in Figure 7. Since the laser bandwidth was broader than the isotope splitting of energy levels of excited Xe atoms, the spectrum in Figure 7 agrees with the known natural abundance of the various Xe isotopes.

The subject of multiphoton ionization has recently been reviewed by Lambropoulos.[41] A great deal of theory has been done on H and rare gases.[22] Many spectroscopic investigations on alkali atoms,[42–45] rare gas atoms,[46–48] and organic molecules[49–53] have been reported. However, very little experimental work on multiphoton transition rates of small molecules or atoms has been reported. In a review article,[14] some of us gave a rough estimate (based on perturbation theory) for the multiphoton ionization cross section. About two orders of magnitude accuracy applies

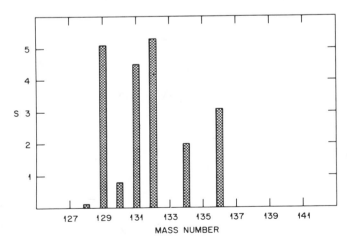

Figure 7. A mass spectrum for various isotopes of Xe produced by the RIS process.

to the following equation:

$$\sigma_n \simeq 3 \times 10^{-18} \left| \frac{3n \times 10^{-7}}{E_1} \right|^{2(n-1)} I^{n-1}$$

In the above, E_1 is the energy of the first excited state in units of eV, I is the laser power density in units of W/cm^2, and n is the number of photons required to ionize the atoms. With multimode lasers, the ionization probability ($\mathscr{F}\tau\sigma_n$, where $\mathscr{F}$ is the photon flux in units of $cm^{-2} sec^{-1}$ and τ is the laser pulse width) should be multiplied by $n!$. The laser power density I was estimated[30] to be $1 \times 10^9 W/cm^2$. If a molecule has an ionization potential lower than 9.8 eV, the ionization probability is 10^{-6} or higher. In most vacuum systems, appreciable ionization of residual gases would occur. An example of a mass spectrum of residual gas ionized by a laser at $\lambda = 250$ nm is shown in Figure 8. The spectrum obtained in Figure 8 is much simpler than the one from electron impact processes. However, the detection of a single rare gas atom would be extremely difficult without a mass filter to reject residual gas ionization.

We now consider the RIS of Xe in more detail. When the power density is less than $10^{10} W/cm^2$ and the laser linewidth is greater than 0.03 Å, the RIS process under consideration for Xe has been described by Eqs. (4) and (5) in the theory of Section 3.

Figure 9 shows the ionization signal as a function of the square of the energy per pulse. We have analyzed the power dependence of the signal in order to infer experimental values for Γ_1 and R as a function of power

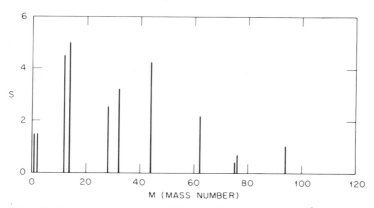

Figure 8. A mass spectrum obtained from residual gas at 1×10^{-6} Torr. The ions were produced by a laser beam through a multiphoton ionization process. The wavelength of the laser is 250 nm and the laser power density is approximately $10^9 W/cm^2$ at the focal point.

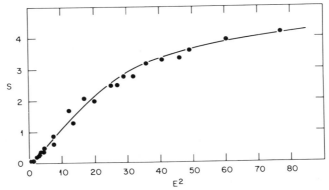

Figure 9. Relative ionization signal S versus the square of the laser energy per pulse at 249.6 nm (10^{-3} J for $E^2 = 100$ in the figure).

density. In the analysis we use a square pulse for the time dependence and assume that the power density at a distance x from the focal point and a distance ρ from beam center is

$$I(\rho, x) = \frac{E}{\tau} A(x) \exp\left\{ - \frac{[\rho - \rho_0(x)]^2}{d^2(x)} \right\}$$

where E is laser energy per pulse, $\rho_0(x) = \rho_{00}x/F$, ρ_{00} is the distance from beam center to the peak intensity before focusing, F is the lens focal length, $d(x) = [(F\theta_{1/2})^2 + (d_0x/F)^2]^{1/2}$, d_0 is the half-width for e^{-1} drop in the intensity of the doughnut-shaped beam before focusing, $\theta_{1/2}$ is the half-angle beam divergence of the initial beam before focusing, and $A(x)$ is determined by

$$A(x) = \left[2\hbar \int_0^\infty \rho \, d\rho \, \exp\left(- \frac{[\rho - \rho_0(x)]^2}{d^2(x)} \right) \right]^{-1}$$

The values of $\theta_{1/2}$, d_0, and ρ_{00} were determined experimentally to make the fit to observation as close as possible.

Thus, if P_I is regarded as a function of power density, the number of ions produced per pulse is

$$N_I = n \int_{-L/2}^{L/2} dx \int_0^\infty 2n\rho P_I(I(\rho, x)) \, d\rho$$

where n is the Xe density in the number of atoms per cubic centimeter and $-L/2 \leqslant x \leqslant L/2$ is the region over which the Xe^+ is seen by the detector ($\sim$0.8 cm). Figure 9 is reproduced very well by the choice $\Gamma_I = 3I$ and $R \simeq 2 \times 10^{-9} I^2$. The two-photon transition rate obtained from this work is somewhat larger than the theoretical estimate by Bichsel et al.[40] The disagreement may be due partly from the strong coupling of Xe(6p) and Xe(7p) states. The difference between two-photon transition rates from ground state Xe to Xe[1/2]$_0$($5p^56p$) and Xe[3/2]$_0$($5p^56p$) should be less than a factor of 2 (cf. Figure 6).

With laser energy at 1 mJ/pulse and the laser beam size focused down to 0.02 cm, we estimate from theory that the intense laser beam should ionize with an effective volume of 3×10^{-4} cm^3. If the Xe ions are accelerated and implanted into a detector and if the implanted atom can stay in the bulk of the detector for a long time (approximately a few days), the number of atoms present in the total cell decreases from nV to $nV - N_I\mu$ in a single laser shot. Here N_I is the number of atoms ionized by the laser and μ is the efficiency of ion implantation. After N_l laser shots, the initial rare gas concentration changes from its initial value n_0 to $n_0 \exp[-N_l\mu \Delta V/V]$, where ΔV is the effective ionization volume. Thus, a number of shots equal to $V/\mu \Delta V$ are required to pump the number of Xe atoms down by one e-fold. Figure 10 shows the decrease in the ionization signal as a function of laser pumping time or the number of laser pulses. We note there is an exponential decrease in signal with the number of laser shots corresponding to $\mu \Delta V \simeq 2 \times 10^{-4}$ cm^3. Comparison with theory

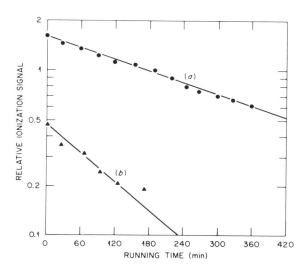

Figure 10. Relative ionization signal versus laser running time (at 10 Hz rate), showing depletion of Xe atoms. Curve a: 48 cm^3 volume. Curve b: 14 cm^3 volume.

suggests that the implantation probability of Xe^+ on the channeltron surface is close to unity at a Xe^+ energy of 3 kV. The results agree with the experimental data of Xe^+ implanting in Al reported by Brown and Davis.[54] The volume ΔV can be determined from the data of both Figures 9 and 10. Since the two methods give about the same ΔV, we estimate that the uncertainty in R is less than a factor of 2.

5. On the Realization of Maxwell's Sorting Demon

In 1871, Maxwell visualized a demon that could see atoms and had sufficient intelligence to perceive the various kinds of atoms or different velocities of atoms. On this basis, the demon could open and close a "door" to accomplish the "sorting of atoms" and store them in a separate chamber. As originally envisioned, such a sorting demon violates the second law of thermodynamics. However, Brillouin[55] pointed out that the demon would have to be equipped with a "flashlight" to see the atoms which more than compensates for the entropy decrease due to the sorting process. Actually, the problem of the demon was resolved much earlier by Leo Szilard[56] by incorporating the dissipation of energy required in the metabolism of an intellectual being. He also showed the connection between a decrease in entropy and a gain of information as well as the central role of memory in such concepts. If a laser beam is introduced to serve as a "flashlight", the demon can sort atoms according to type. Furthermore, if the demon is equipped with a mass spectrometer he can even sort atoms into their isotopes; but, of course, we do not claim that he even approaches the violation of the second law.

Samples containing only a few hundred atoms of Ar, Kr, or Xe can be subjected to our version of Maxwell's sorting demon, whose internal machinery is shown in Figure 11. Pulsed laser beams are tuned to produce the RIS process in rare gas atoms. After ionization and mass analysis, the rare gas ions are accelerated to about 10 kV and implanted into a detector surface such as an electron multiplier. Therefore, a single atom may be simultaneously counted and stored (i.e., remembered). With implantation depths on the order of 100 Å, atoms can be kept in the solid for hours, as shown in Table 2. Thus, the demon has hours in which to count and store on the order of a few hundred atoms.

Since the saturated ionization volume by each laser pulse is usually small ($\leqslant 10^{-3}$ cm^3), the time required to ionize more than 99% of selected atoms in the cell is always longer than a few hours when the volume of the cell is larger than 1 liter. It is necessary to implant the rare gas ions such that the time required for the release of rare gas atoms is much longer than the time required to achieve complete ionization of the atoms. Van

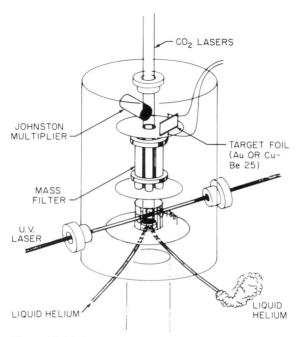

CO$_2$ LASERS

JOHNSTON
MULTIPLIER

TARGET FOIL
(Au OR Cu–
Be 25)

MASS
FILTER

U.V.
LASER

LIQUID HELIUM

LIQUID
HELIUM

Figure 11. Modern experimental arrangement for the realization of Maxwell's sorting demon for counting a few atoms of a selected isotope amid a background of many more atoms of a neighboring isotope.

Liempt[57] derived the probability that a rare gas atom at a depth of x_0 from a surface will escape from the metal in a time t, i.e.,

$$P(t) = 1 - \text{erf}\,[x_0/(4Dt)^{1/2}]$$

where the diffusion constant D is an exponential function of temperature

$$D = D_0\, e^{-E_a/kT}$$

and where E_a is the activation energy for atoms hopping from site to site.

When the probability of gas release, $P(t)$, is higher than 0.23, the time t must be longer than $[(x_0^2/4D_0)\exp(E_a/kT)]$. For rare gas atoms implanted in most metals, E_a is estimated[58,59] to range from 0.3 to 4 eV. If the ions are implanted with an average depth of 100 Å (corresponding to the range of 10 keV Ar^+ ions in Al), the time required for rare gas atoms to be released from metals at room temperature ranges from several hours to several years. An extensive experimental study of the release of rare gases

Table 2. Fraction Retained of ⁸⁵Kr Ions Originally Implanted within the Targets after Successive Heatings at 350–825° C[a]

Heating temperature and duration[b]	5 keV Al(B)	5 keV Al(A)	40 keV Al(B)	40 keV Al(A)	40 keV Mg	10 keV Mo	40 keV Mo	10 keV Pt	40 keV Pt	10 keV Au	40 keV Au
350° C, 1/2 h	0.99	0.96	0.92	0.99	0.96	0.72	0.90	0.64	0.95	0.78	0.93
350° C, 2 h	0.95	0.96	0.92	0.99	0.72	0.72	0.90	0.63	0.95	0.70	0.86
350° C, 24 h	0.95	0.96	0.92	0.97	0.71	0.72	0.90	0.62	0.94	0.60	0.80
500° C, 1/2 h	0.95	0.96	0.92	0.97	0.71	0.72	0.90	0.54	0.91	0.39	0.62
500° C, 2 h	0.95	0.96	0.90	0.97	0.67	0.70	0.90	0.48	0.88	0.28	0.55
500° C, 24 h	0.95	0.96	0.90	0.97	—	0.68	0.86	0.36	0.72	0.16	0.35
825° C, 1/2 h	—	—	—	—	—	0.41	0.71	0.16	0.36	0.03	0.11
825° C, 15 h	—	—	—	—	—	0.36	0.35	0.08	0.23	0.01	0.04

[a] From Ref. 60.
[b] All heatings were performed in a vacuum.

from various metals was done by Lal *et al.*[60] Table 2 shows the fraction of Kr atoms retained in various metals after successive heating. The results agree well with the prediction by the above equations. At temperatures approaching 850° C, nearly all of the Kr atoms could be recovered from gold in a few minutes.

A more challenging problem for the sorting demon is to count a few inert atoms of a given isotope amid a background of many more atoms of another isotope species (such as 1 in 10^{15}). The basic idea now under development is the use of a quadrupole mass spectrometer with the provision that the ionization source utilize the RIS process. However, a small quadrupole mass spectrometer has limited resolution, so that the rejection ratio between adjacent mass peaks may be no more than 1000, for example. Thus, the selected isotope atoms, as well as some of the more abundant isotopes of the same element, will be implanted into a metal such as gold. The remaining rare gas atoms with different mass number will be quickly pumped out of the chamber. Heating of the gold coating up to 800° C will release about 99% of the rare gas atoms[60] (cf. Table 2) in about 30 min. At the end of this cycle, the selected isotope will be enriched by 1000 over the adjacent mass. This cycle can be repeated n times to achieve enrichments of 1000^n.

A commercial (Extranuclear Corporation) quadrupole mass spectrometer with an electron-impact ionizer has a typical sensitivity of 5×10^{-3} A/Torr for the selected mass along with an abundance ratio of 1000. This means that all of the noble gas atoms in a 1-liter volume at a pressure of 10^{-4} Torr can be ionized and mass analyzed in about 1 h. The same processing with the laser would be much longer. Therefore, only after the undesired inert gas isotope atoms are eliminated by the electron enrichment process will the selected isotope be ionized and detected by the RIS process. Ions with selected mass and atomic number will be counted by an electron multiplier operated in a digital mode and time gated with respect to the laser pulse to reduce detector noise.

Finally, in Figure 11 we show an "atom buncher" which functions to reduce the time required for the final laser counting. With the atom buncher, one makes the probability that the inert atom will be in the ionization laser volume much larger than the random probability. The buncher consists of a surface held at liquid He temperature and a CO_2 laser to heat this surface just prior to the pulsing of the detector laser. The length of time that rare gas atoms will remain on metals can be estimated from the following equation:

$$\tau = \tau_0 \exp (E_a/kT_s)$$

where E_a is heat of adsorption, T_s is the temperature of the metal surface, and k is Boltzmann's constant. For heavy rare gas atoms on metals, τ_0 ranges from 10^{-13} to 10^{-20} sec,[61] and E_a from 0.1 to 0.3 eV. If T_s is around 4° K, all of the heavy rare gas atoms adsorbed on the surface will stay there for years. On the other hand, the time that rare gas atoms will remain on metals at room temperature will be shorter than 10^{-8} sec. Thus, all of the free heavy rare gas atoms in the chamber will be adsorbed and frozen on the He-cooled tip within the few seconds required for an atom moving at random in the large volume to make one collision with the cold surface.

A pulsed CO_2 laser can be used to momentarily heat the cold tip where the noble gas atoms are condensed. At the CO_2 laser wavelength, light energy is absorbed very near the surface of a metal (about 500 Å, see Ref. 62). For this case, it has been shown[63] that the temperature of the surface layer will increase to a maximum in about 100 nsec and will decline to near the equilibrium value in a few microseconds. Thus, noble gas atoms will come off in a short duration pulse and will be in the ionization region within a few microseconds. Time correlation is so effective in this case that laser counting times can be reduced to a few minutes, even though the volume of the entire apparatus is quite large.

Since counting the few atoms of choice amid a background of many other atoms of another isotope has to be performed in a closed chamber without pumping during the ionization process, outgassing of the chamber is a concern. However, Reynolds[64] has demonstrated that outgassing rates of closed chambers can be kept lower than 2×10^{-10} Torr/h. Commonly, a small chamber can be kept at less than 10^{-6} Torr for a week without pumping. Since the mass spectrometer can operate up to a pressure of 10^{-4} Torr, even the isotopic enrichment procedures can be completed before the pressure of the chamber reaches the 10^{-4} Torr level.

Another concern of this experiment is the adsorption and desorption of rare gas atoms from the wall of the chamber. Fast et al.[65] showed evidence that the rare gases do not dissolve in or permeate metals under thermal conditions, and that the percentage of time that rare gas atoms can remain on metals at room temperature can be neglected. However, rare gas ions can be implanted into the metal surfaces[66,67] and later be released, producing "memory" effects. The energies of ions are reasonably low during their entire history through a quadrupole mass filter which could make memory effects less pronounced than would be the case in magnetic deflection mass spectrometers.

In spite of concerns of these types, we believe that the Maxwell demon, equipped with a mass filter and modern lasers, can, in effect, be demonstrated. In the next section, therefore, we consider some interesting applications of such a device.

6. Applications

We conclude this chapter by discussing some applications of noble gas detectors. The surprisingly large numbers of possible applications is mainly due to the fact that rare gas atoms can be separated from huge targets in which they are produced by rare nuclear events. The success of Ray Davis, Jr. and his collaborators at Brookhaven National Laboratory in separating a few hundred ^{37}Ar atoms produced by solar neutrino interactions from nearly 400,000 liters of tetrachloroethylene (C_2Cl_4) is proof that such experiments can yield spectacular results. With the advent of laser techniques for the RIS of noble gases which makes it possible to count these atoms even before they have decayed, it is natural that a wide range of interesting proposals would come to our attention. Several of our collaborations involving these are presented here to show actual applications in progress and to provide a tangible basis for the reader to evaluate other possible applications.

Table 3 is a summary of some possible applications and may be used as a guide to the following discussions.

6.1. Baryon Conservation

Baryon conservation is currently regarded as one of the most fundamental problems in physics. Particle physicists, in the past decade,[68] have made great strides in understanding matter at the subatomic level. It is believed that the only truly elementary particles are the quarks and the leptons, yet it has been assumed that the proton obeyed strict baryon number conservation and thus could not decay to quarks or leptons. Recently, however, there is strong evidence in the gauge theories that

Table 3. Some Possible Applications of Noble Gas Detectors, Assuming a Few Atoms Can Be Counted Directly

1. Baryon conservation (e.g., ^{37}Ar or ^{38}Ar from decay of ^{39}K)
2. Solar neutrino flux and neutrino oscillations [^{81}Kr from ^{81}Br$(\nu, e^-)^{81}$Kr]
3. $\beta^-\beta^-$ Decay (e.g., ^{82}Kr from $\beta^-\beta^-$ decay of ^{82}Se)
4. Oceanic circulation (naturally occurring ^{39}Ar)
5. Polar ice caps (naturally occurring ^{81}Kr)
6. Aquifers (naturally occurring ^{81}Kr)
7. Waste isolation (Xe from neutron or photofission of transuranic elements)
8. Pu in soil (^{136}Xe or ^{86}Kr from neutron fission of Pu)
9. Diagnoses of bone diseases [^{37}Ar from ^{40}Ca$(n, \alpha)^{37}$Ar]
10. Fast neutron dosimetry [^{37}Ar from ^{40}Ca$(n,)^{37}$Ar]

baryon conservation is not strictly obeyed. Demonstration of proton decay would have great impact in these already impressive theories and would, generally, have a profound influence on the philosopher's view of the universe.

One of the difficulties in looking for proton decay is that the most likely decay modes are unknown; see, for example, the discussion of Rosen.[69] When a proton decays inside a nucleus, the considerable energy released could be carried away by leptons or photons, none of which interact locally. On the other hand, if a pion is produced, it can fragment the nucleus. An experiment[70] that sets a limit of $>2 \times 10^{30}$ years for the baryon decay depended on the modes in which a muon is produced either directly of indirectly through pion decay. Radiochemical experiments[71] establish a proton stability $>2.2 \times 10^{26}$ years, but they are significant because they are less dependent on the decay modes. The decay of a proton in ^{39}K leads to excited ^{38}Ar which evaporates a neutron about 20% of the time. Thus, ^{37}Ar with a half-life of 35 days is a good indicator of proton decay in K. Accordingly, Fireman[72] measured ^{37}Ar in $KC_2H_3O_2$ in the Homestake mine at a depth of 4400 mwe and found that the production of ^{37}Ar was less than 0.5 atoms per day from 10^{28} atoms of ^{39}K, suggesting proton stability $>2.2 \times 10^{26}$ years.

With the direct counting technique, one can consider the detection of ^{38}Ar (stable, 0.063% abundance) or the detection of ^{37}Ar. However, the concentration of ^{38}Ar is 2000 atoms/cm^3 even at 10^{-8} Torr of residual air. Thus, background due to normal abundance precludes direct detection of ^{38}Ar as a test of baryon conservation. Direct counting of ^{37}Ar in the Fireman experiment may not be quite as bleak. For example, if 1000 tons of $KC_2H_3O_2$ were put at a depth of 10,000 mwe, cosmic ray production of ^{37}Ar would be about 0.1 atom/day. Thus, signals less than 0.5 atom/day could be measured giving a proton stability $>10^{29}$ years. Isotopic interference from ^{36}Ar would be a major problem since there would be 10^4 atoms of ^{36}Ar per cm^3 of the large tank pumped to 10^{-8} Torr. However, a few cycles through the Maxwell's demon should solve the sorting problem. We are not aware of any serious effort to implement this direct counting approach for extending the decay-mode-independent limit on the stability of the proton.

6.2. Solar Neutrino Flux

The flux of solar neutrinos on the earth is regarded as the only means for learning about the processes occurring in the sun's interior. These processes involve the burning of hydrogen to make energy, the transport of energy and matter within the sun, and the emission of radiant energy. In spite of the complexity of the total system when all processes are jointly considered, experts in astrophysics and nuclear physics believe that neutrino emission from the sun should be predictable. Thus, because a radiochemical experiment[73-76] involving the reaction $^{37}Cl(\nu, e^-)^{37}Ar$ showed that the flux was much less than that predicted from the standard stellar model, much attention and effort have been directed toward the "solar neutrino problem"; see, for example, the recent article by Bahcall.[77]

The techniques described here make possible an interesting extension of the Davis type of radiochemical experiment. If the Cl-rich compound is replaced with a Br-rich compound, the reaction $^{81}Br(\nu, e^-)^{81}Kr$ can occur when $E_\nu > 300$ keV and is, therefore, sensitive to the ^7Be neutrinos produced in the stellar atmosphere. The cross section for neutrino capture can be inferred from a measurement[78] of the reverse process, i.e., electron capture in ^{81}Kr. Since the half-life of ^{81}Kr is 2.1×10^5 years, this kind of neutrino experiment is not possible with decay-counting techniques. In contrast, ^{37}Ar has a half-life of 35 days; therefore, it can be counted by decay-counting (proven) or by direct counting (conjecture). In the Br experiment, Davis[79] plans to use nearly 1000 tons of ethylene bromide in which 500 atoms of ^{81}Kr may be produced in 1 year, assuming the standard stellar mode. This is a promising experiment since there is negligible atmospheric ^{81}Kr and the main isotopic interference is stable ^{82}Kr which can be reduced to about 10^7 atoms in the neutrino tank.

6.3. $\beta^-\beta^-$ Decay

Double beta decay is often regarded as a test of weak interaction theory.[80-82] In a 1978 review, Bryman and Picciotto[83] conclude that experimental results show that double beta decay is not primarily a lepton-number-violating neutrinoless process. On the other hand, results are not completely in accord with calculations which assume that the two-neutrino double beta decay process is the only channel for $\beta^-\beta^-$ decay.

Mass-spectrometric detection of ^{82}Kr excesses which accumulated from the $\beta^-\beta^-$ decay of ^{82}Se in large geological deposits of copper selenide have been reported. A half-life for the process $^{82}Se \rightarrow {}^{82}Kr + \beta^-\beta^-$ was found[84] to be $6 \times 10^{19\pm0.3}$ years, and then as $10^{20.16\pm0.16}$ years by Kirsten and Muller.[84,85] Electron sum-energy measurements[86] suggest a half-life $>3 \times 10^{21}$ years. On the other hand, a recent cloud chamber study[87] reported a much shorter half-life ($1.0 \pm 0.4 \times 10^{19}$ years).

Another type of $\beta^-\beta^-$ experiment was suggested to us by Bruce Cleveland and J. D. Ullman.[88] Thus, a carefully evacuated chamber would be filled with a few kilograms of natural Se. If the half-life is 10^{20} years, 1 kg of natural Se (9% ^{82}Se) would give 12 atoms of ^{82}Kr per day. Samples of ^{82}Kr could then be removed daily or weekly and counted as we described above. An apparatus pumped to 10^{-8} Torr would have 30 atoms of ^{82}Kr

per cm^3 due to atmospheric Kr; thus, good vacuum techniques are essential. In spite of practical difficulties of this type, an experiment to measure the double beta decay of ^{82}Se can be viewed as a minor variation of the ^{81}Br$(\nu, e^-)^{81}$Kr solar neutrino experiment.

6.4. Oceanic Circulation

The isotope ^{39}Ar is produced in the atmosphere by cosmic rays and can serve as a steady-state source for the study of oceanic mixing.[89] The chemical inertness of Ar and the 270-year half-life of ^{39}Ar make it an ideal tracer of oceanic currents. One liter of "top water" in the ocean contains about 6000 atoms of ^{39}Ar (with 10^{19} atoms of ^{40}Ar!). However, 6 cycles through a Maxwell demon apparatus should change the situation to about 6000 atoms of ^{39}Ar and 10 atoms of ^{40}Ar; thus, even at depths where the water has been out of contact with the atmosphere for 1000 years, the signal-to-noise ratio should be quite adequate. With the techniques described, one liter of ocean water provides an adequate sample. In contrast, decay counting of ^{39}Ar requires long counting times even when several tons of water are processed to remove the Ar sample.

6.5. Polar Ice Caps and Old Aquifers

Polar ice cap dating can provide valuable information on climatic history.[90] The ideal isotope for dating polar ice caps is ^{81}Kr, whose half-life is 2×10^5 years compared, for example, with ages of about 1×10^5 years found in Greenland ice. However, decay counting of ^{81}Kr in ice requires extremely large samples of ice since even modern ice has only 1400 atoms of ^{81}Kr per kg.

On evaluating the possibility of direct counting of ^{81}Kr in ice samples, we find a rather attractive situation. One kilogram of ice gives a very adequate signal (up to 1400 atoms of ^{81}Kr) with 2.5×10^{14} atoms of ^{82}Kr as the isotopic interference. This should not be too difficult since five cycles of prepurification as described above should eliminate all ^{82}Kr atoms.

A very similar analysis applies to the problem of identifying old aquifers[91] in geological formations.

6.6. Waste Isolation and Pu in Soil

It has already been shown[92] that the RIS technique for detection of Xe can be used to identify materials that contain ^{239}Pu (or other fissionable atoms). For example, a thermal neutron exposure of 10^8 neutrons/cm^3 will produce 3×10^4 atoms of Xe in 1 kg of waste materials containing Pu at the level of 10 nCi/g. Detection of this rather large amount of Xe could produce a rapid scan procedure to determine whether a given batch of material must be stored as hazardous waste.

Similarly, irradiation of a 100-kg soil sample with 10^8 thermal neutrons/cm^2 would enable the detection of even the lowest levels (4 pCi/g) thus far recommended[93] for Pu in soil.

6.7. Diagnosis of Bone Diseases and Fast Neutron Dosimetry

It has been shown[94] that the exposure of an individual to a very modest level of fast neutrons (1 mrad) will produce measurable quantities of ^{37}Ar in the bone via the reaction ^{40}Ca$(n, \alpha)^{37}$Ar. In fact, a healthy man exhales 2300 atoms of ^{37}Ar per minute shortly after this small exposure. Measurement of ^{37}Ar by decay counting was proposed as a method for determining the loss of Ca in bone, and could be thus an indicator of certain types of bone diseases. Recently it was suggested[95] to us that the direct counting approach could offer even greater advantage. We concur in this opinion; furthermore, we suggest that the same technique could be developed to measure neutron exposures to levels much below 1 mrad.

6.8. Conclusion

The above examples of applications of the Maxwell sorting demon are by no means complete. They were selected to illustrate variety and assist the reader in the evaluation of the direct counting of radioactive or stable noble gas atoms for other interesting situations.

Acknowledgments

We should like to acknowledge the valuable discussions with R. D. Willis, G. W. Foltz, B. Lehmann, and G. D. Alton.

References and Notes

1. E. Rutherford, unpublished lecture notes (courtesy of University of Cambridge Library).
2. E. Rutherford and H. Geiger, *Proc. R. Soc. London Ser. A* **81**, 141 (1908).
3. J. J. Thomson, *Rays of Positive Electricity and Their Application to Chemical Analyses*, Green and Co., London (1913).
4. For a more complete perspective on the broader impact of RIS, the reader is referred to a recent article by G. S. Hurst, M. G. Payne, S. D. Kramer, and C. H. Chen, *Phys. Today* **33**(9), 24 (1980).
5. G. S. Hurst, S. D. Kramer, M. G. Payne, and J. P. Young, *IEEE Trans. Nucl. Sci.* **NS-26**, 133 (1979).

6. S. D. Kramer, G. S. Hurst, J. P. Young, M. G. Payne, M. K. Kopp, T. A. Callcott, E. T. Arakawa, and D. W. Beekman, *Radiocarbon* **22**, 428 (1980).

7. G. S. Hurst, S. D. Kramer, and B. E. Lehmann, in *Nuclear and Chemical Dating Techniques*, Ed. L. A. Currie, American Chemical Society Symposium Series, ACS Books (1982).

8. G. S. Hurst, in *Applied Atomic Collision Physics*, Eds. H. S. W. Massey, B. Bederson, and E. W. McDaniel, Academic Press, New York (to be published).

9. G. S. Hurst, M. G. Payne, M. H. Nayfeh, J. P. Judish, and E. B. Wagner, *Phys. Rev. Lett.* **35**, 82 (1975).

10. M. G. Payne, G. S. Hurst, M. H. Nayfeh, J. P. Judish, C. H. Chen, E. B. Wagner, and J. P. Young, *Phys. Rev. Lett.* **35**, 1154 (1975).

11. G. S. Hurst, M. G. Payne, and E. B. Wagner, United States Patent No. 3,987,302, "Resonance Ionization for Analytical Spectroscopy," filed 1974 and granted October 19, 1976.

12. G. S. Hurst, M. H. Nayfeh, and J. P. Young, *Appl. Phys. Lett.* **30**, 229 (1977).

13. G. S. Hurst, M. H. Nayfeh, and J. P. Young, *Phys. Rev. A* **15**, 2283 (1977).

14. G. S. Hurst, M. G. Payne, S. D. Kramer, and J. P. Young, *Rev. Mod. Phys.* **51**, 767 (1979).

15. L. W. Grossman, G. S. Hurst, M. G. Payne, and S. L. Allman, *Chem. Phys. Lett.* **50**, 70 (1977).

16. L. W. Grossman, G. S. Hurst, S. D. Kramer, M. G. Payne, and J. P. Young, *Chem. Phys. Lett.* **50**, 207 (1977).

17. R. V. Ambartzumian, N. P. Furzikov, V. S. Letokhov, and A. A. Puretsky, *Appl. Phys.* **9**, 335 (1976).

18. U. Heinzmann, D. Schinkowski, and H. D. Zeman, *Appl. Phys.* **12**, 113 (1976).

19. A. V. Smith, D. E. Nitz, J. E. M. Goldsmith, and S. J. Smith, *Bull. Am. Phys. Soc.* **24**(9), 1175 (1979).

20. A. V. Smith, J. E. M. Goldsmith, D. E. Nitz, and S. J. Smith, *Phys. Rev. A* **22**, 577 (1980).

21. S. D. Kramer, C. E. Bemis, Jr., J. P. Young, and G. S. Hurst, *Opt. Lett.* **3**, 16 (1978).

22. P. Lambropoulos, *Adv. Atom. Molec. Phys.* **12**, 87 (1976); P. Zoller, *Phys. Rev. A* **19**, 1151 (1979); P. Zoller and P. Lambropoulos, *J. Phys. B* **13**, 69 (1980); H. B. Bebb and A. Gold, *Phys. Rev.* **143**, 1 (1966).

23. P. Zoller, *Phys. Rev. A* **19**, 1151 (1979); P. Zoller and P. Lambropoulos, *J. Phys. B* **13**, 69 (1980).

24. R. S. Hargrove and J. A. Paisner, in *Proceedings of the Topical Meeting on Excimer Lasers* (Optical Society of America, Charleston, 1979).

25. W. R. Garrett and L. D. Mullins, *J. Chem. Phys.* **48**, 4140 (1968).

26. M. G. Payne, W. R. Garrett, and M. Pindzola, *Chem. Phys. Lett.* **78**, 142 (1981).

27. J. Bokor, J. Zavelovich, and C. K. Rhodes, *Phys. Rev. A* **21**, 1453 (1980).

28. M. Pindzola, private communication.

29. R. T. Hawkins, H. Egger, J. Bokor, and C. K. Rhodes, *Appl. Phys. Lett.* **36**, 391 (1980).

30. C. H. Chen, G. S. Hurst, and M. G. Payne, *Chem. Phys. Lett.* **75**, 473 (1980).

31. M. D. Levenson and N. Bloemberger, *Phys. Rev. Lett.* **32**, 645 (1974).

32. R. T. Hodgson, P. P. Sorokin, and J. J. Wyme, *Phys. Rev. Lett.* **32**, 343 (1974).

33. A. H. King, J. F. Youngs, and S. E. Harris, *Appl. Phys. Lett.* **22**, 301 (1973); **28**, 239 (1976).

34. R. Vasudeve and J. C. D. Brand, *Chem. Phys.* **37**, 211 (1979).

35. W. K. Bichsel, P. J. Kelley, and C. K. Rhodes, *Phys. Rev. A* **1817** (1976); **13**, 1829 (1976).

36. D. J. Kligler, J. Bokor, and C. K. Rhodes, *Phys. Rev. A* **21**, 607 (1980).

37. B. A. Bushaw and T. J. Whitaker, in *Proceedings of 20th Annual Technical Symposium of the Society of Photo-Optical Instrumentation Engineers (San Diego, August, 1976)*, Vol. 82, pp. 92–94 (SPIE Publications, Bellingham, Washington).

38. J. E. Bjokholm and P. F. Liao, *Phys. Rev. A* **14**, 751 (1976).

39. C. C. Wang and L. I. Davis, Jr., *Phys. Rev. Lett.* **35**, 650 (1975).

40. W. K. Bichsel, J. Bokor, D. J. Kligler, and C. K. Rhodes, *IEEE J. Quantum Electron.* **QE-15**, 380 (1979).
41. P. Lambropoulos, *Adv. Atom. Molec. Phys.* **12**, 87 (1976).
42. C. B. Collins, S. M. Curry, B. W. Johnson, M. Y. Mirza, M. A. Chellehmalzadeh, J. A. Anderson, D. Popescu, and I. Popescu, *Phys. Rev. A* **14**, 1662 (1976).
43. C. B. Collins, B. W. Johnson, M. Y. Mirza, D. Popescu, and I. Popescu, *Phys. Rev. A* **10**, 813 (1974).
44. G. Wagner and N. R. Isenor, *Can. J. Phys.* **57**, 1770 (1979).
45. M. Crance and M. Aymar, *J. Phys. B* **12**, 3665 (1979).
46. K. Aron and P. M. Johnson, *J. Chem. Phys.* **67**, 5099 (1977).
47. R. N. Compton, J. C. Miller, A. E. Carter, and P. Kruit, *Chem. Phys. Lett.* **71**, 87 (1980).
48. P. Agostini, F. Fabre, G. Mainfray, G. Petite, and N. K. Rahman, *Phys. Rev. Lett.* **42**, 1127 (1979).
49. P. M. Johnson, M. Berman, and D. Zakheim, *J. Chem. Phys.* **62**, 2500 (1975).
50. P. M. Johnson, *J. Chem. Phys.* **64**, 4143 (1976).
51. D. Zakheim and P. M. Johnson, *J. Chem. Phys.* **68**, 3644 (1978).
52. M. A. Duncan, T. G. Dietz, and R. E. Smalley, *Chem. Phys.* **44**, 415 (1979).
53. D. M. Lubman, R. Naaman, and R. N. Zare, *J. Chem. Phys.* **72**, 3034 (1980).
54. F. Brown and J. A. Davis, *Can. J. Phys.* **41**, 864 (1963).
55. L. Brillouin, *J. Appl. Phys.* **22**, 234 (1951).
56. Leo Szilard, *Z. Phys.* **53**, 840 (1929). See also *The Collected Works of Leo Szilard*, Eds. Bernard T. Feld and Gertrude Weiss Szilard, Vol. I, The MIT Press, Cambridge, Massachusetts (1972).
57. J. A. M. van Liempt, *Rec. Tranaux. Chim.* **57**, 871 (1938).
58. F. H. Wohlbier, Ed., *Diffusion and Defect Data*, Vols. 8–10, Trans. Tech. House Publication, Bay Village, Ohio (1968).
59. J. M. Tobin, *Acta Meta.* **5**, 398 (1957).
60. D. Lal, W. F. Libby, G. Wetherill, J. Leventhal, and G. D. Alton, *J. Appl. Phys.* **60**, 3257 (1969).
61. R. G. Wilmoth and S. S. Fisher, *Surf. Sci.* **72**, 693 (1978).
62. H.-J. Hagemann, W. Gudat, and C. Kunz, *Optical Constants from the Far Infrared to the X-Ray Region*: Mg, Al, Cu, Ag, Au, Bi, C, Al_2O_3, Deutsches Elektronen-Synchrotron Report No. DESY SR-74/7 (May 1974).
63. J. F. Ready, *J. Appl. Phys.* **36**, 462 (1965).
64. J. H. Reynolds, *Rev. Sci. Instr.* **27**, 928 (1956).
65. J. D. Fast, H. G. Van Bueren, and J. Philibert, *Diffusion in Metals Symposium*, Bibliotheque Technique Philip Ed., Eindhoven (1957).
66. J. Koch, *Nature* **161**, 566 (1948).
67. L. H. Varnein and J. H. Carmichael, *J. Appl. Phys.* **28**, 913 (1957).
68. For a review of this progress in particle physics, see Roy F. Schwitters, *Sci. Am.* **56** (October 1977); D. Z. Freedman and P. van Nieuwenhuizen, *Sci. Am.* **126** (February 1978); M. L. Perl and W. T. Kirk, *Sci. Am.* **50** (March 1978); L. M. Lederman, *Sci. Am.* **72** (October 1978); K. A. Johnson, *Sci. Am.* **66** (March 1969); M. Waldrop, *C & E News* **44** (January 21, 1980).
69. S. P. Rosen, *Phys. Rev. Lett.* **34**, 774 (1975).
70. F. Reines and M. F. Crouch, *Phys. Rev. Lett.* **32**, 493 (1974).
71. E. L. Fireman, in *Neutrino '77 (Proceedings of the International Conference on Neutrino Physics, Baksan Valley, USSR, 1977)*, Vol. 1, pp. 53–59, Publishing Office (Nauka), Moscow (1978).
72. E. L. Fireman, in *Proceedings 16th International Cosmic Ray Conference, Kyoto, Japan, August 6–18, 1979*, Vol. 13, Physical Society of Japan, Kyoto (1979).

73. R. Davis, Jr., D. S. Harmer, and K. C. Hoffman, *Phys. Rev. Lett.* **20**, 1205 (1968).

74. R. Davis, Jr. in *Proceedings of Neutrino '72 Europhysics Conference, Vol. I, Balatonfured, Hungary, June 11–17, 1972,* OMKD Technoinform, Budapest (1972), pp. 5–22.

75. R. Davis, Jr. and J. M. Evans, in *Proceedings 13th International Cosmic Ray Conference,* Vol. 3, p. 2001, American Institute of Aeronautics and Astronautics Tech. Info. Service, New York (1973).

76. J. K. Rowley, B. G. Cleveland, R. Davis, Jr., and J. C. Evans, in *Neutrino '77 (Proceedings of the International Conference on Neutrino Physics, Baksan Valley, USSR, 1977),* Brookhaven National Laboratory Report BNL-23418.

77. J. N. Bahcall, *Rev. Mod. Phys.* **50**, 881 (1978).

78. C. L. Bennett, M. M. Lowry, R. A. Naumann, F. Loeser, and W. H. Moore, *Phys. Rev. C* **22**, 2245 (1980).

79. Raymond Davis, Jr., private communication.

80. G. F. Dell'Antonio and E. Fiorini, *Supplemento al Volume XVII, Serie X, Del Nuovo Cimento* N. 1, 132 (1960).

81. S. P. Rosen and H. Primakoff, in *Alpha-, Beta-, and Gamma-Ray Spectroscopy,* Vol. 2, Ed. K. Siegbahn, North Holland Publishing Co., Amsterdam (1965).

82. H. Primakoff and S. P. Rosen, *Phys. Rev.* **184**, 1925 (1969).

83. D. Bryman and C. Picciotto, *Rev. Mod. Phys.* **50**, 11 (1978).

84. T. Kirsten, W. Gentner, and O. A. Schaeffer, *Z. Phys.* **202**, 273 (1967).

85. Hans Winfried Muller, Dissertation, Universität Heidelberg (1971).

86. B. T. Cleveland, W. R. Leo, C. S. Wu, L. R. Kasday, A. M. Rushton, P. J. Gollon, and J. D. Ullman, *Phys. Rev. Lett.* **35**, 757 (1975).

87. M. K. Moe and D. D. Lowenthal, University of California at Irvine Report No. UCI-10P19-143 (December 1979).

88. J. D. Ullman, private communication.

89. We are discussing here some ideas which evolved in collaboration with Harmon Craig and R. D. Willis of the Scripps Institution of Oceanography. Furthermore, R. D. Willis is on assignment with the ORNL Photophysics Group for the implementation of the ^{39}Ar counter.

90. H. Oeschger, private communication.

91. S. N. Davis, private communication.

92. L. A. Franks, H. M. Borella, M. R. Cates, G. S. Hurst, and M. G. Payne, *Nucl. Instrum. Methods* **173**, 317 (1980).

93. J. W. Healy, Los Alamos Scientific Laboratory Report LA6741-MS (April 1977).

94. R. E. Bigle, J. S. Laughlin, Raymond Davis, Jr., and J. C. Evans, *Radiat. Res.* **67**, 266 (1976).

95. Raymond Davis, Jr., private communication.

5

Trapped Ion Spectroscopy

Günther Werth

1. Introduction

A single particle, at rest in free space, free of uncontrolled perturbations and at hand for infinitely long times would be the ideal subject for spectroscopic investigations. If the techniques for the preparation of such a particle are available, it becomes feasible to orient the system by a variety of reactions with photons, atoms, and electrons. Line broadening due to Doppler and observation time effects would be minimized.

Although this ideal will ultimately never be reached, the use of ion traps has shown a possible way to approach it to a very high degree: Ions at low density may be confined by electromagnetic fields to a small volume. No wall collisions occur; the rate of collisions with background molecules under UHV conditions is less than 1 per second and the storage time easily exceeds several minutes. Since the introduction of quadrupole traps for mass spectrometric[1-3] and spectroscopic applications,[4] important progress has been made to approach the ideal case even more: Recently, the detection and spectroscopy of single particles has been achieved. Ion cooling reduces the kinetic energy to less than 10^{-4} eV and localizes the particle to a radius of a few μm. The storage time often exceeds several hours.

The basic principles of the ion storage technique have been reviewed by H. G. Dehmelt[5] in 1967 and 1969, and by Todd et al in 1976.[6] In Part B of this book, Hans A. Schuessler[7] described different techniques in stored ion spectroscopy, concentrating on exchange collision processes for reorientation. In this chapter we will give a brief outline of the technique

Günther Werth • Institut für Physik, Universität Mainz, 6500 Mainz, Federal Republic of Germany.

and deal mainly with progress made by laser optical pumping and by radiative cooling of ions. In addition we will mention the use of ion traps for the determination of very low decay rates of metastable states, which is possible because of the long availability of the particles.

2. Storage

Ion traps may be operated either in a static or a rf modes. In both cases three hyperboloids of revolution are used as electrodes, one ring and two endcaps (Figure 1). In the symmetric case we have $r_0^2 = 2z_0^2$, but configurations having $r_0 = z_0$ and cylindrical electrodes have been operated successfully.[8]

2.1. Penning Trap

In the static case we apply a dc voltage U_0 between the ring and endcap electrodes. The resulting potential

$$\Phi = (U_0/2r_0^2)(x^2 + y^2 - 2z^2) \tag{1}$$

leads, for a properly chosen polarity, to a focusing force along the z axis, while the particles are defocused in the x–y direction. Confinement is achieved by superimposing a magnetic field B_z. The ion motion along the z axis remains a harmonic oscillation, unaffected by B_z, with frequency

$$\omega_z^2 = (2e/mr_0^2)U_0 \tag{2}$$

The combined action of the electric and magnetic fields in the x–y plane

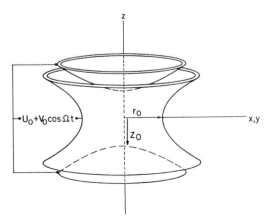

Figure 1. Electrode structure of an ion trap.

leads to a distortion of the cyclotron orbits, whose centers circle around the trap center (magnetron motion). The resulting epicycloide may be characterized by two frequencies

$$\omega_+ = \omega_{c/2} + (\omega_c^2/4 - \omega_z^2/2)^{1/2}$$
$$\omega_- = \omega_{c/2} - (\omega_c^2/4 - \omega_z^2/2)^{1/2} \tag{3}$$
$$\omega_c = (e/m) \cdot B$$

Confinement is assured, if the action of the magnetic field exceeds the defocusing force of the electric field: Then

$$\omega_c^2/2 > \omega_z^2$$
$$B^2 > 4mU_0/er_0^2 \tag{4}$$

Obviously this method of containment is best suited, if a magnetic-field-dependent quantity is to be measured, and as an example, we may cite the determination of the anomalous magnetic moment of the electron by Dehmelt and co-workers.[9]

2.2. rf Trap

No need for a magnetic field exists, if we apply a dc voltage U_0 and a rf voltage $V_0 \cos \Omega t$ to the trap electrodes. The equations of motion for a singly charged ion in the potential

$$\phi = \frac{U_0 + V_0 \cos \Omega t}{2r_0^2} \cdot (x^2 + y^2 - 2z^2) \tag{5}$$

are Mathieu-type equations, which can be written in a canonical form

$$\frac{d^2 u_i}{d\xi^2} + (a_i - 2q_i \cos 2\xi)u_i = 0 \tag{6}$$

$$\xi = \Omega t/2, \qquad a_z = -2a_r = -8eU_0/mr_0^2\Omega^2,$$
$$q_z = 2q_r = 4eV_0/mr_0^2\Omega^2 \tag{7}$$

The solutions of the Mathieu equation are either stable or unstable, depending on the values of a_i and q_i. The solution of (6) for stable ion motion can be written as

$$U_i(\xi) = A \sum_{n=-\infty}^{+\infty} C_{2n} \cos (2n + \beta_i)\xi + B \sum_{n=-\infty}^{+\infty} C_{2n} \sin (2n + \beta_i)\xi \tag{8}$$

where A and B depend on the initial conditions, C_{2n} are functions of a_i and q_i and decrease rapidly for large n, and

$$\beta_i^2 = a_i + q_i^2/2 \qquad (9)$$

If we restrict the description to values of $a, q \ll 1$, (8) reduces to

$$z(t) = z_0[1 + (\beta_z/2) \cos \Omega t] \cos \omega_z t \qquad (10)$$

and similarly for $x(t)$, $y(t)$. This is a superposition of a low-frequency oscillation of the guiding center at $\omega_z = \beta_z \Omega/2$ and a high-frequency micromotion at the driving frequency Ω. A complete description of the equation of motion yields a motional spectrum which contains all the frequencies $n\Omega \pm \omega_z$ $(n = 0, 1, \ldots)$.

Figure 2 shows the boundary of the first region of stability. Included are experimentally found values of the relative ion density, measured by the number of fluorescence photons after resonant laser excitation. Maximum ion density is found near $a_z = -0.03$, $q_z = 0.55$.[10]

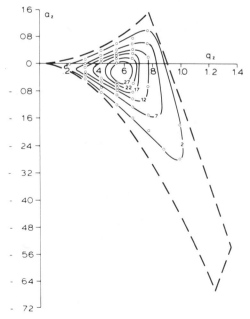

Figure 2. Boundary of the first region of stability of an rf ion trap. Points of equal ion densities, found experimentally, are connected. (From Ref. 10.)

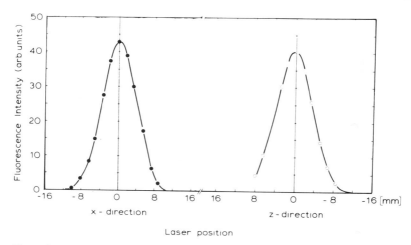

Figure 3. Ion densities in x- and z-directions. Experimental points and fitted Gaussian. (From Ref. 12.)

The ions form a cloud around the trap center (Figure 3). In thermal equilibrium, the spatial density distribution may be described by a Gaussian[11,12] function

$$n(r, z) = n_0 \exp\left[-(x/\Delta x)^2 - (z/\Delta z)^2\right] \qquad (11)$$

The width, Δ, of the distribution is connected to the ion temperature by

$$\Delta x, \Delta z = r_0(kT/eD_{r,z})^{1/2} \qquad (12)$$

D is the maximum depth of the trap potential, given (for $a, q \ll 1$) by

$$eD = \tfrac{1}{8}m\Omega^2(\beta_r^2 r_0^2 + \beta_z^2 z_0^2) \qquad (13)$$

Experimentally, one finds for spherically symmetric potential wells that $\Delta = $ const and $\tfrac{3}{2}kT \simeq 0.1\,eD$. For a well depth of $10\,eV$ we have $T = 6000$ K. Somewhat lower temperatures are obtained if we allow collisions with cold background molecules.[12]

The maximum ion number, N, that can be stored, is determined by space charge compensation of the applied potential, until we have

$$D + V_c = 0 \qquad (14)$$

where V_c is the ionic space charge potential given by

$$\Delta V_c = 4\pi\rho = 4\pi e n \qquad (15)$$

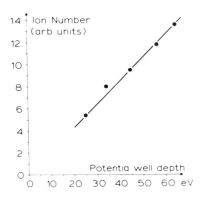

Figure 4. Stored ion number vs. potential well depth.

It follows for spherical potential wells ($\beta_r^2 r_0^2 = \beta_z^2 z_0^2$) and $r_0^2 = 2z_0^2$ that

$$n = m\Omega^2\beta_r^2 r_0^2/16\pi e^2 \tag{16}$$

and from Eq. (13) that

$$n = 3D_z/4\pi e^2 \cdot r_0^2 \tag{17}$$

N follows by integration of Eq. (12) over the active trap volume. Extending the integration to infinity and setting for simplicity $\Delta_x = \Delta_z$ we find

$$n = 3D_z\Delta^3/4\pi e^2 r_0^2 \tag{18}$$

As a numerical example, taken from Ref. 12, we find $\Delta = 0.28r_0$ for a trap having $r_0 = 2$ cm and $D_z = 10$ eV, yielding $N \simeq 1 \times 10^5$, in agreement with experimentally reported values. The linear dependence of the ion number on the potential depth has also been checked experimentally (Figure 4). Collisions with buffer gas atoms influence the dynamics of the ion motion. Of special interest for spectroscopic applications is the presence of light gases. For this case, theoretical considerations by André[13] show that, at low pressures, the time of confinement may be extended by several orders of magnitude, depending on the mass of the colliding particles and the ion energy. This has been confirmed experimentally on different species.[14,15]

2.3. Other Devices

Different ion traps have been developed for special purposes, which make use of the focusing force on charged particles in inhomogeneous electromagnetic fields. Teloy and Gerlich[16] reported an arrangement of metal plates alternatively connected to rf voltages of opposite phase. The time average force, which is exerted on the charged particle, acts in the direction of decreasing field strength. Thus, repulsive "walls" can be formed with the potential essentially constant in between, which can be used for ion containment and guiding.

A static device, already used in 1923 by Kingdon[17] to study electron space charge neutralization by trapped ions, consists of a closed cylinder with a central rod maintained at a negative potential with respect to the grounded cylinder walls. In a plane perpendicular to the rod the potential falls off logarithmically from the center. Ions created at some distance from the center, orbit around the rod in the attractive field and oscillate along its length in the axial well produced by the cylinder ends. Ion trajectories in this potential have been calculated by Hooverman.[18]

The device is especially well suited if the trap has to be used as a cylindrical microwave cavity, and it has been successfully used in the determination of the hyperfine structure of the $2S$ state of $^3He^+$.[19] Recently it has been applied to the investigation of electron capture by highly charged ions.[20]

2.4. Ion Creation and Detection

To be confined, either the ions have to be created inside the well formed by the trapping potentials, or ion beams from external sources have to be cooled down inside the trap. Both methods have been employed in the past. Electroionization of background molecules or atomic beams easily leads to production rates large enough to fill the trap within a few microseconds even at moderate electron currents, if a partial pressure of about 10^{-9} mbar can be maintained. For rare isotopes, however, one may store the ions more effectively if they are created outside the trap and cooled down by inelastic collisions. The residual background pressure of neutral molecules has to be controlled in such a way as to ensure a sufficient collision rate on one side and to avoid ion loss by diffusion processes on the other. Best suited are light buffer gases such as He. Figure 5 shows an example, where Ba^+ ions are formed by surface ionization from a hot Pt filament, placed near the inner surface of one endcap electrode and on the same potential. The total ion current from the filament was 10^{-12} A; sampling time was 10 sec. In our example 2% of the primary ions were confined.

For nondestructive detection of the stored ions, one may most easily use the damping of a weakly excited tank circuit across the trap electrodes, tuned to the z-oscillation frequency. The coherent excitation of the ion cloud, however, leads to energy gain and somewhat reduces the storage

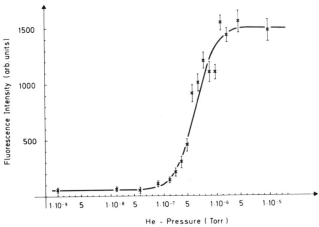

Figure 5. Number of trapped ions, created by surface ionization at an endcap electrode, vs. He buffer gas pressure. (From Ref. 12.)

time. The bolometric technique[21] avoids this disadvantage by measuring the voltage induced in the electrodes by the incoherent motion of the ions.

If the trapped ion has resonance transitions, which are in the reach of available tunable lasers, the observation of fluorescence light from the stored ions offers a detection method, which allows one to distinguish between different isotopes. The sensitivity depends on the spectral range and the particular level structure, which may allow the observation of a wavelength different from that used for the excitation. Using a pulsed dye laser at saturation intensity, 10^3 stored ions would result in about 1 fluorescence count per laser pulse, assuming 1% solid angle and 10% photomultiplier quantum efficiency. Continuous wave laser excitation, however, leads to fluorescence rates of $10^8 \ \text{sec}^{-1}$ per ion which allows visible observation even of a single particle. This has been successfully demonstrated on Ba^+ [22] and Mg^+.[23]

3. Optical Double Resonance Spectroscopy

The almost perturbation-free environment of trapped ions results in extremely long coherence times for long-living states. This advantage over other spectroscopic methods has been used for high-resolution microwave spectroscopy on ground or long-living metastable states, to determine precise hyperfine separations. The first example was the determination of the ground-state hyperfine splitting in $^3He^+$ by Dehmelt and co-workers[24,25] with an uncertainty of 10 Hz in 8.6 GHz. This experiment used spin-dependent collision processes between the stored He^+ ions and a polarized beam of Cs atoms for ion polarization and detection of spin-flip resonances. It has been reviewed in Part B of this book by H. A. Schuessler.[7] An example of work on metastable states is the measurement of the hyperfine structure of the $2S$ state of the $^3He^+$ by Prior and Wang.[16] After creating metastable ions by electron impact on He gas at low pressure, the population difference between two adjacent hyperfine levels was achieved by a rf pulse, which connected one of the metastable states to a fast-decaying $2P$ state. The ions were then irradiated with microwaves of the hyperfine frequency and the transition was monitored by a second rf pulse and the coincident observation of 304-Å decay photons to the ground state. The result of this experiment was $\Delta\nu = 1088.3549807$ (88) MHz $(\partial\nu/\nu = 9 \times 10^{-9})$.

In both experiments on $^3He^+$, methods for ion polarization and resonance detection have been employed, which depend on the particular level structure of the ion and on the size of spin-dependent collision cross sections and which cannot easily be extended to other species.

Application to a larger variety of ions is made possible by the optical pumping technique, which has been successfully used in numerous examples on neutral atoms. The work on optical pumping of ions has been reviewed by E. W. Weber.[26] The experiments described in this paper deal mainly with ions in buffer gases, where densities larger than $10^8 \, cm^{-3}$ are achievable. The precision obtained in these experiments is mainly limited by density shifts to about 1 part in 10^7.[27] Ion densities of less than $10^5 \, cm^{-3}$ in stored ion clouds require different experimental techniques. Absorption measurements of the pumping light beams are impossible, and the observation of fluorescence light requires very effective shielding of stray and background light. Obviously, the use of laser light is best suited for this purpose. However, only some ions have spectral resonance lines within the reach of tunable laser sources available to date. Thus, so far only a few optical double resonance experiments, on Hg^+, Ba^+, and Mg^+, have been reported, which make use of the favorable conditions of ion traps. Each of the three experiments has one characteristic feature, which I will emphasize.

Common to all experiments is the fact that the ion motion is restricted to amplitudes which are small compared to the wavelength of the microwave radiation. For this case, Dicke[28] has shown that the spectrum consists of a non-Doppler shifted frequency plus or minus integral multiples of the ion oscillation frequency. For the special case of a TE_{013} mode, regarding the quadrupole as a cylindrical cavity, the intensities of the sidebands have been calculated by Major and Duchéne.[29] The relative amplitudes of the sidebands contain information about the energy distribution and may serve to evaluate the second-order Doppler correction. Such motional sidebands have been observed in the Hfs spectrum of $^3He^+$.[25]

3.1. Continuous Broadband Light Source: $^{199}Hg^+$

Best suited to exploit the advantages of the ion trap for high-resolution spectroscopy is $^{199}Hg^+$ because of its high mass, which reduces the second-order Doppler effect, its large hyperfine splitting of 40.5 GHz, and its simple spectrum (nuclear spin $I = 1/2$). Several experiments on it have been performed, which exhibit the same basic features.[30–32] The ions were produced by electroionization of the background gas, which consists mainly of the proper isotope after exposure of the carefully cleaned apparatus to a few mg of ^{199}Hg for one hour. Partial pressure was in the 10^{-8}-mbar range. The storage time was of the order of several minutes and was limited by resonant charge exchange collisions with the neutral parent atoms. Two holes were drilled into the electrode to let the pumping light beam pass through. This caused the main deviation from the ideal potential, but did not much affect the storage time, since the ions predominantly move in a small range around the center, far away from the perturbation. The upper endcap electrode was formed by a highly transparent mesh, and the trap center was focused by quartz optics through the mesh and an interference filter to the photocathode of a solar blind photomultiplier.

The $6^2S_{1/2}-6^2P_{1/2}$ resonance line is at 194.2 nm and thus not accessible to laser radiation. However, a fortuitous matching, within the optical Doppler width, of one of the hyperfine components of this line with the corresponding unsplit and isotope shifted line of $^{202}Hg^+$ ($I = 0$) allows easy hyperfine pumping with unpolarized light. Figure 6 shows the pumping scheme. The pumping time constant γ_p depends on the primary photon flux I_γ, the absorption Doppler width $\Delta\nu_D$, and the absorption profile $k(\nu)$[33]:

$$\gamma_p = \frac{k(\nu)}{k_0} \frac{2}{3\Delta\nu_D} \left(\frac{\ln 2}{\pi}\right)^{1/2} \frac{e^2}{4\varepsilon_0 mc} f \cdot I_\gamma \qquad (19)$$

where k_0 is the resonant absorption coefficient and f the oscillator strength. Neglecting the hyperfine splitting in the excited $6P_{1/2}$ state and setting $k(\nu)/k_0 = 1$ and $f = 1$, we have for a Doppler width of 10 GHz (corresponding to an average ion energy of 4 eV)

$$\gamma_p = 10^{-13} I_\gamma \text{ sec}^{-1} \qquad (20)$$

if I_γ is given in photons per cm^2 and per sec. Typical intensities for rf excited spectral lamps are 10^{13} cm^{-2} sec^{-1}, so $\gamma_p \simeq 1$ sec^{-1}. The dominant source of relaxation is charge exchange collisions between the ions and the parent atoms, the cross section being 3×10^{-14} cm^2. At a background pressure of 10^{-8} mbar and an average ion velocity of 2×10^5 cm/sec this

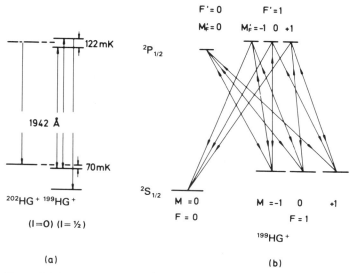

Figure 6. (a) Level structure and (b) optical pumping scheme of Hg^+ ions.

leads to collision rates of $1.8 \, sec^{-1}$. The orders of magnitude for both time constants have been verified experimentally.[34] Other sources of relaxation are magnetic field inhomogeneities, Stark effect, spin exchange between the ions, and ion loss from the trap, but all these are negligible compared to the charge exchange effect. Transitions between the different hyperfine levels are induced by simply irradiating the ions with microwave radiation and detected by the change of the fluorescence intensity observed at right angles to the pumping beam by photon counting techniques. The shape of the resonances may be calculated using the density matrix formalism. It leads to a Lorentzian line shape, whose width, in the case of the magnetic-field-"independent" $F = 1$, $m = 0$–$F = 0$, $m = 0$ transition, is given for weak pumping light and negligible ion loss by $\delta\omega = 9/2(\gamma_p + \gamma_2)$, where γ_2 is the longitudinal relaxation time constant. The expected linewidth of a few hertz for this transition has been obtained experimentally (Figure 7). The rather large straylight background, due to pumping light being scattered at the different apertures, requires averaging times of about 20 min to obtain a signal-to-noise ratio of 10. Some typical numbers are given in Table 1. The statistical uncertainty of the resonance center is 0.1 Hz. The observed resonance has to be corrected to zero magnetic field, using the two $\Delta F = 1$, $\Delta m = \pm 1$ transitions to determine the field. The Breit–Rabi formula leads to $\Delta\nu = 97.2 \cdot H^2$ (Hz) if H is given in Gauss. The total error of the extrapolation is 0.2 Hz. A second source of error is the second-order Doppler effect. At $E_{kin} = 5 \, eV$, one expects a shift of 0.5 Hz. Shifts of this

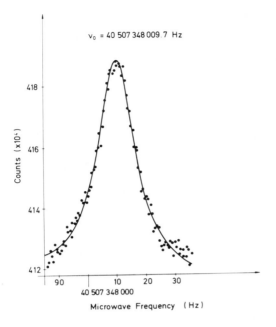

$v_0 = 40\,507\,348\,009.7$ Hz

Figure 7. $F = 0$, $m = 0 - F = 1$, $m = 1$ transition in ^{199}Hg$^+$. The solid line represents a Lorentzian through the experimental points. Averaging time is 50 min. FWHM: 16 Hz. (From Ref. 31.)

order of magnitude may have been observed by variation of the trap potential depth D, which changes the ion energy accordingly.[12] By this method, however, the Doppler shift cannot be separated from a possible hfs Stark shift, since the electric field strength in the trap is changed also. From hfs Stark shift measurements in Cs, one can estimate that its contribution is small ($\delta v/v < 10^{-14}$). The statistical error of a linear extrapolation of the observed transition frequency to $D = 0$ is 0.1 Hz. A systematic error of 1 Hz has to be added to the final value due to uncertainties in the absolute frequency of the Rb clock, which was used as the frequency standard. No attempt has been made to account for a possible light shift, which is assumed to be very small because of the broad overlap of the pumping and absorption light profiles.

Table 1. Typical Data of the Trapped Hg$^+$ Experiment

Number of trapped ^{199}Hg$^+$ ions	$10^5 - 10^6$
Storage time	ca. 30 sec
^{199}Hg partial pressure	10^{-8} mbar
Primary photon flux at 194 nm	10^{-12} sec^{-1}
Stray background count rate	1.6×10^5 sec^{-1}
Additional count rate at resonance	1.5×10^3 sec^{-1}

Table 2. Contribution of Errors to the Determination of the ^{199}Hg$^+$ Hfs Separation

$V_{Hfs} = 40\,507\,347\,997.8$	
Errors	
Statistical uncertainty	0.1 Hz
Magnetic field extrapolation	0.2 Hz
Electric field extrapolation	0.1 Hz
Absolute frequency calibration	1 Hz

Table 2 gives the result of the measurement, including different sources of error, taken from Ref. 31.

3.2. Pulsed Laser Excitation: Ba$^+$

The $6^2S_{1/2}-6^2P_{1/2}$ and $6^2S_{1/2}-6^2P_{3/2}$ resonance transitions of the Ba$^+$ ion at 493 and 455 nm are easily accessible by laser light. The increase in signal-to-noise ratio compared to the lamp-excited Hg$^+$ experiment compensates for the somewhat smaller hfs separation and mass. In addition, the excited P levels decay with a probability of $1:7.6$ into low-lying metastable $5D$-states, which allows the detection of fluorescence light at a wavelength different from that of the excitation. Experiments on both stable odd isotopes, ^{135}Ba$^+$ and ^{137}Ba$^+$, have been performed.

For the optical pumping, a nitrogen pumped pulsed tunable dye laser was used whose spectral width (1 GHz) was less than the ground-state hfs splitting of the Ba$^+$ isotopes (8.05 and 7.18 GHz, respectively). The laser was tuned to one of the hfs components of the resonance lines at 493.4 or 455.4 nm (Figure 8), and the fluorescence light perpendicular to the

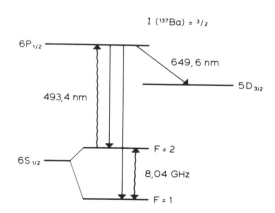

Figure 8. Energy levels and pumping scheme of ^{137}Ba$^+$. For ^{135}Ba$^+$ it differs only by the size of the Hfs splitting (7.18 GHz).

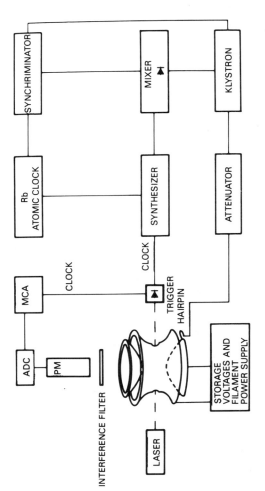

Figure 9. Experimental setup for the precision laser optical double resonance experiment on Ba$^+$.

laser beam was monitored. Individual fluorescence light quanta could not be resolved in time ($\tau_{1/2} = 8$ nsec) and were integrated by an analog-to-digital converter, which was gated for a few lifetimes. Figure 9 shows the experimental setup.

The microwave transitions, frequency stabilized and referenced to a Rb atomic standard, were induced between two laser pulses, whose width was 6 nsec at a maximum repetition rate of 50 Hz. The advantage of using the pulsed scheme is the complete absence of light shifts. The line shape, however, now deviates substantially from a Lorentzian and contains sidebands depending on the laser repetition rate, the spatial distribution of the laser output, and the microwave power: If we neglect collisions, the coherence time for one ion is now given by the time T between two consecutive laser pulses, which are at saturation intensity. However, since the laser beam diameter is smaller than the ion cloud radius and since the ions perform many oscillations between two laser pulses, there is a finite chance for an ion not to be excited after 2, 3, or even more laser pulses. For the $F = 1$, $m = 0$ and $F = 2$, $m = 0$ states, which depend on the magnetic field only in second order, one finds

$$p(\omega) = \frac{A^2}{(\omega - \omega_0)^2 + A^2} \sum_{k=1}^{n} a_k \sin^2 \left\{ k \frac{T}{2} [(\omega - \omega_\delta)^2 + A^2]^{1/2} \right\} \quad (21)$$

where ω_0 is the resonance frequency and A the induced transition rate.

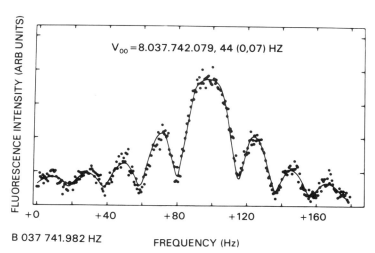

$V_{00} = 8.037.742.079, 44\ (0,07)\ HZ$

B 037 741.982 HZ

FREQUENCY (Hz)

FLUORESCENCE INTENSITY (ARB UNITS)

Figure 10. "Field-independent" Hfs transition in ^{137}Ba$^+$. Side-bands are due to the pulsed laser excitation. Laser repetition rate 16.6 Hz. Width of the central maximum: 25 Hz.

The a_k depend on the ratio of the ion distribution width and laser beam diameter and decrease for increasing k. Five parameters were needed to fit, with sufficient accuracy, the whole resonance pattern observed in the experiment. Figure 10 shows an example at a laser repetition rate of 16 Hz. Each data point represents an average of 40 laser pulses. At the present level of accuracy no influence of collisions on the line shape can be seen. Even at laser repetition rates as low as 1 Hz, the line shape given by Eq. (21) describes the observations sufficiently well (see Figure 11). The statistical uncertainty of the line center is less than 0.02 Hz, but the total averaging time increases to 20 min due to the low laser repetition rate. The observed resonance frequency has to be corrected to zero magnetic and electric fields. The first is done according to the Breit–Rabi formula, with the magnetic field strength determined by the observation of the remaining $\Delta F = 1$, $\Delta m = 0$, ± 1 transitions. The accuracy was limited by

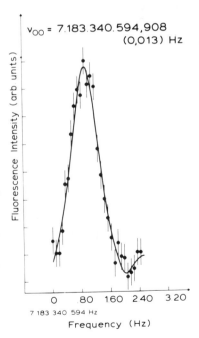

Figure 11. Hyperfine resonance in ^{135}Ba$^+$. Laser repetition rate was 1 Hz, FWHM = 0.87 Hz.

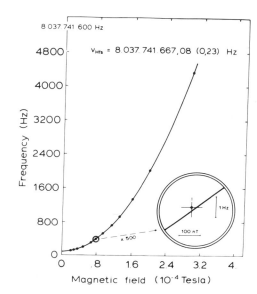

Figure 12. Magnetic field dependence of the $F = 1$, $m = 0 - F = 2$, $m = 0$ transition in $^{137}Ba^+$.

the magnetic field inhomogeneity. Figure 12 shows the result for $^{137}Ba^+$. The electric field dependence contains a possible hfs Stark effect and—more important—a second-order Doppler shift, since the ion temperature is a linear function of the depth of the potential well created by the electric fields. Experimentally, we find

$$\Delta\nu/\nu = -2.3\,(1.0) \times 10^{-11}\,\bar{E}_{kin}$$

The final results for the $^{137}Ba^+$ and $^{135}Ba^+$ isotopes are given in Table 3.

Table 3. Result of Hfs Measurements for Ba^+ Isotopes[a]

	$^{135}Ba^+$	$^{137}Ba^+$
Hyperfine separation	7 183 340 234.90	8 037 741 667.69
Statistical uncertainty	≪0.10	<0.10
Magnetic field extrapolation	0.40	0.23
Electric field extrapolation	0.40	0.27
Total error	0.57	0.36

[a] All frequencies are given in Hz.

3.3. Ion Cooling: $^{25}Mg^+$

While, in ion storage experiments, uncontrolled perturbations due to collisions or interaction with external fields can be made extremely small, the main limitation of the accuracy is given by the second-order Doppler effect. In the experiments on Hg^+ and Ba^+ the ion temperature was of the order of several thousand Kelvin, giving rise to a fractional frequency shift of some parts in 10^{11}, even for the heavy ions. This shift may be almost completely eliminated by motional sideband cooling of the ions, which was first proposed by Wineland and Dehmelt[37] and successfully demonstrated by Wineland et al.[38] and Neuhauser et al.[39] The basic features of the cooling process may be understood if we consider the optical electric field of a plane wave, incident along the x axis, seen by the harmonically bound ion: We have

$$\bar{E}_{ion} = \bar{E}_0 \sin (kx - \omega t)$$

and

$$x = x_0 \sin (\Omega_x t + \phi) \tag{22}$$

where x is the amplitude of the ion oscillation, Ω_x is its frequency, and k and ω are the wave vector and frequency of the incident radiation. Choosing $\phi = 0$ we get

$$\bar{E}_{ion} = \bar{E}_0 \sin (kx_0 \sin \Omega_x t - \omega t) \tag{23}$$

This can be expanded in terms of Bessel functions and gives rise to a spectrum consisting of a center frequency ω and symmetrical sidebands at the frequency $\omega \pm m\Omega$, m integral. When the natural linewidth of the transition is less than the spacing between the Doppler generated sidebands, we can tune the incident radiation to one of the resolved lower sidebands. Thus the ion absorbs photons of energy $\hbar (\omega_0 - m\Omega_x)$. The average energy of the re-emitted photons, however, is $\hbar\omega_0$, leading to a net cooling of the ions. The cooling rate dE/dt is proportional to the ratio of the energy difference $m\hbar\Omega_x$ to the total energy $\hbar\omega_0$, the laser intensity I, the cross section σ for resonant photoabsorption, and the amplitude of the sideband, which is proportional to the Bessel function:

$$dE/dt = (m\Omega_x/\omega_0)I\sigma J_m^2 (kx_0) \tag{24}$$

We have neglected recoil effects and have assumed complete overlap of the laser beam and the ion cloud. Even at moderate laser powers

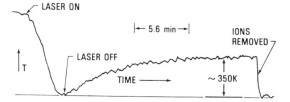

Figure 13. Demonstration of ion cooling and subsequent heating on Mg$^+$ ions (from Ref. 38).

($\sim$mW/cm^2) this leads to rather large cooling rates of 0.1 eV/sec. For details of the calculation and a quantitative description of the cooling process, the reader may be referred to Refs. 40 and 41.

The experimental demonstration of the cooling process has been performed on Mg$^+$ in a Penning trap[38] and on Ba$^+$ in a rf trap.[39] In both cases, temperatures lower than 1 K have been obtained (Figure 13). At the same time, the sensitivity of the optical detection has been increased to the extent that one single trapped particle could be detected and even be observed visually.[42]

The first application to high-resolution spectroscopy resulted in a determination of the ground-state hyperfine splitting of ^{25}Mg$^+$.[43] The ions (nuclear spin $I = 5/2$) were stored in a Penning trap, whose magnetic field of 1.24 T had been set to a value where the first derivative of the ($m_I = -3/2$, $m_J = 1/2$) to ($m_I = -1/2$, $m_J = 1/2$) transition frequency with respect to the magnetic field was zero. Ion cooling and optical pumping were achieved by using circular polarized light from a frequency doubled cw dye laser, whose frequency was tuned slightly below the ($^2S_{1/2}$, $m_I = -5/2$, $m_J = -1/2$) to ($^2P_{3/2}$, $m_I = -5/2$, $m_J = -3/2$) transition. Although other hyperfine levels were off by at least 9.3 GHz, optical pumping occurred in the Lorentzian wings of these states. At the light intensities used, these weak pumping rates were about 1 sec^{-1}, much faster than any other relaxation rate between ground-state sublevels, which were estimated to be >1 h at a background pressure of 10^{-9} mbar. The Zeeman transition at 292 MHz was detected by the change in the fluorescence intensity after the light had been blocked, while the microwaves were on, to avoid any light shift. Using the Ramsay method[44] of rf pulses separated in time, one obtains an oscillating line shape with a central minimum of the width $(2T)^{-1}$, where T is the time between two rf pulses. Linewidths as small as 12 mHz were obtained with $T = 41.4$ sec (Figure 14). This corresponds to a line Q of 2.4×10^{10}. At times longer than 40 sec the fluorescence rate decreased due to heating of the ions while the light was off, which increased the Doppler width. The final result for the A value, which was obtained from the observed frequencies using the Breit–Rabi formula, is $A = -596\,254\,376\,(54)$ Hz. The error reflects, in about equal parts, the statistical

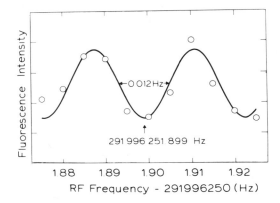

Figure 14. Narrow Zeeman resonance in $^2Mg^+$ (from Ref. 43).

uncertainty of determining the line center and possible magnetic field drifts and inhomogeneities. The number of ions used in this experiment varied between 10 and 100, and the temperature was about 20 K. At this temperature, the fractional frequency shift due to second-order Doppler effect is -1.1×10^{-13}, which is negligible, as are collisions with ions and neutral molecules. Also, Stark shifts of the hyperfine levels are too small to be observable, since the maximum electric field in this experiment was 21 V/cm, and the rms field is much lower.

3.4. Possible Applications as Frequency Standards

The inherent possibilities of narrow resonance lines and very small line shifts in hyperfine transitions have led to early proposals for frequency standard applications.[45,46] The short-term frequency stability (Allen variance $\delta f / f$) of this system, operated in feedback mode as a frequency standard, can be estimated. When the frequency of the microwave excitation is chopped between the steepest points of the Lorentzian curve, the Allen variance is given by

$$\frac{\delta f}{f} = \frac{3}{4} \frac{1}{Q} \frac{1}{\text{SNR}} \tag{25}$$

where Q is the quality factor of the resonance and SNR the signal-to-noise ratio. For the case of $^{199}Hg^+$ this leads to

$$\delta f / df = 1.5 \times 10^{-11} / \sqrt{t}$$

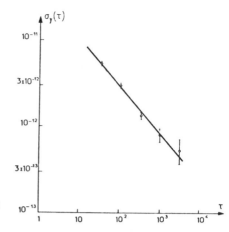

Figure 15. Frequency stability of a ^{199}Hg$^+$ ion trap (from Ref. 34).

The first experimental results[47,48] indicate that the frequency stability is comparable to commercially available frequency standards (Figure 15) and, with possible improvements in signal-to-noise ratio and elimination of the second-order Doppler effect, may well serve as a device comparable to primary cesium standards.

4. Lifetime of Metastable States

A necessary requirement for the determination of radiative decay times of metastable levels is the fact that the particle spends a time in the apparatus which is at least of the same order as the lifetime. During this time collisional deexcitation should not take place or has to be determined. These requirements limit conventional methods, like beam foil or Hanle signals, to cases where the lifetime does not exceed several milliseconds. Electromagnetic confinement of ions, however, offers storage times which may exceed several hours. The background pressure may vary over many orders of magnitude up to 10^{-12} mbar, thus allowing the variation of collisional quenching rates. The possible extension of the measurement of very slow decay rates is of basic interest in simple atomic systems like He$^+$ and Li$^+$ and may have metrological applications in heavy systems like Ba$^+$. The basic principles of the different measurements are similar: Because of the slow decay rate and the small number of trapped particles, it is usually impossible to observe the decay of the metastable state directly. Instead, one either monitors the number of particles remaining in the metastable state at a given time after the excitation, or one determines the ground-state population as a function of time, assuming that for a given total number of ions the remaining particles populate the metastable state, if no other long-living state is involved. An example of the first method is the determination of the $2\,^3S$ lifetime in Li^{+}.[49] The relevant energy levels are shown in Figure 16. Li^{+} was produced by electron impact from an atomic beam, and 2.10^5 metastable ions were stored in a rf quadrupole trap for about 20 sec at 10^{-9} mbar residual pressure. A cw dye laser excites the $2\,^3S_1-2\,^3P_1$ transition at 548.5 nm. The $^3P_1$ state, mixed by spin–orbit interaction to the $n\,^1P_1$ states, decays partly into the ground state by emission of a 20.2-nm photon. At a laser power of ~50 mW the entire metastable population is depleted in a time of 0.3 sec. By varying the delay time between creation and detection, the time dependence of the metastable population is measured. However, at a residual pressure of 10^{-9} torr the $2\,^3S_1$ decay is mainly determined by quenching collisions, the mean lifetime being ~5 sec. Variation of the background gas composition and pressure allows a separate determination of the radiative and the collisional decay. The final result is $\tau = 58.6 \pm 12.9$ sec, which has to be compared with the theoretical value of 49.1 sec.

In the case of the $5D_{3/2}$ state of Ba^{+},[50] the ions are excited by pulsed laser radiation tuned to the $6S_{1/2}-6P_{1/2}$ transition at 493.4 nm. Via spontaneous decay, the metastable $D_{3/2}$ state is populated (Figure 9). The P–D fluorescence at 694.8 nm is monitored and is a measure of the ground-state population. The metastable lifetime is determined in the following way: The first of the laser pulses at $t = 0$ finds all the ions in the ground state. Some of the excited particles decay within a few nanoseconds into the metastable state. Since the mean lifetime is large compared to the time between two consecutive laser pulses (20–100 msec), the second laser pulse finds less ions in the ground state and so on, until equilibrium is reached. The observed fluorescence intensity S_n decays accordingly. If n denotes the number of laser pulses, N the total number of ions, T the time between

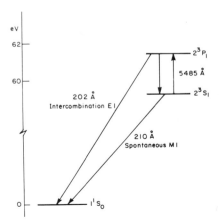

Figure 16. Level scheme of Li^{+}.

two laser pulses, γ the metastable decay rate, and c a constant, which contains the detection efficiency, the fraction of the stored ion cloud, which is excited by the laser pulse, and the P–S/P–D branching ratio, we find for the fluorescence intensity at the nth laser pulse,

$$S_n/N = (c \cdot e^{-\gamma T})^n + (1 - e^{-\gamma T}) \frac{1 - (ce^{-\gamma T})^n}{1 - ce^{-\gamma T}} \qquad (26)$$

For $n \to \infty$ this goes to the equilibrium value

$$S_\infty/N = (1 - e^{-\gamma T})/(1 - c\, e^{-\gamma T}) \qquad (27)$$

The decay rate γ is the sum of the radiative decay and the deexcitation by collision with neutral background atoms. Ion losses from the trap as additional relaxation mechanism can be neglected, since the storage time was many hours. P–D Stark mixing by the electric storage field contributes less than 1% to the observed decay rates. While the residual background pressure was kept below 10^{-9} mbar, He was used as a buffer gas. Figure 17 shows the variation of the fluorescence decay curves at different He pressures. The linear extrapolation to $p = 0$ leads to a radiative lifetime of 17.5 ± 4 sec. The theoretical value, using Coulomb approximation with spin–orbit interaction,[51] is 45.4 sec. A similar discrepancy was found in an experiment on Sr$^+$, which was almost identical to the one described above. Experimentally, an upper limit for the $5D_{3/2}$ lifetime of 120 msec has been found,[52] while the calculation predicts $\tau = 0.25$ sec.[51] Proposals exist[53] to use the extremely long lifetime of the Ba$^+$ $5D$ states for optical metrology by stabilization of a laser to the forbidden S–D transition, which would be of extremely narrow natural width.

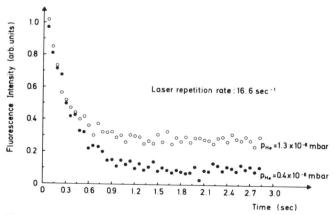

Figure 17. Decay of stored ion fluorescence at consecutive laser pulses. Decay curves for 2 He buffer gas pressures are shown. (From Ref. 50.)

5. Conclusion

Although the number of ions investigated using the ion storage technique has been small up to now, the method has already been proven to be a powerful tool for high-precision spectroscopy. Compared to measurements on ions in buffer gas atmospheres, the uncertainty has been reduced by more than two orders of magnitude and is now well below 1 part in 10^{10}. This is supported by recent experiments on Yb^{+} [55] and Be^{+} [56] which have been performed after the completion of this article. The use of tunable lasers for optical pumping will extend the number of investigated ions in the near future and has already found an extension to molecular ion spectroscopy.[54] The successful demonstration of laser sideband cooling to temperatures below 1 K indicates that one might be able to eliminate virtually all possible lineshifts in rf spectroscopy to the order of 10^{-15}. In addition, the sensitivity of the optical detection allows spectroscopy on a single particle, which opens applications to the investigation of fundamental processes in the interaction of light with atoms as well as the extension to very rare isotopes.

References

1. W. Paul, O. Osberghaus, and E. Fischer, Forschungsber. Wirtschafts- und Verkehrsmin. NRW, No. 415 (1958).
2. E. Fischer, Z. Phys. **156**, 1 (1959).
3. R. F. Wuerker, H. Shelton, and R. V. Langmuir, J. Appl. Phys. **30**, 342 (1959).
4. H. G. Dehmelt and F. G. Major, Phys. Rev. Lett. **8**, 213 (1962).
5. H. G. Dehmelt, Adv. At. Mol. Phys. **3**, 53 (1967); **5**, 109 (1969).
6. J. F. J. Todd, G. Lawson, and R. F. Bonner, in Quadrupole Mass Spectrometry (P. Dawson, ed.), Elsevier, Amsterdam (1967).
7. H. A. Schuessler, Progress in Atomic Spectroscopy (W. Hanle and H. Kleinpoppen, eds.), Part B, Plenum, New York (1979).
8. M. N. Benilan and C. Audoin, Int. J. Mass. Spectrosc. Ion Phys. **11**, 421 (1973).
9. R. S. van Dyck, P. B. Swinberg, and H. G. Dehmelt, Phys. Rev. Lett. **38**, 310 (1977).
10. R. Iffländer and G. Werth, Metrologia **13**, 167 (1977).
11. R. D. Knight and M. H. Prior, J. Appl. Phys. **50**, 3044 (1979).
12. H. Schaaf, U. Schmeling, and G. Werth, Appl. Phys. **25** (1981).
13. J. André, J. Phys. (Paris) **37**, 719 (1976); Proceedings of the 2nd Frequency Standards and Metrology Symposium, Copper Mountain, p. 325 (1976).
14. J. P. Scherman and F. G. Major, NASA Report No. X-524-71-343 (1971).
15. R. Iffländer and G. Werth, Proceedings of the 2nd Frequency Standards and Metrology Symposium, Copper Mountain, p. 347 (1976).
16. E. Teloy and D. Gerlich, Chem. Phys. **4**, 417 (1974).
17. K. H. Kingdon, Phys. Rev. **21**, 408 (1923).
18. R. H. Hooverman, J. Appl. Phys. **34**, 3505 (1963).
19. M. H. Prior and E. C. Wang, Phys. Rev. Lett. **35**, 29 (1975).

20. C. R. Vane, M. H. Prior, and R. Marrus, *Phys. Rev. Lett.* **46**, 107 (1981).
21. H. G. Dehmelt and F. Walls, *Phys. Rev. Lett.* **21**, 127 (1968).
22. W. Neuhauser, M. Hohenstatt, P. Toschek, and H. Dehmelt, *Phys. Rev. Lett.* **41**, 233 (1978).
23. D. J. Wineland and W. M. Itano, *Phys. Lett.* **82A**, 75 (1981).
24. H. G. Dehmelt and F. G. Major, *Phys. Rev.* **170**, 91 (1968).
25. H. A. Schuessler, E. N. Fortson, and H. G. Dehmelt, *Phys. Rev.* **187**, 5 (1969).
26. E. W. Weber, *Phys. Rep.* **32**, 123 (1977).
27. H. Ackermann, G. zu Putlitz, J. Schleusener, F. v. Sichart, J. Vetter, G. W. Weber, and S. Winnike, *Phys. Lett.* **44A**, 515 (1973).
28. R. H. Dicke, *Phys. Rev.* **89**, 472 (1953).
29. F. G. Major and J. L. Duchéne, *J. Phys. (Paris)* **36**, 953 (1976).
30. F. G. Major and G. Werth, *Phys. Rev. Lett.* **20**, 1155 (1973).
31. H. D. McGuire, R. Petsch, and G. Werth, *Phys. Rev. A* **17**, 1999 (1978).
32. M. Jardino, F. Plumelle, M. Desaintfuscien, and C. Audoin, *Proceedings of the 10th International Congress of Chronometry*, Genéve (1979).
33. Mitchel and Zemansky, *Resonance Radiation and Excited Atoms*, Cambridge Univ. Press, Cambridge (1961).
34. M. Jardino and M. Desaintfuscien, *IEEE Trans. Instr. Meas.* **29**(3) (1980).
35. R. Blatt and G. Werth, *Z. Phys.* **299A**, 93 (1981).
36. W. Becker, R. Blatt, and G. Werth, Proc. 1st ECAP, Heidelberg (1981).
37. D. J. Wineland and H. G. Dehmelt, *Bull. Am. Phys. Soc.* **20**, 637 (1975).
38. D. J. Wineland, R. E. Drullinger, and F. L. Walls, *Phys. Rev. Lett.* **40**, 1639c (1978).
39. W. Neuhauser, M. Hohenstatt, P. Toschek, and H. Dehmelt, *Phys. Rev. Lett.* **41**, 233 (1978); *Appl. Phys.* **17**, 123 (1978).
40. D. J. Wineland and W. M. Itano, *Phys. Rev. A* **20**, 1521 (1979).
41. J. Javanainen, *Appl. Phys.* **23**, 175 (1980).
42. W. Neuhauser, M. Hohenstatt, P. E. Toschek, and D. H. Dehmelt, *Phys. Rev. A* **22**, 1137 (1980).
43. W. M. Itano and D. J. Wineland, *Phys. Rev. A* **24**, 1364 (1981).
44. N. F. Ramsay, *Molecular Beams*, Oxford Univ. Press, London (1956).
45. F. G. Major, NASA Report No. X-521-69-167 (1969).
46. H. A. Schuessler, NSF Report No. PQ 19 166-000 (1969).
47. G. Werth, *Proceedings of the 10th International Congress of Chronometry*, Genéve (1979).
48. M. Jardino, M. Decaintfuscien, R. Barillet, R. Barillet, J. Viennet, P. Petit, and C. Audoin, *Appl. Phys.* **24**, 107 (1981).
49. R. D. Knight and M. H. Prior, *Phys. Rev. A* **21**, 179 (1980).
50. R. Schneider and G. Werth, *Z. Phys.* **A293**, 103 (1979).
51. S. Warner, *Mon. Not. R. Astron. Soc.* **139**, 115 (1968).
52. A. Osipowicz, Diplomarbeit Mainz, 1980 (unpublished).
53. W. Neuhauser, M. Hohenstatt, P. E. Toschek, and H. Dehmelt, in *Spectral Line Shapes*, (B. Wende ed.), p. 1045 (1981).
54. F. J. Grieman, B. H. Mahan, and A. O'Keefe, *J. Chem. Phys.* **72**, 4246 (1980).
55. R. Blatt, H. Schnatz, and G. Werth, *Phys. Rev. Lett.* **48**, 1601 (1982).
56. W. M. Itano, D. J. Wineland, H. Hemmati, J. C. Bergquist, and J. J. Bollinger, *8th Intern. Conf. on Atomic Phys.*, Göteborg (1982).

6

High-Magnetic-Field Atomic Physics

J. C. Gay

1. Introduction

1.1. Historical Survey

In 1845, Faraday[1] discovered a connection between optics and magnetism, which up until that time had been two entirely distinct fields in physics. This was the magnetic rotation in atomic vapors of the plane of polarization of light traveling along a magnetic field, a process strikingly different from natural rotation because of the peculiar symmetries of the interaction with the magnetic field. Later, Lorentz[2] developed the classical theory of the normal Zeeman effect, discovered in 1896.[3] This was the object of many studies which stimulated the development of the technology of high fields[4,5] and led to the quantum mechanics of spinning particles.[6] Magnetic fields still continue to be one of the finest tools for investigating the atomic and molecular structure.[7–9]

The possible importance of diamagnetism for weakly bound electronic systems was first theoretically established by Landau.[10] The experimental evidence was found in 1939 by Jenkins and Segre[11] on sodium atoms, and the existence of a Landau regime in the exciton spectrum was demonstrated in the 1950s.[12] In 1969, Garton and Tomkins[13] first demonstrated, on barium, the existence close to the ionization limit of a new signature of the atomic spectrum which was neither the Coulombic nor the Landau

J. C. Gay • Laboratoire de Spectroscopie Hertzienne de l'Ecole Normale Supérieure, Associé au CNRS LA 18, Tour 12-E01-4, place Jussieu, 75230 Paris, Cedex 05.

177

one. This was later called by A. R. P. Rau[14] the "strong field mixing regime" and was the origin of new experimental and theoretical developments on diamagnetism. The strong field mixing regime forms the main subject of this review, though some aspects of Zeeman studies at strong fields are also discussed.

1.2. Outlook on the Production of High Fields

Two classes of devices which are presently used have been developed by Kapitza[4] and Bitter.[5]

1.2.1. Production of DC Fields

Superconducting magnets allow the production of fields up to 25 T. Bitter and Wood coils[15,16] allow the production of fields up to 30 T, but they require huge electrical intensities and cooling with high-pressure water flow. This needs special installations such as those developed at Bitter Magnet Laboratory (MIT), Braunschweig (RFA), and SNCI (Grenoble).

1.2.2. Production of Pulsed Magnetic Fields

The practical limitations of the previous methods are due to the huge magnetic pressure which is exerted on the low mechanical resistivity copper disks of the coils. Considerable increase of the field strength to 60 Tesla is possible in pulsed operation due to limited heating and constraints on the coils. Second, imploded fields may be produced through trapping of the flux of a classical coil in a conductive cylinder, the radius of which is reduced by implosion through the action of electromagnetic forces or classical chemical exploders. Fields up to 2500 Tesla[17] have been produced for some microseconds.

1.3. Other Sources of Magnetic Fields

Some possibilities of reaching very large field values exist but have not yet been used. These are connected with accelerator techniques, such as passing a high-energy ionic beam through a macroscopically strong electromagnetic field. Fields up to 10^7 T can be produced in the rest frame of the beam. Highly inhomogeneous ultrahigh fields of up to 10^{13} Tesla could be generated by the internuclear motion in heavy ion collisions at MeV energies,[18] and magnetohydrodynamical processes in plasmas are responsible for fields of up to 10^8 T in pulsars.

1.4. Atomic Diamagnetism: A Cross-Disciplinary Problem with Wide Implications in Physics

The fundamental aspects of the problem of atoms in magnetic fields are connected with diamagnetism and basically with what happens to the motion of the electron when submitted to the joint actions of the Coulomb and magnetic fields. This is really an unsolved problem, while numerous experiments on atoms are renewing our current understanding of the phenomena.

The interest of the problem comes firstly from its fundamental simplicity, making it really a prototype of a wide class of unsolved questions. It is a nonseparable problem with two forces of different symmetries and of comparable strengths acting on the electron. Associated with this are the problems of stability of the classical trajectories, of proper semiclassical quantization, and of adiabatic invariants.[19-21] In addition, the two limiting cases are the only two exactly separable situations known in classical or quantum physics associated with Bohr–Sommerfeld and Landau quantizations. Secondly, this problem has wide application in various branches of physics—in solid state physics of excitonic systems,[12] in surface physics of two-dimensional electron layers,[14] in astrophysics of pulsars and white dwarfs,[22,23] in heavy ion collisions,[18] in laser physics,[24] and of course in plasma physics and fusion.[25] Further developments in fundamental QED problems are expected. In all these domains, there is some need of a quantum solution to the problem of an atom in a magnetic field. The recent experimental progress in atomic physics in the field of Rydberg states is in that sense a major step towards this achievement.

2. Contents of the Review

A general presentation of the problem of atoms in magnetic fields is given in Section 3. Some applications of Zeeman scanning techniques at strong fields are described in Section 4. They essentially concern some striking contributions from anticrossing experiments to the theory of atomic and molecular spectra. Some applications to the study of atom–atom interactions are also examined. The remaining part of the review deals with fundamental developments on diamagnetism which do not necessarily imply the use of strong or ultrastrong magnetic fields as this problem has a wealth of strange and surprising aspects and applications. The Landau regime is not as well known as the atomic structure theory (see Chapter 1 of Part A) and is briefly described in Section 5 with some emphasis on radiative properties. In section 6 we present some of the theoretical aspects of diamagnetism concerning both laboratory and ultra-high-field conditions and a review of unsuccessful quantum theoretical attempts together with some prospects. A review of recent experiments on atomic Rydberg states is given in Section 7 and important consequences are deduced. Section 8 deals with astrophysical and QED applications, and applications of the ultra-high-field limit of the problem to excitons, electron traps, and surface physics. Then a rapid examination of the undeveloped situation in molecules and molecular ions is performed.

3. The Atom in a Magnetic Field

3.1. Hamiltonian and Symmetries

3.1.1. General Expression

For the sake of simplicity, we only consider the one- or two-particle problem in a B field. The Hamiltonian is then given by

$$H_0 = \frac{1}{2m}(\mathbf{p} - q\mathbf{A})^2 + V(r) \tag{1}$$

where $\mathbf{A}$ is the vector potential, V the Coulomb potential, and q the electron charge. We will use the notation $e = |q|$. The momentum $\mathbf{p}$ in the sense of Lagrange satisfies

$$\mathbf{p} = -i\hbar\mathbf{\nabla}_r \tag{2}$$

while the momentum of the velocity is

$$\mathbf{\Pi} = \mathbf{p} - q\mathbf{A}(r) \tag{3}$$

as expected from Eq. (1). Contrasting to the $\mathbf{\Pi}$ operators, the $\mathbf{p}$ do not have in general any physical meaning but are proper artefacts for the calculations.[26,27]

3.1.2. Field-Dependent Terms in the Hamiltonian

Upon expanding (1) one obtains two classes of field-dependent terms, one linear, the other quadratic in $\mathbf{A}$. They are

$$H_z = \frac{e}{2m}(\mathbf{p} \cdot \mathbf{A} + \mathbf{A} \cdot \mathbf{p}) \tag{4}$$

and

$$H_D = \frac{e^2}{2m} A^2(r) \tag{5}$$

The first term is the well-known Zeeman Hamiltonian while the second one is the diamagnetic term. Usually the gauges are transverse satisfying $\nabla \cdot \mathbf{A} = 0$, and allowing straightforward simplification of (4).

3.1.3. Choices of Gauges for a Uniform Field

Practical choices of gauge are restricted to the Landau one satisfying, e.g.,

$$\begin{aligned} A_x &= -B \cdot y \\ A_y &= A_z = 0 \end{aligned} \tag{6}$$

and to the symmetric gauge

$$\mathbf{A} = \tfrac{1}{2}(\mathbf{B} \wedge \mathbf{r}) \tag{7}$$

These choices lead obviously to different expressions of H_z and H_D and to different mathematical expressions of the constants of motion, though of course the physical results are unchanged. The wave function expressions in (6) and (7) differ only in the unessential $\exp[iq(B/2)xy]$ phase factor. The Landau gauge obviously exhibits a clear translational invariance apparently unsuitable to atomic studies such as in Zeeman effect.

3.1.4. Symmetric Gauge Expressions

The choice of (7) seems more appropriate for atomic physics applications as it exhibits a clear rotational symmetry. Writing $\mathbf{L}$ the angular momentum in the sense of Lagrange

$$\mathbf{L} = \mathbf{r} \wedge \mathbf{p} \tag{8}$$

one obtains the expression of the Zeeman Hamiltonian

$$H_z = \frac{e}{2m} \mathbf{B} \cdot \mathbf{L} \tag{9}$$

which is a constant over the whole spectrum whatever the degree of excitation of the electron. The diamagnetic term takes the form

$$H_D = \frac{e^2 B^2}{8m}(x^2 + y^2) \tag{10}$$

and, obviously, strongly depends on the degree of excitation of the electron.

The angular momentum associated with the velocity is

$$\mathbf{l} = \mathbf{r} \wedge \mathbf{\Pi} \tag{11}$$

and does possess an obvious physical meaning.[27]

3.1.5. Nonseparability of the Two-Particle Problem

The question of the separability of the equations in the center-of-mass frame has been the subject of many investigations. As shown by Lamb,[28] the two-particle problem in a magnetic field is really nonseparable since the *energies* and *variables* of the reduced particle and of the center of mass are always coupled.[29,30] But some kind of separability exists, the consequences of which are not completely explored as demonstrated by the numerous contradicting papers[31-33] dealing with the effect of the proton rest mass on the quasi-Landau spectrum.

The Hamiltonian for the two-particle system is given by

$$H = \frac{1}{2m_1}(\mathbf{p}_1 + e\mathbf{A}_1)^2 + \frac{1}{2m_2}(\mathbf{p}_2 - e\mathbf{A}_2)^2 + V_{12}(r) \tag{12}$$

Using, for instance the classical equation of motion,[29] one can show that the operator

$$\mathbf{P} = \mathbf{p}_1 + \mathbf{p}_2 + e(\mathbf{A}_1 - \mathbf{A}_2) + e\mathbf{B} \wedge (\mathbf{r}_2 - \mathbf{r}_1) \tag{13}$$

is a constant of motion. Then, the velocity of the center of mass is no longer a constant of motion, and $\mathbf{P}$ depends on both the center of mass and the reduced coordinates.

In the reduced coordinates and in the symmetric gauge, the expression of $\mathbf{P}$ which does not coincide with the momentum for the two-particle system in the magnetic field is just

$$\mathbf{P} = -i\hbar\nabla_{\mathbf{R}} - \frac{e}{2}\mathbf{B} \wedge \mathbf{r} \tag{14}$$

Performing a translation on the wave function

$$\Psi_{\mathbf{P}}(\mathbf{r}_1, \mathbf{r}_2) = \exp\left\{i\left[\mathbf{P} + \frac{e}{2}(\mathbf{B} \wedge \mathbf{r})\right]\frac{\mathbf{R}}{\hbar}\right\} \cdot \Psi_{\mathbf{P}}(\mathbf{r}) \tag{15}$$

and upon substitution in (12), one obtains

$$H_{\mathbf{P}} = \frac{p^2}{2\mu} + \frac{e}{2\mu}\beta \cdot \mathbf{B} \cdot \mathbf{L} + \frac{e^2 B^2}{8\mu}(x^2 + y^2) + \frac{e}{M}(\mathbf{P} \wedge \mathbf{B}) \cdot \mathbf{r} + V_{12}(r) + \frac{P^2}{2M} \tag{16}$$

where $\mathbf{p}$ and $\mathbf{L}$ are the canonical operators associated with the reduced particle of mass $\mu = m_1 m_2/(m_1 + m_2)$ and $\beta = (m_2 - m_1)/(m_2 + m_1)$. In (16) $\mathbf{P}$ is the eigenvalue of the operator (14). Then equation (16) is that of a particle of mass μ, of gyromagnetic factor $\beta \cdot (e/2\mu)$ evolving in a magnetic field of strength $\mathbf{B}$ and in a motional Stark field $(e/M)(\mathbf{P} \wedge \mathbf{B})$ crossed to $\mathbf{B}$. The energies associated with the motion of the center of mass and the motion of the reduced particle are no longer conserved independently, which means that the problem is nonseparable, though soluble.[30] Complete solution of the one-electron atom problem in a magnetic field really needs the solution of a one-particle problem in crossed electric and magnetic fields. Complete solution of (16) is needed to conclude about the effect of the proton rest mass on the quasi Landau spectrum.[34]

3.1.6. Symmetries and Constant of the Motion

We now consider the mass of the nucleus to be infinite.

The only constants of motion are L_z and parity, and (S^2, S_z) since the spin is not involved in the diamagnetic term. In comparison to the usual atomic situation, l^2 and the Lenz vector components no longer commute with H.

L_z is a constant of motion in the symmetric gauge but not in general in the other gauges as is readily seen in the Landau gauge, for example with the Hamiltonian

$$H = \frac{p^2}{2m} + eByp_x + \frac{e^2 B^2}{2\mu}y^2 + V(r) \tag{17}$$

This is due to the fact that the expression for L_z in the symmetric gauge just coincides with that of the operator M_z which is a constant of motion and gauge invariant. The expression for M_z is just

$$M_z = l_z - \frac{eB}{2}(x^2 + y^2) \tag{18}$$

where l_z is defined through (11). M_z and the expression for L_z in Landau gauge no longer coincide.

3.2. The Various Magnetic Regimes in Atomic Spectra

Corresponding to the various terms in (1), there are special characteristics and alterations of the atomic spectra and approximate sets of constants of motion.[26]

3.2.1. Zeeman Regimes

For low-lying excited states, the diamagnetic term is negligible. The dominant term is then the Zeeman one which is proportional to B and small compared to the Coulomb interaction. For a many-electron system, various regimes exist, following the respective sizes of the Zeeman term and of the fine or hyperfine Hamiltonians. Some nonlinearities of the pattern as a function of B occur which are associated with partial decoupling of $(\mathbf{L}, \mathbf{S})$ or $(\mathbf{I}, \mathbf{J})$. All these regimes, Zeeman, Paschen–Bach, and Back–Goudsmit have been extensively studied and do not have any mystery. They are the basis of the analysis of Section 4.

3.2.2. Diamagnetic Regimes

The more fundamental aspects of the problem are connected with diamagnetism. The diamagnetic Hamiltonian depends strongly on the degree of excitation of the electron's motion and is of the order of

$$H_D \sim \frac{e^2 B^2}{8m} n^4 a_0^2 \tag{19}$$

It will then totally dominate the Zeeman contribution in high Rydberg states and, in some conditions, the electrostatic contribution to the binding energy.

Four characteristic regimes of diamagnetism have been recognized, though they are really the expression of the same feature.

The inter-l mixing regime[26] occurs for

$$\frac{2R}{n^3} \gg H_D \sim \frac{e^2 B^2}{8m} n^4 a_0^2$$

l^2 is not a constant of motion, while the radial Coulomb quantum number is still defined. Energies behave as B^2 with some deviations associated with quantum defects in nonhydrogenic situations.

In the inter-n mixing regime for which $2R/n^3 \sim H_D$, the atomic spectrum is completely altered through mixing of the various adjacent manifolds.

The third regime is the "strong field mixing regime"[14] when the electrostatic and magnetic contributions to the energies are comparable and $2R/n^3 \sim \hbar\omega_c$.[34]

Finally, when the Lorentz force completely overwhelms the Coulomb one, far into the continuum, the Landau regime of the atomic spectra is reached. This is discussed in Section 5.

Due to the diamagnetic interaction, the atomic spectrum in a magnetic field has signatures which are, successively, Coulomb- or Landau-like or some intermediate mixture of these. The passing from one regime to the other is the subject of a part of this review.

4. Applications of Zeeman Effect at Strong B Fields

We will, in this section, focus our attention on some special applications of strong B fields in the investigation of simple atomic and molecular structures and in the analysis of the dynamics of atom–atom interactions. Only classical aspects of the Zeeman effect are involved.

4.1. High-Field Anticrossing Experiments

Anticrossing (AC) spectroscopy is one of the simplest, and earliest developed, applications of Zeeman techniques to the determination of parameters of the atom,[9,35-38] either isolated or interacting with other atoms[39] or with the radiation field.[40] Together with the Hanle effect[7] and with optical pumping[8] this formed the basis for investigations in atomic physics before the dye laser era.

4.1.1. Generalities on AC Spectroscopy

The main features and general theory of AC spectroscopy have been described in Chapter 13 of Part A. Here we will recall only the leading ideas and emphasize the particular characteristics of the studies at strong fields.

The AC technique at strong fields turned out to be especially useful in the last decade for solving difficult problems of atomic and molecular spectroscopy such as those connected with the positions of triplet states in the H_2 molecule,[41] or for high-precision measurements of energy intervals in helium atoms.[37] In both cases, the need for tunability and for high field strengths was evident as the energy intervals were of the order of 10 cm^{-1}, requiring the use of fields of about 10 to 20 T. In parallel with the development of new spectroscopic techniques, the anticrossing method is also converging toward the study of the diamagnetism in Rydberg states.

The elementary theory of AC signals can be derived using a two-state approximation of the problem. These two states $|a\rangle$ and $|b\rangle$ are eigenstates of an unperturbed hamiltonian H^0. The associated energies $E^{a,b}$ have a crossing point at $B = B_0$. Usually, the states are coupled due to a symmetry-breaking interaction V which may be a small perturbation to H^0 or an effective potential deduced from a dressed atom formalism.[40] Then the energy levels at B_0 anticross and the eigenstates of H are mixed combinations of $|a\rangle$ and $|b\rangle$. V may be an internal perturbation due to the peculiarities of the atomic structure (spin–orbit coupling–Fermi contact interaction) or an external one associated, e.g., with motional stark field.[42] The relationship with collision theory (Landau–Zener problem, hypergeometric models, diabatic and adiabatic basis, e.g.) is obvious.[43–46]

A general analysis, including optical excitation and detection of the AC signals, involves the use of a density matrix formalism with a master equation[47] for describing the evolution under spontaneous decay, optical pumping, collisions, and multiple scattering.[48] After polarization averaging the total optically detected signal is given by[37,49]

$$I = I_0\left(\gamma_a\lambda_a\beta_a + \gamma_b\lambda_b\beta_b - \frac{2\bar{\gamma}(\lambda_a/\gamma_a - \lambda_b/\gamma_b)(\beta_a/\gamma_a - \beta_b/\gamma_b)|V_{ab}|^2}{4|V_{ab}|^2 f_\gamma + \hbar^2\bar{\gamma}^2 + (E_a - E_b)^2}\right) \quad (20)$$

The γ_i's are radiative decay probabilities and β the quantum efficiency. $\bar{\gamma}$ is the mean value of the radiative decay rates and $f_\gamma = \bar{\gamma}^2/\gamma_a \cdot \gamma_b$. The field varying part of the signal depends on the differences of the steady state populations of levels $|a\rangle$ and $|b\rangle$. The observation of AC signals then requires V_{ab} to be nonzero and several other conditions which have been extensively discussed.[37,49,50]

The main characteristics of the AC signals in the simple situation of (20) are the width and amplitude of the Lorentzian. The width is given by

$$\Delta B \, \alpha \, (4|V_{ab}|^2 f_\gamma + \hbar^2\bar{\gamma}^2)^{1/2} \quad (21)$$

with an obvious similarity to magnetic resonance signals. Various regimes can then occur—a saturation one for which the amplitude does not depend on $|V_{ab}|$, a sub-Doppler regime when $\hbar^2\bar{\gamma}^2 \gg 4|V_{ab}|^2 \cdot f_\gamma$ for which the width is just proportional to $\bar{\gamma}$. Of course, each situation needs a proper discussion both with respect to the atomic structure and to experimental conditions.

4.1.2. Magnetically Tuned Anticrossing Experiments

We will discuss some of the common features of the apparatuses and selected results obtained with this technique.

(a) *Sketch of the Experimental Setup.* A typical setup[51] is shown in Figure 1. The sources of magnetic field are Bitter coils producing fields of up to 20 tesla. The inhomogeneity is one part in 10^5 in a 1-cm-diam sphere. Control of the field is achieved with the use of locking loops connected to Hall or NMR probes. The production of excited atomic or molecular species is achieved through a discharge or through electron bombardment in sealed or pumped cells. The optical AC fluorescence signals are monitored with a combination of optical filters, spectrometer and photomultiplier, and computer processed.

Severe limitations and problems of this technique are connected with the instabilities of the discharge in the magnetic field. Gas pressure and electric field values are then limited by the requirements of stability and are not always compatible with the optimum conditions for obtaining accurate results on "free atom" structures. Alternatively, it may be more convenient to use electronic bombardment where the perturbations are smaller, but the AC signals are somewhat weaker than in discharge conditions.

(b) *Atomic Anticrossings—The Helium Atom.* Experiments at strong fields can be seen as the continuation of those initiated by Dilly and Descoubes[52] in low, highly homogeneous fields. The experiments have allowed important accurate determinations of the atomic structure for the simplest of many-electron systems for which abundant literature exists.[53-55] We will recall here the basic theoretical features of helium which are needed within the scope of this review.

(i) Singlet–triplet systems of energy levels. The spectrum is organized into two ladders associated with singlet ($S = 0$) and triplet ($S = 1$) states. Due to the Pauli principle the ground state is a singlet state. Spin–orbit interactions between the triplet and singlet states are always small correc-

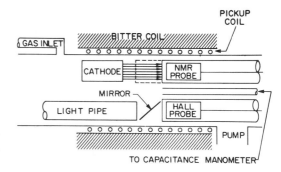

Figure 1. Typical setup for anticrossing experiments (Refs. 37, 51) at strong fields.

tions, and for a given (nL) the energy difference between them is essentially due to exchange interaction. The relative position of the two ladders is known with poor accuracy from some intercombinations lines, one of the most famous being the $\lambda = 591$ Å.

(ii) The singlet–triplet nD levels provide us with a situation for which the AC technique allows accurate determinations of the spin–orbit interaction, V_{so}, of the mixing of the wave functions and of the zero field energy difference.[37] The symmetry-breaking part of the Hamiltonian is the V_{so} interaction which does not commute with S. Four anticrossings are then allowed between Zeeman substates of the singlet and triplet manifolds. They occur at exactly the same field value if one neglects the role of second-order corrections to the Hamiltonian which are extensively discussed in Refs. 51, 56, and 57.

(iii) Anticrossings on $n\,^1D$ and $n\,^3D$ states which are unequally populated by electronic bombardment are easy to detect and appear as an increase of the optical fluorescence from $n\,^3D$ levels or a decrease on the $n\,^1D$ levels. Typical signals from the $3D$ states are shown in Figure 2. Experiments have been performed on the whole ensemble of $n(^1D-^3D)$ couples with $n < 20$.[37,50,51,57] Several experiments have also been carried out on ^{3}He in which the hyperfine structure is clearly resolved.[37,50]

Basic information on the atomic structure has been deduced. The positions of the anticrossings combined with some theoretical analysis of second-order effects determine quite accurately the zero-field separations of the levels. These are listed in Table I of Ref. 37. The linewidth gives the value of the spin–orbit interaction between singlet and triplet states. One then straightforwardly deduces the purity of the states in zero field. The technique is especially accurate for the determination of the $n(^1D-^3D)$ spacings, as it is Doppler free. The intervals have been shown to follow with n, the theoretical model of Chang and Poe.[54] Nevertheless, the accuracy depends on the need and importance of second-order corrections. In addition, the absolute energy of each level cannot be accurately related to that of the $2\,^3S$ and $1\,^1S$ ground state of each ladder as this involves optical measurements with only Doppler accuracy. Doppler-free two-photon techniques will probably help to overcome these limitations,[58] allowing absolute calibrations.[59]

(iv) Extensions of the method to other atomic states, either in helium or in other atoms, are limited. For P states the energy spacings greatly exceed the field possibilities, while for F, G, and H states the simplicity of the analysis for D states no longer holds. Other classes of anticrossings exist such as those due to spin–spin interaction.

Of course, this unique series of high-accuracy measurements (see also Section 4.1.3) is now to be complemented by several classes of high-resolution experiments,[55] especially those involving Doppler-free two-photon spectroscopy.

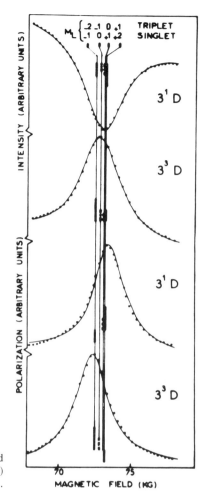

Figure 2. Typical signals of the total and polarized fluorescence intensities from the $n = 3(^1D-^3D)$ anticrossing of energy levels in helium (Ref. 51).

(c) *Molecular Anticrossings. H_2 Molecule.* Anticrossings between states of different multiplicities are possible for simple molecules in exactly the same way as for He. The best illustrations of this is the simplest neutral molecule H_2. The experiments on H_2 are among the greatest achievements of the anticrossing method in settling over 40 years of controversy[41] over the relative positions of the triplet manifold and of the singlet ground state which resulted in values of the excitation energy differing by hundredths of cm^{-1}.[37,60] The anticrossing experiment of Jost and Lombardi[41] in 1974, definitely solved the problem in a rather elegant and simple manner.

(i) A striking analogy exists between the singlet–triplet structure of He and that of H_2 molecules. The main difference, of course, lies in the axial symmetry of the electrostatic interaction and in the rotational and vibrational energy structures of the levels, though to a first approximation, one can assume that the mean internuclear distance is fixed at values of the order of those of the H_2^+ ground state.

(ii) A large number of anticrossings between singlet and triplet states exists for which the zero field spacings are smaller than 10 cm^{-1}. This is especially true for rotational states of the $3d$ configuration. Selection rules are then $\Delta S = 1$; $\Delta \Lambda = 0$; $\Delta v = 0$; $\Delta N = 0$. Spin–orbit and hyperfine Fermi contact interactions play an essential role in the AC experiments performed on H_2 and D_2, though a wide variety of phenomena can occur in molecules.[61,67] A typical diagram[37,68,69,70] in Figure 3 shows these two categories of AC signals for the $(G(3d)\ ^3\Sigma_g^+, G(3d)\ ^1\Sigma_g^+)$ pairs of singlet–triplet states of the para-D_2 molecule together with the recorded intensities from the singlet and triplet states. Of course, following the nature of the perturbation, selection rules differ considerably, as is exemplified in the figure.[37,50]

(iii) The extracted data for H_2 and D_2 are essentially the zero-field spacings between triplet and singlet states, Landé factors and the strengths of the interactions from the widths of AC signals. For example, in Figure 3, the Fermi contact interaction, being weaker than V_{so}, is responsible for the sharper lines. Numerous experiments have been performed on H_2[59,67] and D_2[68,70] each of which yields a value for the excitation energy of the lowest triplet state which is now known with an accuracy of 0.2 cm^{-1}.

(d) *Other Molecular Studies.* AC studies on H_2 have opened the way to a wide extension of the method to molecular systems, allowing important progress in the determination of structures, though the analysis is far more difficult than in the atomic spectrum. Singlet–triplet AC have been observed in the He_2 molecule,[71] and doublet–quartet AC in CN,[72] NO,[73] and the O_2^+ ion,[50,73] some of them exhibiting double structures associated with Λ-type doubling. Some experiments on HD and $^3He_2$ are as yet unsuccessful.[50]

Molecules are certainly the future of this technique, but it is outside the scope of the review to discuss this in greater detail.

4.1.3. Electric Resonance and Motional Stark Field Spectroscopy

An inherent feature of vapor phase experiments at strong fields is the existence in the atom rest frame of the motional Stark field as shown in Eq. (16), which in turn produces a velocity-dependent Stark effect in the atomic structure. This was earlier recognized as a powerful tool at low fields, for the analysis of H_2 dissociation fragments in electron bombardment experiments.[74] Recent developments at the MIT Bitter Magnet Lab make it really a new sub-Doppler tool for investigating the atomic structure.

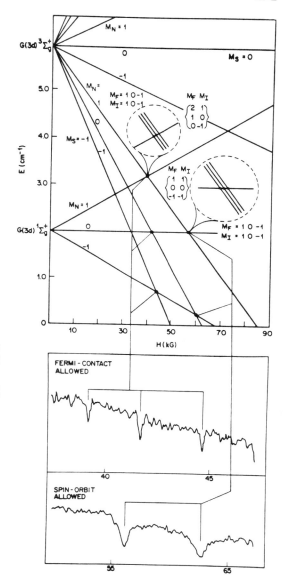

Figure 3. Anticrossings of the $G(3d)$ $^1\Sigma_g^+$, $v = 0$, $N = 1$ and $G(3d)$ $^3\Sigma_g^+$, $v = 0$, $N = 1$ states in the para-D_2 molecule (Refs. 37, 68). At the top of the figure is the Zeeman energy level diagram. The inserts show the details of the hyperfine structure which is not resolved in the experiments. The recordings (bottom) have been obtained by monitoring the $^1\Sigma_g^+$ optical emission. The signals due to the Fermi contact interaction are narrower than those due to V_{so}, indicating a smaller perturbation.

Similar features in the laser domain have also been demonstrated,[75,76] which involve the existence of velocity-dependent light shifts.

(a) *Basic Features of the Method.* The method is an extension of the basic ideas of rf magnetic resonance to electric resonance with lasers. Unfortunately, high-power IR lasers (as for example CO_2 lasers) are not tunable, though some considerable advance in the domain of IR tunable dye lasers have been made.[77-79] Levels have to be tuned to the optical resonance using a magnetic field, and most situations of interest require the use of high fields.

Figure 4 illustrates the main ideas of the technique.[37] The setup is conceptually quite similar to AC apparatuses, with the addition of a CO_2 laser as a radiation source. The CO_2 laser provides about 100 lines with a spacing of 35 GHz, and a field tunability of 15 T allows the observation of about 10 coincidences with the $7\,^1S \to 9\,^1P$ transition in helium.

The zero-field separations can then be determined with a rather good accuracy of 3×10^{-7}.[37] The transitions have Gaussian Doppler-broadened line shapes with a width of about 150 MHz.

In addition due to the mixing of states through motional Stark effect, the detection of forbidden transitions, such as $7\,^1S \to 9\,^1S, 9\,^1D$ is possible in fields. In contrast with the zero-field allowed electric dipole transitions, they have motional Stark field line shapes as shown in Figure 5. The line shape is characterized on one side by an abrupt cutoff of a width close to the radiative one and on the other side by a long exponential tail. These

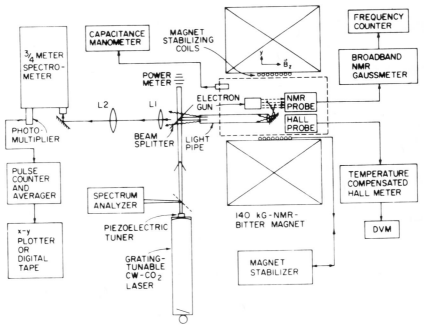

Figure 4. Diagram of the apparatus used (Ref. 81) for electric resonance experiments and sub-Doppler motional Stark field spectroscopy.

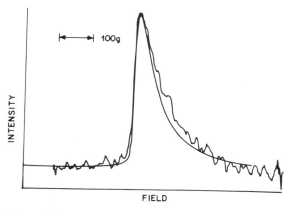

Figure 5. Motional Stark line shape for the $8\,{}^3D$–$8\,{}^1F$ anticrossing. The continuous curve has been computed following the hypothesis outlined in Ref. 82.

are the basic features of a sub-Doppler spectroscopic tool, though the exponential tail could be a nuisance.[80,81]

(b) *Selection Rules for Stark-Induced AC.* The pairs of levels observed in a Stark-induced anticrossing must differ in L by one. Selection rules for the motional Stark field perpendicular to **B** are then $\Delta M_L = \pm 1$. An obvious choice is the $8\,{}^3D$–$8\,{}^3F$ system in helium,[82] though the linewidth is quite large whatever the n value. A better choice is to study the AC signals forbidden in first order occurring as a conjunction of Stark mixing and fine-structure coupling. Such a pair of levels could be, for example, the $(M_S = 0;\, M_L = 0)\,{}^3F$ and $(M_S = 1;\, M_L = 0)\,{}^3D$ sublevels of helium which are mixed through spin–spin, spin–orbit, and Stark interactions. The widths of such AC signals are of the order of 10 to 100 gauss as the mixing is very small (10^{-3}). A wide variety of mechanisms is possible, e.g., mixed effects of diamagnetism and motional Stark field.[82–84]

(c) *Line-Shape Analysis.* In some situations,[75,80] the resonance conditions describing the absorption of a photon in the rest frame of the atom depends quadratically on the components of the atomic velocity. This is especially true under the action of the motional Stark field on the atomic structure, where one gets

$$\nu - \nu_0 = \frac{\nu_0}{c} \cdot v_y + k(v_x^2 + v_y^2) \qquad (22)$$

k being just proportional to B^2 and to the difference of the polarizabilities of the levels. The Doppler term is responsible for symmetric shifts while the quadratic one can produce an unidirectional frequency shift. The resonance conditions will not be fulfilled for all values of the velocity. Then a sharp cutoff will exist in the line shape, meaning that some of the atoms will never interact with the laser beam. The center of the line is shifted from ν_0 by a quantity of the order of $(1/k)\nu_0^2/c^2$. Of course, according to the sign of the difference of the polarizabilities, the cutoff will occur for frequencies lower or higher than ν_0.

The actual line shape is obtained by taking account of the velocity distribution of the atoms and the homogeneous Lorentzian line shape of width Γ. Usually, the cutoff will acquire a finite width of the order of Γ, while on the other side and far from the center of the line, the decrease will be exponentially slow.[79,82,84] Such a line shape is shown in Figure 6.

(d) *Experimental Results on Helium.* These classes of experiments have been performed with CO_2 lasers on the $4\,^3S$–$4\,^3P$; $7\,^1S$–$9\,^1P$; $8\,^3S$–$12\,^3P$ intervals in helium. Some of the determinations have achieved an accuracy

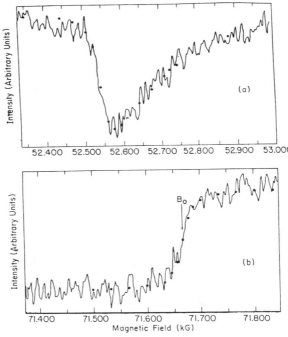

Figure 6. Some recordings of motional Stark line shapes in CO_2 laser resonance between the $7\,^1S$–$9\,^1P$ helium states (Ref. 81). The top curve (a) corresponds to the $7\,^1S$–$M_L = 0$, $9\,^1P$–$M_L = -1$ transition while the curve (b) corresponds to the $7\,^1S$–$M_L = 0$, $9\,^1P$–$M_L = 1$ transition.

of better than 10^{-6} as for example the $7\,^1S-9\,^1P$ zero field spacing.[37,82] The extension of the method to $n\,(^1S-^1P)$ intervals has also been performed.

The Stark-induced $(n\,^3D-n\,^{1,3}F)$ anticrossings have also been observed for $n = 4$ to 9. The results of measurements are in good agreement with theories of Chang and Poe,[54] and MacAdam and Wing.[85]

4.1.4. Conclusions

Anticrossing techniques at strong fields are therefore of importance for the simplest systems such as He and H_2 and have allowed major determinations. In helium the method is a useful complement to other techniques allowing confirmation to 1% of the variational Brueckner–Goldstone theories.[54] Better determinations in the future will probably result from the use of two-photon Doppler-free spectroscopy,[86] then overcoming the actual limitations of the AC spectroscopy where additional information is needed.

4.2. Dynamics of Atom–Atom Interactions

An important parameter in collision studies at thermal energies is the adiabaticity parameter X[45,87]

$$X = \omega \cdot T_c \tag{23}$$

where T_c is the mean duration of the collision or the correlation time of the interatomic potential. ω is a characteristic atomic frequency, Larmor frequency,[89] isotopic shift,[90–92] fine or hyperfine constant.[48,93,94] X is of course a statistical version of the well-known Massey parameter.[87] According to the value of X, the effects of the collision on an atomic excited state may be completely different:

- If $X \ll 1$, the sudden approximation is valid and the evolution due to ω during the collision is negligible.
- If $X \gg 1$, the evolution is quasiadiabatic for the observables associated with ω.
- If $X \simeq 1$, the atomic evolution takes place under the combined actions of the atomic Hamiltonian and of the time-dependent interaction potential which have different symmetries but are of comparable strengths. This results in considerable modifications of both the symmetries and magnitude of the relaxation matrix. When it is possible to vary the X value by several orders of magnitude, reliable comparison between experiments and theory may be made.

4.2.1. Role of the Magnetic Field in Collisions

The role of magnetic fields in collision processes at thermal energies is twofold and concerns both intensity and symmetry effects.

(a) *Intensity Effects.* The magnetic field appears as a convenient tool for the continuous variation of the energy difference between Zeeman substates. These aspects of the field action[89] can be described using the general formalism of relaxation theory,[95–97] specifically with the Fourier transform of the correlation function of the time-dependent interaction potential, the width of which, in frequency units, is just T_c^{-1}. Experimental and theoretical studies of the excitation transfer between Zeeman sublevels allow a sensitive test of the interaction potential. Of course, field tunabilities of several T_c^{-1} are needed, which in general means field strengths of several teslas.

(b) *Symmetry Effects.* The role of the field is also to break down the statistical isotropy of the relaxation processes,[98–100] which looks like a supersymmetry in the Liouville space. In the presence of the B field, the relaxation process still possesses rotational invariance around the field direction, and invariance *in the product* of time reversal and reflections in planes parallel to the field.[89,101,102] The breakdown of the zero-field supersymmetry results in new effects which can provide new information on the potential and on the mechanism of the collision process.

(c) *Anticrossings of Molecular Curves.* Long-range collisional interactions are usually of electrostatic nature and exhibit well-known molecular symmetries along the internuclear axes of the colliding atoms. The effect of the magnetic field is also to induce a breakdown of the previous Stark degeneracies which results in anticrossings of the quasi-molecular energy levels at some values of the internuclear distance. This is shown in Figure 7. The effects are large enough during the transient formation (10^{-12} sec) of the quasimolecule to induce important effects after statistically averaging over all the processes.[89,103,104]

In this sense, the present situation looks like that occurring in strong-field mixing regimes,[14] involving here a time-dependent Stark Hamiltonian and the Zeeman effect. The main difference with situations of atomic spectroscopy is that we detect in collision processes the modifications of the relaxation matrix. It is then a dynamical approach of the strong-field mixing regimes allowing only statistical determinations.

(d) *Basis of the Experimental Studies.* The principle is to perform selective optical excitation of one Zeeman sublevel of an atomic excited state using conventional light sources or tunable dye lasers.[103,105,106] Collisionally transferred excitation is recorded by monitoring the fluorescence intensities. The rates of transfer and variations with the field are then straightforward to deduce. The collision processes of interest are those involving long-range interactions. They give rise to large collision cross sections and collision times, compatible with significant studies using dc fields under laboratory conditions. In addition, theoretical models are simpler to set up in these conditions and are usually quite accurate descriptions.

Resonant Holtzmark collisions involve the interaction of an excited atom in a resonance level with a similar one, in the ground state.[107,112] The interaction potential is mainly dipole–dipole interaction. As the cross sections are large (10^{-12} to 10^{-13} cm^2), the collision times are long (10^{-11} to 10^{-12} sec) and fields of no more than 5 T are sufficient for a complete study of the correlation function.

Nonresonant collisions are the processes where an excited atomic species A collides with a buffer gas X. The interaction potential is not well known, the long-range part being usually R^{-6} van der Waals interaction.[48] Cross sections are smaller than 10^{-13} cm^2, meaning that huge dc field values are needed for a complete study of the correlation function, which is in general not possible under laboratory conditions. On the other hand, such situations allow the striking demonstration of the strong anisotropic behavior of the relaxation process and provide a way of determining the anisotropy of the interaction potential.[103]

4.2.2. Theoretical Approach of Thermal Collisions in a Strong Magnetic Field

General discussions of the properties of thermal collisions can be found in various review papers[48,112] and in Chapters 28 and 29 of Part B. Here we focus our attention on the role of a magnetic field in such processes, especially concerning symmetry properties and detailed balance. Situations for resonant collisions will be somewhat peculiar in comparison with the general case since the system is invariant under the exchange of atoms.

(a) *Symmetries of the Processes in **B** Fields.* In zero field, the collision matrix associated with a given collision process in space, defined by the impact parameter **b** and the velocity **v**, allows all collision matrices associated with other orientations of (**b**, **v**) to be deduced through straightforward rotation. This is no longer true in the presence of the **B** field. However, it is still possible to relate collision matrices associated with different collisions in space which give rise to simple properties of the relaxation matrix after angular averaging. More general quantal arguments are also valid and symmetries can be derived from rotational invariance around the field direction—invariance in parity operation, invariance *in the product* of reflection in planes containing the field direction and time reversal operation.[89,101] This last operation looks like a dynamical symmetry for the system.

(i) *Establishing the symmetries.* Derivations may be found in Refs. 89, 101, and 103, using both semiclassical approaches, as well as quantal treatment which is the only way of satisfying the laws of conservation of quantized momentum and energy in the external field.[113] The derivations assume that the interaction potential is of an electrostatic type and, therefore, invariant under time reversal and plane reflections around the internuclear axes. A general form of the interaction potential is

$$V(R) = f(R)(U^{(2)} \otimes T^{(2)})^{(0)} \qquad (24)$$

Where $U^{(2)}$ is a second-order rank tensor built on the external orbital variables (nuclear positions) and $T^{(2)}$ a second-order rank tensor built on the internal variables of the two-atom system.

The definition of the relaxation matrix and relaxation coefficients is, following Baranger's[99] notation,

$$g^{\alpha\alpha'} = \langle\!\langle \alpha | \langle M \rangle | \alpha' \rangle\!\rangle \qquad (25)$$

Where $|\alpha\rangle\!\rangle$ indicates the Zeeman dyadic or irreducible tensorial basis.

(ii) *Symmetries for nonresonant collisions.* The relations for A^*–X collisions are summarized in Table 1 for a $J = 1$ excited state. Further generalizations can be found in Ref. 89. The origins of the relations are successively (1) rotational invariance; (2) invariance in time reversal times plane reflections; (3) hermiticity of the density matrix; (4) unitarity of the collision matrix. These relations really assume that both the perturbations and the coupling with the surrounding are weak. Relation (5) connects the parameters of the relaxation for the situations in which the potential energy curves Σ and Π of the quasimolecule are inverted, but it has limited validity in practical situations.[104]

(iii) *Symmetries for resonant collisions.* The main difference comes from the additional invariance under exchange of atoms, leading to symmetrization and antisymmetrization of the results of Table 1. One then gets the

Table 1. Symmetries of the Relaxation Matrix in A^*–X Nonresonant
Collisions

$g_{qq'}^{kk'} = \delta_{qq'} g_q^{kk'}$	$g^{(rr')(pp')} = \delta_{r-r',p-p'} g^{(rr')(pp')}$	(1)
$g_q^{kk'} = g_q^{k'k}$	$g^{(rr')(pp')} = g^{(pp')(rr')}$	(2)
$g_q^{kk'} = g_{-q}^{kk'*}$	$g^{(rr')(pp')} = g^{(r'r)(p'p)*}$	(3)
$g_0^{0k} = g_0^{k0} = 0$	$\sum_p g^{(rr)(pp)} = 0$	(4)
$g_q^{kk'}(+) = (-)^{k+k'} g_q^{kk'}(-)$	$g^{(rr')(pp')}(+) = g^{(-r-r')(-p-p')*}(-)$	(5)

Table 2. Symmetries in A^*–A Resonant Collisions[a]

$g_{qq'}^{kk'}(i) = \delta_{qq'} g_q^{kk'}(i)$	$g^{(rr')(pp')}(i) = \delta_{r-r',p-p'} g^{(rr')(pp')}$ (1)
$g_q^{kk'}(i) = g_q^{k'k}(i)$	$g^{(rr')(pp')}(i) = g^{(pp')(rr')}$ (2)
$g_q^{kk'}(i) = g_{-q}^{kk'}(i)^*$	$g^{(rr')(pp')}(i) = g^{(r'r)(p'p)*}(i)$ (3)
$g_q^{kk'}(i) = (-)^{k+k'} g_q^{kk'}(i)$	$g^{(rr')(pp')}(i) = g^{(-r-r')(-p-p')*}$ (5)

[a] The $g^{(rr')(pp')}$ are matrix elements of the relaxation matrix in Liouville space and represent the transfer rates between the $|Jr, Jr'\rangle\rangle = |Jr\rangle\langle Jr'|^+$ and $|Jp\rangle\langle Jp'|$ Zeeman dyadic states. $g_q^{kk'}$ are also matrix elements of $\langle M \rangle$ but in a tensorial irreducible representation T_q^k.

relations of Table 2 with similar interpretations, but now (5) is a new symmetry property which limits the possible anisotropy of the relaxation.

(b) *Quantal Derivations and Detailed Balance.* Time reversal invariance does not exist in the presence of a magnetic field. Then the usual demonstration of detailed balance breaks down, but the demonstration of Messiah[114] turns out to be valid if one uses instead of the time reversal operator K, the operator $A = K \cdot \Sigma$, where Σ represents reflections in planes parallel to the field. Provided the coupling with the surrounding is weak, one obtains the following generalization, which is valid in a magnetic field[48,49]:

$$g^{(\alpha_b)(\alpha_a)} = e^{-[E(\alpha_b)-E(\alpha_a)]/kT} g^{(A\alpha_a)(A\alpha_b)} \tag{26}$$

Where (α_a), (α_b) are, respectively, the initial and final states of the systems and $(A\alpha_a)$, $(A\alpha_b)$ the results of their transformations in the operation A. $E(\alpha_i)$ is the energy associated with state (α_i). Equation (26) reduces to relation (2) of Tables 1 and 2 when $\hbar\omega \ll kT$.

At this stage, one must emphasize the fundamental difference which exists between these expressions and the usual expression of detailed balance in zero field, owing to the fact that the time reversal operation K and the plane reflections Σ are no longer separately proper symmetry operations for the process in the presence of fields. This means that the relations†

$$g^{(Jm)(Jp)} = g^{(Jp)(Jm)} = g^{(J-p)(J-m)} \tag{27}$$

for population transfer between Zeeman sublevels, valid in zero field, no longer holds in finite fields, where in general

$$g^{(Jm)(Jp)} \neq g^{(J-p)(J-m)} \tag{28}$$

† NOTE: The notation $g^{(Jp)(Jm)}$ is an abbreviation of $g^{(Jp,Jp)(Jm,Jm)}$ for the population transfer rate between the (Jp) and (Jm) Zeeman sublevels.

which is not a consequence of detailed balance as has been inferred by Baylis in Chapter 28 of Part B. The only relation generalizing detailed balance in the field is

$$g^{(Jp)(Jm)} = e^{-(p-m)\hbar\omega/kT} g^{(Jm)(Jp)} \tag{29}$$

which is far different from (28) even if $\hbar\omega \ll kT$.

Of course a quantal justification of other relations is straightforward.

(c) *Field Dependence of the Relaxation Matrix.* Previous analysis allows the solution of the problem of symmetries without any calculations and almost independently of the model adopted for the collision process.

Explicit dependences of the $\langle M \rangle$ relaxation matrix on the B field strength are obtained either by the use of a correlation function formalism in the general framework of the relaxation theory or by solving the semi-classical Schrödinger equation. This needs a large number of additional hypotheses about trajectories, close-range corrections to the potential, and so on,[89,101–103] though in general these are of little importance for resonant collisions involving R^{-3} dipole–dipole interaction.

(i) *Correlation function analysis.* Writing H_0 as the two-atom Hamiltonian and $V_{12}(t)$ as the time-dependent interaction potential, a perturbational solution of Schrödinger's equation produces the following expression of the collision matrix[114,115]

$$S = T(\exp[-i \int e^{iH_0 t} V_{12}(t) e^{-iH_0 t} dt]) \tag{30}$$

where T is the chronological time-ordering operator. One then deduces straightforwardly, after analytical angular averaging, the expression for the matrix elements of the relaxation operator in Liouville space which in general are just expressed in terms of the Fourier transform of the symmetric correlation function of the interaction potential[89,101,102,116]

$$f(\eta) = \int_{-\infty}^{+\infty} \int_{-\infty}^{+\infty} e^{i\eta(x-x')} \mathrm{Tr}\,[V(x)V(x')^+]\,dx\,dx' \tag{31}$$

Relaxation and transfer rates involve $f(\eta)$ while collisional shifts involve the Cauchy principal value of $f(\eta)$.[101]

Convenient cutoff procedures[115,117] allow us to obtain the rates in closed form after summation over impact parameters and velocity averaging.[101,102,116] For R^{-3} dipole–dipole interactions, for example, the cross sections are entirely expressed in terms of modified Bessel functions[89] depending on the detuning ω. The analogy with Fourier analysis is then obvious. Usually, the physical quantities of interest involve several competing physical processes[106,116] in the two-atom systems when the spin is nonzero. These processes are associated with different values of the detuning ω^i and have different rates of variation with the field. Scanning the field allows one to demonstrate the existence of these terms which are masked in zero field.[106]

(ii) *Exact semiclassical solutions.* The numerical solution of the Schrödinger equation allows more accurate predictions to be made in some special situations especially concerning anisotropy effects[101,102,104] for R^{-3} and R^{-6} interactions. But usually, for excitation transfer between Zeeman sublevels, a correlation function analysis at second order in the potential is sufficiently accurate.

(iii) *Striking manifestations of the anisotropy.* Usually, anisotropy effects appear at third order in the interaction potential and require an extension of Eq. (31). For example, with R^{-6} van der Waals interaction, one can show through perturbation theory that the excitation transfer rates from $m = 0$ to $m = \pm1$ are different proving the failure of (27), though (28) and (29) are still verified.[89] Equivalently, one proves the existence of a nonzero coupling term between longitudinal alignment and orientation in the vapor.[104]

A more precise understanding of these features can be obtained from a numerical study of the collision process. For a $J = 1$ atomic state, the associated quasi-molecular states are two (uncoupled) Π states and one Σ state. Because of the symmetries of the electrostatic interaction, the two Π states are degenerate for all internuclear distances, and the potential energy curves are symmetric under the $t \rightarrow -t$ operation. As shown in Figure 7, in the presence of a magnetic field, the fundamental Kramer's degeneracy no longer exists. The potential energy curves are now characteristic of a "strong field mixing regime" between the electrostatic and magnetic fields. A plot of the associated transition probabilities from $m = 0$ to $m = \pm1$ is shown in Figure 7 exhibiting the fundamental anisotropy of the process when the atom is strongly interacting with the two fields.[89,103] The results of calculations for R^{-6} van der Waals interactions are shown in Figure 8 as a function of the field.[102,104] The rates of transfer between $m = 0$ and $m = \pm1$ are always different and behave completely differently in the field. These are not consequences of the detailed balance as these coefficients are never related by any symmetry relation, except relation (5) of Table 1 which allows one to assert that the general dependences interchange when one shifts from one sign of the anisotropic part of the potential to the other. The situation of Figure 8 corresponds to $E_\Sigma < E_\Pi$ in zero field. This is then a way to determine the relative positions of E_Σ and E_Π. In contrast, for resonant collisions between identical atoms, the (u, g) separability of Schrödinger's equation ensures that the $m = 0 \rightarrow m = \pm1$ transfer rates are always identical.[101]

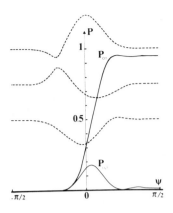

Figure 7. Correlations between transition probabilities and potential-energy curves during the collision. The situation corresponds to R^{-6} anisotropic van der Waals interaction with a sign such that $E_\Sigma > E_\Pi$ in zero magnetic field. The field belongs to the collision plane and is parallel to the second bisector plane of $(\mathbf{b}, \mathbf{v})$, where $\mathbf{b}$ is the impact parameter and $\mathbf{v}$ the relative velocity of the colliding atoms (Ref. 89). The strength of the field corresponds to a Zeeman splitting of about $2T_c^{-1}$, and the impact parameter is about four times the Weisskopf radius. At $\Psi = -\frac{1}{2}\pi$ $(t = -\infty)$ the system is in the $m = 0$ Zeeman substate. The $P_{0\pm1}$ full curves represent the probabilities of finding the system in the $m = \pm1$ substates during the collision. The dashed curves (arbitrary units) represent the eigenvalues of the time-dependent Hamiltonian. At $\Psi = \pm\frac{1}{2}\pi$ they correspond to the pure Zeeman case. Potential-energy curves, anticrossing partly, explain the strong value of P_{01} at the end of the collision process. (From Ref. 103.)

4.2.3. Experimental Results for Strong B Fields

(i) In convenient experimental conditions, the ratio of the detected fluorescence intensities is just proportional to the transfer rates g^{mm_0}/Γ.[118] The variation of the $m_0 = 3/2 \rightarrow m = -3/2$ transfer rate with the B field in resonant $Na^*(3^2P_{3/2}) - Na(3^2S_{1/2})$ collisions is shown in Figure 9.

A huge decrease of the cross section by a factor of 20 is exhibited (the zero-field cross section is about 8400 Å^2), illustrating the role of the adiabaticity in the transfer.[106,116] In addition, the very good agreement

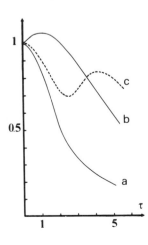

Figure 8. Anisotropy of the collision process in the field for R^{-6} van der Waals interaction (Ref. 89). Field variations (reduced to the zero-field value) of the $m = 0 \rightarrow m = \pm1$ and $m = \pm1 \rightarrow m = \mp1$ transfer rates. The parameter τ is just proportional to $Bp^{1/5}v^{-6/5}$, where p is the anisotropy parameter of the potential related to the product of the polarizability of the perturbers and to the mean value of $\langle r^2 \rangle$ for the $J = 1$ excited atom: (a) $m = \pm1 \rightarrow m = \mp1$, (b) $m = 0 \rightarrow m = -1$, and (c) $m = 0 \rightarrow m = 1$. The situation corresponds to a zero-field position of potential-energy curves such that $E_\Sigma < E_\Pi$. The inverse situation, $E_\Sigma > E_\Pi$ is obtained by interchanging only the signs of m in (b) and (c) because of the scaling laws for the problem. (From Ref. 103.)

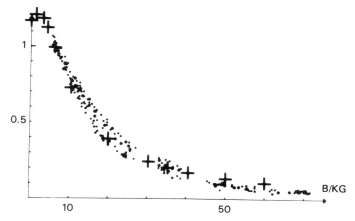

Figure 9. Variations with the field of the collisional transfer rate $m = -3/2 \rightarrow m = 3/2$ in resonant Na $(3\,^2P_{3/2})$–Na $(3\,^2S_{1/2})$ collisions (Refs. 106, 116). The theoretical values (+) are from a correlation function analysis (Ref. 89).

with a correlation function analysis allows us to confirm that the processes occur at 2ω and are associated with reversal of the spin during excitation transfer. This, then, allows a test of the form of the interaction and also of the electric dipole selection rules. This ensemble of results is extensively discussed in Refs. 89, 103, 105, 116, and 118.

(ii) The best illustrations of anisotropy effects are probably those obtained on Hg$(6\,^3P_1)$–rare gas collisions in fields of up to 20 tesla.[119] Optical excitation of the $m = 0$ Zeeman substate is performed and the fluorescence intensities reemitted from $m = +1$ and $m = -1$ substates are detected. This gives at low mercury densities (10^{10} at/cm^3) and low rare gas pressures ($\leqslant 1$ Torr)

$$I_{\sigma^-}/I_{\sigma^+} = g^{(-1-1)(00)}/g^{(11)(00)}$$
$$I_{\sigma}/I_{\pi} = (g^{(-1-1)(00)} + g^{(11)(00)})/2\Gamma \tag{32}$$

where Γ is the radiative decay rate of the $6\,^3P_1$ state of mercury.

Figure 10 clearly shows that the transfer rates behave *quite differently with the field* and that the *behavior strongly depends on the nature of the rare gas atom*. The qualitative understanding of the dependences shown in Figure 10 has been demonstrated in several ways.[89,102,104] Theoretical curves for Xe and Kr perturbers could be plotted and are in qualitative agreement with the respective experimental behavior, confirming that for heavy rare gas perturbers the interaction with Hg takes place at long range, where $E_\Sigma > E_\Pi$.

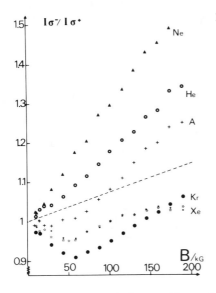

Figure 10. Anisotropy of the excitation transfer in Hg ($6\,^3P_1$)–rare gas collisions. Continuous optical excitation of the $m = 0$ Zeeman sublevel is performed in conditions where Hg densities are 5×10^{10} at/cm³ (Ref. 119). Optical detection of fluorescence intensities from the $m = 1$, 0, -1 Zeeman sublevels allows the determination of $I_{\sigma+}$, $I_{\sigma-}$, I_{σ}, I_{π}. The ratio $I_{\sigma-}/I_{\sigma+}$ is just $g^{(-1-1)(00)}/g^{(11)(00)}$. The results show that the transfer rates $g^{(11)(00)}$ and $g^{(-1-1)(00)}$ are always different for $B \neq 0$. They also behave very differently following the nature of the rare gas atom. The results are consistent with theoretical calculations for heavy rare gas atoms (Refs. 103, 119) and allow the determination of the sign of the anisotropic part of the potential in Hg*–X collisions. In addition, the model allows the qualitative explanation of the strange behavior observed for neon and helium (Refs. 89, 104) through the effect of close-range repulsive parts of the potential.

Obviously, these results have nothing to do with detailed balance (dashed curve) as expected and are equivalent to the creation of orientation from longitudinal alignment.

In contrast, the behavior with (He, Ne) indicates that $E_\Sigma < E_\Pi$ which means that short-range interactions are dominant for the light rare gas perturbers. In addition, the critical field will be stronger for He than for Ne in agreement with the low-field experimental behavior. Results for argon disagree with one or other of the predictions. The processes take place near the crossing of Σ and Π energy curves. The mean anisotropy of the potential is zero and $g^{(-1-1)(00)} \sim g^{(11)(00)}$. Then at higher fields, the role of the repulsive region again becomes dominant.[103]

These results clearly confirm the strong anisotropy of the relaxation in a magnetic field and allow the determination of the sign of the anisotropic part of the interaction potential.

(iii) Of course the previous features have nothing to do with the detailed balance which is expressed by Eq. (29). Experimental verification of the detailed balance has been performed in conditions of high mercury densities (10^{13} at/cm³) where total collisional redistribution occurs. $I_{\sigma-}/I_{\sigma+}$ is then no longer given by (32) but by

$$\frac{I_{\sigma-}}{I_{\sigma+}} = \frac{g^{(-1-1)(11)}(g^{(11)(00)} + g^{(-1-1)(00)}) + g^{(-1-1)(00)}g^{(00)(11)}}{g^{(11)(-1-1)}(g^{(11)(00)} + g^{(-1-1)(00)}) + g^{(11)(00)}g^{(00)(-1-1)}} \approx e^{-2\hbar\omega/kT} \quad (33)$$

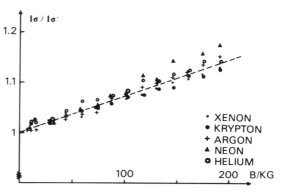

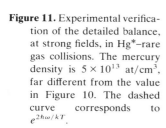

Figure 11. Experimental verification of the detailed balance, at strong fields, in Hg*–rare gas collisions. The mercury density is 5×10^{13} at/cm^3, far different from the value in Figure 10. The dashed curve corresponds to $e^{2\hbar\omega/kT}$.

which, as shown in Figure 11 is consistent with experimental results. Even though $g^{(-1-1)(00)}$ and $g^{(11)(00)}$ are different due to inner mechanisms of the collision processes, this is not detailed balance and does not violate obvious laws of physics.[119] Anisotropy effects and detailed balance are then two aspects of the action of the magnetic field on the vapor and are really different.

4.2.4. Conclusions

The use of strong magnetic fields allows the analysis of the behavior of the cross sections as a function of the energy difference between the levels and quasi-direct recording of the correlation function of the interaction potential, a test of the validity of the potential, of selection rules and symmetry properties.[89] Finally, this is a direct confirmation of the very special mechanisms of the action of the field in inelastic collisions. Of course, this is another aspect, from a statistical point of view, of the importance of the strong-field mixing concept.[14]

We will now examine the more fundamental aspects of the problem of atoms in a strong magnetic field, i.e., the aspects associated with diamagnetism.

5. Landau Regime for Loosely Bound Particles

We will recall here some of the aspects of the Landau regime concerning energies, wave functions, and interaction with the radiation field. They are of importance for the understanding of atomic diamagnetism for which it is one limiting case. The situation here is that the Lorentz force completely overwhelms all the other forces in the problem.

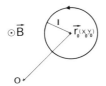

Figure 12. Constants of the classical motion for a "free" electron in a **B** field. These are basically the radius Γ of the circle, the position $r_0(x_0, y_0)$ of the center of the circle, and the velocity v_z along the **B** field.

5.1. Classical Mechanics Aspects

Classical trajectories are circular helices described at the cyclotron frequency ω_c. The constants of the classical motion are indicated in Figure 12 and are the coordinates (x_0, y_0) of the center of the circle, the distance r_0, the radius Γ of the circle, and the angular momentum l_z defined through Eq. (11).[27] The motion along B is free and decoupled from the transverse motion. Ways of generating the various forms of the Jacobi classical functions are given in Ref. 120.

5.2. Quantum Mechanical Aspects

The classification of the spectrum is more explicit using the symmetric gauge. The constants of the motion are readily deduced from their classical expressions. Defining a_c, the cyclotron radius, one gets

$$[x_0, y_0] = ia_c^2 = i\frac{\hbar}{qB} \tag{34}$$

meaning that the center of the trajectory is no longer defined. Among various possibilities, two compatible sets of observables are of interest. They are $\{H, L_z, r_0^2, \Gamma^2\}$ and $\{H, x_0\}$ or $\{H, y_0\}$. The first choice corresponds to wave functions with rotational symmetry while the second one corresponds to wave functions with translational invariance.

5.2.1. Fock Representations of Landau States

From the partial equivalence with a two-dimensional symmetric harmonic oscillator, one can define the set of Fock operators associated with the creation and the annihilation of elementary circular excitation.[27] That is

$$a_\pm = \frac{1}{\sqrt{2}}(a_x \pm ia_y) \tag{35}$$

where $a_{x,y}$ are the canonical definitions of the ladder operators for a one-dimensional oscillator.[27] One gets straightforwardly

$$H = (n_+ + \tfrac{1}{2})\hbar\omega_c + \hbar^2 k_z^2/2m$$
$$L_z = (n_+ - n_-)\hbar = M\hbar \qquad (36)$$
$$n_\pm = a_\pm^+ a_\pm$$

This shows that for particles with negative charges, the transverse part of the energy only depends on the number of right-handed circular excitations in the system, exhibiting an infinite degeneracy on n_- (or M). M varies from n_+ to $-\infty$ in a given Landau manifold. The associated spectrum is shown in Figure 13. A possible choice of the transverse part of the wave function is

$$|n_+, n_-\rangle = \frac{1}{[n_+ \cdot n_-]^{1/2}} (a_+^+)^{n_+} (a_-^+)^{n_-} |0, 0\rangle \qquad (37)$$

where $|0, 0\rangle$ is the ground state of the two-dimensional symmetric oscillator,[27] and the form of $\langle \rho, \phi | n_+, n_-\rangle$ is shown in Figure 13,[120] making evident the connections with the Jacobi functions.[120]

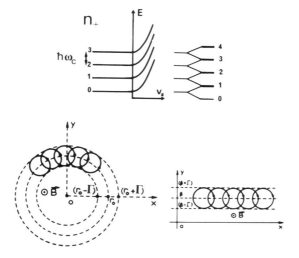

Figure 13. Pure Landau spectrum (with spin effects) and schematic representation of the wave functions with rotational and translational symmetries corresponding to two possible descriptions of the degeneracies in a Landau manifold.

5.2.2. Wave Function with Translational Invariance

The wave functions associated with the $\{H, y_0\}$ representation are easy to deduce from (37). Classification in a Landau band is done with the continuous parameter β which is the eigenvalue of y_0. Representation of such states is given in Figure 13. Their expression is just

$$|n_+, y_0\rangle = \sum_{n_-}^{\infty} i^{n_-} \phi_{n_-}^* (K\sqrt{2}y_0)|n_+, n_-\rangle \tag{38}$$

where ϕ_{n_-} is proportional to one-dimensional oscillator wave functions. Of course, such an expression appears quite naturally using the Landau gauge. L_z and parity are not defined in this representation.[121]

5.2.3. Other Choices

Other choices are possible. In a given Landau manifold one can build classical states in the sense of Glauber,[27,122] on left-handed circular excitations, for which the fluctuations $\Delta x_0 \cdot \Delta y_0$ take the minimum value $a_c/\sqrt{2}$. Quasiclassical states of the Landau spectra can also be built[27] using standard techniques for the two-dimensional harmonic oscillator.

Of course, all these results do not depend on the gauge and are still valid in the Landau gauge. Other descriptions are possible[123,124] though the present one is the more general and convenient.

5.3. Spin Effects

Both S^2 and S_z commute with the Hamiltonian. Energies in a nonrelativistic approximation are given by

$$E = (n_+ + \tfrac{1}{2})\hbar\omega_c + g_e M_s \frac{\hbar\omega_c}{2} + \frac{\hbar^2 k_z^2}{2m} \tag{39}$$

where g_e is the anomalous Landé factor for the spin. Degeneracies for spin-up and spin-down are removed due to this anomalous correction, while degeneracies with respect to M and parity are removed by the introduction of small perturbations.

5.4. Absorption and Emission of Radiation in the Landau Spectrum

This is an important question for atomic physics applications, though here we consider a limiting case of the real situation. The Hamiltonian of the interaction with the radiation field is just

$$H^{\text{int}} = +\frac{e}{m}\mathbf{\Pi} \cdot \mathbf{A}_R \tag{40}$$

where $\mathbf{A}_R$ is the vector potential of the radiation field.

5.4.1. Electric Dipole Selection Rules

Using Fock expressions for the wave functions and operators, it is straightforward to show that π transitions will not be possible inside a given (n_+, M) Landau band, as they would only affect the free longitudinal motion. In contrast, σ transitions will not affect the longitudinal motion. The only possibilities are transitions in which

$$(n_+, n_-, k_z) \rightarrow (n_+ + 1, n_-, k_z) \tag{41}$$

that is, the absorption of a photon of energy $\hbar\omega_c$, of polarization ε^+. The absorption or emission of ε^- photons is impossible in first order. Of course, such a feature is connected with the fact that the electron can only interact with waves rotating in the same sense around the B field. It is analogous to well-known features of magnetic resonance or propagation of helicon waves in plasmas.

One can conclude that the Landau spectrum has very anisotropic properties concerning the interaction with radiation.

5.4.2. M_1, E_2, and Two-Photon Transitions

Further development leads to the interaction Hamiltonian for M_1 and E_2 transitions, and not to Eq. (40). A large number of possibilities now exist, especially those involving mixed discrete–continuous transitions, but continuous transitions involving free motion along the B field are forbidden here.

Two-photon transitions also involve many processes, discrete, continuous, or mixed ones, but they are not of interest in general as they are competing with one- or two-step resonant transitions due to the degeneracies in the Landau ladder. Anyway, in the atomic physics situation, these degeneracies are removed and these processes will regain all their proper interest. Moreover, the longitudinal motion will no longer be free and continuous π transitions and mixed discrete–continuous ones can take place (see Section 6).

5.4.3. General Remark

The previous discussion implicitly supposes that one deals with a neutral system. If not, for charged systems such as the electron in a **B** field, the standard treatment based on multipolar expansions of the Hamiltonian does not apply. Of course, in such a case, making the electric dipolar approximation does not have any physical meaning. A convenient description of the interaction of the Landau electron with E.M. fields will be achieved considering the complete Hamiltonian

$$H = \frac{1}{2m}[\mathbf{p} - q(\mathbf{A} + \mathbf{A}_R)]^2$$

and using, e.g., perturbative approaches in $\mathbf{A}_R$ or convenient redefinition of the unperturbed basis. A first order treatment will imply to deal with a term equivalent to (40) in which the spatial dependance is $e^{i\mathbf{kr}}$, but cannot be reduced to 1 (the so-called "electric dipolar approximation" as used in Section 5.4.1). In the situations of interest, $e^{i\mathbf{kr}}$ is acting as a Glauber displacement operator D on the Landau wave function. Selection rules are no longer then as restrictive as described above.

Indeed, the derivations of the previous sections suppose that the Landau electron is tightly bound to an ionic core—the atomic physics conditions—so as to ensure the neutral character of the system and the existence of a "center." In addition, such an interaction must be weak enough and not strongly modify the wavefunctions and energy diagram described in Section 5.2. One then deals with a limiting case of the real ultra-strong field situation for atoms. It is likely that part of the properties described above will disappear due to the breakdown of the degeneracies in the Landau ladder when the electrostatic force is taken into account.

6. Theoretical Concepts of Atomic Diamagnetism

The interest of atomic diamagnetism and the wide generality of the implied physical mechanisms result from the experimental demonstration of the simplicity of the spectrum in situations where all intuitive arguments suggest, on the contrary, strong complexity. For example, in the Rydberg spectrum, near threshold, the action of both the Coulomb and magnetic fields may be of the same importance so that neither l^2, A_z (the z component of the Lenz vector), nor the operators x_0, y_0, Γ^2, r_0^2 defined in Section 5, are constants of the motion. Without an additional constant of the motion, the spectrum of the nonseparable Hamiltonian must be formidably complicated. Why it is not, is till now an elusive question which has received indirect answers through approximate semiclassical quantization of the motion.

6.1. Phenomenological Description of the Atomic Spectrum

In atomic units, the Hamiltonian in the symmetric gauge takes the form

$$H = p^2 + \gamma L_z + \frac{\gamma^2}{4}\rho^2 - \frac{2}{r} \tag{42}$$

with

$$\gamma = (a_0/a_c)^2 = \hbar\omega_c/2R = B/B_c \tag{43}$$

where a_0 and a_c are the Bohr and cyclotron radii, respectively. B_c is the critical field value of 2.35×10^9 gauss for which the Larmor frequency equals the Rydberg constant. In laboratory conditions for atomic physics one has $\gamma \ll 1$.

The various field-dependent terms in (42) are, respectively, the Zeeman one and the diamagnetic one ($\gamma^2\rho^2/4$). The Zeeman term is associated with the well-known Lorentz force. Changing to Larmor frame allows one to be rid of this contribution as it exactly cancels the Coriolis term in the transformation. In the Larmor frame, there is still a remaining force, the diamagnetic one with the expression $\gamma^2\boldsymbol{\rho}/2$, and axial symmetry. It is the one responsible for the effects of diamagnetism in the atomic spectrum, which must be compared to the Coulomb force for classifying the various regimes in the spectrum.

6.1.1. Weak-Field Regime of Atomic Physics [$\gamma \ll 1$; $\hbar\omega_c \ll R$; $B \ll B_c$]

As shown in Figure 14, there are roughly three characteristic regions in the atomic spectra, which are exhibited quite clearly using a Bohr model[125] or semiclassical arguments.[126,127] The appearance and the classification of the spectrum depend only on the order of magnitude of the diamagnetic force with respect to the Coulomb force, the ratio of which f_L/f_C is just $n^3\gamma$.[14,34] In the first region, $n^3\gamma \ll 1$, and the Coulombic energy levels are faintly perturbed.

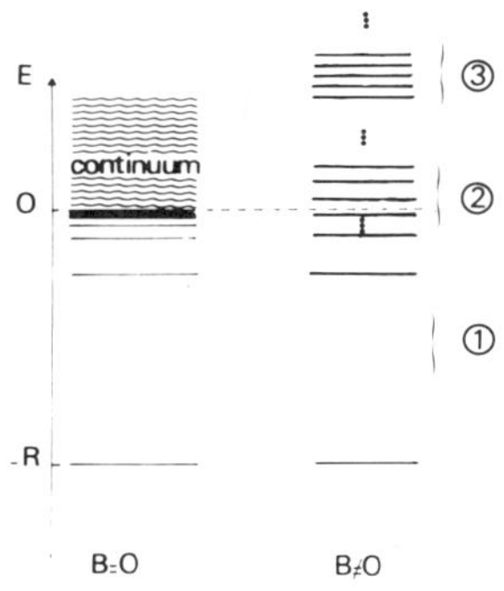

Figure 14. The three regions of the atomic spectrum in the weak-field limit ($\gamma \ll 1$) derived from a Bohr or semi-classical model. They correspond, respectively, to the Coulomb region (1), the strong-field mixing (2) with $3\hbar\omega_c/2$ spacing of the levels, and the Landau region (3) ($\hbar\omega_c$ spacing) for the motion of the electron.

The strong field mixing regime occurs for $n^3\gamma \simeq 1$ for which the total energy of the electron is nearly zero.[14] This also means that the classical frequencies $2R/n^3$ and $\hbar\omega_c$ are equal.[34] The spacing of energy levels is then about $\frac{3}{2}\hbar\omega_c$ as proved experimentally and with numerous kinds of arguments.[14,34] The quasi-Landau region is reached for $n^3\gamma \gg 1$. This is fulfilled for positive energies of the electron and results in a partial separation of the atomic continuum into a discrete structure. The spacing of energy levels is then close to $\hbar\omega_c$. Indeed, these levels above the zero field ionization limit are more likely resonances as the motion of the electron is not completely bound in the field direction.

6.1.2. Ultra-High-Field Regime ($\gamma \gg 1 - F_C \ll F_L$)

In this case, the Lorentz force greatly exceeds the Coulomb one. The spectrum is mainly of the Landau type with the removal of all degeneracies in M due to the one-dimensional Coulomb potential acting along $\mathbf{B}$ field.[128,129,130]

6.2. Semiclassical WKB Approach

WKB methods play an important historical role in this problem. They allowed the first theoretical explanation of the $\frac{3}{2}\hbar\omega_c$ feature of the spectrum of Garton and Tomkins,[126] but they do not allow any accurate prediction of the positions of the resonances. the origin of these limitations lies in the nonseparability of the three-dimensional Hamiltonian (42) in any of the 13 sets of coordinates of $R(3)$.[131] Bohr–Sommerfeld quantization conditions apply for one-dimensional problems in the form

$$\oint p_i \, dq_i = (n_i + \gamma_i)h \qquad (44)$$

where γ_i is the Maslov index[132,133] and allow exact treatments of the Coulomb or Landau limits of the present problem. Between these limiting cases, Eq. (44) is of no use, and one must develop the procedure of Einstein, Brillouin, and Keller[19,134,135] to quantize the nonseparable problem. A simpler way is to make some approximation in (42) in order to allow separation. For example, a crude one is to put $z = 0$ in the Hamiltonian, assuming that the longitudinal motion is not excited.[126] This leads, in cylindrical coordinates, to a one-dimensional problem for which (44) applies:

$$\int_{\rho_1}^{\rho_2} \{[(E_1 - \gamma^2\rho^2)/4] + 2/\rho - (|M| + 1/2)^2/\rho^2\}^{1/2} \, d\rho = (N_r + \alpha)\pi \qquad (45)$$

where $E_\perp$ is the electronic energy including paramagnetic terms. Such an expression, which is referred to hereafter as the "two-dimensional WKB approach," gives simply, for $E_\perp = 0$, the $\frac{3}{2}\hbar\omega_c$ spacing, while the relation between n and γ is just $n^3\gamma = 1.56$. Of course, (45) suffers from various short-comings. It gives the proper Coulomb limit for $\alpha = 1/2$ and the Landau limit for $\alpha = 1/4$. No absolute calibration of the spectrum can be done through (45). Of course, a correct "three-dimensional WKB" formula has yet to be established!

6.3. Generalized Semiclassical Methods

Proper semiclassical quantization of nonseparable problems involves a study of the classical motion of the electron as in the EBK quantization procedures on invariant tori.[136,137] It is suitable especially in the present situation, for which it has been inferred that a hidden approximate dynamical symmetry seems to exist.[138,139] However, no rigorous proofs have been given, nor has any study of the stability of the classical motion been successful.

"Three-dimensional WKB calculations" are of course essential for taking account of the effects of the longitudinal motion on the spectrum. Recent analyses by Fano[140–142] through the use of an appropriate set of coordinates are of this kind.

Other methods in $R(4)$ allow the problem to be reduced to that of a two-dimensional symmetric anharmonic oscillator, which is appropriate in a search for adiabatic invariants and in studying the classical motion.[143]

Finally, a very important work exists[144,145] in which the problem is dealt with, under conditions of the adiabatic approximation and with $\gamma \gg 1$. A generalization of Eq. (45) is considered, assuming that the classical turning points ρ_1 and ρ_2 depend on the variable z. One then deduces proper adiabatic potentials for the longitudinal motion. It seems that the region of validity of the deductions made in Ref. 144 is somewhat wider than $\gamma \gg 1$, possibly extending into the atomic physics domain. In fact one gets

$$E = (2N_r + 1)\gamma \left[1 - \frac{2}{(\gamma)^{1/2}} \frac{1}{(2N_r + 1)^{3/2}} + \frac{1}{\sqrt{2}} \frac{(n + 1/2)}{\gamma^{1/4}(2N_r + 1)^{7/4}} \right] \quad (46)$$

for $z \simeq 0$, and

$$E = \gamma(2N_r + |M| + M + 1) - \frac{1}{(n + \delta n)^2} \quad (47)$$

for excited motion along z, where N_r is the radial quantum number and n is a quantum number associated with the longitudinal motion. The physical meaning of the two formulas is obvious. For weakly excited longitudinal motion, the electron sees a harmonic potential well, the parameters of which depend on those of the transverse motion. On the other hand, for fairly excited longitudinal motion, one gets the combination of a Landau spectrum and of a one-dimensional Rydberg spectrum. From (46), it is easy to deduce that, for $E \simeq 0$, the spacing of the lines is $\frac{3}{2}\hbar\omega_c$. Structures due to the longitudinal motion have a spacing of $\hbar\omega_c/2\sqrt{2}$ for the bands near the zero-field atomic threshold ($E = 0$). Then, three "fine-structure components" will occur between two main resonances of spacing $\frac{3}{2}\hbar\omega_c$. In addition, all the frequencies are in general noncommensurable, thus explaining the "chaotic" character of some of the experimental data. The lower end of the bands described through Eq. (46) will be very important in optical excitation giving the stronger lines of the spectrum. Thus this is really the first model to predict quasi-Landau behavior and strong field mixing regimes near threshold, though the results have not so far been used in this context.

6.4. Quantum Mechanical Approach

Semiclassical methods are still being developed and are an essential tool for the interpretation of the spectrum, but they do not give any reliable answer to the crucial problem of wave functions. The need for accurate expressions of the wave functions is particularly clear from the fact that in the experiments optical absorption techniques are used which are sensitive to the short-range part of the wave function, while semiclassical methods deals with the long-range asymptotic part. Important properties of quasi-Landau states like autoionization, optical properties, and collisions, cannot be predicted through semiclassical methods.

Fully quantum theoretical approaches are under development. This is, of course, a tremendous task with only limited attempts so far,[139,146,147] some of which seem rather successful, e.g., those using a Sturmian basis.[148,149] Some attempts in the ultra-high-field limit ($\gamma \gg 1$), in the framework of the adiabatic approximation, are successful and may provide a convenient way to investigate the situation in Rydberg states which is still open.

6.4.1. Ultra-High-Field Quantum Solution ($\gamma \gg 1$)

The transverse motion of the electron is described with Landau-type wave functions given in Section 5. The L_z degeneracy of the spectrum is removed by the Coulomb perturbation. One gets a one-dimensional

Schrödinger equation for the longitudinal part of the wave function

$$\left[\frac{p_z^2}{2m} + V_m^{n_+}(z)\right] f_m^{n_+}(z) = E_m^{n_+} f_m^{n_+}(z) \qquad (48)$$

This, of course, assumes that the nonadiabatic $V^{n_+ n_+}$ coupling terms are negligible compared to the V^{n_+} ones, which are just the average of the Coulomb potential in the transverse Landau wave function. This is basically the method used by Schiff and Snyder.[26] Further approximations allow equation (48) to be reduced to a Whittaker equation for a one-dimensional Coulomb potential of the form $(\rho_m^{n_+} + |z|)^{-1}$. The spectrum looks like a Rydberg spectrum of doublets[129,130] associated with the symmetric and antisymmetric solutions along z. The energies are then[150,151]

$$E = \hbar\omega_c(n_+ + m_s + \tfrac{1}{2}) + E_\nu^\mp \qquad (49)$$

with

$$E_\nu^+ = -\frac{R}{\nu^2}\left(1 - \frac{2}{\nu}\log\frac{a_0}{\rho_m^{n_+}}\right) \qquad \text{(even states)}$$

$$E_\nu^- = -\frac{R}{\nu^2}\left(1 - 4\frac{\rho_m^{n_+}}{a_0}\right) \qquad \text{(odd states)}$$

In addition, there exists one deep level of even parity, the energy of which, from variational procedures, is just[150,152]

$$E_{m,\nu=0}^{n_+} = -\frac{8R}{\pi}\log^2(\gamma\varepsilon_m^{n_+}) \qquad (50)$$

and is minimal for $m = 0$ and $n_+ = 0$. The origin of such a high binding energy in the ground state lies in the compression of the orbit by the Lorentz force to dimensions of the order of the cyclotron radius $a_c \ll a_0$. Collapse on the Coulombic center is now possible.

6.4.2. Intermediate-Field Cases ($\gamma \lesssim 1$)

No reliable theories exist in this regime, except for low-lying atomic levels through numerical solutions of Schrödinger's equation.[153–161] The adiabatic approximation breaks down, but the symmetries of the wave functions in the Landau regime are still of interest. Furthermore, the ultra-high-field solution is of interest near the threshold of each Landau band. But the lower one-dimensional Coulombic states of equations (49) and (50) recondense and have strong interactions with the other series as is seen from Eq. (46) using the semiclassical three-dimensional approach. Near the bottom of each band, the structure of the longitudinal motion is no longer Coulombic but oscillator like. The interaction between the various channels is also clear from the fact that the bottom of the bands follows the strong-field mixing predictions and that the transverse motion is no longer of Landau type. No valuable quantum theoretical treatments have been attempted in this regime.

6.4.3. Diamagnetism from the Low-Field Coulomb Limit

The other types of quantum theoretical approaches are those related to the low-field Coulomb limit. A weak field is in any case sufficient for completely altering some parts of the Rydberg spectrum and of the continuum, and, of course, is responsible for a total breakdown of the O (4, 2) supersymmetry.[162]

Eigenfunctions of (42) at low fields are proper linear combinations of spherical harmonics at fixed M and parity values. We call them $|n, K, M, \Pi\rangle$, where K is just a label.[139] In our conventions, $K = 1$ is associated with the fastest rise of the energy with the B field within a given manifold (see Figure 15) and retains the most part of the spherical harmonics of the lowest l value compatible with M and Π. These states, in contrast to spherical harmonics, are eigenvectors of the Hamiltonian (42) under inter-l mixing conditions ($\gamma^2 n^7 \ll 1$) and do not depend on the field value in this limit. They are obtained through straightforward diagonalization of H_D inside the hydrogenic manifold of quantum number n, and the associated energies behave as B^2. As shown in Figure 16 for $n = 17$,[139] it clearly appears that the symmetries of $|n, K, M, \Pi\rangle$ states are far different from those in the usual spherical basis.[163] From nodal surface representations, one can easily imagine how the wave functions are built up through the merging of cones and spheres of the hydrogenic limit and the merging of planes and cylinders of the ultra-high-field limit. A general noncrossing rule of these nodal surfaces applies.[139]

In addition, Figure 16 exhibits quite clearly that K states are roughly organized into two classes. The first one, for small K values, is associated with low excitation of the longitudinal motion out of the $z = 0$ plane. The second one, for high K values, is associated with fairly excited motion along the B field and looks like the class of states which are deduced in the ultra-high-field limit. These simple calculations, therefore, strongly support the view expressed by (46) and (47) using three-dimensional semiclassical approaches.

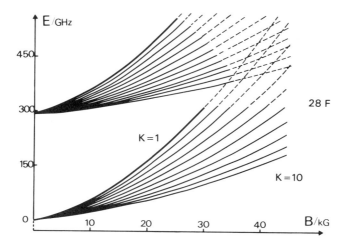

Figure 15. Inter-l mixing regime. Diamagnetic structure of the $(n = 28, M = 3,$ odd) hydrogenic manifold. "Forbidden" lines, labeled with K, appear due to the breakdown of l^2 as a good quantum number. n is still an approximate constant of the motion.

Under inter-n mixing conditions, the $\{|nKM\Pi\rangle\}$ states interact as a consequence of the breakdown of n as a good quantum number. Eigenvectors are then $|n\tilde{K}M\Pi\rangle$ states which are obtained with limited accuracy since the diagonalization of H_D needs to take into account continuum states! One can show that the anticrossing rules are general between $|n\tilde{K}M\Pi\rangle$ states, but that the anticrossing coupling is exponentially small between $|n\tilde{K} = 1\rangle$ and $|(n + 1)\tilde{K}_{max}\rangle$ states which is a possible clue to the existence of a hidden dynamical symmetry.[138,139] More powerful methods have now to be used for concluding.†

6.5. A Brief Review of Other Methods

Numerous other techniques have been used which essentially produce the exact energies and wave functions of the 13 first levels of hydrogen as a function of the field.[153–155] Of course, the problem in Rydberg states is almost an open one.

---

† NOTE: Recent developments based on a group theoretical analysis of this regime[243–247] have been able to fully understand these features.

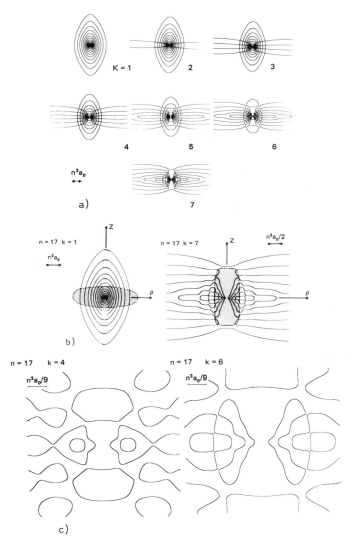

Figure 16. Nodal surface representation of the $n = 17$, odd parity, $M = \pm 3$, K states which are eigenfunctions of the Hamiltonian at low fields (Ref. 139). Depending on K, the symmetries of the states are very different (a). The $K = 1$ state is concentrated rather near the $Z = 0$ plane. The $K_{\max} = 7$ state corresponds to excited motion along **B** field (the shaded area corresponds to values of the wave functions $|\Psi| > 3.5 \times 10^{-7}$) (b). A general noncrossing rule of the nodal surfaces is fulfilled (c).

6.5.1. Variational and Perturbational Techniques

Choices of various sets of coordinates have been made: spherical coordinates,[164] parabolic ones,[165] spheroidal wave functions,[155] and the use of the ultra-high-field basis,[158] leading to no results for Rydberg states. Ground-state energies have been deduced through the use of trial wave functions of a mixed Landau–Coulomb type[150] which give the best results and analytical expressions for the wave function with one variational parameter.

6.5.2. Group Theoretical Method

Such methods are the more promising in view of solving the problem for Rydberg states if a dynamical symmetry really exists.

(a) *Use of a Sturmian Basis.* In such an approach connected with some aspects of group theory, the hydrogenic spectrum is described through the use of discrete complete sets of nonorthogonal Sturmian functions,[166] allowing a partial representation of the continuous spectrum. Applications to this problem were first initiated by Edmonds,[167] and recent advances by C. W. Clark and K. T. Taylor[148] allowed the first simulation of the Rydberg spectrum in fields, in fairly good agreement with semiclassical predictions and with experimental results. As shown in Figure 17, the quasi-Landau phenomena display well-defined series of discrete lines, some of which are dominant in the spectrum.

(b) *Group Theory.* Various attempts have been made through group theoretical methods, especially by Bednar.[168] Nevertheless, the problem of dynamical symmetry is still open. Recent advances using $R(4)$ methods demonstrate the equivalence with a two-dimensional symmetric anharmonic oscillator,[143] which perhaps may turn out to be a better way of investigating the inner symmetries of the problem.

(c) *Dilation Analyticity.* A dilatation transformation $r \to re^{i\theta}$ in the Hamiltonian (42)[169,170] will permit convenient studies of the Landau resonances and the computation of their widths, i.e., autoionization rates.[171] Such problems have also been approached through three-dimensional semiclassical methods.[145] The interest of the method, which is not widely used for the present problem,[34,171] lies in the exhibition of resonances in the complex half-plane as shown in Figure 18 for a similar situation.

6.5.3. Correlation Diagram and Conjectures.

The basic idea for establishing the correlation diagram between the low- and high-field limits, is the conservation of L_z and of the parity. This is, of course, insufficient. Some additional conjectures involve the conserva-

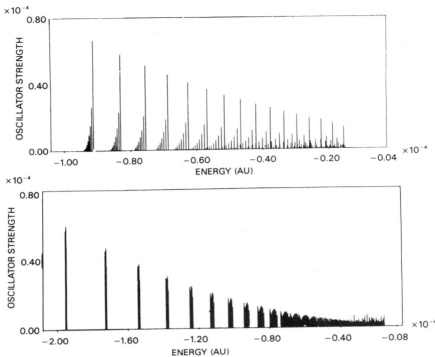

Figure 17. Numerical simulation of the Rydberg spectrum in a field (Ref. 148) for $\Delta m = 1$ and $\Delta m = 0$ transitions using a Sturmian approach. In one situation (a) ($\Delta m = 1$), dominant lines exist in the spectrum (associated with $K = 1$ states at low fields) which are obeying the semiclassical quantization conditions. The situation for $\Delta m = 0$ (b) is much more complicated with regard to oscillator strengths, but individual components of the series are still obeying the semiclassical predictions.

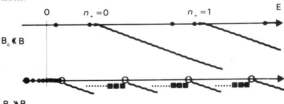

Figure 18. Schematic behavior of the Hamiltonian spectrum under a dilatation transformation $r \rightarrow re^{i\theta}$. The bound states (●) below the first quasi-Landau threshold are invariant to the transformation while the continua rotate 2θ out of the real axis. Rydberg resonances (■), associated with each quasi-Landau manifold, are then exposed (their positions are purely arbitrary, for the purpose of illustration). The upper part of the figure refers to the spectrum in the ultra-high-field limit, when the adiabatic separation of the longitudinal and transversal motions is valid.

tion of the number of nodal surfaces[172,173] or of adiabatic invariants.[174] These rules in general contradict each other and in addition are not fulfilled in the low-field limit, as may be seen from the nodal surface representations shown in Figure 16.[139] The existence of anticrossings between energy levels, even though they are weak, is an additional proof of the failure of these conjectures. Of course, a fundamental advance would be the discovery of an approximate dynamical symmetry in the problem.

6.6. Dipole Selection Rules and Short-Range Corrections to the Coulomb Potential

6.6.1. Role of Non-Coulombic Corrections

Short-range corrections to the Coulomb potential may affect the strength of anticrossings between energy levels in the sense that they are still weak but that the exponential singularity disappears, thereby implying the breakdown of the possible hidden symmetry.[138,139] Of course, the characteristics of the strong-field mixing regime (the $n^3 B = C^{\text{ste}}$ law at threshold), associated with the asymptotic behavior of the wave functions, will not be sensitive to these short-range corrections.

6.6.2. Oscillator Strengths

Oscillator strengths are of major importance for the general appearance of the quasi-Landau spectrum when optical excitation is used.[146] The short-range aspects of the wave functions and of the potential are then involved in all experimental work. This may hinder any detailed analysis of the data if situations are too far from the hydrogenic one. For example, the inter-l-mixing regime is modified completely for atoms with important quantum defects,[11] both with respect to the oscillator strengths and the energies. Under these nonhydrogenic conditions diamagnetic shifts are obtained through first-order perturbation theory[26,123]:

$$\Delta E_D = \langle nlm_l m_S | H_D | nlm_l m_S \rangle$$

$$= \frac{e^2 B^2}{8m} a_0^2 \cdot \frac{n^2 [5n^2 + 1 - 3l(l+1)][l^2 + l - 1 + m_l^2]}{(2l-1)(2l+3)} \tag{51}$$

Of course, at higher fields, when $(2R/n^3) \cdot \delta \ll H_D$ (δ the quantum defect), one partly recovers the behavior in the hydrogenic situation, but oscillator strengths will always be perturbed, adding spurious effects to the inner perturbation associated with the field action.

6.6.3. Qualitative Arguments on Electric Dipole Selection Rules for Highly Hydrogenic Systems

Due to the lack of any realistic approximation to the wave functions in fields, it is only possible to develop qualitative arguments on this point which is probably now the most important one for the understanding of the physics in Rydberg states. Some useful guides are, of course, the numerical simulations of Clark[148] and the ultra-high-field limit selection rules. In contrast to this last situation, it appears that all transitions ε^+, ε^-, and π are now possible. Mixed transitions, involving changes in both the transverse and longitudinal motions, are also possible as a consequence of the breakdown of the adiabatic approximation. The longitudinal states, associated with different Landau bands, are not orthonormalized, in analogy to the situation known from molecules. The spectrum will then exhibit both discrete structures and continua.[145]

In practical situations, optical excitation of Rydberg states takes place from a low-lying atomic state having approximately the zero-field hydrogenic symmetry. Preferrential excitation of quasi-Landau states concentrated near the origin and especially near the $z \simeq 0$ plane will then occur, meaning that the bottom of the Landau bands will be seen dominantly in the optical absorption spectrum. Especially the components of the $K = 1$ series (see Section 6.4.3) will be dominant under these conditions. However, the actual situation is somewhat more complicated and depends on the parity $(-1)^{l-M}$ of the states along the B axis, as shown both in numerical simulation and in experiments.[140,149,175]

In the intermediate-field case, many lines must exist in absorption due to the completely dissimilar symmetries of the low-lying Coulombic state and the highly excited quasi-Landau state.

6.7. The Hydrogen Atom in Crossed (E, B) Fields

As quoted in Section 3.1.5, the real problem to solve in a B field is a class of one-particle problems in crossed (E, B) fields. This will furnish a solution to the important question of the effects of the proton rest mass on the Landau spectrum, a source of large controversy.[31,33] In addition, the intrinsic effects of a crossed E field on the Landau spectrum seems of considerable interest both for the physics of Rydberg states and for astrophysical applications.[127,176–179] A major prediction of the various models is the existence of a new anisotropic non-Coulombic potential well at large distances between the electron and the proton as shown in Figure 19. This indeed produces various types of possible effects which are not necessarily equivalent and are discussed in Chapter 41. A total understanding of the physics of atoms in crossed (E, B) fields requires a quantum

Figure 19. Schematic representation (in a one-dimensional approximation) of the potential along the **E** field, in the crossed (**E**, **B**) field problem. This allows the prediction of the existence of a one-dimensional anisotropic non-Coulombic potential well at large distances between the electron and the proton, occurring under the joint action of the Coulomb (1), diamagnetic (2), and dipole electric (3) contributions.

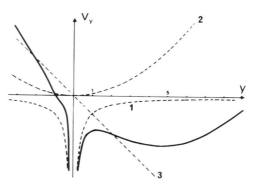

mechanical approach which is far beyond our reach owing to the lack of any constant of motion. Many models have been developed using classical or semiclassical arguments, but some questions as to their validity always remained.

6.7.1. Classical and Semiclassical Arguments

From the cycloidal trajectories of the electron and the proton in crossed (**E**, **B**) fields when the Coulomb potential is turned off, there is a clear tendency to form a bound system of the mean extension y_0

$$y_0 = (E \cdot M)/(e \cdot B^2) \tag{52}$$

(with $M = m_1 + m_2$) along the electric field and of the drift velocity $\mathbf{E} \wedge \mathbf{B}/B^2$. In the presence of the Coulomb field, the equations are no longer separable especially with respect to the energies of the reduced particle and of the center of mass.[29] If the Coulomb contribution is weak enough, equivalent to an excited motion of the reduced particle, the equation of motion will show some stationary behavior for conditions close to those associated with the classical trajectory of the center of mass for $y \simeq y_0$. A finite amount of the total energy, viz., $\frac{1}{2} M E^2/B^2$, will be "frozen" to ensure the motion of the center of mass. Since only the total energy of the system is conserved, this will result in the appearance of a second, non-Coulombic, potential well in the effective energy of the reduced particle, with a fairly anisotropic shape along **E**, and located near $y \simeq y_0$. If in addition $y_0 \gg a_0$, the Coulomb potential is almost negligible and the problem is equivalent to that of a one-dimensional oscillator centered at y_0, the frequency of which is just $\omega_c/2$. One is then led to the existence of a new class of states with enhanced stability compared to the Rydberg situation and with a spacing of energy levels of $\hbar\omega_c/2$ rather than $3/2\hbar\omega_c$. In addition, these resonances will have permanent dipole moments along **E**.[177-179]

6.7.2. Form of the Two-Particle Hamiltonian in Crossed (E, B) Fields

The Hamiltonian in the notation of Section 3 is

$$H_{12} = \frac{\pi_1^2}{2m_1} + \frac{\pi_2^2}{2m_2} + V_{12}(r) + e\,\mathbf{E}(\mathbf{r}_1 - \mathbf{r}_2) \tag{53}$$

It is straightforward to show that $\mathbf{P}$, as defined in (13) and (14), is the only constant of motion. Upon transformations similar to those of Section 3, one is led to an equation very similar to (16) with an additional term due to the $\mathbf{E}$ field:

$$H_\mathbf{P} = \frac{p^2}{2\mu} + \frac{e}{2\mu}\beta\mathbf{B}\cdot\mathbf{L} + \frac{e^2}{8\mu}B^2(x^2 + y^2)$$
$$+ V_{12}(r) + \frac{e}{M}(\mathbf{P}\wedge\mathbf{B})\mathbf{r} + e\,\mathbf{E}\cdot\mathbf{r} + \frac{P^2}{2M} \tag{54}$$

A further transformation of (54) and of the wave function

$$\mathbf{P}' = \mathbf{P} + \frac{M}{B^2}(\mathbf{B}\wedge\mathbf{E})$$

$$\mathbf{y}_0 = \frac{\mathbf{B}\wedge\mathbf{P}'}{eB^2} = \frac{M\cdot\mathbf{E}}{eB^2} + \frac{\mathbf{P}\wedge\mathbf{B}}{eB^2} \tag{55}$$

$$\Psi(\mathbf{r}) = \Phi(\mathbf{r} - \mathbf{y}_0)\,e^{i\gamma\mathbf{r}\cdot\mathbf{P}'/2\hbar}$$

leads to Eq. (56) in which all the operators are expressed in the new frame of coordinates and referred to the new origin at y_0:

$$H = \frac{p^2}{2\mu} + \gamma\beta\,\frac{\mathbf{B}}{B}\cdot\mathbf{l} + \frac{e^2B^2\rho^2}{8\mu} + V(r + y_0) + \frac{P^2}{2M} - \frac{P'^2}{2M} \tag{56}$$

This is the one-particle problem in a $\mathbf{B}$ field but in a non-central potential well. The last term is just

$$\frac{P^2}{2M} - \frac{P'^2}{2M} = -\mathbf{P}\cdot\frac{\mathbf{B}\wedge\mathbf{E}}{B^2} - \frac{M}{2}\frac{E^2}{B^2} \tag{57}$$

6.7.3. Remarks on the Conceptual Importance of the Nonseparability

(i) The solution of the two-particle problem in a $\mathbf{B}$ field implies the solution of an infinite set of Schrödinger equations for the eigenvalues of the operator **P**, the definition of which involves both the velocity of the center of mass and the coordinates of the reduced particle. Choosing atomic beam conditions in order to cancel motional Stark effects will reduce the effects of the center-of-mass motion on the spectrum, but such cancellation can never be total, as coupling between the velocity of the center of mass and the coordinates of the reduced particle will always exist.

(ii) Due to this coupling, cooling and trapping of the atoms is of interest as it will modify the statistics of the external variables and hence the spectrum of the reduced particle. Quantum effects associated with the external degree of freedom could perhaps be seen in the optical quasi-Landau spectrum, at low temperatures. Anyway, the purest conditions of observation will be achieved, from (54), for $P \simeq \mathbf{0}$, and thus, *in traps*.

6.7.4. Various Regimes in Crossed (**E**, **B**) Fields

Because of the lacking of any further separability of the equations (54) and (56), discussions are necessarily qualitative.

(i) From (56), one is led to the discussion of Section 6.7.1. The effect of the **E** field is to lower the effective potential of the reduced particle when conditions are close to the classical ones. One assumes here that the rotational energy *near* $y \simeq y_0$ is negligible. The depth of the well and conditions for its existence are easy to derive.[34,179] Taking λ as the strength of the electric field in atomic units, one gets the condition for the existence:

$$\frac{M}{\mu} \cdot \lambda \gg \gamma^{4/3} \quad \text{or} \quad \frac{E}{B^{4/3}} > \frac{3}{M}\left(\frac{\hbar^2 e}{a_0}\right)^{1/3}$$

The depth of the well is then (in units of R)

$$\Delta V \simeq (\lambda^2/\gamma^2)(M/\mu) \tag{58}$$

which is just the kinetic energy term associated with the drift velocity [see (57)]. This is equivalent to a one-dimensional shifted harmonic oscillator of frequency $\hbar\omega_c/2$.

(ii) From (54), neglecting the *rotational energy near* 0, one is led to the situation of Figure 19 where the combination of the diamagnetic, Coulomb, and dipole electric terms may create a second quasi-harmonic potential well. Conditions for the existence and depth of the well are then

$$\lambda \gg \gamma^{4/3}$$
$$\Delta V' \simeq \lambda^2/\gamma^2 \tag{59}$$

Equation (59) differs from (58) and involves only the motion of the electron.[127] This expression is only valid if the distance y_0' is sufficiently small to justify the neglect of the rotational energy term in (54). Especially if $y_0' \simeq y_0$ the analysis does not hold and one turns back to the features of formula (58). Of course, many intermediate regimes may occur between (59) and (58), and no rigorous treatment has yet been possible.

(iii) A third possible approach to Eqs. (58) and (59) is through semiclassical methods. Phenomenologically, the problem is a three potential one which may lead to strong field mixing phenomena for the Coulomb–diamagnetic terms, for the Coulomb–electric dipole term (Stark resonances),[127,181] and to a third regime which is characterized by a spacing $\hbar\omega_c/2$ of the energy levels and new quantization laws. A suitable approach is to use the separability of the equations in parabolic coordinates for the Stark problem,[178] leading to the $\hbar\omega_c/2$ signature of the spectrum near threshold.

(iv) Other ways of generating $\hbar\omega_c/2$ spacings exist using parity breaking in the spectrum[182,183] and equating $2R/n^3$ and the Larmor frequency (which here represents paramagnetic terms).

6.7.5. Conclusions

All these models or arguments are far from being clear cut! But they all show that something new must happen in the theory of atoms in crossed $(\mathbf{E}, \mathbf{B})$ fields, which would considerably modify the description of Rydberg states. The study of this problem in the Landau gauge seems of special interest due to the translational invariance it retains. Suitable wave functions then seem to be those defined in Section 5 as type-II wave functions, shifted at y_0.

Under ultra-high-field conditions, the non-Coulombic potential well can support the ground state of the stystem.[177]

7. Experimental Advances in Atomic Diamagnetism

Since 1977, numerous experiments have been performed on atomic diamagnetism.[146,147,184–191] Some of the current studies are providing major advances, proving that the quasi-Landau phenomena cause well-defined series of discrete lines from the Coulomb to the Landau regions and not only broad modulations of the oscillator strengths.[147,190] This explains why the "two-dimensional" semiclassical approximation of section 6.2 is rather successful in the interpretation of the data. Conclusions can then be drawn concerning the structure of the quasi-Landau spectrum, thus opening the way to a quantum approach to the problem. We will review here the more striking features of these experiments.

7.1. Basis of the Experiments

Several experimental choices are of special importance in this problem, as they lead to the simplest interpretation of the data. They are the choice of the light source, the method of experimental detection, and the important choice of the atomic element. Each choice in general excludes the others, making the various experimental attempts rather complementary, but not perfect.

7.1.1. The Ideal Experimental Situation

It would be, of course, highly desirable to perform the experiments on a collimated beam of low-velocity hydrogen atoms (or in traps), conveniently excited by means of a single-mode dye laser of well-controlled line shape and small line width. Scanning both the field and the wavelength is a necessity for the experiments since the problem depends on the two parameters, the electron's energy and the magnetic field. This would recreate the purest and simplest theoretical situation concerning the fundamental mechanisms of the field action. Unfortunately, this is far from being possible in high Rydberg states of hydrogen.

Such an ideal situation would be free from any effects connected with short-range corrections to the Coulomb potential, especially those associated with oscillator strengths.[146,190] In addition, if the beam is parallel to the magnetic field and is as monokinetic as possible, one can minimize the effects of the external degree of freedom on the spectrum[147] (see Section 6.7), although these effects are inherently always present, but presumably weak. Of course, it is of fundamental importance to reduce the effects associated with motional Stark fields as they will broaden the lines or completely alter the signature of the spectrum; but it is conceptually important to point out that, even in beam experiments, one can never get rid of the effects of the nonseparability.

Beam experiments allow accurate determinations of the parameters of Rydberg states. However, they do not allow any study of n values greater than 80 owing to low atomic densities, and in general need huge optical powers which restrict the optical purity of the lines, except if one uses synchronization of a pulsed dye laser on a cw single mode one.[192-194]

Spectral purity and tunability are important parameters in that they allow resolution of the entire Landau spectrum and of its "fine structure." Furthermore, experiments at high n values in the Rydberg spectrum are of interest. For example, for $n = 140$, the quasi-Landau regimes occur for field values of 500. And it is highly desirable to study accurately the regions near threshold under such conditions. Vapor phase experiments, in contrast to beam experiments, allow the production of highly excited ($n \sim 160$) Rydberg states under low perturbation conditions[190] using totally controlled single-mode cw dye lasers. Of course, the motional Stark field cannot be eliminated completely thus requiring the choice of heavy atoms, low temperatures, and low field conditions. Thus cell experiments are of interest if they are combined with high-resolution techniques and performed on elements having highly hydrogenic behavior.

7.1.2. Actual Experiments

In the pioneering experiment of Garton and Tomkins,[13] use was made of conventional high-resolution absorption techniques. The light source was a high-pressure hydrogen tube producing a continuous spectrum of uv radiation. The resolution was achieved with the use of the Argonne spectrometers. The Landau spectrum was observed on Ba atoms and appeared as broad modulations.

More recent experiments at Bell Lab[186] and at Argonne[185] have been performed at higher resolution with pulsed dye lasers of several gigahertz widths, but under high pressure conditions with heat pipe devices and on atomic states of elements departing strongly from the hydrogenic behavior. This hinders any possible interpretation of the important problem of the "fine structure" of the spectrum. In addition, the intensity information is not reliable due to fluctuations of the dye laser line shapes and intensities.

Two recent experiments[146,147,150] have allowed major advances in this problem. One of them uses atomic beam techniques and excitation with pulsed dye lasers of gigahertz widths on sodium nD states with some departures from the hydrogenic behavior. Only the bound spectrum below threshold[147] has been studied. The other experiment uses highly selective cw single-mode dye laser excitation of cesium states in a vapor phase experiment at low pressures. Highly hydrogenic $M = \pm 3$, odd parity states of cesium with quantum defects smaller than $\delta = 0.033$ are produced. Studies were done, usually, by scanning the magnetic field. Variable resolution, as required, is achieved by varying the sweep rate of the magnetic field with time. In these experiments, Rydberg states as high as $n = 162$ were detected, and a complete study of the phenomena from the Coulomb to the Landau limits was realized for the first time, exhibiting the $n^3 B = C^{\text{ste}}$ and $nB = C^{\text{ste}}$ quantization laws associated with the strong field mixing and Landau regimes.[191] Motional Stark field effects for this heavy element are small.

7.2. The Discrete Nature of the Quasi-Landau Spectrum

Fundamentally new pictures of the structure of the Landau spectrum can be deduced from these two experiments which use complementary techniques scanning either the field[190] or the energy.[147] It is shown that the strong-field mixing phenomena cause series of well-defined sharp discrete lines from the Coulomb to the Landau limit. The dominant series is associated with $K = 1$ states as defined in Section 6.4.3 in the low-field inter-l mixing regime for highly hydrogenic species.[139,146] These states are concentrated near the $z = 0$ plane, obeying the "two-dimensional WKB quantization", and are characterized by the famous $\frac{3}{2}\hbar\omega_c$ spacing near threshold. Numerical simulations of the spectrum with a Sturmian basis confirm these features,[148] linking the low-field quantum theoretical approach and the approximate semiclassical method. In addition, other discrete series, obeying approximately the same quantization laws, appear in these experiments and represent the so-called "fine structure."

7.2.1. Landau Spectrum in Energy

A schematic view of the set up of the beam experiment is shown in Figure 20. Stepwise excitation of Na states with $|M| \leq 2$ and even parity is achieved by the use of pulsed YAG pumped dye lasers with 2 GHz line widths. This allows the recording of the quasi-Landau spectrum as a function of the parity along the z axis, since the distributions of oscillator strengths are expected to be rather different[149,189,195] for different parities. One of the major results of these experiments is shown in Figure 21 where the various recordings, obtained by scanning the laser frequency, are plotted against the B field. Approximate quantum theoretical calculations, shown in comparison, provide a first rough interpretation of the spectrum.

In Figure 22 we show a recording obtained on Sr atoms near and above the threshold, exhibiting the $\frac{3}{2}\omega_c$ spacing of the resonances. It was obtained in a cell experiment in low-resolution conditions.[185,195] Comparison with the same features in the Ba spectrum allows one to emphasize the very important role of close-range corrections to the Coulomb potential when optical excitation is used.[187]

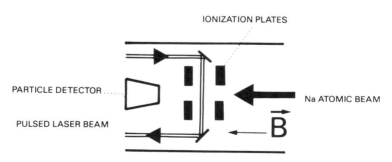

Figure 20. Apparatus in the beam experiment on Na atoms (Refs. 147, 175).

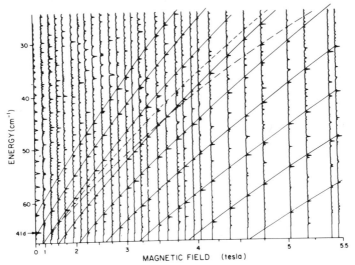

Figure 21. The quasi-Landau spectrum of Na atoms below threshold ($M = 2$, even parity states). The full curves are approximate numerical quantum calculations (Ref. 147).

7.2.2. Landau Spectrum in the Magnetic Field

A schematic view of the experiments on cesium F states is shown in Figure 23.[146] One of the aims of these experiments was to search for the purest situation concerning the structure of the states, as hydrogenic as possible, and to use ultra-high-resolution optical excitation. This was achieved by using the hybrid resonance pumping scheme of Collins *et al.*[196,197] which results in the population of high-lying nF Rydberg states with quantum defects $\delta = 0.033$ for n values as high as 160. The detection of the ions is achieved by the use of a thermoionic detector. Continuous 120-GHz single-mode frequency scans of the 10-MHz-width laser line are possible and highly linear scans of the 8T superconductive magnet with Nb–Ti windings can be made.

In the field, the pumping scheme leads essentially to the excitation of the $M = \pm 3$ odd parity states. Then, the diamagnetically mixed states will

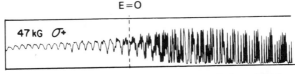

Figure 22. The quasi-Landau spectrum of Sr exhibiting at threshold the $\frac{3}{2}\hbar\omega_c$ spacing of energy levels (Refs. 185, 188).

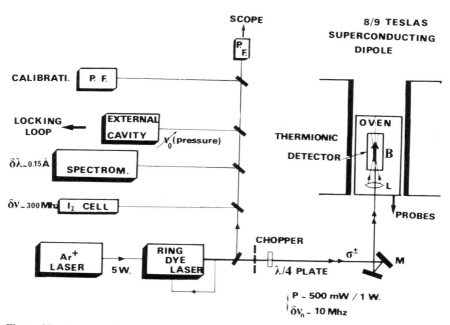

Figure 23. Apparatus for the cesium experiment ($M = \pm 3$, odd parity states), using a cw single-mode dye laser.

have quantum defects smaller than 0.033, which is one of the purest quasihydrogenic situations so far investigated. For example, Figure 24 shows a plot with the field of the diamagnetic structure of the $n = 50$ F state, exhibiting the components $K = 1$ to $K = 14$ ($l = 3$ to $l = 29$) of the hydrogenic manifold.[130]

The important feature of these studies performed over electron energies ranging from the Coulomb to the Landau limit, is the existence of series of discrete lines, the dominant of which are associated with the $K = 1$

Figure 24. Diamagnetic structure of the ($n = 50$, $M = \pm 3$; odd parity) states of cesium (Ref. 190) seen at fixed laser frequency, scanning the field. 14 components of the quasihydrogenic manifold are seen on this recording, corresponding to (perturbed) hydrogenic states from $l = 3$ to $l = 29$.

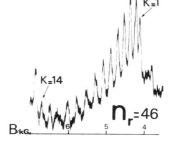

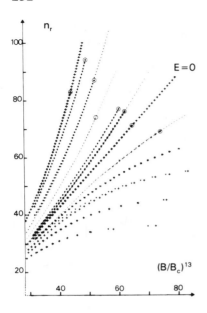

Figure 25. Experimental and theoretical plots of the radial quantum number n_r versus $\gamma^{-1/3} = (B/B_c)^{-1/3}$ at various energies, for the dominant discrete series of the spectrum ($K = 1$ states). Theoretical points (crosses or dashed lines) are from formula (45) with $\alpha = 1/2$. Experimental points are for energies, respectively, of 121.5 (top line), 99.5, 65.7, 37.3, 17.2, 11.5, 0, -10.9, -20, -29.3, -43.6, -66.6 cm^{-1} (lower line) from the ionization limit. At the threshold ($E = 0$) the curve is a straight line, showing the $n^3 \times B = 1.56$ quantization law. For positive and negative energies of the electron, the results curve away in opposite directions towards the Landau and Coulomb regions.

states in the inter-l mixing regime. For this series, a complete study of the various regimes of quantization has been performed experimentally, as shown in Figure 25, from the Coulomb to the Landau regimes.[190] Especially at threshold, the $n^3B = C^{\text{ste}}$ law is fulfilled while, in Figure 26, the $nB = C^{\text{ste}}$ law, characteristic[191] of the Landau regime, is seen for energies 100 cm^{-1} above threshold and n values of 90. One must remark

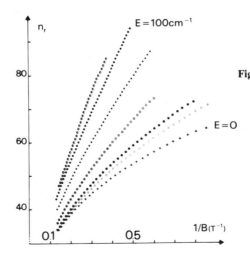

Figure 26. Reaching the Landau limit of the atomic spectra. A plot (for the $K = 1$ discrete lines) of the values of n_r as a function of $1/B$ showing, for energies above threshold, that the Landau limit is almost reached for $n_r \simeq 90$ and $E \sim 100$ cm^{-1}; the $n_r \times B = C^{\text{ste}}$ quantization law, characteristic of the Landau regime, is almost fulfilled under these conditions. At higher fields, the results curve away towards the strong field mixing regime characterized by the $n^3 \times B = C^{\text{ste}}$ quantization law at the threshold.

that in any situation, the increase of the field strength always results in a quantization law which tends towards the $n^3 B = C^{\text{ste}}$ law, characteristic of the strong field mixing regime.[14]

Finally, good overall agreement exists between the data and the predictions of the approximate two-dimensional WKB method.

7.3. Conclusions

From these experiments it seems clear that the quasi-Landau phenomena result from the superposition of various interacting series involving discrete lines below threshold. Each of them roughly follows the general predictions of the two-dimensional WKB approximation with small differences in the characteristics of the spacings. Due to the role of the optical excitation, states concentrated near the $Z \simeq 0$ plane will be favored in the process. This has to be compared with the predictions of (46) and (47) through three-dimensional calculations.[144,145] A plot from (46) with an oscillatorlike fine structure is compared with experimental results in Figure 27 for $E = 0$, and allows the qualitative understanding of the important decrease which occurs after each dominant line.[191] In addition, some chaotic character of the spectrum is certainly associated with the noncommensurability of the frequencies for the various series as noted in numerical simulations.[148] Obviously, the role of states described through (47) will be negligible in optical excitation, but the band structure, which emerges from this model, clearly indicates that the *spacing of energy levels* at threshold will really be distributed from $\frac{3}{2}\hbar\omega_c$ to $\hbar\omega_c$ for the whole ensemble of series. Physical reality will then probably be much more complicated than apparent in optical absorption experiments. A quantum theoretical model, allowing description of the properties of these states, has yet to be developed, and further generalization to problems of spontaneous autoionization in fields will be of interest.[198]

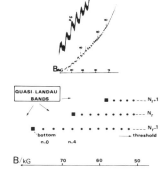

Figure 27. A comparison of theoretical predictions from the "three-dimensional semiclassical" theory (Refs. 144, 145) and experimental results above threshold. This allows the tentative understanding of the strong decrease following the resonance as a physical effect caused by leaving a Landau band, which no longer contributes to the signal at higher fields (Ref. 191).

Of course, the experimental reproduction of the $\frac{1}{2}\hbar\omega_c$ spacing, observed in the lithium experiments[159] and associated with motional Stark effects,[182] will be of special interest when an electric field of known strength is applied perpendicular to **B**.

8. Ultra-High-Field Regime and Quantum Electrodynamics

The ultra-high-field regime under conditions where $R \ll \hbar\omega_c$ very often implies the use of a relativistic solution of the problem, especially for astrophysical purposes. However, numerous discrepancies and disagreements exist among these various attempts[200-206] even in the framework of the adiabatic approximation. The ultra-high-field limit is also of interest for situations in which the effective Rydberg or coupling constant, R^*, is small as, for example, for electrons in traps or excitons, and may be reached at laboratory field strengths.

8.1. Relativistic Solution for $\gamma \gg 1$

The relativistic Landau spectrum is straightforwardly derived from the Dirac equation[202,207,208] and takes the form

$$E = \pm (c^2 P_z^2 + m^2 c^4 + 2K \hbar\omega_c m c^2)^{1/2} \tag{60}$$

where K is the Landau quantum number (including spin).

The relativistic Coulomb problem in a **B** field is solved for the ground state by the use of the adiabatic approximation. This produces a one-dimensional Dirac equation for a Coulomb-like potential.[209] The energy of the ground state is given with asymptotic validity for $B \gg B_c/\alpha^2$ (α the fine structure constant) by[210]

$$E/mc^2 = \cos\,[2Z\alpha \log\,(2\lambda d) + 2\,\mathrm{Arg}\,\Gamma(-2iZ\alpha) - \pi$$
$$- 2\,\mathrm{Arg}\,\Gamma(-Z\alpha\,(E/\lambda mc^2 + i))] \tag{61}$$

with $d = (B_c/B\alpha^2)^{1/2}$ and the Compton wavelength $\lambda = (1 - E^2/m^2 c^4)^{1/2}$. When $Z\alpha \ll 1$, one recovers the nonrelativistic solution of formula (50).

Equation (61) neglects Darwin terms and $(g - 2)$ corrections[211] and it displays the very important problem of close-range corrections, of nuclear origin, to the Coulomb potential.[207,212] Derivation of (61) explicitly involves boundary conditions to be fulfilled in the vicinity of the nucleus. The parameter d in (61) may then be understood in some situations as a mean nuclear radius. Possible interference terms between Coulomb and nuclear potentials may exist.

8.2. Spontaneous Decay of the Neutral Vacuum at Strong B Fields

A possible fundamental application of diamagnetism in the ground state is connected with the neutral vacuum decay in high magnetic fields.

8.2.1. Neutral Vacuum Decay at High Z Values

Such a process has so far attracted interest especially in heavy ion collisions and is extensively described in Chapter 5 of Part A of this book. Let us just recall that within the framework of the Dirac equation, in zero magnetic field, the ground state of the united hydrogenlike atom fails to reach the boundary of the lower Dirac continuum since for $Z > 137$ the energy suddenly becomes imaginary.[211] This is just the expression, in quantum mechanics, of the forbidden character of collapse on the center with a purely Coulombic potential.

The latter is not a physical effect, however, and when allowance is made for the finite nuclear dimensions, it is found that the ground state reaches the lower boundary of the positron continuum for $Z \simeq 169$.[212] Decay of the neutral vacuum may then occur, but a large number of problems are still unsolved.

8.2.2. Neutral Vacuum Decay in Strong B Fields

The action of strong B fields on the electron's motion, as in pulsars and white dwarfs, completely modifies the previous analysis of the neutral vacuum decay.[34,209,210,213] The Lorentz force compels the electron to explore regions of dimensions $a_c \ll a_0/Z$ of the Coulomb potential closer to the Coulomb center. The binding energy of the electron is considerably enhanced by the B field and is more rapidly increasing with Z than in the pure Coulomb case. Neutral vacuum decay may then occur for values of Z lower than 169, depending on the B value.[209,210]

From (61) one can deduce a provisional estimate of the critical curve $Z_{cr}(B)$ for which diving occurs ($E = -mc^2$). These models do not include any corrections of nuclear origins to the potential. Due to the field action, collapse on the center is possible here without taking account of these corrections.

Of course, the electron spin is of extreme importance in this problem. It allows zero Landau energy for the transverse motion with spin down [neglecting $(g-2)$ corrections]. If such a condition was not fulfilled, the logarithmic term in the binding energy would be overwhelmed by the positive $\hbar\omega_c/2$ contributions and diving would not be possible. Also, in these circumstances, positron emission will probably occur with complete spin polarization which has to be compared with the equivalent roles played by the two spin polarizations when no magnetic field exists. One must also mention that positron emission will really occur if nonzero coupling terms exist between the discrete diving state and the positronic continua having partially separated in the field into a discrete structure.[207]

8.2.3. Pulsars, White Dwarfs, and Heavy Ion Collisions

The previous analysis may have some interest in conditions of pulsars and white dwarfs for which fields of about 10^9 to 10^{12} gauss are suspected to exist. Observation of transitions between the two first Landau bands, at keV energies, have been reported in pulsars.

Generation of intense magnetic fields will also occur in collisions between heavy ions at MeV energies.[18] Fields of 10^{17} gauss will be generated by the nuclear motion at internuclear distances of 20 fermi, but they will be highly inhomogeneous. Huge electrostatic fields will probably dominate the physics of the processes, but the possibility of important magnetic effects is nevertheless still an open problem.[214-216] A realistic approach will have to take into account the quasimolecular character of the collision, and will have to solve, as a model, the relativistic H_2^+ problem in the presence of an inhomogeneous magnetic field. This implies the study of a two-center Dirac equation, and the stability of the three-body problem is an open question. But, as a clue, the enhancement of the ground state binding energy of H_2^+ with the field does not seem to be as important as in the single center problem.[217-220]

8.2.4. Crossed (**E**, **B**) Field Effects

As mentioned in Section 6, the non-Coulombic potential well may acquire a depth far deeper than the Coulomb one and then may support the ground state of the atomic system. New classes of exotic atoms of interest for the physics of pulsars and heavy ions may then exist in which the ground state will possess permanent electric dipole moments and a strong anisotropy, of interest in chemical reactivity.[23,176,221]

Also, some aspects of dielectronic recombination in megagauss fields, in plasmas, and in collisions at MeV energies, have been considered,[222-224] exhibiting singular features which are certainly the manifestation of the underlying Landau behavior of the atoms or ions. The high-energy applications of the ultra-high-field situation will certainly turn out to be as spectacular as what is now known in Rydberg states.

8.3. Radiation, Chemical Reactivity, and Miscellaneous Questions in Strong Magnetic Fields

(i) The ultra-high-field solution has wide applications in astrophysical problems. Of course, diagnostics of the field strengths through optical methods has been the subject of a large number of investigations. The Kemp model[22] in 1970 was a first attempt to interpret the polarization of the continuous spectrum from magnetic white dwarfs. It is a theory of magnetoemission from a thermal source which is modeled as an assembly of radiating harmonic oscillators, subjected to a magnetic field. Since the problem of the harmonic oscillator in a B field can be solved exactly,[27] complete analytical calculations are possible, though this is far from the real atomic problem.[225-228] The polarization rate allows the determination of the field strength. Such a model has now been abandoned for more realistic ones based on synchrotron radiation which have been presented in the work of Ginzburg.[229]

Also connected to these topics are all problems involving high-energy electron beams, interacting with static or electromagnetic fields, such as in free electron lasers, which will not be discussed here but will have major importance in the near future.

In the purely QED domain, the search for a modification of the polarization of the neutral vacuum, induced by magnetic fields, is also in progress, using huge fields produced with accelerator techniques or in pulsed conditions.[230]

(ii) In ultra-high-field conditions, atomic shells have very strange properties[231] which are exemplified in the low-field case in the nodal surface diagrams of Figure 16. In addition to the obvious anisotropy, the quasi-infinite degeneracy of the first Landau band allows a completely different filling with electrons compared with the zero-field atomic physics situation.[23,232,233] Atoms will have a tendency to form bands of complex linear molecules or crusts as in pulsars.[22] Chemical reactivity will then be completely different, and atomic species will have strong electronic affinity resulting in bound negative ions.[234,235] It is not clear whether such behavior can survive at laboratory field strengths, in Rydberg states, but as an indication, the very anisotropic aspects of the wave functions will certainly lead to unusual properties in collisions.

Thomas–Fermi statistical models for heavy ions show, of course, unusual features compared with the zero-field situation.[231-233] Enhanced stability of molecules through the weakening of repulsive close-range interactions is also predicted together with the creation of new species of bound molecular ions.[235] This may be of interest in the situation of biexcitons,[236,237] at laboratory field strengths.

8.4. Laboratory Fields and Weakly Bound Particles

The ultra-high-field limit is, of course, of interest for some laboratory applications because of the scaling law for the $\gamma^* = \hbar\omega_c/2R^*$ parameter— for excitons and biexcitons, for electron layers in surface physics, and for electrons in traps.[238,239,240]

Excitons are quasi-particles formed by an electron–hole pair during the absorption of a photon. The interaction, in some respect, is of Coulomb type but with a smaller effective Rydberg constant R^*. Practically, conditions are those of the ultra-high-field limit, which explains the earlier development of this subject in solid state physics.[12,240] Due to the complications associated with anisotropy in the solid, complete equivalence with the atomic physics situation is somewhat exceptional. Interesting predictions on the molecular biexcitonic system are certainly to be considered experimentally.[237]

Electrons trapped by the image-induced potential at the surface of liquid helium also behave as a quasihydrogenic one-dimensional system with a small effective Rydberg constant[14,178] and allow complete studies of all the regimes of diamagnetism. Applications to the realization of some kinds of barrier junctions in crossed $(\mathbf{E}, \mathbf{B})$ fields seem possible.[180,238]

To end, $(g - 2)$ measurements for electrons in traps, which are described in Chapter 22 of Part B, are among the more beautiful QED applications of the techniques in high fields. The quadrupolar field, which is superimposed in order to bind the motion along the $\mathbf{B}$ field, is a small perturbation of the diamagnetic term. The spectrum of electrons in traps is then analogous to that of the ultra-high-field solution. Radiative transitions between the $(n_+, S_z = +1/2)$ and $(n_+ + 1, S_z = -1/2)$ Landau states have allowed the measurement of the $(g - 2)$ anomaly with an exceptional accuracy.[241]

The combination of the trap techniques and high-resolution laser spectroscopy has allowed high-precision laser photodetachment spectroscopy in high fields,[242] thus opening the way to studies of chemical reactivity in high fields (see Figure 28). Further development seems especially promising with regard to state selection in such reactions, when the atomic or molecular continua change to discrete Landau states, and with regard to the modification of the inner mechanisms of the chemical process.

9. Conclusions

In this review, we have dealt both with traditional developments of Zeeman techniques at strong fields and with fundamental concepts of diamagnetism, which are in rapid evolution. Developments in atomic experimental physics and prospects have been intentionally mixed, allowing the prediction of much exotic behavior of the atom which may turn out to be of major importance for the applications in plasma physics. Further developments in QED using accelerator techniques will certainly be of interest in the near future.

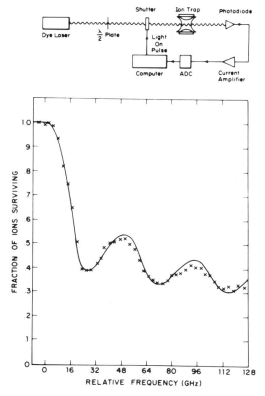

Figure 28. Apparatus of the laser photodetachment experiment in high fields (upper part). The lower part shows some typical experimental results, viz., the number of surviving S^- ions in a 15.7-kG magnetic field when the laser frequency is swept from threshold of the S^- $(^2P_{3/2}) \to S$ $(^3P_2) + e^-$ reaction. The curve clearly exhibits the resonances associated with the successive Landau bands for the motion of the free electron (solid curve: theoretical predictions).

To close, the wide implications of the problem of the strongly magnetized hydrogen atom in various domains of physics and its obvious conceptual importance concerning theoretical methods of classical and quantum mechanics largely justify the experimental and theoretical efforts now being undertaken in atomic physics to obtain some insights towards a solution.

Since the writing of this review, further progress has been achieved in the understanding of diamagnetism. Classical mechanics studies have been developed by several groups, permitting the discovery of an adiabatic invariant[243] in the inter-l-mixing Coulomb regime, which is $\Lambda = 4A^2 - 5A_z^2$, where $\mathbf{A}$ is the Lenz vector. It allows a complete classification of the states in this limit and the understanding of the exponentially small anticrossings in the energy diagram. The clarification of the meaning of Λ through group theory on the Fock hypersphere[244] allows one to understand the very different symmetries of the wave functions in the inter-l-mixing regime and suggests some analogy between the problem of diamagnetism and other important problems in physics. Also, numerical simulations have shown that classical dynamics becomes chaotic above a critical energy, thus restricting the possibility of a dynamical symmetry in this problem.[245] A more detailed analysis of these recent advances can be found in Refs. 246 and 247.

Acknowledgments

Several stimulating discussions on some of the topics of this review with B. Cagnac, C. W. Clark, D. Delande, U. Fano, A. Omont, L. R. Pendrill, W. B. Schneider, and K. T. Taylor are gratefully acknowledged. Some of the figures have been kindly supplied by J. Castro, C. W. Clark, T. A. Miller, and F. S. Tomkins.

References

1. A. Sommerfeld, *Lectures in Theoretical Physics*, Academic Press, New York (1964).
2. H. A. Lorentz, *K. Acad. Wet. Amsterdam Proc.* **8**, 591 (1906).
3. P. Zeeman, *Phys. Z.* **10**, 217 (1909).
4. P. Kapitza, *Proc. R. Soc. London Ser. A* **115**, 658 (1927).
5. F. Bitter, *Rev. Sci. Instr.* **7**, 482 (1936).
6. Uhlenbeck and Goudsmit, *Naturwissenschaften* **13**, 953 (1925).
7. W. Hanle, *Z. Phys.* **41**, 164 (1927).
8. J. Brossel and A. Kastler, *C.R. Acad. Sci. Paris* **229**, 1213 (1949).
9. F. O. Colegrove, P. A. Franken, R. R. Lewis, and R. A. Sands, *Phys. Rev. Lett.* **3**, 420 (1959).
10. L. Landau, *Z. Phys.* **64**, 629 (1930).
11. F. A. Jenkins and E. Segre, *Phys. Rev.* **55**, 52 (1939).
12. H. Hasegawa, *Physics of Solids in Intense Magnetic Fields* (E. D. Haidemenakis, Ed.) Plenum, New York (1969); J. G. Mavroides, *Optical Properties of Solids* (F. Abeles, Ed.) North-Holland, Amsterdam (1972).
13. W. R. S. Garton and F. S. Tomkins, *Astrophys. J.* **158**, 839 (1969).
14. A. R. P. Rau, *Phys. Rev. A* **16**, 613 (1977).
15. D. B. Montgomery, *Rep. Prop. Phys.* **26**, 69 (1963).

16. M. F. Wood, *Cryogenics* **2**, 297 (1962).
17. Sakharov and Lyndrev, *Dokl. Akad. Nauk. SSR* **165**, 65 (1965).
18. W. Greiner, B. Müller, and G. Soff, *Phys. Lett.* **69A**, 1, 27 (1978).
19. A. Einstein, *Ver. Deut. Phys. Ger.* **19**, 82 (1917).
20. I. C. Percival, *Adv. Chem. Phys.* **36**, 61 (1977).
21. M. P. Strand and W. P. Reinhardt, *J. Chem. Phys.* **70**, 3812 (1979).
22. J. C. Kemp, *Astrophys. J.* **162**, 169 (1970).
23. R. O. Mueller, A. R. P. Rau, and L. Spruch, *Nature Phys. Sci.* **234**, 31 (1971).
24. C. K. N. Patel and Y. Yafet, *Appl. Phys. Lett.* **34**, 318 (1975).
25. P. Bakshi, G. Kalman, and A. Cohn, *Phys. Rev. Lett.* **31**, 1576 (1973).
26. L. I. Schiff and H. Snyder, *Phys. Rev.* **55**, 59 (1939).
27. B. Diu, C. Cohen-Tannoudji, and F. Laloe, *Mécanique Quantique*, Hermann, Paris (1977).
28. W. E. Lamb, *Phys. Rev. A* **85**, 259 (1952).
29. L. P. Gorkov and I. E. Dzyaloshinskii, *JETP* **26**, 449 (1968).
30. B. P. Carter, *J. Math. Phys.* **10**, 788 (1968).
31. R. F. O'Connell, *Phys. Lett.* **70A**, 389 (1979).
32. M. S. Ryvkin, *Sov. Phys. Dokl.* **20**, 192 (1975).
33. G. Wunner, H. Ruder, and H. Herold, *Phys. Lett.* **79A**, 159 (1980).
34. J. C. Gay, *Comments At. Mol. Phys.* **9**, 97 (1980).
35. H. J. Beyer and H. Kleinpoppen, *J. Phys. B* **4**, L129 (1971).
36. M. Glass and J. P. Descoubes, *C.R. Acad. Sci. Paris* **273**, B721 (1971).
37. T. A. Miller, in *Proceedings of the International Conference on Solids and Plasmas in High Magnetic Fields*, MIT, Boston (1978); *J. Magnetism Magnetic Mater.* 11 (1978).
38. T. A. Miller and R. S. Freund, *Adv. Mag. Reson.* **9**, 49 (1977).
39. J. Meunier and A. Omont, *C.R. Acad. Sci. Paris* **262B**, 190 (1966).
40. C. Cohen-Tannoudji and S. Haroche, *J. Phys. (Paris)* **30**, 153 (1969).
41. R. Jost and M. Lombardi, *Phys. Rev. Lett.* **33**, 53 (1974).
42. M. Glass-Maujean, *J. Phys. B* **11**, 431 (1978).
43. E. C. G. Stueckelberg, *Helv. Phys. Acta* **5**, 369 (1932).
44. L. D. Landau, *Phys. Z. Sowj.* **2**, 46 (1932).
45. E. E. Nikitin, *Comments At. Mol. Phys.* **2**, 59 (1970).
46. D. S. F. Crothers, *Adv. Phys.* **20**, 86 (1971).
47. J. P. Descoubes, thesis, University of Paris, Paris (1967).
48. A. Omont, *Prog. Quantum Electron.* **5**(2), 69 (1976).
49. M. Glass, thesis, University of Paris, Paris (1974); M. Glass, *Comments At. Mol. Phys.* **7**, 83 (1977).
50. R. Jost, thesis, University of Grenoble, Grenoble (1978).
51. J. Derouard, R. Jost, M. Lombardi, T. A. Miller, and R. S. Freund, *Phys. Rev. A* **14**, 1025 (1976).
52. J. P. Descoubes, *Physics of the One and Two Electron Atoms* (F. Bopp and H. Kleinpoppen, Eds.), North-Holland, Amsterdam (1969).
53. Y. Accad, C. L. Pekeris, and B. Schiff, *Phys. Rev. A* **4**, 516 (1971).
54. T. N. Chang and R. T. Poe, *Phys. Rev. A* **10**, 1981 (1974).
55. K. B. MacAdam and W. H. Wing, *Phys. Rev. A* **12**, 1464 (1975).
56. T. A. Miller, R. S. Freund, F. Tsai, T. J. Cook, and B. R. Zegarski, *Phys. Rev. A* **9**, 2474 (1974).
57. T. A. Miller, R. S. Freund, and B. R. Zegarski, *Phys. Rev. A* **11**, 753 (1975).
58. E. Giacobino, F. Biraben, G. Grynberg, and B. Cagnac, *J. Phys. (Paris)* **38**, 623 (1977).
59. F. Biraben, E. De Clercq, E. Giacobino, and G. Grynberg, *J. Phys. B* **13**, L685 (1980).
60. H. M. Crosswhite, *The Hydrogen Molecule Wavelength Tables of Gerhard Heinrich Dieke*, Wiley-Interscience, New York (1972).

61. T. A. Miller and R. S. Freund, in *Colloque International du CNRS* No. 217 (Paris, 1974).
62. R. Jost, M. Lombardi, and J. Derouard, in *Colloque International du CNRS* No. 242 Paris, 1975).
63. T. A. Miller and R. S. Freund, *J. Chem. Phys.* **61**, 2160 (1974).
64. T. A. Miller and R. S. Freund, *J. Chem. Phys.* **63**, 256 (1975).
65. T. A. Miller and R. S. Freund, *J. Chem. Phys.* **62**, 2240 (1975).
66. T. A. Miller and R. S. Freund, *J. Mol. Spectrosc.* **63**, 193 (1976).
67. R. J. Freund, T. A. Miller, R. Jost, and M. Lombardi, *J. Chem. Phys.* **68**, 1983 (1978).
68. R. Jost, M. Lombardi, J. Derouard, R. S. Freund, T. A. Miller, and B. R. Zegarski, *Chem. Phys. Lett.* **37**, 507 (1976).
69. T. A. Miller, R. S. Freund, and B. R. Zegarski, *J. Chem. Phys.* **64**, 1842 (1976).
70. T. A. Miller, B. R. Zegarski, and R. S. Freund, *J. Mol. Spectrosc.* **69**, 199 (1978).
71. T. A. Miller, R. S. Freund, B. R. Zegarski, R. Jost, M. Lombardi, and J. Derouard, *J. Chem. Phys.* **63**, 4042 (1975).
72. T. A. Miller, R. S. Freund, and R. W. Field, *J. Chem. Phys.* **65**, 3790 (1976).
73. R. Jost and M. Lombardi, *J. Phys. Colloq.* **C1-39** (1978).
74. L. Julien, M. Glass-Maujean, and J. P. Descoubes, *J. Phys. B* **6**, L196 (1973).
75. C. Cohen-Tannoudji, F. Hoffbeck, and S. Reynaud, *Opt. Commun.* **27**, 71 (1978).
76. S. Reynaud, thesis, University of Paris, Paris (1981).
77. M. Leduc and C. Weisbuch, *Opt. Commun.* **26**, 78 (1978).
78. L. F. Mollenauer, *Opt. Lett.* **5**, 188 (1980).
79. L. F. Mollenauer, in *Methods of Experimental Physics* (C. L. Tang, Ed.) Academic Press, New York (1983).
80. D. M. Larsen, *Phys. Rev. Lett.* **39**, 878 (1977).
81. M. Rosenbluh, T. A. Miller, D. M. Larsen, and B. Lax, *Phys. Rev. Lett.* **39**, 874 (1977).
82. G. C. Neumann, B. R. Zegarski, T. A. Miller, M. Rosenbluh, R. Panock, and B. Lax, *Phys. Rev. A* **18**, 1464 (1978).
83. R. Panock, M. Rosenbluh, B. Lax, and T. A. Miller, *Phys. Rev. A* **22**, 1041 (1980).
84. R. Panock, M. Rosenbluh, B. Lax, and T. A. Miller, *Phys. Rev. A* **22**, 1050 (1980).
85. K. B. MacAdam and W. H. Wing, *Phys. Rev. A* **15**, 678, (1977).
86. G. Grynberg and B. Cagnac, *Rep. Prog. Phys.* **40**, 791 (1977).
87. N. F. Mott and H. S. W. Massey, *The Theory of Atomic Collisions* Oxford University Press (1965).
88. A. Gallagher, in *Physics of the One- and Two-Electron Atoms* (F. Bopp and H. Kleinpoppen, Eds.) North-Holland, Amsterdam (1969).
89. J. C. Gay and W. B. Schneider, *Phys. Rev. A* **20**, 879 (1979).
90. J. C. Gay and A. Omont, *J. Phys. (Paris)* **35**, 9 (1974).
91. J. C. Gay, *J. Phys. (Paris)* **35**, 813 (1974).
92. J. C. Gay, *J. Phys. B* **10**, 1269 (1977).
93. F. Masnou and R. McCaroll, *J. Phys. B* **7**, 2230 (1974).
94. J. C. Gay and W. B. Schneider, *Z. Phys. A* **278**, 211 (1976).
95. A. Abragham, *Principles of Nuclear Magnetism*, Oxford University Press, New York (1961).
96. C. Cohen-Tannoudji, Lectures 1977–78 College de France (Paris).
97. A. Omont, thesis, Paris (1967).
98. U. Fano, *Rev. Mod. Phys.* **29**, 74 (1957); *Phys. Rev. A* **131**, 259 (1963).
99. M. Baranger, *Phys. Rev. A* **111**, 481, 484 (1958).
100. U. Fano and J. H. Macek, *Rev. Mod. Phys.* **45**, 553 (1973).
101. J. C. Gay, *J. Phys. (Paris)* **37**, 1135 (1976).
102. J. C. Gay, *J. Phys. (Paris)* **37**, 1165 (1976).
103. J. C. Gay, thesis, University of Paris, Paris (1976).

104. J. C. Gay, *J. Phys. Lett.* **36**, L239 (1975).
105. J. C. Gay and W. B. Schneider, *J. Phys. Lett.* **36**, L185 (1975).
106. J. C. Gay and W. B. Schneider, *Phys. Lett.* **62A**, 403 (1977).
107. A. Omont and J. Meunier, *Phys. Rev.* **169**, 92 (1968).
108. M. I. Diakonov and V. I. Perel, *Sov. Phys. JETP* **21**, 227 (1965).
109. P. R. Berman and W. E. Lamb, *Phys. Rev.* **187**, 221 (1969).
110. R. Friedberg, S. R. Hartmann, and J. T. Manassah, *Phys. Rep.* **3**, 101 (1973).
111. Y. A. Vdovin and V. M. Galitskii, *Sov. Phys. JETP* **69**, 5 (1967).
112. E. E. Nikitin, in *The Excited State in Chemical Physics* (J. W. McGowan, Ed.), Wiley, New York (1975).
113. E. Gerjuoy, *Case Stud. At. Phys.* **3**, 1 (1972).
114. A. Messiah, *Quantum Mechanics*, Wiley, New York (1962).
115. C. J. Tsao and B. Curnutte, *J. Quant. Spectrosc. Radiat. Transfer* **2**, 41 (1962).
116. J. C. Gay and W. B. Schneider, *Phys. Rev. A* **20**, 905 (1979).
117. P. W. Anderson, *Phys. Rev.* **76**, 647 (1949).
118. J. C. Gay and W. B. Schneider, *Phys. Rev. A* **20**, 894 (1979).
119. J. C. Gay and A. Omont, *J. Phys. Lett.* **37**, L69 (1976).
120. A. Durand, *Mécanique Quantique*, Dunod, Paris (1970).
121. L. Landau and E. Lifchitz, *Mécanique Quantique*, Mir, Moscow (1966).
122. R. J. Glauber, *Phys. Rev.* **131**, 2766 (1963).
123. R. H. Garstang, *Rep. Progr. Phys.* **40**, 105 (1977).
124. D. Ter Haar, *Selected Problems in Quantum Mechanics*, Academic Press, New York (1964).
125. R. F. O'Connell, *Astrophys. J.* **187**, 275 (1974).
126. A. F. Starace, *J. Phys. B* **6**, 585 (1973).
127. A. R. P. Rau, *J. Phys. B* **12**, L193 (1979).
128. R. J. Elliot and J. Loudon, *J. Phys. Chem. Solids* **15**, 196 (1960).
129. H. Hasegawa and R. E. Howard, *Phys. Chem. Solids* **21**, 179 (1961).
130. L. K. Haines and D. M. Roberts, *Am. J. Phys.* **37**, 11, (1969).
131. P. M. Morse and H. Feshbach, *Methods of Theoretical Physics*, McGraw-Hill, New York (1953).
132. V. P. Maslov, *Théorie des Perturbations et Méthodes Asymptotiques*, Dunod, Paris (1972).
133. C. Jaffe and W. P. Reinhardt, *J. Chem. Phys.* **66**, 1285 (1977).
134. J. B. Keller, *Ann. Phys. (NY)* **4**, 180 (1958).
135. I. Percival and D. Richards, *Adv. Atom. Mol. Phys.* **11**, 1 (1975).
136. W. P. Reinhardt, in *Electronic and Atomic Collisions* (N. Oda and K. Takayanagi, Ed.), North-Holland, Amsterdam (1980).
137. V. I. Arnold, *Méthodes Mathématiques de la Mécanique Classique*, Mir, Moscow (1976).
138. M. L. Zimmerman, M. M. Kash, and D. Kleppner, *Phys. Rev. Lett.* **45**, 1092 (1980).
139. D. Delande and J. C. Gay, *Phys. Lett.* **82A**, 393 (1981).
140. U. Fano, in *Colloque International du CNRS* No. 273 (S. Feneuille and J. C. Lehmann, Eds.) (Paris, 1977).
141. U. Fano, *J. Phys. B* **13**, L519 (1980).
143. D. Delande and J. C. Gay (unpublished); D. Delande, thesis, University of Paris, Paris (1981).
144. A. G. Zhilich and B. S. Monozon, *Sov. Phys. Solid State* **8**, 2846 (1967).
145. B. S. Monozon and A. G. Zhilich, *Sov. Phys. Semicond.* **1**, 563 (1967).
146. D. Delande and J. C. Gay, *Phys. Lett.* **82A**, 399 (1981).
147. J. C. Castro, M. L. Zimmerman, R. G. Hulet, and D. Kleppner, *Phys. Rev. Lett.* **45**, 1780 (1980).
148. C. W. Clark and K. T. Taylor, *J. Phys. B* **13**, L737 (1980).
149. C. W. Clark, *Phys. Rev. A* **24**, 605 (1981).

150. A. R. P. Rau and L. R. Spruch, *Astrophys. J.* **297**, 671 (1976).

151. A. R. P. Rau, R. O. Mueller, and L. Spruch, *Phys. Rev. A* **11**, 1865 (1975).

152. C. Angelie, thèse 3e cycle (Orsay, 1978).

153. E. R. Smith, R. J. W. Henry, G. L. Surmelian, R. F. O'Connell, and A. K. Rajagopal, *Phys. Rev. D* **6**, 3700 (1972).

154. H. C. Praddaude, *Phys. Rev. A* **6**, 1321 (1972).

155. A. F. Starace and G. L. Webster, *Phys. Rev. A* **19**, 1929 (1979).

156. D. Cabib, E. Fabri, and G. Fioro, *Solid State Commun.* **9**, 1517 (1971).

157. Y. Yafet, R. W. Keyes, and E. N. Adams, *J. Phys. Chem. Solids* **1**, 137 (1956).

158. J. Simola and J. Virtamo, *J. Phys. B* **11**, 3309 (1978).

159. S. M. Kara and M. R. C. McDowell, *J. Phys. B* **13**, 1337 (1980).

160. J. Killinbeck, *J. Phys. B* **12**, 25 (1979).

161. S. C. Kanavi and S. H. Patil, *Phys. Lett.* **75A**, 189 (1980).

162. M. Bander and C. Itzykson, *Rev. Mod. Phys.* **38**, 330–346 (1966).

163. T. Hamada and Y. Nakamura, *Publ. Astron. Soc. Jpn* **25**, 527 (1973).

164. H. S. Brandi, *Phys. Rev. A* **11**, 1835 (1975).

165. W. Eckardt, *Solid State Commun.* **16**, 233 (1975).

166. M. Rotenberg, *Ad. Atom. Mol. Phys.* **6**, 233 (1970).

167. A. R. Edmonds, *J. Phys. B* **6**, 1603 (1973).

168. M. Bednar, *Phys. Rev. A* **15**, 27 (1977).

169. J. Aguilar and J. M. Combes, *Commun. Math. Phys.* **22**, 269 (1971).

170. W. P. Reinhardt, *Int. J. Quantum Chem.* **10**, 359 (1976).

171. S. I. Chu, *Chem. Phys. Lett.* **58**, 462 (1978).

172. W. H. Kleiner, Progress Report, Lincoln Lab. (1958).

173. M. Shinada, O. Akimoto, H. Hasegawa, and K. Tanaka, *J. Phys. Soc. Jpn* **28**, 4 (1970).

174. C. Angelie and C. Deutsch, *Phys. Lett. A* **67**, 357 (1978).

175. R. R. Freeman, *Proceedings of the 7th International Conference on Atomic Physics*, MIT (1980).

176. A. D. Jannussis, A. D. Leodaris, and G. N. Brodimas, *Phys. Lett.* **71A**, 205 (1979).

177. G. P. Drukarev and B. S. Monozon, *Sov. Phys. JETP* **34**, 509 (1972).

178. A. R. P. Rau, *Bull. Am. Phys. Soc.* **23**, 10 (1979).

179. J. C. Gay, L. R. Pendrill, and B. Cagnac, *Phys. Lett.* **72A**, 315 (1979).

180. R. F. O'Connell, *Phys. Lett.* **60A**, 481 (1977).

181. R. R. Freeman, N. P. Economou, G. C. Bjorklund, and K. T. Lu, *Phys. Rev. Lett.* **41**, 1463 (1978).

182. H. Crosswhite, U. Fano, K. T. Lu, and A. R. P. Rau, *Phys. Rev. Lett.* **42**, 963 (1979).

183. A. R. P. Rau and K. T. Lu, Contributed paper at the 7th International Conference on Atomic Physics (MIT, 1980).

184. M. L. Zimmerman, J. C. Castro, and D. Kleppner, *Phys. Rev. Lett.* **40**, 1083 (1978).

185. R. J. Fonck, F. L. Roesler, D. H. Tracy, K. T. Lu, F. S. Tomkins, and W. R. S. Garton, *Phys. Rev. Lett.* **39**, 1513 (1977).

186. N. P. Economou, R. R. Freeman, and P. F. Liao, *Phys. Rev. A* **18**, 2506 (1979).

187. K. T. Lu, F. S. Tomkins, and W. R. S. Garton, *Proc. R. Soc. London Ser. A* **364**, 421 (1979).

188. R. J. Fonck, D. H. Tracy, D. Wright, and F. S. Tomkins, *Phys. Rev. Lett.* **40**, 1366 (1978).

189. C. D. Harper and M. D. Levenson, *Opt. Commun.* **20**, 107 (1977).

190. J. C. Gay, D. Delande, and F. Biraben, *J. Phys. B* **13**, L729 (1980).

191. D. Delande, C. Chardonnet, and J. C. Gay, *Opt. Commun.* **42**, 25 (1982).

192. J. Pinard and S. Liberman, *Opt. Commun.* **20**, 344 (1977).

193. F. Trehin, thesis, University of Paris, Paris (1979).

194. F. Trehin, F. Biraben, B. Cagnac, and G. Grynberg, *Opt. Commun.* **31**, 76 (1979).

195. R. J. Fonck, F. L. Roesler, D. H. Tracy, and F. S. Tomkins, *Phys. Rev. A* **21**, 861 (1980).
196. C. B. Collins, B. W. Johnson, M. Y. Mirza, D. Popescu, and I. Popescu *Phys. Rev. A* **10**, 813 (1974).
197. L. R. Pendrill, D. Delande, and J. C. Gay, *J. Phys. B* **12**, L603 (1979).
198. J. P. Grandin and X. Husson, *J. Phys. B Lett.* **14**, 433 (1981).
199. K. T. Lu, F. S. Tomkins, H. M. Crosswhite, and H. Crosswhite, *Phys. Rev. Lett.* **41**, 1034 (1978).
200. R. Cohen, J. Lodenquai, and M. Ruderman, *Phys. Rev. Lett.* **25**, 467 (1970).
201. M. L. Glasser and J. Kaplan, *Phys. Lett.* **53A**, 373 (1975).
202. C. Angelie and C. Deutsch, *Phys. Lett.* **67A**, 353 (1978).
203. J. T. Virtamo and K. A. U. Lindgren, *Phys. Lett.* **71A**, 329 (1979).
204. M. Sminada, *Phys. Lett.* **74A**, 401 (1975).
205. V. B. Pavlov-Verevkin and B. I. Zhilinskii, *Phys. Lett.* **75A**, 279 (1980).
206. J. T. Virtamo and J. T. A. Simola, *Phys. Lett.* **66A**, 371 (1978).
207. A. I. Akhiezer and V. B. Berestetskii, *Quantum Electrodynamics*, Wiley, New York (1965).
208. I. I. Rabi, *Z. Phys.* **49**, 507 (1928).
209. V. P. Krainov, *Sov. Phys. JETP* **37**, 406 (1974).
210. J. C. Gay and D. Delande, *Proc. XIe I.C.P.E.A.C.* (S. Datz, Ed., 1981).
211. J. J. Sakuraï, *Advanced Quantum Mechanics*, Addison-Wesley, Reading, Massachusetts (1967).
212. Y. B. Zeldovich and V. S. Popov, *Sov. Phys. Usp.* **14**, 673 (1972).
213. B. B. Kadomtsev, *Sov. Phys. JETP* **31**, 945 (1970).
214. K. Smith, H. Pfeitz, B. Müller, and W. Greiner, *Phys. Rev. Lett.* **38**, 592 (1977).
216. G. Soff, W. Greiner, W. Betz, and B. Müller, *Phys. Rev. A* **20**, 169 (1979).
217. C. P. de Melo, R. Ferreira, H. S. Brandi, and L. C. M. Miranda, *Phys. Rev. Lett.* **37**, 676 (1976).
218. R. K. Badhuri, Y. Nogami, and C. S. Warke, *Astrophys. J.* **217**, 324 (1977).
219. C. S. Lai, *Can. J. Phys.* **55**, 1013 (1977).
220. J. M. Peek and J. Katriel, *Phys. Rev. A* **21**, 413 (1980).
221. R. O. Mueller, A. R. P. Rau, and L. R. Spruch, *Phys. Rev. A* **11**, 789 (1975).
222. W. A. Huber and C. Bottcher, *J. Phys. B* **13**, L399 (1980).
223. A. Ohsaki, I.S.A.S. Research note 107, Tokyo (1980).
224. A. Ohsaki, I.S.A.S. Research note 117, Tokyo (1980).
225. A. K. Rajagopal, G. Chanmugam, R. F. O'Connell, and G. L. Surmelian, *Astrophys. J.* **177**, 713 (1972).
226. G. Chanmugam, R. F. O'Connell, and A. K. Rajagopal, *Astrophys. J.* **175**, 157 (1972).
227. G. Chanmugam, R. F. O'Connell, and A. K. Rajagopal, *Astrophys. J.* **177**, 719 (1972).
228. A. Rich and W. L. Williams, *Astrophys. J.* **190**, 117 (1974).
229. V. Ginzburg, *Physique théorique et Astrophysique*, MIR, Moscow (1978).
230. E. Iacopini, P. Lazeyras, M. Morpugo, E. Picasso, B. Smith, G. Stefanini, E. Zavattini, R. Desalvo, and E. Polacco, in *Proceedings of the 12th E.G.A.S. Conference* (Pisa 1980-E.P.S. Abstracts).
231. R. O. Mueller, A. R. P. Rau, and L. Spruch, *Phys. Rev. Lett.* **26**, 1136 (1971).
232. B. B. Kadomtsev and V. S. Kudryatev, *Sov. Phys. JETP* **62**, 144 (1972).
233. B. Banerjee, D. M. Constantinescu, and P. Rehak, *Phys. Rev. D* **10**, 2384 (1976).
234. R. J. W. Henry, R. F. O'Connell, E. R. Smith, G. Chanmugam, and A. K. Rajagopal, *Phys. Rev. D* **9**, 329 (1974).
235. J. Avron, I. Herbst, and B. Simon, *Phys. Rev. Lett.* **39**, 1068 (1977).
236. L. N. Labzowsky and Y. E. Lozovik, *Phys. Lett.* **40A**, 281 (1972).
237. Y. E. Lozovik and A. V. Klyuchnik, *Phys. Lett.* **66A**, 282 (1978).

238. R. F. O'Connell, *Physica* **579B**, 1 (1980).

239. M. Altarelli and N. O. Lipari, *Phys. Rev. B* **9**, 1733 (1974).

240. R. F. O'Connell, *Phys. Rev. A* **17**, 1984 (1978).

241. H. G. Dehmelt, in *Proceedings of the 7th International Conference on Atomic Physics*, MIT (Boston, 1981).

242. W. A. M. Blumberg, R. M. Jopson, and D. J. Larson, *Phys. Rev. Lett.* **40**, 1320 (1978).

243. E. A. Solov'ev, *JETP Letters* **34**, 265 (1981).

244. D. R. Herrick, *Phys. Rev. A* **26**, 323 (1982).

245. M. Robnik, *J. Phys. A* **14**, 3195 (1981).

246. J. C. Gay, in *Photophysics and Photochemistry in the V.U.V.*, (D. Reidel, N.Y., 1983).

247. J. P. Connerade, J. C. Gay, and S. Liberman, *Physique Atomique et Moléculaires près des Seuils d'Ionisation en Champs Intenses*, Proceedings of the Colloque CNRS 334, Orsay (1982).

7

Effects of Magnetic and Electric Fields on Highly Excited Atoms

CHARLES W. CLARK, K. T. LU, AND ANTHONY F. STARACE

1. Introduction

The long and glittering history of the study of atoms in external electric and magnetic fields dates from the late 19th century work of Zeeman[1] and the early 20th century works of Stark[2] and Paschen and Back.[3] The birth of quantum mechanics was followed by the pioneering studies of diamagnetic effects of Van Vleck[4] and Jenkins and Segré,[5] and by the prediction of Landau resonances[6] in free-electron spectra of solids. More recently, observations of highly excited atoms in external fields,[7-10] made possible by the advent of high-resolution spectrometers, superconducting magnets, and lasers, have led to a revitalization of atomic spectroscopy. In particular, the observation of field-induced resonances in an otherwise smooth continuum[11-13] has compelled theory[14-16] to deal with a class of phenomena involving competing forces of different symmetries and comparable magnitudes. Fragmentary advances of theoretical understanding have in turn pointed to possibilities of using external fields as probes of atomic structure. In this sense the present lines of investigation adhere to

CHARLES W. CLARK • SERC Daresbury Laboratory, Daresbury, Warrington WA4 4AD, England. K. T. LU • Argonne National Laboratory, Argonne, Illinois 60439. ANTHONY F. STARACE • Behlen Laboratory of Physics, The University of Nebraska, Lincoln, Nebraska 68588-0111. Work supported by Office of Basic Energy Sciences, Division of Chemical Sciences, U.S. Department of Energy, under contract No. W-31-109-Eng-38; also, for Dr. Starace's work, under contract No. EY-76-S-02-2892. Present address for Dr. Clark: Atomic and Plasma Radiation Division, National Bureau of Standards, Washington, D.C.

the spirit of Professor Hanle's pioneering work, and we are honored to be able to dedicate this article to him on the occasion of his eightieth birthday.

This review describes primarily recent theoretical developments on highly excited atoms in uniform external fields, and complements the experimental review of Gay in the present volume. The Zeeman and Stark effects on low-lying atomic states, on the other hand, constitute a mature field of study which has been reviewed previously by Garstang,[17] Kollath and Standage,[18] Bayfield,[19] and Kleppner.[20] Most theoretical work on highly excited states in laboratory strength fields has heretofore focused on the prototype system of atomic hydrogen, and accordingly hydrogen receives special emphasis in this article. For nonhydrogenic atoms we review theoretical work using the framework of quantum defect theory. Magnetic and electric effects are treated separately and in combination. For magnetic fields of astrophysical magnitude, on the other hand, the competition between external and atomic forces becomes important for low-lying states. Recent developments in the understanding of hydrogen in such fields are reviewed here, and qualitative aspects of general atomic structure in such fields are briefly discussed.

As indicated above, the principal difficulty in dealing with highly excited atoms in external fields arises from the simultaneous presence of separate strong forces of different symmetry. For hydrogen in a magnetic field, the two forces involved are the spherically symmetric Coulomb interaction between the electron and the proton, and the cylindrically symmetric interaction between the electron and the magnetic field. Specifically, the potential consists of two terms: the Coulomb potential, $-1/r$, and the diamagnetic potential, proportional to $r^2 \sin^2 \theta$, where θ is the angle between the magnetic field axis and the electron position vector. These two in combination yield a Schrödinger equation which is nonseparable in any coordinate system. When one of the potentials is significantly smaller than the other, for instance at small r where the Coulomb potential is dominant, the problem can be solved by perturbation theory. This is the case for low-lying states in laboratory magnetic fields. For highly excited states, however, as the principal quantum number n increases, the electron moves to larger distances r where the strength of the diamagnetic potential becomes comparable to the Coulomb binding. As a result, one observes that as n increases there is a transition from a primarily Rydberg-like spectrum, through a region in which there is a breakdown of the zero-field Rydberg classification of states, to a new spectral regime of quasi-Landau resonances. This is also the case for nonhydrogenic atoms, but the spectrum of hydrogen is distinctive because of the zero field degeneracy of levels. We consider recent classical, semiclassical, and quantum mechanical approaches to the understanding of the spectrum. A partial classification of the hydrogen spectrum in the transition region is discussed.

The same competition between external and atomic forces occurs for highly excited states of hydrogen in a uniform electric field. However, the Schrödinger equation is separable in parabolic coordinates. Solutions to the separated equations are not available in closed form, so that approximate analytical or numerical means must be employed. For nonhydrogenic atoms, this separability of the equations of motion is broken near the ionic core. Thus again, although the dynamics of electron motion at large distances are the same for all atoms, the problem of hydrogen is unique. We discuss recent developments in analytical and numerical approaches to its solution.

The differences between hydrogen and other atoms arise from the presence of a non-Coulombic interaction between the excited electron and the residual ion core. In the absence of external fields, the effects of this interaction are conveniently characterized by quantum defects.[21-24] When external fields are present, quantum defects may still be used to describe this short-range interaction, because the magnitude of the external fields is negligible in comparison to the internal atomic fields. Thus we employ quantum defect theory as a tool for analyzing nonhydrogenic spectra in external fields. If one knows the analytic solutions of the Schrödinger equation in the region outside the core where only the net Coulomb and external field potential(s) are present, then the quantum defects may be used to construct a linear combination of these solutions that represents the electron wave function everywhere outside the core. For the diamagnetic problem such analytic solutions are not yet available, and so the present utility of a quantum defect approach is essentially restricted to perturbative treatments, which we shall review here. For the Stark effect, on the other hand, such solutions can be obtained in parabolic coordinates. Since, however, the usual quantum defect theory requires knowledge of the wave function on a spherical boundary enclosing the core, the matching of parabolic coordinate solutions to spherical boundary conditions is nontrivial. The solution to this matching problem has recently been obtained, and we shall review its main features.

As well as reviewing the Zeeman and Stark effects as distinct phenomena, we shall also treat cases in which both fields are simultaneously present. For the magnetic problem such a study is necessary because the center-of-mass motion of the atom induces an electric field in the atomic rest frame. Such motional Stark fields have been observed to have pronounced effects on the spectra of light atoms. We discuss present theoretical understanding of this phenomenon. This motional Stark effect is a special case of the general problem of crossed magnetic and electric fields, whose principal features we also review.

For laboratory strength fields the competition between external and atomic forces becomes significant only in highly excited states. For very high fields, such as may occur in astrophysics, this competition is important for low-lying excited states and even the ground states of atoms. In fact, for magnetic fields of the magnitude expected to exist on the surfaces of neutron stars, the magnetic field dominates the electronic motion in such a way as to radically change atomic structure. We shall review some of the general properties of atoms in such fields.

2. Diamagnetic Effects in the Hydrogen Atom

Effective use of a general quantum defect formulation can only be made if one has some knowledge of the solutions of the equations of motion for the electron in the region outside the atomic core. When the electron wave function in the exterior region is adequately represented by a perturbed Coulomb wave function, matching it to boundary conditions imposed near the atomic core is straightforward. This will be shown in Section 4.4. below. However, when the magnetic and Coulomb potentials become comparable in strength, perturbation theory is unsatisfactory and a direct solution to the equations of motion must be sought. At present no general method for obtaining such solutions is known. Thus most previous theoretical effort has been directed towards identifying and elaborating qualitative properties of the electron motion in the exterior region, and, recently, towards obtaining accurate wave functions for hydrogenic atoms. In this section we will review the development of the theory along these lines. Some progress towards a general theory is evident, but it has not reached the stage where a definitive treatment is possible.

2.1. The Equations of Motion

We consider first the classical Lorentz equation for the motion of an electron and a proton in a magnetic field:

$$m_e \ddot{\mathbf{r}}_e = -\frac{e^2 \mathbf{r}}{r^3} - \frac{e}{c} \dot{\mathbf{r}}_e \times \mathbf{B}$$

$$m_p \ddot{\mathbf{r}}_p = \frac{e^2 \mathbf{r}}{r^3} + \frac{e}{c} \dot{\mathbf{r}}_p \times \mathbf{B} \tag{1}$$

where $\mathbf{r} = \mathbf{r}_e - \mathbf{r}_p$. If we take $M = m_e + m_p$, $\mathbf{R} = (m_p \mathbf{r}_p + m_e \mathbf{r}_e)/M$, and $\mu = m_p m_e/(m_e + m_p)$ as is usually done to separate center-of-mass and internal motions, we find that

$$M \dot{\mathbf{R}} + \frac{e}{c} \mathbf{r} \times B = \boldsymbol{\pi} \tag{2}$$

where $\boldsymbol{\pi}$ is a constant. Thus $\boldsymbol{\pi}$ is to be identified with the net momentum of the system; the kinetic momentum $M\dot{\mathbf{R}}$ is not constant in time, but is coupled to the electric dipole moment $-e\mathbf{r}$ of the system. By combination of (1) and (2) the equation of motion for $\mathbf{r}$ is readily shown to be

$$\mu\ddot{\mathbf{r}} = -\frac{e^2\mathbf{r}}{r^3} - \frac{e}{c}\left(\sigma\dot{\mathbf{r}} - \frac{e}{Mc}\mathbf{r}\times\mathbf{B}\right)\times\mathbf{B} - \frac{e}{Mc}\boldsymbol{\pi}\times\mathbf{B} \tag{3}$$

where $\sigma = (m_p - m_e)/M \simeq 1$. The rightmost term of (3) indicates the presence of a uniform electric field in the center-of-mass frame, induced by the motion of that frame across the magnetic field lines. In the remainder of this section we shall take $\boldsymbol{\pi}\times\mathbf{B} = 0$. The effect of the motional electric field in the general case is discussed in Section 5.2 below. The second rightmost term may be simplified by writing the equations of motion in a coordinate frame rotating about the magnetic field axis: the angular frequency of rotation being either of $\omega_{\pm} = (eB/\mu c)(\sigma \pm 1)/2$, corresponding, respectively, to a counterclockwise rotation at approximately the electron cyclotron frequency $eB/m_e c$ or a clockwise rotation at approximately the proton cyclotron frequency $eB/m_p c$. Either choice gives equivalent results, but since $|\omega_-| \ll |\omega_+|$ we shall express the equations of motion in the slowly rotating frame. They are

$$\mu\ddot{\mathbf{r}} = -\frac{e^2\mathbf{r}}{r^3} - \frac{e}{c}\mathbf{r}\times\mathbf{B} \tag{4}$$

Thus the classical equations of motion in the rotating frame are equivalent to those for an electron of reduced mass μ moving in a fixed Coulomb potential in the presence of a magnetic field. We shall hereafter discuss just this idealized problem, its relation to realistic cases being taken as understood from the above arguments. In calculation of energies it must be remembered that the rotation of the coordinate frame produces a current which gives a slight paramagnetic energy; this appears in quantum mechanics as an adjustment of the electron's Landé g factor.

The formal separation of the center-of-mass motion in quantum mechanics is straightforward. Finite nuclear mass corrections to the Hamiltonian for a hydrogen atom in a uniform magnetic field were derived to first order in the electron–proton mass ratio of Lamb.[25] A physically intuitive derivation of these corrections has also been given by Bethe and Salpeter.[26] Interest in exotic atoms such as muonium and positronium, which have much larger mass ratios, led Carter[27] to solve the nuclear motion problem exactly for neutral two-body systems. Recently the problem has been reexamined by a number of authors.[28–34] In particular,

Avron et al.[30] have examined the general question of the separability of center-of-mass motion for N charged particles in a uniform magnetic field.

The Hamiltonian for atomic hydrogen in a uniform magnetic field B directed along the z axis is

$$H = \sum_{i=1}^{2} \frac{[\mathbf{p}_i - q_i \mathbf{A}(\mathbf{r}_i)]^2}{2m_i} - \frac{e^2}{|\mathbf{r}_1 - \mathbf{r}_2|} \tag{5}$$

where the index 1 refers to the electron and the index 2 refers to the proton, $-q_1 = q_2 = e$, and the vector potential is chosen in the Landau gauge, $A_y = Bx$, $A_x = A_z = 0$. Carter[27] has shown that the total wave function described by the Hamiltonian in (5) may be written in terms of the center-of-mass coordinate $R = (X, Y, Z)$ and the relative coordinate $r = (x, y, z)$ as

$$\Psi(\mathbf{R}, \mathbf{r}) = \exp i(\mathbf{P} \cdot \mathbf{R} + eByX)\psi(\mathbf{r}) \tag{6}$$

where $\mathbf{P}$ is the center-of-mass momentum and $\psi(\mathbf{r})$ is described by the following reduced Hamiltonian in the relative coordinate $\mathbf{r}$:

$$h = \frac{1}{2\mu}\left(\mathbf{p}_\mu + \frac{e}{c}\mathbf{a}\right)^2 - \frac{e^2}{r} - \frac{e\mathbf{r}}{c} \cdot \left(\frac{\mathbf{P}}{M} \times \mathbf{B}\right) + \frac{P^2}{2M} + \frac{e^2 B^2}{2Mc^2}(x^2 + y^2) \tag{7}$$

In Eq. (7), μ is the reduced mass, $\mathbf{p}_\mu$ is the momentum operator for the reduced mass particle, and $\mathbf{a}$ is a new vector potential defined by

$$\nabla \times \mathbf{a} = \left(\frac{m_2 - m_1}{m_1 + m_2}\right)\mathbf{B} = \mathbf{b} \tag{8}$$

The first two terms in (7) represent the usual Hamiltonian for a particle of mass μ moving in the Coulomb field and in the magnetic field given in (2.8); the third term represents the interaction of the reduced mass particle with the motional electric field $\mathbf{E} = (\mathbf{P}/M) \times \mathbf{B}$ arising from the center-of-mass velocity $(\mathbf{P}/M)$; the fourth term is the kinetic energy of center-of-mass motion; and the last term is a harmonic oscillator potential. As pointed out by O'Connell,[31] Eq. (7) may be greatly simplified by means of a judicious choice of gauge for the vector potential $\mathbf{a}$: $a_x = -by/2$, $a_y = bx/2$, $a_z = 0$. In this gauge and switching to the center-of-mass coordinate system (i.e., setting $\mathbf{P} = 0$) one finds that h reduces to

$$h = \frac{p_\mu^2}{2\mu} - \frac{e^2}{r} + \frac{e}{2\mu c} gBL_z + \frac{e^2}{8\mu c^2} B^2(x^2 + y^2) \tag{9a}$$

where

$$g \equiv \frac{m_2 - m_1}{m_1 + m_2} \tag{9b}$$

The reduced Hamiltonian h in Eq. (9) is idential to that for a particle of mass μ moving in a Coulomb field as well as a uniform magnetic field B except for the presence of the factor g in the linear Zeeman term instead of the usual factor unity. Interestingly, $g/\mu = m_1^{-1} + m_2^{-1}$ is Lamb's[25] correction for center-of-mass motion.

In summary, then, in the center-of-mass coordinate system the effect of a finite nuclear mass of the Hamiltonian for a hydrogen atom in a uniform magnetic field is to replace the electron mass by the reduced mass and to multiply the linear Zeeman term by the factor g in Eq. (9b).

2.2. Solutions near the Ionization Threshold

2.2.1. Classical and Semiclassical Approaches

We consider first the treatment of the Zeeman effect by classical and semiclassical methods. An intrinsic limitation of any such approach is that it cannot offer an accurate account of the electron interaction with the ionic core. However, as we have stressed above, this interaction can often be dealt with in terms of a few quantum defect parameters. The major task set to any theoretical treatment of this problem is to give a good description of the electron motion in the region outside the ionic core; and in this region the quasiclassical criterion—that the variation of the potential over an electron de Broglie wavelength be sufficiently small—is largely satisfied for the magnetic field strengths and excitation energies of current experimental interest. Thus classical methods may be expected to give at least correct qualitative information on the observable quantities which do not depend strongly on the particular nature of the core. The most prominent of these, the nonintegral spacing of photoabsorption resonances near ionization thresholds was indeed first explained in semiclassical terms.[35,36] The rather simple physical picture developed then has been elaborated upon, but has been essentially retained in all subsequent theoretical work.

The classical equation of motion for an electron in the field of an infinitely massive point nucleus of charge $+Z$ and a uniform magnetic field is (in c.g.s.—Gaussian units)

$$m_e \ddot{\mathbf{r}} = -\frac{Ze^2 \mathbf{r}}{r^3} - \frac{e}{c} \dot{\mathbf{r}} \times \mathbf{B} \tag{10}$$

The correction for finite nuclear mass is insignificant for the kilogauss field strengths we treat in this section, as pointed out by O'Connell.[31] We shall hereafter take the magnetic field to point along the Cartesian z axis, $\mathbf{B} = B\hat{z}$. As pointed out by Gajewski,[37] who seems to have performed the first systematic numerical study of the classical problem, Eq. (10) can be reduced to a parameter-free form by appropriate scaling of the space and time coordinates. With

$$\tau = \omega t$$
$$\mathbf{R} = \zeta \mathbf{r} \tag{11}$$

where $\omega = eB/m_e c$ is the cyclotron frequency and $\zeta = (Zm_e c^2/B^2)^{-1/3}$, Eq. (10) becomes equivalent to

$$\ddot{\mathbf{R}} = -\mathbf{R}/R^3 - \dot{\mathbf{R}} \times \hat{z} \tag{12}$$

The solutions to Eq. (12) are determined solely by the initial conditions $\mathbf{R}(\tau = 0)$, $\dot{\mathbf{R}}(\tau = 0)$; if solutions for all such initial conditions are known, solutions of Eq. (10) appropriate to any values of Z and B are obtained from Eq. (11). Some appreciation of the magnitude of the scaling parameters can be had from the observation that the characteristic length $L = \zeta^{-1}$ is the radius of a sphere containing magnetic field energy $(B^2/8\pi) \cdot (4\pi/3)L^3$, which is comparable to the rest mass energy $m_e c^2$ of the electron. For the kilogauss fields discussed in this section, the cyclotron frequency ω is typically less than one hundred thousandth of the orbital frequency of the electron in the ground state of a hydrogen atom.

Two elementary constants of the motion are apparent in Eq. (12); an energy, $\varepsilon = \frac{1}{2}\dot{\mathbf{R}}^2 - 1/R$; and an effective z component of angular momentum, $\lambda_z = \hat{z} \cdot \mathbf{R} \times \dot{\mathbf{R}} - \frac{1}{2}(\hat{z} \times \mathbf{R})^2$. It is convenient to recast Eq. (12) in Hamiltonian form, utilizing ε and λ_z. With R and $\theta = \cos^{-1}(\mathbf{R} \cdot \hat{z})$ denoting the usual polar coordinates of the vector $\mathbf{R}$, it is readily shown that Eq. (12) is equivalent to

$$\frac{1}{2}\dot{R}^2 + \frac{1}{2}R^2\dot{\theta}^2 + \frac{\lambda_z^2}{2R^2\sin^2\theta} - \frac{1}{R} + \frac{\lambda_z}{2} + \frac{1}{8}R^2\sin^2\theta = \varepsilon \tag{13}$$

This is recognizable as the equation of motion of a particle in a combined spherical Coulomb potential $-1/R$ and a cylindrical harmonic oscillator potential $\frac{1}{8}(R\sin\theta)^2$. The constant $\lambda_z/2$ on the left-hand side of Eq. (13) corresponds to the linear Zeeman shift of quantum theory, which depends only on the z component of angular momentum.

(a) *Classical Theory of Planar Motion.* No general solution to Eq. (13) has yet been discovered, nor are any additional constants of the motion known (though, as described below, there is some evidence for the existence of an approximate constant of the motion). However, one class of solutions of Eq. (13)—those with initial conditions $\theta = \pi/2$, $\dot{\theta} = 0$—can be carried out in closed form. Such solutions describe electron motions which are always confined to the plane $z = 0$ which contains the nucleus and lies perpendicular to the direction of the field. It is clear that in this case the solution to Eq. (13) can be carried out in terms of elliptic integrals.

The relevance of this class of solutions to the spectroscopy of highly excited states in kilogauss fields is not immediately evident. In photoabsorption experiments the Rydberg electron emerges from the atom in a more or less spherical wave, and one would expect population of such planar motions to be quite improbable. Moreover, the planar orbits tend to be unstable with respect to small excursions out of the plane. It will be seen, however, that such motions, even though unstable, are of great importance in the photoabsorption spectrum. It is this feature which links the theory of the quadratic Zeeman effect with broader questions of dynamics which are raised below.

Without immediately addressing the question of orbital stability, we shall now examine the properties of the solutions of Eq. (13) for which $z = 0$ always. Because of the smallness of the scaling parameter ζ—it is about $(1350a_0)^{-1}$ for hydrogen in a 50-kG field—the quadratic term $\frac{1}{8}R^2 \sin^2 \theta$ can be disregarded in determining the low-energy solutions, which are then essentially the orbits of a two-dimensional hydrogen atom. At high energies, on the other hand, the quadratic term becomes the most important part of the potential and the solutions must go over to those of the two-dimensional harmonic oscillator. The relevance of this equation to the interpretation of spectroscopic data, which covers an intermediate range of energies, can, however, only be determined by finding the allowed quantum-mechanical energy levels.

(b) *Semiclassical Generalization*: *Interpretation of the Quasi-Landau Resonances*. Edmonds[35] first did this by applying the appropriate Bohr–Sommerfeld quantization rule:

$$\oint P_\rho \, d\rho = (n + 1/2)h \tag{14}$$

with P_ρ being the classical momentum conjugate to the two-dimensional radial coordinate $\rho = (x^2 + y^2)^{1/2}$. By numerical integration of Eq. (14) he found that, for hydrogen in a field of 24 kG, the energy levels near the ionization threshold are uniformly spaced, the separation between adjacent levels being approximately $1.58 \, \hbar\omega$. Moreover, Eq. (14) can be evaluated in an arbitrary plane $z = $ const (though, strictly speaking, the orbits are not solutions to the classical equations of motion unless $z = 0$). This results in a set of energy levels which vary smoothly with z, going over to the harmonic oscillator levels as $z \to \infty$. It may also be remarked that the energy levels so calculated vary weakly with z for small z. Thus it seems plausible to associate those prominent spectral features which are observed to be spaced by an energy of $\sim 1.5 \, \hbar\omega$ (Figure 1), with electron orbits which are largely confined to small values of z.

Of course, the actual position of an allowed energy level will depend strongly on the interaction of the Rydberg electron with the atomic core, which is not given realistically in this model. However, when cast in the form of a quantum defect, this residual interaction generally varies smoothly with energy near the ionization limit. This is because a small change in the kinetic energy of the Rydberg electron at large r results in a much smaller proportional change in its kinetic energy near the residual core. Thus, as in the ordinary field-free quantum defect theory, the density of states in energy $\partial n/\partial E$ is determined principally by the form of the potential at large distances. Differentiation of Eq. (14) gives an expression for the density of states confined to the plane. Starace[36] computed the resulting integral numerically and obtained the results shown in Figure 2. The basic result of interest can be seen from the expression

$$\frac{\partial n}{\partial E} = \frac{2}{\pi \hbar \omega} \int_{-1}^{1} \frac{dz \, (z + \gamma)}{[(1 - z^2)(z^2 + 4\gamma z + 6\gamma^2 + 1 - 4\varepsilon)/(\rho_2 - \rho_1)^2]^{1/2}} \quad (15)$$

where $\varepsilon = 8E/(m_e \omega^2)$, ρ_2 and ρ_1 are the greater and lesser of the two classical turning points, and $\gamma = (\rho_2 + \rho_1)/(\rho_2 - \rho_1)$. The inner turning point ρ_1 is (for the kilogauss fields of interest) determined solely by the Coulomb potential and the centrifugal barrier, and so is of the order of unity. The outer turning point ρ_2 is, on the other hand, determined by the relative magnitude of the Coulomb and quadratic potentials, and for energies near threshold, $\rho_2 \simeq 2/\zeta$. Thus, γ is very nearly equal to 1, with only weak dependence on the z component of angular momentum $m\hbar$ and the magnetic field strength; for $m = 0$ we have $\gamma = 1$ independent of the field. If we take $\gamma = 1$ and consider the case $E = 0$, the integral in Eq. (15) can be evaluated by elementary means and yields

$$\frac{\partial n}{\partial E} = \frac{2}{3\hbar\omega} \quad (16)$$

When $E \gg \hbar\omega$, Eq. (15) reduces to the result for the two-dimensional oscillator

$$\frac{\partial n}{\partial E} = \frac{1}{\hbar\omega} \quad (17)$$

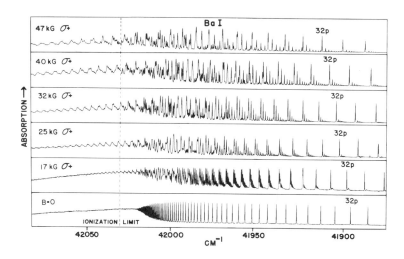

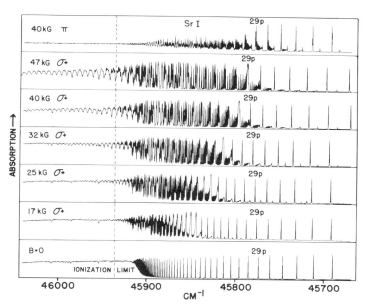

Figure 1. (a) Microdensitometer recording of the plate transmission of the Ba principal series with magnetic field strengths $B = 47, 40, 32, 25$, and 17 kG (σ^+ polarization) and $B = 0$ (from Ref. 11). (b) Microdensitometer recording of the plate transmission of the Sr principal series. The top spectrum is that for π-polarization with $B = 40$ kG. The next five spectra are for σ^+ polarization with $B = 47, 40, 32, 25$, and 17 kG. The last spectrum is for $B = 0$. (From Ref. 11.)

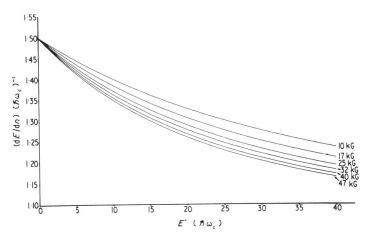

Figure 2. WKB results for the quasi-Landau level separations $\partial E/\partial n$ [cf. Eq. (15)] plotted vs. energy above threshold for various magnetic field strengths. Both axes are in units of the cyclotron energy $\hbar\omega$. Note that $\partial E/\partial n = 1.5$ at threshold regardless of field strength. (From Ref. 36.)

The density of states increases monotonically as the energy increases from the ionization threshold. The experimental spectra show rather broad features, with some secondary structure, above the ionization limit (cf. Figure 1). However, it is possible to determine the positions of the centers of the major peaks without too much ambiguity. Garton et al.[38] have done this and found that their measurements of the quasi-Landau energy spacings are in quite good agreement with Starace's[36] results for $(\partial E/\partial n)$ using the WKB approximation of Eq. (15), as shown in Figure 3. Setting $\gamma = 1$ again, the total number N_T of bound energy levels in the plane with given angular momentum $m\hbar$ can be determined as

$$N_T = \frac{1}{2} + \frac{2^{-1/3}\Gamma(1/3)}{[\Gamma(2/3)]^2} \frac{m_e\omega}{\hbar\zeta^2} \tag{18}$$

For example, $N_T \approx 43$ for hydrogen in a 50-kG field. The only dependence on m comes through the ratio γ. This number N_T, if correctly interpreted, is also in rough agreement with "experimental" data, as will be seen below.

(c) *Three-Dimensional Orbits.* The quasi-classical approach thus accounts for some systematic features of the experimental spectra. It cannot, however, provide any definite prediction of line widths or the distribution of oscillator strength, and it is clear from the data that there are many more lines of spectroscopic importance than can be accounted for in this

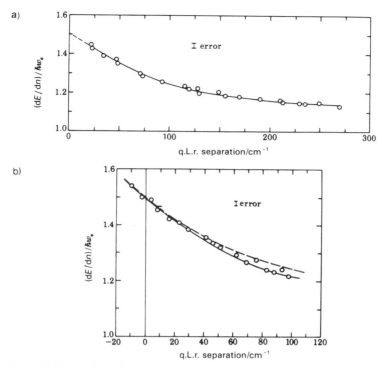

Figure 3. Comparison of the measured quasi-Landau energy level separations of Garton *et al.* (Ref. 38), shown by open circles, with predictions of the WKB result for $(\partial E/\partial n)/\hbar\omega_c$ of Starace (Ref. 36). (a) Barium in a 46.9-kG field; WKB result [cf. Eq. (15)] indicated by the solid curve. (b) Strontium in a 48.3-kG field; WKB result [cf. Eq. (15)] indicated by the dashed curve. The solid curve in (b) shows the trend of the experimental data. The difference between the WKB result and experiment is about twice the estimated experimental error in the case of Sr. The abscissas indicate the energy above the threshold in cm^{-1}.

way. An alternative, purely classical, approach to the problem has been taken by Edmonds and Pullen,[39] who do not restrict the electron's motion to the plane. They proceed from an equation similar to Eq. (13), with, however, a different choice of scaling parameters: the unit of distance being $2/\zeta$ and the unit of time chosen as $4/\omega$. Their unit of energy is twice the electrostatic potential between a proton and electron separated by a distance ρ_0. In these units the magnitude of Planck's constant depends on the strength of the field, but typically $\hbar \approx 10^{-2}$ for fields in the range 10–50 kG. Since classical and quantum mechanics coincide when $\hbar \to 0$, this magnitude may be taken as a rough measure of the validity of a classical treatment.

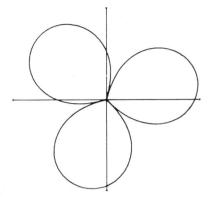

Figure 4. The classical planar orbit for an electron with zero energy in combined Coulomb and magnetic fields. (From Ref. 39.)

Edmonds and Pullen have done considerable numerical investigation of the classical solutions. One novel feature of their approach is the use of a regularization transformation of the space and time coordinates, of a type familiar in celestial mechanics, which removes the singularity in the Coulomb potential near the nucleus and thus greatly improves the numerical stability of the calculation. Their interpretation of the results is founded on the application of the correspondence principle. If the classical motion is periodic with angular frequency ω_{cl}, and if ω_{cl} does not vary appreciably over a given energy interval, then the spacing of quantum mechanical energy levels within that interval must be $\Delta E = \hbar\omega_{cl}$. A trefoil classical planar orbit of zero energy and small angular momentum is shown in Figure 4. The angular frequency of this motion is $\omega_{cl} = \frac{3}{2}\omega$. Edmonds and Pullen report that the angular frequency of classical orbits of this type does not vary by more than 5% throughout the energy range (in their units) $-0.1 < E < 0.01$. Thus the correspondence principle can be invoked with reasonable confidence, and yields the same result as the WKB treatment for the threshold spacing.

However, for any given energy there are many orbits which are not periodic so that this analysis does not apply. By computing surfaces of section[40] for the classical trajectories, Edmonds and Pullen have obtained a comprehensive catalogue of the types of orbit which exist in different ranges of energy. For energies $E < -1/2$ in their units, they find that nearly all motions are regular (in the sense of Percival[41]), i.e., they can be regarded as the resultant of superposed periodic motions. As the energy rises above this value, an increasing volume of phase space is filled by irregular orbits and at the ionization limit $E = 0$ the entire volume of phase space seems to be occupied by irregular trajectories. Though none of these are strictly periodic, some consist of motions in which the electron returns to the region near the proton a number of times before drifting away. These may be imagined to have important quantum mechanical analogs, i.e., states with small but nonnegligible oscillator strength which might be associated with the fine structure in the photoabsorption cross sections seen above the ionization threshold. A very qualitative but plausible method of assessing the quantum mechanical significance of such orbits has been put forward. In photoabsorption the electron will be ejected at some angle relative to the plane perpendicular to the field. If it is placed in a quasi-periodic orbit, it will return at some later time to the neighborhood of the nucleus. The inverse of the distance of closest approach upon return to the nucleus may be supposed to be a measure of both the stability of the quasiperiodic motion and of its associated oscillator strength. Figure 5 shows the spectrum of this inverse impact parameter as a function of the ejection angle, for a class of quasiperiodic orbits calculated by Edmonds and Pullen.[39] Firm quantitative results have not however been produced in this framework.

Edmonds and Pullen[39] have also provided a classical interpretation of the effect of the motional electric field on the spectrum near threshold. This field has not been of obvious importance in experiments on heavy atoms. In photoabsorption by Li vapor at a temperature of $\approx 800°$ K,

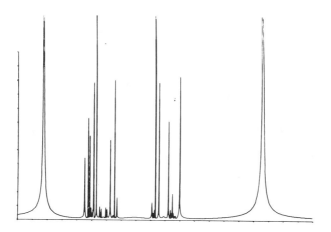

Figure 5. A representation of periodic classical orbits at zero energy. The abscissa is one of the initial regularized momenta of the system, corresponding to the initial angle of the orbit with respect to the plane $z = 0$. The ordinate is the inverse of the impact parameter on the return journey. (From Ref. 39.)

however, it has been observed[13] that the spacing between major features is $\frac{1}{2}\hbar\omega$ rather than the semiclassical value $\frac{3}{2}\hbar\omega$. This difference was attributed by Crosswhite *et al.*[15] to the presence of the motional electric field, using arguments summarized in Section 5 of this paper. Edmonds and Pullen[39] have shown that a motional Stark field of the appropriate magnitude splits the trefoil orbit of Figure 4 into another three-lobed periodic orbit, only one lobe of which passes through the nucleus (Figure 6). As the frequency of return is thus 1/3 the value without the electric field, the correspondence principle gives a $\frac{1}{2}\hbar\omega$ spacing.

NOTE (added in proof): Since this article was written, another investigation of the quadratic Zeeman effect by classical methods was reported by Robnik.[138] Though it was applied to fields of astrophysical strength, this work employed methods similar to those of Edmonds and Pullen and yielded results consistent with theirs. In particular, Robnik also observed a transition to occur between regular and irregular motion at a critical energy.

(d) *Classical and Semiclassical Perturbation Theory: Coupled Motion of Runge–Lenz and Angular Momentum Vectors.* An elegant treatment of the classical problem by perturbation theory has been recently carried out by Goebel and Kirkman.[42] If the magnetic term of Eq. (12) is suppressed,

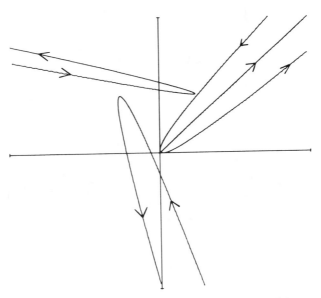

Figure 6. The disruption of the trefoil orbit of Figure 4 by a motional electric field. (From Ref. 39.)

we recover the ordinary Kepler equation. This exhibits seven constants of the motion: the energy ε, the angular momentum vector $\boldsymbol{\lambda} = \mathbf{R} \times \dot{\mathbf{R}}$ and the Runge–Lenz vector $\boldsymbol{\alpha}$

$$\boldsymbol{\alpha} = (-2\varepsilon)^{-1/2}[(\dot{\mathbf{R}} + \boldsymbol{\lambda} - \mathbf{R})/R] \tag{19}$$

where $\varepsilon < 0$ for bound orbits. These are not all independent since

$$\boldsymbol{\lambda} \cdot \boldsymbol{\alpha} = 0, \qquad \boldsymbol{\lambda}^2 + \boldsymbol{\alpha}^2 = (-2\varepsilon)^{-1} = N^2 \tag{20}$$

Goebel and Kirkman calculate the average value of the diamagnetic potential $\frac{1}{8}R^2 \sin^2 \theta$ over an unperturbed Kepler orbit and find that

$$U = \langle \tfrac{1}{8}R^2 \sin^2 \theta \rangle = \frac{N^2}{16} [5(\alpha^2 - \alpha_z^2) + \lambda^2 + \lambda_z^2] \tag{21}$$

If U is then regarded as an effective potential, equations of motion for the vectors $\boldsymbol{\lambda}$ and $\boldsymbol{\alpha}$ may be obtained. If we substitute U for $\frac{1}{8}R^2 \sin^2 \theta$ in Eq. (13), we obtain an effective classical Hamiltonian function

$$H = H_c + \frac{\lambda_z}{2} + U \tag{22}$$

where H_c is the ordinary Coulomb Hamiltonian. Recall that the classical equation of motion for a function $f(q, p)$ of canonical coordinates q and their conjugate momenta p is

$$f = \{f, H\} = \sum_i \frac{\partial f}{\partial q_i} \frac{\partial H}{\partial p_i} - \frac{\partial f}{\partial p_i} \frac{\partial H}{\partial q_i} \tag{23}$$

the brackets being the familiar Poisson brackets. As is well known, they are the classical analog of the quantum mechanical commutation brackets.[43] The equations of motion which result for $\boldsymbol{\lambda}$ and $\boldsymbol{\alpha}$ by applying Eqs. (22) and (23) are described as follows. Since $\boldsymbol{\lambda}$, $\boldsymbol{\alpha}$ are constants of motion for the Coulomb problem, $\{\boldsymbol{\lambda}, H_c\} = \{\boldsymbol{\alpha}, H_c\} = 0$. The term $\lambda_z/2$ results in a uniform precession of $\boldsymbol{\lambda}$ and $\boldsymbol{\alpha}$ about the direction of the field, with frequency $1/2$ with respect to the scaled time τ (recall the quantum mechanical description of angular momentum as the generator of rotations). Since U is quadratic in the components of $\boldsymbol{\lambda}$ and $\boldsymbol{\alpha}$, its contribution to their time derivatives is an asymmetric bilinear form; in fact, the equations of motion resulting from U are isomorphic to the Euler equations for a rotating four-dimensional top. We shall not reproduce the equations in detail here, but will only give a brief summary of the properties of their solutions. They exhibit four constants of motion: λ_z, U, and the two conditions of Eq. (20). Thus working with this system of equations is the classical analog of degenerate perturbation theory in quantum mechanics (discussed below), since from Eq. (20) N may be readily identified with the principal quantum number n.

The qualitative behavior of the orbits computed in classical perturbation theory is as follows. Recall that, in field free hydrogen, $\boldsymbol{\alpha}$ is directed along the major axis of the elliptical orbit and its magnitude is proportional to the eccentricity of the orbit. For low-energy perturbed orbits, the major axis oscillates in a plane parallel to the field, and stays near the z axis, i.e., α_z remains either always positive or always negative. For higher energies the orbital axis oscillates about the $z = 0$ plane; the highest energy orbit of fixed N is confined to that plane. In addition, there is uniform precession of the orbital axis about the direction of the field.

Goebel and Kirkman[42] have carried out semiclassical quantization of the perturbed system by requiring the resulting action integrals to be multiples of Planck's constant. The resulting energy spectrum resembles that of a double-well potential problem. Low-lying states are doubly degenerate, being confined to either of the wells. The highest state lies on the barrier separating the wells.

NOTE (added in proof): After the completion of this article, another treatment by classical secular perturbation theory was reported by Soloviev.[139] He obtains equations of motion equivalent to those of Geobel and Kirkman for the Runge–Lenz and angular momentum vectors. Their associated quantum mechanical spectrum is surveyed by applying Bohr–Sommerfeld quantization to the component of angular momentum perpendicular to the magnetic field [i.e., our Eq. (14) with ρ replaced by θ]. The form of the quantization integral suggests a double-well potential for the low-lying states; however, for the highest states in the spectrum, the effective potential takes the form of a single well which confines the electron orbit to the angular range $\theta = \pi/2 \pm \sin^{-1}(1/\sqrt{5})$, i.e., about the plane $z = 0$. This is consistent with the non-perturbative treatment described in the next subsection, and with the numerical evidence discussed in Section 2.2.2d.

(e) *Wave Propagation along a Potential Ridge.* In the results of Goebel and Kirkman[42] we see the localization of some orbits in the plane $z = 0$. Fano[44,45] has recognized this aspect as being common to a large class of problems in which electron motion takes place in the presence of a rising "ridge" of a potential surface. The general semiclassical theory he has developed for such systems remains yet to be applied in detail to the quadratic Zeeman problem. Nevertheless, even in its present stage of development this theory provides some account of the mechanism responsible for the novel results appearing in large brute-force quantum mechanical calculations. Since many of its implications have not been worked out, a full review here of this theory would be premature. A brief discussion of its elements may, however, help fill in the background for quantum mechanical considerations.

In the atomic units which will be employed in the remainder of this section, the quadratic Zeeman Hamiltonian is

$$H = \tfrac{1}{2}\mathbf{p}^2 - Z/r + \tfrac{1}{2}\beta^2 r^2 \sin^2 \theta \qquad (24)$$

with $\beta = \omega/2$ a.u. and θ the polar angle of the electron position vector as defined previously. The linear Zeeman shift is here understood to be included in the energy eigenvalue. At fixed electron–nuclear distance r, the potential in Eq. (24) is at a maximum in the plane $\theta = \pi/2$. The full potential surface (Figure 7) takes the form of a ridge straddling this plane. The azimuthal motion of the electron is trivial, so we need consider only the two degrees of freedom associated with the coordinates r and $\xi = \theta - \pi/2$. This choice of the angular parameter ξ is motivated by the implication from spectroscopic data that motions involving only small displacements from the ridge at $\xi = 0$ are of greatest importance. Then by expanding the classical Hamilton–Jacobi equation appropriate to the system (24), and applying a linearization transform analogous to that employed by Wannier,[46] Fano identifies two bundles of trajectories $\xi_{\pm}(r)$. Trajectories of diverging type $\xi_+(r)$ move gradually off the ridge as r increases; the converging trajectories ξ_- remain on the ridge. Individual trajectories within a bundle are described by a scale parameter T which is determined

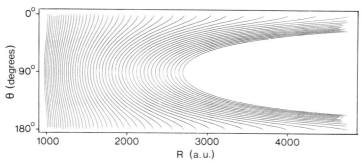

Figure 7. Equipotential surfaces for hydrogen in a 47-kG field. Abscissa: electron–proton distance in atomic units; ordinate: angle between electron position vector and magnetic field. The equipotential lines are spaced by one fifth of the cyclotron energy, the rightmost corresponding to zero energy. (From Ref. 60.)

by initial conditions; specifically, the diverging and converging trajectories with a particular value of T are given, to lowest order, by

$$\xi_{T\pm} = Tr^{\zeta_\pm(r)} \qquad (25)$$

When the two exponents $\zeta_\pm(r)$ do not vary rapidly with r, they are the roots of a quadratic equation; in general they are the solutions to a simple Riccati differential equation. The radial dependence of these exponents is indicated in Figure 8.

Since $\zeta_+ \to 0$ at small r, a bound electron excited by a photon will necessarily move to large r on a diverging trajectory. As the classical turning point is approached, the diverging exponent ζ_+ begins to increase rapidly, and some fractional transfer of the electron wave function to a converging trajectory will occur. At the turning point the diverging trajectory veers sharply away from the ridge; the converging trajectory, on the other hand, undergoes elastic reflection. Thus that fraction of the electronic wave function on the converging trajectory travels back towards the nucleus; that on the diverging trajectory is scattered to large angles ξ, leading to ionization if there is sufficient energy. The standing waves which are associated with the structures seen in photoabsorption may then be visualized as being built up by repeated passes from small to large r along the appropriate trajectories. The actual magnitude of the coupling between converging and diverging trajectories has not yet been established, so it is

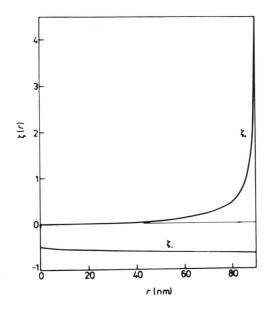

Figure 8. Characteristic exponents of diverging and converging trajectories for hydrogen in a 94-kG field, as a function of electron–proton distance in units of 10^{-9} m. (From Ref. 45.)

not presently possible to make any definite statement about the energies or widths of the bound levels and continuum resonances. However, this formulation provides a plausible mechanism for the generation of standing wave motion along the ridge. Some qualitative conclusions about its further development can be inferred from the "experimental" data provided by large quantum mechanical calculations.

2.2.2. The Quantum Theoretical Treatment

(a) *General Aspects.* We continue on the trail of the Zeeman effect for moderate field strengths (10–100 kG), though the bulk of quantum theoretical work has been done for much stronger fields (10^4–10^9 kG). These two cases are sufficiently different to warrant a separate section on the strong fields. For although in both cases one must deal with a nonseparable Hamiltonian not usually amenable to a perturbative treatment, in the strong field case the density of levels in the energy region of interest is very much smaller than that arising in the case of a moderate field. Thus, any calculation of a standard type applied to the case of moderate field strength, is necessarily large, and is bound to produce many solutions of little relevance to conventional spectroscopy.

We return to Eq. (24), which gives the Hamiltonian for electron motion in the region $r > r_0$, where r_0 is the size of the residual core. For a field of 47 kG, $\beta = 10^{-5}$, which we shall take to be a typical value in the remainder of this section. It is clear that for a significant distance beyond the core, $r_0 < r < r_1$, the magnetic potential is negligible in comparison to the Coulomb and centrifugal potentials; r_1 being, say, several hundred Bohr radii. Then in the region $r_0 < r < r_1$ the wave function may be expanded in the form

$$\psi(\mathbf{r}) = \sum_\alpha A_\alpha \sum_i \Phi_i U_{\alpha i}[f_i(r) \cos \pi\mu_\alpha - g_i(r) \sin \pi\mu_\alpha] \qquad (26)$$

Here the terms in the summation over i are those determined by the MQDT treatment of the field-free atom (though they may be slightly modified by the Zeeman effect on the core) as defined subsequently in Section 4; in the case of hydrogen, Φ_i reduces to a spherical harmonic of the electron position vector, f_i is the appropriate Coulomb wave function, and $\mu_\alpha \equiv \mu_l = 0$. The A_α, on the other hand, are determined by boundary conditions imposed at large r; that is, they determine the superposition of eigenchannels at small r which leads to a wave function vanishing at large r (or one which describes electron escape along the z axis, if the energy is sufficient). They will be dependent on the energy and the eigenchannel parameters, in a manner which has not yet been generally determined.

(b) *Degenerate Perturbation Theory in Spherical Coordinates.* We shall now discuss the computation of the A_α in the context of perturbation theory, which is applicable when the states considered do not extend far beyond $r = r_1$. The calculations involved are relatively trivial but, at least for hydrogen, bring out a number of features which do appear to persist beyond the strict range of validity of perturbation theory.

Hydrogen in the absence of a field has bound states grouped in degenerate manifolds with given principal quantum number n. The nth manifold is n^2-fold degenerate (neglecting the electron spin, which is for our purposes a constant of the motion[47]), containing all states with orbital momenta $l < n$. Taking $V = \frac{1}{2}\beta^2 r^2 \sin^2\theta$ as a perturbation, it is found that the matrix elements of V within an n manifold are of order $\beta^2 n^4$. The energy difference between adjacent manifolds is of order n^{-3}. Thus if $\beta^2 n^7$ is small, perturbation theory can be carried out within a single n manifold; on the other hand, it is apparent that as n becomes large a significant number of manifolds may be coupled by the magnetic interaction. The density of states and the strength of the perturbation increase together with n. We shall for the moment consider only the cases in which perturbation theory can be carried out within a degenerate manifold.

As $m = \mathbf{l} \cdot \hat{z}$ is a constant of the motion, one treats separately each submanifold of states with fixed m. Each of these is further split by the conservation of parity, so that for each manifold of fixed n, m one has a separate calculation to perform for states of even and odd parity: roughly equal numbers of states of each parity $[\approx (n - |m|)/2]$ are involved. For fixed n, m, and parity the perturbation $\frac{1}{2}\beta^2 r^2 \sin^2\theta$ is diagonalized. The resulting eigenstates are independent of the field strength, and the perturbed energies scale with field strength simply as β^2.

As $\sin^2\theta$ is a combination of tensors of rank 0 and 2 with respect to rotations, the matrix elements $\langle nl'm|V|nlm\rangle$ necessarily vanish when $|l - l'| > 2$. This enables the perturbation matrix to be written in tridiagonal form. The angular matrix elements $\langle l'm|\sin^2\theta|lm\rangle$ are determined straightforwardly by angular momentum algebra; the radial matrix elements, which are independent of m, are given by[17]

$$\langle nl'|r^2|nl\rangle = \frac{n^2}{2Z^2}[5n^2 + 1 - 3l(l + 1)] \qquad l' = l$$

$$= \frac{5n^2}{2Z^2}\{(n^2 - l_>^2)[n^2 - (l_> - 1)^2]\}^{1/2} \qquad l' = l \pm 2$$

(27)

The energies and eigenfunctions of hydrogenic degenerate perturbation theory can be determined routinely for n values of up to several thousand. (Though, for hydrogen in a 50-kG field, degenerate perturbation theory ceases to be valid at around $n = 20$). Methods alternative to straightforward matrix diagonalization have been put forward by Killingbeck[48] and Avron et al.[49] These, however, appear to be more appropriate to the consideration of low-lying states in very strong fields, and so are dealt with in Section 2.2.3.

We give here only a brief summary of the results obtained in numerical solution of the perturbation equations. It is found that the highest and lowest eigenvalues are nearly equally spaced in energy, in agreement with the semiclassical results of Goebel and Kirkman.[42] If the oscillator strengths for dipole transitions from, e.g., the hydrogen $1s$ ground state to perturbed Rydberg states are computed, a significant difference between the spectra of states with different values of m is observed. If the parity of the perturbed manifold is equal to $(-1)^{|m|}$, the oscillator strength tends to be concentrated in a single state (see the lowest line cluster of Figure 10). For parity equal to $(-1)^{|m|+1}$, on the other hand, the oscillator strength tends to be spread more or less uniformly among all the perturbed states (Figure 9). Since the total oscillator strength associated with a given manifold must be equal to the field-free value of the oscillator strength—this is the classical concept[4] of "spectroscopic stability"—the spectrum of states in the latter category will be more diffuse. It should be recalled that these states necessarily have a node in their wavefunctions in the plane $z = 0$. Thus they are less confined by the magnetic field than the states with parity $(-1)^{|m|}$, which necessarily display an antinode along $z = 0$ (the lowest clusters of Figure 10 are an example). It will be seen below that in fact the wave function of the strongest line of that manifold of states attains its maximum in the plane.

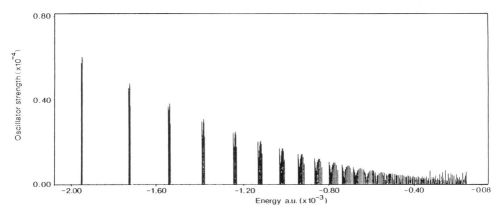

Figure 9. Partial photoabsorption spectrum of ground state hydrogen in a magnetic field of 47 kG. The light is polarized linearly along the field axis (π polarization). The abscissa is the absolute energy of the final states involved; the ordinate, the oscillator strength. The lowest cluster of lines describes the perturbed $n = 23$ manifold. (From Ref. 59.)

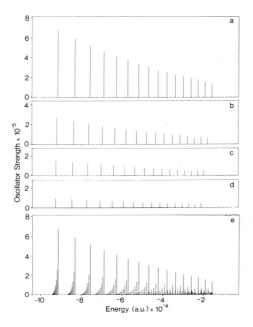

Figure 10. The σ polarization absorption spectrum of hydrogen in the same conditions as of Figure 9. Frame (e) shows the net oscillator strength distribution in this energy region; frames (a)–(d) depict the principal, second, third, and fourth series, respectively. (From Ref. 60.)

For nonhydrogenic atoms similar calculations may be carried out, though the zero-field degeneracy of levels is broken by the presence of quantum defects. One is then faced with evaluating the radial matrix elements of Eq. (27) with respect to principal quantum numbers $n^* = n - \delta$ which are nonintegral. This can be done to very good accuracy by using only the asymptotic form of the Coulomb wavefunction [from Eq. (26)], a method analogous to that employed by Bates and Damgaard[50] for the computation of oscillator strengths. The diagonal matrix elements of Eq. (27) turn out to be nearly equal to those obtained simply by substituting n^* for n; the off-diagonal elements are not so easy to summarize succinctly (see, however, Picart et al.[51]), but their calculation presents no real difficulty. This method has been applied by Crosswhite et al.[15] to portions of the lithium spectrum, by Edmonds and Kelley to barium, and by Clark and Taylor to magnesium (both unpublished). The magnesium spectrum is of some particular interest in that in the range $n \cong 5 - 12$ the quantum defects for p and f states differ by nearly unity,[52,53] so the zero-field degeneracy is almost recovered.

(c) *Alternative Perturbation Treatment in Parabolic Coordinates: The Role of the Runge–Lenz Vector.* An alternative formulation of degenerate perturbaton theory for hydrogen reveals several striking new features, and indeed nearly allows the perturbation equations to be solved in closed form as shown in recent work by Goebel and Kirkman.[42] As is well known, the field-free equations of motion for hydrogen are separable in the parabolic coordinates $\xi = r + z$ and $\eta = r - z$. Their eigensolutions take the form

$$|n_1 n_2 m\rangle = f_{n_1}(\xi) f_{n_2}(\eta) \frac{e^{im\phi}}{(2\pi)^{1/2}}$$

$$n = n_1 + n_2 + |m| + 1 \tag{28}$$

where n_1, n_2 are, respectively, the number of nodes in the ξ and η components of the wave function, and assume all positive values consistent with Eq. (28); the f's are products of Laguerre polynomials and exponentials.[54,55] Since the parity operator P interchanges ξ and η, these wave functions are generally not eigenfunctions of parity; rather

$$P|n_1 n_2 m\rangle = (-1)^m |n_2 n_1 m\rangle \tag{29}$$

In the parabolic coordinates the diamagnetic perturbation takes the form

$$V = \tfrac{1}{2}\beta^2 r^2 \sin^2 \theta = \tfrac{1}{2}\beta^2 \xi\eta \tag{30}$$

so that, in contrast to the spherical system, V is both linear and symmetric in the separate coordinates. Consequently, within the degenerate manifold $n_1 + n_2 = n - |m| - 1 = \text{constant}$

$$\langle n_1 n_2 m | V | n_1' n_2' m \rangle = \langle n_2 n_1 m | V | n_2' n_1' m \rangle$$

$$\langle n_1 n_2 m | V | n_1 + i n_2 - i m \rangle = 0 \qquad \text{for } |i| > 1 \tag{31}$$

The specific forms of the matrix elements $\langle n_1 n_2 m | \xi\eta | n_1 + i n_2 - i m \rangle$ are[54]

$$\langle n_1 n_2 m | \xi\eta | n_1 + i n_2 - i m \rangle$$
$$= \tfrac{1}{2}n^2\{3[n^2 - (n_1 - n_2)^2] + 1 - m^2\} \qquad \text{for } i = 0 \tag{32}$$
$$= 2n^2[(n_1 + 1)(n_1 + 1 + |m|)n_2(n_2 + |m|)]^{1/2} \qquad \text{for } i = 1$$

Thus the Hamiltonian matrix for the full manifold $n, m = \text{const}$ can be written in tridiagonal form with respect to the parabolic basis of Eq. (28). A further reduction can be obtained by transforming the parabolic basis functions to functions with definite parity:

$$|n_1 n_2 m \pm\rangle = 2^{-1/2}\{|n_1 n_2 m\rangle \pm (-1)^m |n_2 n_1 m\rangle\} \tag{33}$$

It is then easily shown that the Hamiltonian matrix in the basis of Eq. (33)

splits into two uncoupled submatrices, one for the even parity states $|n_1 n_2 m + \rangle$ and one for the odd states $|n_1 n_2 m - \rangle$; and, from Eq. (31), the even and odd Hamiltonian matrices are tridiagonal and nearly identical. Specifically, when $n - |m|$ is even so that the even and odd matrices are of the same size, the even and odd matrices differ only in the diagonal element associated with the state for which $|n_1 - n_2| = 1$. When $n - m$ is odd, the Hamiltonian matrix for states with parity $(-1)^m$ has an additional row and column (associated with the state for which $n_1 = n_2$); otherwise, it is identical with the matrix for states with parity $(-1)^{m+1}$. If these discrepancies in the single diagonal element (or the additional dimension) were not present, the perturbed spectrum would be doubly degenerate: each energy being associated with both an even and an odd state. This is the result of the semiclassical treatment of Goebel and Kirkman.[42] Thus the difference between the even and odd Hamiltonians may be viewed as a result of quantum mechanical tunneling, which breaks the degeneracy of the semiclassical double-well problem. This effect of tunneling is most pronounced for orbits lying near the potential ridge $z = 0$ of Figure 7, as can be seen by considering the Runge–Lenz vector. In quantum mechanics this is an operator whose z component is diagonal in the basis of Eq. (28), with eigenvalues proportional to the difference $n_2 - n_1$. The differences in even and odd Hamiltonian matrices are associated with states which assume the smallest possible values of $|n_2 - n_1|$: that is, those describing orbits whose major axes are most nearly perpendicular to z. It may also be noted that the operator $\sigma = (\alpha_z^2)^{-1/2} \alpha_z$ which gives the sign of the z component of the Runge–Lenz vector $\boldsymbol{\alpha}$, interchanges the even and odd states of the basis (33):

$$\sigma |n_1 n_2 m \pm \rangle = \pm |n_1 n_2 m \mp \rangle \tag{34}$$

Since this transformation almost preserves the form of the Hamiltonian matrices, σ may be said to almost be a constant of the motion. As in the semiclassical perturbation theory, the lower-energy orbits will tend to be confined to the valleys of the potential surface around $\theta = 0$ and π, i.e., σ will tend to remain either positive or negative. As energy increases, however, the conservation of σ ceases to hold true. Note particularly that for a planar orbit, $\alpha_z = 0$ and so σ is indeterminate.

Goebel and Kirkman[42] have made considerable analysis of the solutions of the equations of degenerate perturbation theory in the parabolic basis. They note that since from Eq. (31) the equation for the eigenvector components is of the form of a three-term recursion relation, it is the finite-difference analog of a second-order differential equation. They are able to solve approximately the corresponding differential equation, in limiting cases, by a WKB approach. Moreover the recursion relations reduce, in some asymptotic limits, to those for standard[56] orthogonal polynomials: the Hermite polynomials when $n \gg |m|$ and when the energy is high; Meixner polynomials for low energy and $n \gg |m|$; and Krawtchouk polynomials when $n - |m| \ll n$. In these limits oscillator strengths for transitions from the hydrogen ground state can also be computed. At the time of writing of this article, however, a detailed comparative study of these approximations has not been published.

Thus both semiclassical and quantum mechanical perturbation treatments indicate that some states become localized upon the magnetic potential ridge. It seems plausible from the discussion of planar motion given above—and it will be shown explicitly in the following section—that such states are of considerable importance in the experimental photoabsorption spectra. It may be appropriate to remark here that such localization arises quite generally in systems described by tridiagonal Hamiltonian matrices.[57]

Some simple relevant systems are a chain of coupled oscillators, or the Huckel model for the pi-electron spectrum of an aromatic molecule. If the coupling is uniform, one obtains wave functions which are entirely delocalized, and a band of energy levels in which the energy depends quadratically on the wavenumber near the bottom and the top of the band. If the coupling is nonuniform, or a substitutional impurity is introduced in the molecule, the energy spectrum is perturbed and the wave functions in the region of greatest perturbation become partially localized about the impurity. A typical energy band which arises in the perturbative treatment of the quadratic Zeeman effect can be seen in the lowest clusters of lines in Figures 9 and 10. There the energies are very nearly equally spaced at the edges of the band. States on the lower and upper edges of the band are then necessarily localized in regions of minimum and maximum potential.

NOTE (added in proof): Since the submission of this article, Herrick[140] has published a comprehensive treatment of degenerate perturbation theory. He found that (within a given n manifold) the matrix elements (32) of the diamagnetic potential are equivalent to those of the operator U of Eq. (21). Therefore, in the context of perturbation theory, U can be regarded as the exact diamagnetic potential [a different substitution, which omitted some terms of Eq. (21), was proposed by Labarthe[141]]. It turns out that the resulting equations of motion can be solved in closed form in momentum space in terms of Lamé functions of Jacobian elliptic coordinates, a fact which was also noted by Soloviev.[139]

(d) *Matrix Diagonalization Treatments.* The many suggestive features arising in perturbative treatments have been given somewhat more concrete expression in the course of more accurate nonperturbative calculations. Evidence for the existence of a quasi-constant of the motion in the hydrogen Zeeman spectrum has been presented by Zimmerman, Kash, and

Kleppner[58] and by Clark and Taylor.[59] In light of the work of Edmonds and Pullen[39] described above, it is apparent that the validity of such a conservation law will depend on the energy of the motion, and that it must cease to apply as the ionization threshold is approached. Further development has, however, linked the quasi-constant of the motion with the asymptotic forms of the wave function on the potential ridge.[60] It seems probable that the regularities observed in the asymptotic forms will be preserved well beyond the energy at which the approximate conservation rule fails.

The mechanics of the calculations have been given in detail elsewhere,[45,59] so only a brief summary need be provided here. Both Zimmerman, Kash, and Kleppner[58] and Clark and Taylor[60] diagonalized the Hamiltonian of Eq. (24) in a finite basis; Zimmerman *et al.* employing a basis of discrete hydrogenic wave functions and Clark and Taylor a basis of Sturmian functions. An advantage of the hydrogenic basis is that the effect of a quantum defect can be incorporated fairly readily. For the specific treatment of the hydrogen problem, however, the Sturmian basis appears preferable for a number of reasons, and we shall give it primary attention here.

A Sturmian basis was first employed in this problem by Edmonds[61] to treat a strong field (10^4 kG) case. In a general form, the Sturmian radial functions are

$$S_n^{(\zeta)}(r) = \left[\frac{(n-l-1)!}{2 \cdot (n+l)!}\right]^{1/2} e^{-\zeta r/2} (\zeta r)^{l+1} L_{n-l-1}^{(2l+1)}(\zeta r) \tag{35}$$

where L is a Laguerre polynomial.[56] In the diagonalization procedure these are taken in combination with spherical harmonics

$$\psi(\mathbf{r}) = \sum_{n,l} \psi_{nl} S_{nl}^{(\zeta)}(r) Y_{lm}(\hat{r}) \tag{36}$$

and the expansion coefficients ψ_{nl} are determined. For a given l, the S_{nl} form a complete set of functions; this is not the case with any set of discrete hydrogenic wave functions. The scale parameter ζ may take any positive value. When $\zeta = 2/n^*$, the radial functions S_{n^*l} coincide, up to a normalization factor, with the hydrogen radial functions for the states with principal quantum number n^*. In Edmonds' original paper $\zeta = 2$, since it was desired to examine low-lying states. The flexibility in choice of ζ allows one to "center" a calculation in a given energy region, and for computation of high Rydberg wave functions in moderate fields one employs values of n^* of the order of 40.

Another very significant advantage of the Sturmian basis is the simplicity of matrix elements. In particular, the radial matrix elements of the magnetic potential obey the selection rule $|\Delta n| \leq 3$. Since the angular selection rule $|\Delta l| \leq 2$ is also in force, the Hamiltonian matrix is very sparse, and can be written in banded form. There is a small price to be paid for achieving this sparsity, in that the S_{nl} are orthogonal over a weight function equal to the Coulomb potential $1/r$ rather than unity. Thus one has to contend with the generalized eigenvalue probem

$$H\psi = \varepsilon B\psi \qquad (37)$$

where ε is the energy eigenvalue and B the matrix of overlap between Sturmian functions. Since the overlap between Sturmian functions vanishes when $|\Delta n| > 1$, B can always be put in tridiagonal form. Efficient algorithms exist[62] for the solution of Eq. (37) when, as here, H and B are banded matrices. On a sufficiently large computer, like the CRAY-1, determination of the eigenvalues and eigenvectors of the system (37) can be done routinely for Sturmian bases with up to 1500 elements.

When the eigenvectors have been computed, the determination of transition oscillator strengths is straightforward. Figures 9 and 10 show computed oscillator strengths for dipole transitions from the ground state of hydrogen in a field of 47 kG. The energy scale has been chosen to span the region from where perturbation theory breaks down up to a few units of cyclotron frequency below the ionization threshold. The two spectra shown would be observed in photoabsorption of light in different states of polarization. The spectrum of Figure 9 is associated with states with magnetic quantum number $m = 0$, which are produced by absorption of photons polarized along the direction of the magnetic field; that of Figure 10, for which $m = 1$, is generated by absorption of photons with circular polarization in the plane perpendicular to the field. Since all states have odd parity, those with $m = 0$ have wave functions which vanish on the ridge $\theta = \pi/2$, whereas the states with $m = 1$ have an antinode on the ridge. Thus it should be possible to describe some of the lines in the $m = 1$ spectrum in terms of the two-dimensional semiclassical solutions of Edmonds[35] and Starace.[36]

In both spectra one sees the presence at low energy of line clusters with a well-defined shape. These can be identified with the manifolds of hydrogenic states with fixed principal quantum number n, and are adequately described by degenerate perturbation theory. As the energy increases the clusters begin to interpenetrate, and this simple hydrogenic description ceases to be appropriate. Note, however, that the clusters tend to pass through one another without significant perturbation of their individual members. This is more readily apparent in the $m = 1$ spectrum than in the $m = 0$ because of the monotonic distribution of oscillator strength among the lines of a given cluster. This near absence of mutual perturbation among lines with comparable oscillator strengths and nearly equal energies is the signature of a quasiconservation law. It is a complementary aspect of the curve crossings noted by Zimmerman *et al.*[58] in their plot of energy levels as a function of magnetic field strength. Thus it is appropriate to regard the spectra, in this energy region, as the superposition of a number of essentially independent line series. This is analogous to the decomposition of a complicated molecular band spectrum into lines corresponding to definite rovibrational transitions, though at the moment the intuitive physical basis for such a description is not obvious. The identification of the line series in the spectrum is not ambiguous, however; in Figure 10 the first four series, in order of spectral prominence, are drawn out in separate frames. We call the set of strongest lines the first or principal series, the second, third, and fourth series following in order. Within each series one sees a regular energy spacing between members, and a decrease in oscillator strength as the energy increases. At higher energies some irregularities in the oscillator strength are apparent; these occur when members of separate series nearly coincide in energy.

The spacing between members of a series is in good agreement with the predictions of the semiclassical theory,[35,36] as is shown in Figure 11 for the principal series. The ordinate is the radial quantum number N (at low energies $N = n - 2$, where n is the principal quantum number) which can be regarded as a continuous function of energy in a WKB treatment. The value of N to be associated with a given line of the principal series is simply determined by its position within the series. It should be noted that the absolute energies of the principal lines are very near those given by the WKB theory at high energies; thus, the total number of principal lines

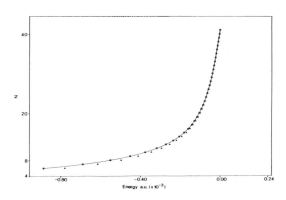

Figure 11. Comparison of the energies of the principal lines of Figure 10e with the two-dimensional WKB theory of Refs. 35 and 36. Full curve: the WKB integral as a continuous function of energy; the circles are placed at points where the action is a multiple of Planck's constant. Triangles denote the positions of principal series lines. The ordinate is the number of nodes of the wave function in the plane $z = 0$. (From Ref. 59.)

in the spectrum may be expected to be close to the N_T of Eq. (18). Note also that the oscillator strength in the principal series decreases with increasing N roughly as $N^{-2.4}$.

An examination of the wave functions indicates the reason for this agreement between two-dimensional WKB and three-dimensional quantum mechanical predictions. Figure 12 shows the radial wave functions for the principal lines of Figure 10, evaluated in the plane $\theta = \pi/2$, i.e., along the ridge of the Coulomb-magnetic potential surface of Figure 7. At small r the wave functions are all nearly proportional, as would be the case in the absence of the magnetic field and as is consistent with Eq. (26). At large r the wave function for each successive state of the series has an additional node. The system of functions looks very much like that which would result from a computation of the eigenfunctions of a one-dimensional Schrödinger equation. Thus in this three-dimensional calculation the confinement of the wave functions to the ridge region is seen to take place. Plots of wave functions for the higher series along the ridge show similar regularities, though not to as great a degree. In this spectrum the amount of confinement is in fact closely correlated with the magnitude of the oscillator strength.

This is seen from the variation of the wave functions in the direction across the ridge, i.e., the variation in θ with r held fixed. A systematic description of the angular dependence of the wave functions at small r is rather difficult when there is no predominance of a single angular momentum component, as is the case for the higher energy lines of each series. This is implied by Eq. (26), in which the amplitude associated with each angular momentum component will oscillate with increasing r as a Coulomb function. In the diffraction picture of Fano as well, one must view a standing wave at small r as being built up of a large number of diverging and converging trajectories. We consider now the angular behavior of the wave

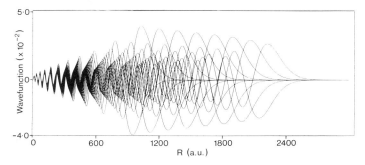

Figure 12. Wave functions for the principal series lines of Figure 10a in the plane $z = 0$. (C. W. Clark and K. T. Taylor, unpublished.)

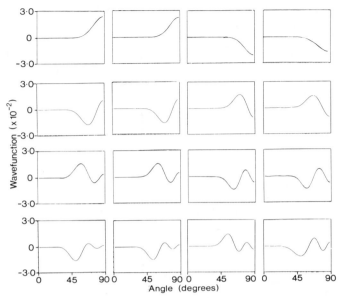

Figure 13. Wave functions along the turning surface for various states of Figure 10. First row: first, fifth, tenth, and sixteenth states of the principal series; second through fourth rows: the corresponding states of the second through fourth series, with the exception of the lower right-hand corner being the fifteenth state of the fourth series. The relationship between distance and angle is determined from Eq. (38). (From Ref. 60.)

functions at large r, specifically along the classical turning surface. Such a surface is the locus of the equation

$$-1/r + \tfrac{1}{2}\beta^2 r^2 \sin^2 \theta = \varepsilon \tag{38}$$

where ε is the energy of the state involved. These are just the various equipotential surfaces of Figure 7. Figure 13 shows wave functions for states of different series, plotted along their respective turning surfaces. The abscissa is actually the angle θ, the value of r to be inferred from Eq. (38) (the dependence on the azimuthal angle ϕ is trivial). For the states of low energy the value of r is nearly constant; at higher energies the variation of r with θ can be seen from Figure 7. In the top row are wave functions for the first, fifth, tenth, and sixteenth members of the principal series, as shown in Figure 10. The same order is taken for the members of the second and third series shown in the next two rows, respectively; the first, fifth, tenth, and fifteenth members of the fourth series are shown in the bottom row, the sixteenth member being strongly perturbed by a state from the second series (cf. Figure 10). An anticipated, the wave functions of the principal series are localized about the ridge, this behavior actually being established well before the turning surface is reached. The states of the higher series become progressively delocalized, and in a regular manner. The principal series wave functions have no nodes across the ridge; the second series wave functions have two nodes (recall that the wave functions are necessarily symmetric about $\theta = \pi/2$); the third series, four nodes; and the fourth series, six nodes. States with odd numbers of nodes occur only in the $m = 0$ spectrum. Thus the series number serves as an index of the degree of excitation across the ridge at large distances. However, it should be kept in mind that the motion across the ridge is intrinsically unstable, since it is always possible for the electron to fall into the valleys. Thus the results shown here are entirely at variance with any adiabatic theory; since, in a quasi-separable problem, the states of high energy which can reach the top of the ridge must also have many nodes in the valleys. It seems that this simple picture which obtains at large distances must instead be due to a particular form of interference between diverging and converging trajectories; it is more appropriate to consider these angular wave functions as a form of diffraction pattern, than as analogs to a harmonic oscillator. It is not improbable that similar phenomena will be seen in other problems involving motion about a potential ridge, for example in low-energy electron impact ionization, but this specific example does not provide enough evidence to make firm general predictions.

A physical basis for the independent series representation has been established, at least in this energy region: the members of a given series are distinguished by the degree of excitation along the potential ridge, and the series themselves are characterized by the degree of excitation across the ridge. At low energies this is a vague way of listing the properties of a quasi-constant of the motion. It remains to be seen whether such a viewpoint retains its utility as the energy increases towards, and goes beyond, the ionization limit. The calculations indicate that the members of the principal series tend to retain their form, and it seems apparent that this series will extend well into the continuum to produce the most prominent features of the photoionization spectra. The secondary features of those spectra may well be attributed to the persistence of the second, and higher, series. Since the wave functions of members of the higher series get more deeply into the valleys, it would be expected that the secondary features will be damped out more rapidly as the energy increases. This is in accordance with experimental findings. Unfortunately it is now only possible to speculate upon, rather than to review, the properties of the continuum wave functions for this problem. Their calculation does not seem feasible by any method employed thus far.

The hydrogen atom is a system with a number of unusual properties, and many of the results reviewed here will not apply directly to the spectra of more complex atoms. For instance, the equal partitioning of oscillator strength in the clusters of the $m = 0$ spectrum of Figure 9 is a result of the zero-field l degeneracy, and so must be changed appreciably in the presence of a non-Coulombic core. Calculations of states with larger values of m, e.g., those which are involved in Balmer transitions, have revealed many near degeneracies among states of opposite parity.[47] These must also disappear when finite quantum defects are introduced. Even a superficial examination of the spectra of a number of atoms, e.g., along the second column of the Periodic Table, reveals many significant variations of detail; almost all of which must be attributed to the changes in the atomic core.

Nevertheless there are a number of aspects of the solution to the hydrogenic problem which must have general applicability. The spacing of the quasi-Landau resonances near threshold is a universal phenomenon (albeit sometimes obscured by the motional Stark field, or by the weakness of the transition oscillator strength); and in many instances it is possible to pick out sets of lines which resemble the independent series in hydrogen. The degree to which the knowledge of these regularities can be used to build a general and practical theory remains a question for the future.

2.2.3. Bound States in Strong Fields

The previous sections of this review have discussed the properties of the Rydberg spectra of atomic hydrogen in high laboratory magnetic fields, particularly near the ionization threshold. In this section we review recent theoretical treatments of the ground and lowest excited levels of atomic hydrogen in magnetic fields stronger than those normally obtainable in the laboratory. Of particular interest are field strengths of the order of 10^7 gauss, which is typical of white dwarf stars, and of 10^{12} gauss, which is typical of neutron stars. The astrophysical applications of the theory have been discussed in detail by Garstang.[17] (Note that correspondingly large effects are found for exciton spectra in laboratory-sized magnetic fields, $10^3 \, \mathrm{G} \le B \le 10^5 \, \mathrm{G}$, due to the small effective mass of the electron.[63]) While we present here all recent theoretical work on hydrogenic states of low excitation in strong magnetic fields, the emphasis is on theories appropriate for magnetic fields of strength $B \le 10^9$ gauss in which the Coulomb field is the dominant influence on the electronic motion. General aspects of atomic structure and of atomic scattering processes in extremely high magnetic fields $B \gg 10^9$ gauss—in which the diamagnetic interaction is the dominant influence on the electronic motion—are discussed in Section 6.

(a) *Quasiseparation of Electronic Motion Using an Adiabatic Approximation.* The Hamiltonian in Eq. (9) is difficult to treat since the effective potential is not separable. The only good quantum numbers are parity and the axial component of orbital angular momentum m. Equation (9) may, however, be cast into a form susceptible to an adiabatic separation of variables.[14,64,65] To this end we write the wave function $\psi(\mathbf{r})$ for the reduced mass particle as

$$\psi(\mathbf{r}) = \frac{\chi_m(r, \theta)}{r} \frac{e^{im\phi}}{(2\pi)^{1/2}} \tag{39}$$

and substitute Eq. (39) in Eq. (9). We obtain in atomic units (i.e., $m = e = \hbar = 1$) the following equation for $\chi_m(r, \theta)$:

$$\left[\frac{1}{\mu} \frac{d^2}{dr^2} + \frac{2}{r} + 2E' - \frac{\Lambda_m(\alpha r^2, \theta)}{\mu r^2} \right] \chi_m(r, \theta) = 0 \tag{40}$$

where

$$\Lambda_m(\alpha r^2, \theta) \equiv -\frac{1}{\sin\theta} \frac{\partial}{\partial\theta} \left(\sin\theta \frac{\partial}{\partial\theta} \right) + \frac{m^2}{\sin^2\theta} + \alpha^2 r^4 \sin^2\theta \tag{41}$$

In Eqs. (40) and (41) α is a strength parameter defined as

$$\alpha \equiv \frac{B}{2c} = (2.127 \times 10^{-10} \text{ a.u./G}) B(\text{G}) \tag{42}$$

and in Eq. (40) the prime on the energy E' indicates that the linear Zeeman energy is subtracted from the total energy E, i.e.,

$$E' \equiv E - \alpha m (g/\mu) \tag{43}$$

When the operator $\Lambda_m(\alpha r^2, \theta)$ defined in Eq. (41) is considered to depend only parametrically on the quantity αr^2, it has as eigenstates the oblate spheroidal angle functions $g_{m\nu}(\alpha r^2, \theta)$:

$$\Lambda_m(\alpha r^2, \theta) g_{m\nu}(\alpha r^2, \theta) = U_{m\nu}(\alpha r^2) g_{m\nu}(\alpha r^2, \theta) \tag{44}$$

where ν labels the eigenstates and $U_{m\nu}(\alpha r^2)$ are the corresponding eigenvalues. In the limit that $\alpha r^2 \to 0$ (due either to $B \to 0$ or $r \to 0$), the index ν becomes the ordinary orbital angular momentum quantum number l, $U_{m\nu}(\alpha r^2)$ becomes $l(l + 1)$, and $g_{m\nu}$ becomes the associated Legendre polynomial $P_l^m(\cos\theta)$. For finite values of the parameter αr^2, each oblate spheroidal angle function may be written as a linear combination of associated Legendre polynomials. More specifically, as αr^2 increases from zero, each oblate spheroidal angle function with index ν loses its identity to the single associated Legendre polynomial with $l = \nu$ and must be represented by an increasingly larger sum over Legendre polynomials with other orbital angular momenta $l \neq \nu$ and the same parity. Thus the oblate spheroidal angle functions include implicitly a large amount of magnetic field distortion of the atom's spherical symmetry which spectroscopically is observed as "l-mixing." Due to the parametric dependence of the oblate spheroidal angle functions on the radial coordinate, they do not represent eigenstates for the angular part of the wave function $\chi_m(r, \theta)$ in Eq. (40). However, in an adiabatic approximation this is a good first approximation in many cases, as shown below.

In an exact treatment, the wave function $\chi_m(r, \theta)$ must be expanded as a linear combination of oblate spheroidal angle functions with radially dependent coefficients:

$$\chi_m(r, \theta) = {\sum_{\nu'}}' h_{m\nu'}(r) g_{m\nu'}(\alpha r^2, \theta) \qquad (45)$$

The prime on the summation indicates that either even or odd $\nu \geq |m|$ are included in the summation depending on whether the party of the state is even or odd. Substituting Eq. (45) into Eq. (40), multiplying from the left by $g_{m\nu}$, and integrating over θ gives the following set of coupled differential equations for the radial function $h_{m\nu}(r)$[64]:

$$\left[\frac{1}{\mu} \frac{d^2}{dr^2} + \frac{2}{r} + 2E' - \frac{U_{m\nu}(\alpha r^2)}{\mu r^2} \right] h_{m\nu}(r)$$

$$+ 2 {\sum_{\nu'}}' \left(g_{m\nu}, \frac{\partial g_{m\nu'}}{\partial r} \right) \frac{d}{dr} h_{m\nu'}(r)$$

$$+ {\sum_{\nu'}}' \left(g_{m\nu}, \frac{\partial^2 g_{m\nu'}}{\partial r^2} \right) h_{m\nu'}(r) = 0 \qquad (46)$$

where

$$\left(g_{m\nu}, \frac{\partial^n g_{m\nu'}}{\partial r^n} \right) \equiv \int_0^\pi g_{m\nu}(\alpha r^2, \theta) \frac{\partial^n}{\partial r^n} g_{m\nu}(\alpha r^2, \theta) \sin \theta \, d\theta \qquad (47)$$

Whether or not an adiabatic approximation is reasonable depends on the strength of the first and second derivative coupling terms in Eq. (46). Use of the Hellman–Feynman theorem[66,65] shows that they are large only near avoided crossings in the "potential" curves $U_{m\nu}(\alpha r^2)$. These curves are shown for the odd ν-values 1, 3, 5, 7, and 9 in Figure 14. In general, the locus of avoided crossings between these curves lies along the curve defined by $U_{m\nu}(\alpha r^2) = m^2 + \alpha^2 r^4$ as this defines the classical turning point for the oblate spheroidal angle functions for motion in r along $\theta = 90°$ [cf. Eqs. (41) and (44)]. That is, at any larger value of $\alpha^2 r^4$ the "kinetic energy" of the oblate spheroidal angle functions, given by the first term on the right in Eq. (41), becomes negative since the potential energy, $m^2 + \alpha^2 r^4$, on the right of Eq. (41) for $\theta = 90°$ becomes greater than the eigenvalue $U_{m\nu}$ of Λ_m [cf. Eq. (44)]. Hence tunneling behavior must set in—unless, that is, a transition is made to a higher state ν with a larger $U_{m\nu}$, in which case the particle can propagate outward in αr^2 until it reaches the turning point of the new potential.

Examination of Figure 14 shows that the first avoided crossing, that between the curves $\nu = 1$ and $\nu = 3$, occurs near $\alpha r^2 = 3$. For a magnetic field B of order 10^5 G, which is a typical strength for laboratory magnetic fields, this corresponds to an r value of 375 bohr; for B of order 10^7 G, this corresponds to an r value of 37.5 bohr; and for B of order 10^9 G, this corresponds to an r value of 3.75 bohr. Thus for hydrogenic states having a radial extent less than these values of r, one may quite accurately represent the wave function for the state in terms of a single oblate spheroidal angle function,

$$\chi_{m\nu}(r, \theta) \approx h_{m\nu}(r) g_{m\nu}(\alpha r^2, \theta) \tag{48}$$

where $h_{m\nu}(r)$ is obtained by solving Eq. (46) in the adiabatic approximation in which the coupling terms are set equal to zero.

Such adiabatic calculations have been carried out by Starace and Webster[64] for the 1s, 2s, and 2p levels of atomic hydrogen. The calculated energies—which may be shown to be rigorous lower bounds on the true

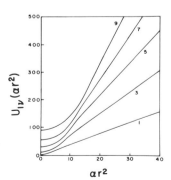

Figure 14. Oblate spheroidal eigenvalues $U_{m\nu}(\alpha r^2)$ for $m = 1$ and $\nu = 1, 3, 5, 7$, and 9 plotted as a function of the parameter αr^2.

energy for the lowest states of each symmetry—are shown in Figure 15 for the $1s$ and $2p$ levels and compared with the best available variational upper bounds on these same energies.[67] A detailed comparison of the calculated binding energy of the $1s$ level with other theoretical results is shown in Table 1. The adiabatic oblate spheroidal angle function treatment gives results that agree very well with more elaborate calculations for magnetic fields up to about 10^9 G.

Even when an adiabatic approximation is not valid, the oblate spheroidal angle function representation may be useful. The reason is that the eigenvalue curves in Figure 14 indicate how many oblate spheroidal adiabatic functions need to be coupled at any value of the parameter αr^2. Of course, due to the variation of the locus of avoided crossings with the *square* of the parameter αr^2, the number of adiabatic solutions which must be coupled increases rapidly with increasing r but less rapidly with increasing magnetic field strength B. Thus the method appears most useful for treating the lowest levels of an atom, even in high fields for which nonadiabatic couplings must be considered, rather than for treating Rydberg states in high magnetic fields due to their large spatial extensions in r. This state of affairs is not peculiar to the oblate spheroidal representation, but is a feature of any theoretical representation based upon spherical symmetry. All such theoretical methods have difficulty representing the cylindrically symmetric magnetic field distortion of the electronic orbit far from the nucleus of the atom.

We should mention that a number of other adiabatic treatments have been given for atomic hydrogen in a uniform magnetic field.[68–73] All of these, however, employ cylindrical coordinates, thus applying only to very high field strengths $B > 10^9$ G. Furthermore these methods give a poor description of the electron wave function in the neighborhood of the origin, where optical absorption takes place. An attractive feature of the adiabatic approach in spherical coordinates presented above is that the electron wave

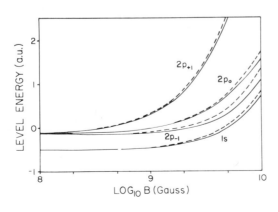

Figure 15. Upper and lower bounds for the hydrogenic $1s$ and $2p$ level energies in a uniform magnetic field. Solid lines: adiabatic results of the oblate spheroidal treatment of Starace and Webster (Ref. 64); dashed lines, best variational upper-bound results (cf. Ref. 64, Tables III–V).

Table 1. Binding Energy of the Hydrogenic Ground State in a Uniform Magnetic Field

| α[a] (a.u.) | B (10^9 G) | Present adiabatic results | | Variational results[b] | | | Ed | Eigenfunction expansion and perturbation results[b] | | | | Other adiabatic results[b] |
		Upper	Lower	L	PR	SHSOR		CFF	P	Ek	GP	BB
0.00213	0.01	0.5021	—	—	—	—	0.5021226	—	—	—	—	—
0.00638	0.03	0.5063	—	—	—	—	0.5063408	—	—	—	—	—
0.05	0.235	0.5475	0.5474	0.54752	0.54752	—	—	0.54754	0.54752	—	0.54753	—
0.1	0.470	0.5905	—	—	0.59036	—	—	0.59038	—	—	0.59038	—
0.15	0.705	0.6296	0.6280	0.62918	0.62908	—	—	0.62920	—	—	0.62919	—
0.2	0.940	0.6657	0.6622	—	0.66438	—	—	0.66462	—	—	0.66461	—
0.3	1.41	0.7307	0.7217	—	0.72687	0.73	—	0.72747	—	0.707	0.72753	—
0.4	1.88	0.7887	—	—	0.78122	0.78	—	0.78228	—	0.757	0.78250	—
0.5	2.35	0.8418	0.8167	0.831	0.82956	0.83	—	0.83120	0.83116	0.802	0.83166	0.86
1	4.70	1.061	—	1.02	1.01762	1.0254	—	1.02225	1.02221	0.998	1.02526	1.02
1.5	7.05	1.238	—	1.16	1.15701	1.11	—	1.16456	1.16452	1.106	1.17091	1.16
2.5	11.8	1.533	—	1.378	1.3676	1.38	—	1.3804	—	1.304	1.3931	1.38
12.5	58.8	3.444	—	2.366	—	2.40	—	—	—	—	2.424	—
50	235	8.21	—	3.728	—	4	—	—	—	3.174	3.798	—

[a] $\alpha = (2.12715 \times 10^{-10}$ a.u./G)B(G).

[b] Letter key to results of other authors—L: Larsen, Ref. 75; PR: Pokatilov and Rusanov, Ref. 76; SHSOR: Smith et al., Ref. 80(a), as reported in Ref. 80(b); Ed: Edmonds, Ref. 61; CFF: Cabib, Fabri, and Fiorio, Ref. 88; P: Praddaude, Ref. 89; Ek: Ekardt, Ref. 91; GP: Galindo and Pascual, Ref. 92; BB: Baldereschi and Bassani, Ref. 68.

function is accurately represented near the origin no matter how high the field strength. However, the matching of the spherically symmetric wave function near the origin onto wave functions having the cylindrical symmetry more appropriate far from the origin remains an unsolved theoretical problem.

(b) *Other Recent Theoretical Developments.* Besides the adiabatic approximations, numerous studies for atomic hydrogen in a uniform magnetic field have been carried out using either variational[68,74–87] or eigenfunction expansion[59,61,69,88–92] methods. Variational methods, of course, are most useful for calculating upper bounds on the energy of the lowest state of a given symmetry. Upper bounds on excited energy levels may also be obtained by variational methods, but only at the cost of greater computational labor.[93] Eigenfunction expansion calculations carry out an exact or approximate diagonalization of the Hamiltonian, but often require the use of very large numbers of basis states to achieve accurate results.

Very recently attention has been focused on mathematically rigorous perturbative treatments of atomic hydrogen in a uniform magnetic field. Killingbeck[48] has separated the quadratic Zeeman term in the Hamiltonian into so-called s and d components, as follows:

$$\frac{e^2 B^2}{8\mu c^2}(x^2 + y^2) = \left(\frac{e^2 B^2}{8\mu c^2}\right)\left[\tfrac{2}{3}r^2 P_0(\cos\theta) - \tfrac{2}{3}r^2 P_2(\cos\theta)\right] \qquad (49)$$

Here $P_0(\cos\theta)$ and $P_2(\cos\theta)$ are Legendre polynomials: $P_0 = 1$ and $P_2 = (\tfrac{3}{2}\cos^2\theta - 1/2)$. Initially Killingbeck drops the second term in Eq. (49) and solves the remaining Hamiltonian exactly; since $P_0(\cos\theta) = 1$ the potential has only radial dependence. The resulting energies are rigorous upper bounds on the correct energies and in fact turn out to be very accurate for magnetic fields less than 10^9 G. Treating then the second term in Eq. (49) perturbatively, Killingbeck is able to obtain rigorous upper and lower bounds on the $1s$ level of hydrogen which agree to four digits at 10^9 G!

Asymptotic perturbation expansion formulas for the ground-state energy of atomic hydrogen in a uniform magnetic field have been studied by Avron and co-workers.[49,94,95] In one such study, Avron *et al.*[49] used the Bender–Wu formulas[95] to obtain analytic expressions (to order $1/n$) for the coefficients E_n in the following expansion for the ground-state energy:

$$E(B) = \sum_{n=0}^{\infty} E_n \left(\frac{B^2}{8}\right)^n \qquad (50a)$$

Their result is

$$E_n = (-1)^{n+1}\left(\frac{4}{\pi}\right)^{5/2}\left(\frac{8}{\pi^2}\right)^n \left(2n + \frac{1}{2}\right)!\left[1 + O\!\left(\frac{1}{n}\right)\right] \qquad (50b)$$

In another study, Avron, Herbst, and Simon[94] obtain a detailed asymptotic expression for the ground-state binding energy of hydrogen in an extremely large magnetic field. It has been known for some time now that in extremely strong magnetic fields the hydrogenic electron is forced so close to the nucleus that its binding energy increases as $\ln^2 B$.[70,77,96] Avron *et al.*[49] obtain instead a new implicit asymptotic formula for the binding energy whose numerical predictions are in excellent agreement with earlier work.

3. Stark Effect in the Hydrogen Atom

The standard treatment of the Stark effect in the hydrogen atom is in parabolic coordinates.[26,97] The Schrödinger equation for an electron in a Coulomb and a uniform electric field is separable into one-dimensional WKB-type equations in the ξ and η coordinates. Various techniques have been developed for calculating the Stark effect for atoms of low excitation.[98–104] Recent review articles[18–20] have covered these works. Renewed interest has been generated by the recent experimental observations of regular intensity modulations in the photoionization cross section of atoms in an external electric field near and above the zero-field ionization limit.[12,105–107] The polarization dependence of these features has been discussed[108] and confirmed by a numerical calculation of the Stark photoionization process in hydrogen by Luc-Koenig and Bachelier,[109] who have adapted a method developed by Blossey[110] for Wannier excitons. Semi-analytic theories for this effect have also been developed.[111,112] Below we review this recent work on photoionization of atoms in a uniform electric field. Recent advances in high-order perturbative calculations will be briefly discussed.[113–115]

3.1. Semiclassical Treatments: Photoionization Cross Section

Following Landau and Lifshitz,[97] the Schrödinger equation for combined Coulomb and Stark potentials, $-e^2/r + e\mathbf{F}\cdot\mathbf{r}$, is separable in parabolic coordinates $\xi = r + z$, $\eta = r - z$, and $\phi = \tan^{-1}(y/x)$. The two separated one-dimensional equations for ξ and η are

$$\frac{d^2\chi_1(\xi)}{d\xi^2} + \left[\frac{1}{2}\varepsilon - 2\left(\frac{m^2-1}{8\xi^2} - \frac{\beta_1}{2\xi} + \frac{F}{8}\xi\right)\right]\chi_1(\xi) = 0 \qquad (51)$$

$$\frac{d^2\chi_2(\eta)}{d\eta^2} + \left[\frac{1}{2}\varepsilon - 2\left(\frac{m^2-1}{8\eta^2} - \frac{\beta_2}{2\eta} - \frac{F}{8}\eta\right)\right]\chi_2(\eta) = 0 \qquad (52)$$

and the separation parameters β_1 and β_2 satisfy $\beta_1 + \beta_2 = 1$. The energy

ε is regarded as a parameter which has a definite value, and the β_1 and β_2 as eigenvalues of corresponding equations. These quantities are determined by solving Eqs. (51) and (52) as functions of ε and field strength F. The condition $\beta_1 + \beta_2 = 1$ then gives the required relation between ε and F, i.e., the energy as a function of the external field F. The eigenstate wave function is represented by

$$|\varepsilon, F; n_1 m\rangle = N_{\varepsilon n_1} \chi_1(\xi) \chi_2(\eta) \, e^{im\phi} (2\pi)^{-1/2} \tag{53}$$

where n_1 is the parabolic quantum number representing the number of nodes in the ξ-mode and $m = 0, \pm 1, \pm 2 \ldots$. The $\chi_1(\xi)$ and $\chi_2(\eta)$ are Coulombic for small values of ξ and η, and are normalized to $\xi^{m/2}$ and $\eta^{m/2}$ as $(\xi, \eta) \to 0$. The total wave function is normalized by $N_{\varepsilon n_1}$ to satisfy the orthonormality condition.

For low excited states and moderate electric field strengths, one can use a perturbation method to evaluate the Stark shifts.[102,104] When the electron energy increases, the binding energy of the electron can be comparable with the Stark energy shift. Also, the electron can ionize due to the quantum tunneling created by the Stark field. Figure 16 shows the potential energy in Eq. (51) and (52)

$$U_1(\xi) = \frac{m^2 - 1}{8\xi^2} - \frac{\beta_1}{2\xi} + \frac{F}{8}\xi \quad \text{and} \quad U_2(\eta) = \frac{m^2 - 1}{8\eta^2} - \frac{\beta_2}{2\eta} - \frac{F}{8}\eta$$

The dependence of β_1 and β_2 on energy ε are fixed by the Bohr–Sommerfeld

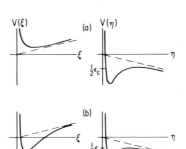

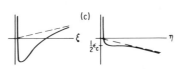

Figure 16. Qualitative plots of the potentials $U_1(\xi)$ [denoted here by $V(\xi)$] and $U_2(\eta)$ [denoted here by $V(\eta)$] in Eqs. (51) and (52) for $m > 1$, $F \geqslant 0$, and sample values of $\beta_1 = 1 - \beta_2$: (a) $\beta_1 \approx -0.1$, (b) $\beta_1 \approx 0.4$, and (c) $\beta_1 \approx 0.9$. The dotted lines represent the Stark potentials $\frac{1}{4}F\xi$ and $-\frac{1}{4}F\eta$. The top of the potential hump in η, $\frac{1}{2}\varepsilon_c$, and the potential coalesce in (c) where $\beta_2 \approx \beta_{crit} \sim 0.1$. (From Ref. 112.)

quantization rule

$$\int_{\xi_1}^{\xi_2} \{2[\tfrac{1}{4}E - U_1(\xi)]\}^{1/2} \, d\xi = (n_1 + \tfrac{1}{2})\pi$$

$$\int_{\eta_1}^{\eta_2} \{2[\tfrac{1}{4}E - U_2(\eta)]\}^{1/2} \, d\eta = (n_2 + \tfrac{1}{2})\pi$$

where n_1 and n_2 are parabolic quantum numbers. These integrals can be solved numerically. For a given energy level E_n, according to the classical picture, there is a critical electric field F_c,[26] where $F_c = E^2/4\beta_2 = 1/16n^4\beta_2$, such that ionization can take place only if E_n exceeds this value. However, in reality, ionization can take place for field strengths lower than this because of the quantum mechanical tunneling. One can see this effect readily by examining the potentials in Figure 16. The potential $U_1(\xi)$ increases as $\xi \to \infty$, so that the eigenvalues β_1 correspond to a bound mode and $\chi_1(\xi)$ decays exponentially. In contrast, there is a potential barrier in the η-mode and the wave function $\chi_2(\eta)$ is oscillating at large η. Thus, the η-mode corresponds to ionization and tunneling. The asymptotic expression for $\chi_2(\eta)$ can be obtained by combining an independent pair of Airy functions at large η[103,112]:

$$\chi_2(\eta) = A_{n_1,m}(\varepsilon, F)\left[\frac{2}{\pi k(\eta)}\right]^{1/2} \sin\{[\Delta(\eta) + \tfrac{1}{4}\pi] + \delta_{n_1 m}(\varepsilon, F)\} \quad (54)$$

Here $\delta_{n_1 m}(\varepsilon, F)$ is the total phase accumulated over the interval $\eta = 0$ to $\eta = \infty$ and $A_{n_1 m}(\varepsilon, F)$ is the asymptotic amplitude and $\Delta(\eta)$ is the WKB phase integral. An energy level of the atom to which an electron is excited in the presence of a Stark field will reveal itself with a finite width corresponding to an ionization yield. This decay of atomic resonances due to the Stark effect involves two coupled motions, the bounded ξ-mode and the continuum η-mode, with a coupling through the separation parameter, $\beta_1 + \beta_2 = 1$. It thus resembles the familiar picture of autoionization.[108] The amplitude A and phase shift δ can be used to fit a simple Breit–Wigner formula for the Stark resonance, which can be characterized by a resonance position ε_{n_2} and a width Γ_{n_2}[103,108,112]

$$\delta_{n_1 m}(\varepsilon, F) = \delta_{n_1 m}(\varepsilon_{n_2}, F) + \tan^{-1}[\Gamma_{n_2}/2(\varepsilon - \varepsilon_{n_2})]$$

and

$$A_{n_1 m}^2(\varepsilon, F) = A_{n_1 m}^2(\varepsilon_{n_2}, F)[(\varepsilon - \varepsilon_{n_2})^2 + \tfrac{1}{4}\Gamma_{n_2}^2]$$

(55)

The width Γ_{n_2} is related to the ionization rate, and calculations of these rates in hydrogen in the vicinity of E_c [102–104] as well as near and above the zero-field limit [109,112] are available.

The photoionization cross section of the hydrogen atom can be obtained easily by replacing the energy-normalized wave function $|\varepsilon, F; n_1 m\rangle$ in Eq. (53) by the zero-field hydrogen wave function $|\varepsilon, 0; n_1 m\rangle$, [112] since the ground-state wave function $|g\rangle$ is concentrated near the nucleus, where the Stark field is negligible. The usual selection rules apply: (1) $\Delta m = 0$ for light polarization parallel to the field, and (2) $\Delta m = \pm 1$ for perpendicular light polarization. The photoionization cross section is [111,112]

$$\sigma_F^m(\varepsilon) = \left(\frac{4\pi^2}{137}\hbar\omega\right) d_{lm}^2(\varepsilon) \sum_{n_1=0}^{\infty} N_{\varepsilon n_1 m}^2(\varepsilon)\alpha_{lm}^2(\beta_1, \nu) = \sum_{n_1} \sigma_{F,n_1}^m(\varepsilon) \quad (56)$$

where $\hbar\omega = E_i - E_f$ and $\nu = (-\varepsilon/13.6\ \text{eV})^{1/2}$. The magnitude of the photo-ionization cross section depends on three factors: (1) the radial dipole integral

$$d_{lm}(\varepsilon) = \int_0^{\infty} dr\, F_l^*(\varepsilon, r) r_m R(r)$$

where $R(r)$ and $F_l(\varepsilon, r)$ are wave functions of the ground state and final state, respectively, and where r_m equals z for $m = 0$ and $\sqrt{\frac{1}{2}}(x \pm iy)$ for $m = \pm 1$; (2) the normalization factor of Eq. (53), $N_{\varepsilon n_1 m}^2(\varepsilon)$; and (3) the geometric factor $\alpha_{lm}(\beta_1, \nu)$, which represents the projection of the confluent hypergeometric factors of $|n_1, m\rangle$ with $F = 0$, onto the associated Legendre polynomials $P_l^m(\cos\phi)$. [116,117] The dipole factor $d_{lm}(\varepsilon)$ can be obtained from the intensities of zero-field spectral lines.

The polarization dependence of photoionization cross sections have been carried out [118] and the results are shown in Figure 17. They are in good agreement with exact numerical calculations by Luc-Koenig and Bachelier [109] and with the early findings of the intensity modulations near the zero-field ionization limit for π-polarization (i.e., the $m = 0$ mode). [12] To understand this effect, it is more transparent to transform the eigenstate for a given energy from parabolic coordinates to spherical polar coordinates

$$|\varepsilon, n_1 m\rangle = N_{n_1 m}(\varepsilon) \sum_l F_l(r) Y_{lm}(\theta, \phi)\alpha_{lm}(\beta_1, \nu) \quad (57)$$

The spherical harmonics $Y_{lm}(\theta, \phi)$ are nonzero for $\theta = 0°$ only for the $m = 0$ component of any arbitrary l. This condition can be fulfilled provided

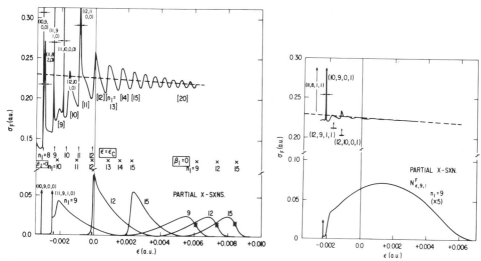

Figure 17. Photoionization cross section (in a.u.) of the ground state of atomic hydrogen in a Stark field $F = 77\,\text{kV/cm}$ vs. energy ε (in a.u.) for different light polarizations. (a) The upper part shows the total cross section from Eq. (56) for π-polarization. The dashed curve indicates $\sigma_0(\varepsilon)$ for $F = 0$. The lower part shows the partial cross sections. (b) The total cross section for σ-polarization. The parabolic quantum numbers (n, n_1, n_2, m) for $\varepsilon < 0$ and the n_1-channel corresponding to the peaks for $\varepsilon \geq 0$ are marked in the figure. (Courtesy of D. A. Harmin.)

that the electron's energy is above the potential barrier created by the Coulomb and Stark field. The barrier will attain its maximum for $\theta = 0°$. However, only that portion of the electron's motion parallel to the field axis, i.e., the z axis, and having $m = 0$, has a high probability of being scattered by the infinite barrier toward the core and thus gains intensity by overlapping extensively with the ground-state wave function. For a given n_1 manifold, the $m = 0$ mode corresponds then to the electron's motion being confined and the charge distribution being stretched along the field axis. Indeed, one can understand this by examining the results shown in Figure 17 in parabolic coordinates. Each intensity maximum (or resonance) for energy $\varepsilon > 0$ in the photoionization cross section corresponds to the energy for a given n_1 manifold in the partial cross section such that $\beta_1 \sim 1$ and $\beta_2 \sim 0$. The value $\beta_1 \sim 1$ corresponds to the maximum distribution of the excited wave function along the z axis, and this effect is most enhanced for final states with $m = 0$. When the electron's energy is below the potential barrier, $\varepsilon < 0$, its motion is more conditioned by the barrier penetration effect. In other words, the condition to produce $m = 0$ intensity modulation will not be so well satisfied. Photoionization calculations from the unperturbed $3p$ state in hydrogen show pronounced intensity modulations, in

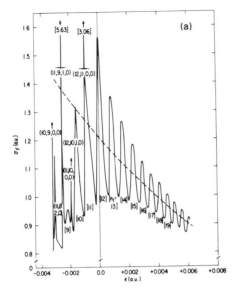

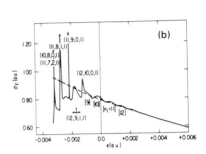

Figure 18. Photoionization cross sections (in a.u.) of the $3p$ state of atomic hydrogen in a Stark field $F = 77$ kV/cm vs. energy ε (in a.u.) for different light polarizations. (a) $\pi\pi$: Transition from $3p$, $m = 0$ to $\varepsilon s + \varepsilon d$, $m = 0$. (b) $\sigma^{\pm}\pi$ (or $\pi\sigma^{\pm}$): Transition from $3p$, $m = 1$ or 0 to εd, $m = 1$. (Courtesy of D. A. Harmin.)

particular in the π–π transitions where both intermediate and final states have $m = 0$. Figure 18 shows the results.[118] Experimental works on hydrogen Rydberg states are underway.

3.2. Perturbative Treatment: Excited Rydberg States

The perturbative treatment of the Stark effect in atoms has been revived recently, partly because of the advance of higher-order perturbation theory, notably the Bender–Wu theory,[113] and partly because of the use of the Stark effect on excited states to detect highly excited atoms. In particular, a connection between the Bender–Wu theory for high-order Rayleigh–Schrödinger perturbation coefficients for the two-dimensional rotationally symmetric anharmonic oscillator and the behavior of resonances in the hydrogen Stark problem has been made.[114,115] The energy of the Stark effect in hydrogen can be expressed in terms of arbitrarily high orders of Rayleigh–Schrödinger perturbation theory [115]

$$E(n_1, n_2, m, F) \sim \sum_N E^{(N)}_{n_1 n_2 m} F^N \qquad (58)$$

The expression in Eq. (58) is known as the Rayleigh–Schrödinger series and $E_{n_1 n_2 m}^{(N)}$ is a Rayleigh–Schrödinger coefficient. The coefficients are expressed in terms of parabolic quantum numbers, n_1 and n_2, and the magnetic quantum number m. F is the electric field strength. The precise asymptotic behavior of $E_{n_1 n_2 m}^{(N)}$ for high-order perturbation theory in N up to 150 has been carried out.[115]

4. Nonhydrogenic Atoms in External Fields

4.1. Introduction

In the following, we discuss magnetic and electric effects of nonhydrogenic atoms in terms of quantum defect theory (QDT). The electron–ion core interaction of nonhydrogenic atoms is represented by the Coulomb potential, $-1/r$, plus a short-range electrostatic interaction $v(r)$. In the far zone, $r \to \infty$, the interaction is dominated by the pure Coulomb potential and the wave function is known analytically. In the near zone, $r \lesssim r_0$, the electron's interaction with the short-range non-Coulombic potential $v(r)$ is characterized by a parameter. This parameter is fixed by the boundary condition at $r = r_0$ for the wave function in the far zone and is identified as the quantum defect or phase shift. This is the underlying principle of quantum defect theory.[21–24]

The magnetic and electric potentials are proportional to r^n, with $n = 1$ for electric and $n = 2$ for diamagnetic potentials. For nonhydrogenic atoms in external fields, the effective potential has the form $-1/r + v(r) + a_n r^n$. The external fields dominate in the far zone whereas $v(r)$ dominates in the near zone. The situation thus appears applicable for a QDT approach. However, the symmetries in these two configuration spaces are different. It is cylindrical symmetry in the far zone where external fields dominate and spherical symmetry in the near zone where the electrostatic field dominates. This difference of symmetry in different configuration spaces requires a transformation between two different coordinate systems. The Stark effect of nonhydrogen atoms provides such an example. The pure Coulomb plus Stark potential is separable in parabolic coordinates and an analytically known wave function exists. Thus the eigenfunctions of a nonhydrogenic atom can be represented by a linear superposition of hydrogenic Stark wave functions in parabolic coordinates. The mixing coefficients are expressed in terms of a reaction matrix and fixed by the boundary conditions in the near zone. The two configuration spaces are related by a local frame transformation.[119] The scattering effect of the short-range electrostatic interaction $v(r)$ of the electron with the ion core is expressed in terms of a quantum defect (or phase shift) which represents the eigenvalue of the reaction matrix in the near zone. Fano has succeeded in expressing the photoionization cross section of nonhydrogenic atoms in a Stark field for *all energies* in terms of quantum defect and frame transformation parameters.[119]

However, the situation for the diamagnetic problem is not so simple. We have seen in the previous sections that the Coulomb plus diamagnetic potentials are not separable in any coordinate system. That is, we do not have analytic solutions to represent the far zone region. Thus a unified channel approach for atomic diamagnetism is not yet possible; nonetheless, if the diamagnetic potential can be treated as a perturbation, then the quantum defect treatment can be readily made. In evaluating the matrix elements of external field potentials, it is only important to know the radial functions for large r, well outside of the inner core. We should expect to obtain a good approximation for the radial integrals by using the zero-field hydrogenic wave functions with indices n and l and putting $n \equiv \nu$, as discussed in Section 2.2.2b. In fact, Schiff and Snyder[69] have worked out the diamagnetic shifts for the alkalis using this approach. This method leads to the normal diamagnetic scaling law as a quartic power of ν, the effective quantum number, along a single Rydberg series.

For perturbed Rydberg spectra, in particular when the excited electron interacts with an open-shell ion core, a multichannel quantum defect theory (MQDT) is required.[120] We shall begin with a brief discussion of MQDT suitable for the alkaline earths, such as Ba.

For an atom in combined external magnetic and electric fields, the relevant Hamiltonian is

$$H_{\mathrm{ME}} = \mu_0(g_e \mathbf{L} + g_s \mathbf{S}) \cdot \mathbf{B} + \frac{e^2}{8mc^2}(\mathbf{r} \times \mathbf{B})^2 - \frac{e}{c}(\mathbf{V} \times \mathbf{B}) \cdot \mathbf{r} - e\mathbf{r} \cdot \mathbf{F} \quad (59)$$

The first term is the paramagnetic Zeeman potential. It is independent of radial excitation and depends only on spin and orbital angular momentum. The shifts and splittings of spectral lines are the same for all levels of a channel for a given magnetic field B. The Landé g-factor is also characteristic for all levels belonging to an unperturbed Rydberg channel. These simple systematics are no longer true for perturbed spectra.[24] The last three terms in Eq. (59) are the diamagnetic, motional Stark, and Stark potentials, respectively. Since these three terms depend on the extent of radial excitation, the shifts and splittings vary from level to level. One can control experimental conditions such that only one term in Eq. (59) is important at a time. We shall discuss the electric and magnetic effects separately for perturbed Rydberg atoms in this section.

4.2. Quantum Defect Theory of Rydberg Spectra

Since the most complete diamagnetic spectra are on alkaline earth atoms,[11,38] i.e., Ba and Sr, we hereby outline a two-channel MQDT model which is suitable for the alkaline earths in order to illustrate the major aspects of the formulas of the quantum defect theory.[120] The formulas can be easily generalized to cases of more than two channels as well as simplified to a single-channel case, such as for the alkalis.

A channel is defined, according to the MQDT, as a set of discrete and continuum states of an ion–electron complex which differ only in the energy of the excited electron. A channel is specified by the orbital, spin, and fine-structure quantum numbers of the ion, the orbital and spin angular momenta of the electron, and the coupling of the electron to the ionic core. For example, all states specified by Ba $6s(^2S_{1/2})\varepsilon s$, $J = 0$ for discrete and continuum values of ε form a singly excited electronic channel, while the set of states specified by Ba $5d(^2D_{3/2})\varepsilon d_{3/2}$, $J = 0$ comprise a doubly excited electronic channel.

In the case of the Ba atom, the singly excited Rydberg channels $6snl$ converge to the first ionization limit I_1, corresponding to the $6s$ level of Ba$^+$; whereas the doubly excited channels $5dnl$ converge to the second ionization limit I_2, corresponding to the $5d$ levels of Ba$^+$. At each level position we define two effective quantum numbers, ν_1 and ν_2, such that

$$E_n = I_1 - \mathrm{Ry}/\nu_1^2$$
$$= I_2 - \mathrm{Ry}/\nu_2^2 \tag{60}$$

where Ry is the Rydberg constant. The quantum defect μ is defined by $\mu = n - \nu_1$. Equation (60) establishes a functional relationship between ν_1 and ν_2:

$$\nu_1 = \nu_2 \left[1 - \nu_2^2 \left(\frac{I_2 - I_1}{\mathrm{Ry}} \right) \right]^{-1/2} \tag{61}$$

The quantum defect μ in a multichannel case is not required to be a smoothly varying function of energy. Instead, it can vary from level to level due to interchannel interactions. The zero-field potential between the excited electron and the ion core has the following property:

$$V(r) = -\frac{e^2}{r} + v(r) \qquad \text{for } r < r_0$$

$$= -\frac{e^2}{r} \qquad \text{for } r > r_0 \tag{62}$$

The wave function for the scattering state $e + Ba^+$ can be written in the general form

$$\psi = \sum_{ij} \Phi_i [f(\nu_i, r)\delta_{ij} - g(\nu_i, r)R_{ij}]b_j \qquad \text{for } r > r_0 \qquad (63)$$

where f and g are the regular and irregular Coulomb wave functions, respectively, with $\nu_i = \nu_1$ or ν_2. R_{ij} is the reaction matrix which characterizes the short-range non-Coulombic potential, and the b_j are the mixing coefficients which are determined by application of boundary conditions. An energy-dependent R-matrix would lead to energy-dependent eigenvalues μ_α and eigenvectors $U_{i\alpha}$, where

$$R_{ij}(\varepsilon) = \sum_\alpha U^\dagger_{i\alpha}(\varepsilon) \tan \pi\mu(\varepsilon) U_{\alpha j}(\varepsilon) \qquad (64)$$

For a two-channel case

$$U_{i\alpha} = \begin{pmatrix} \cos\theta & \sin\theta \\ -\sin\theta & \cos\theta \end{pmatrix}$$

In general $U_{i\alpha}$ is an orthogonal matrix which relates the asymptotic channels i applicable when the excited electron is far from the core to the close-coupling channels α applicable when the excited electron penetrates the core. To account for energy-dependent effects we expand the channel mixing angle θ and the eigen-quantum defect μ_α in energy using the first two terms of a Taylor series expansion

$$\theta(\varepsilon) \cong \theta_0 + \varepsilon\theta^{(1)}$$

with

$$\theta^{(1)} \equiv \frac{\partial\theta}{\partial\varepsilon}$$

A similar expression applies for μ_α. To obtain discrete energy levels, we impose the boundary condition that $\psi \to 0$ as $r \to \infty$, where ψ is the wave function given in Eq. (63). This leads to the consistency relation

$$F(\nu_1, \nu_2) = \det |U_{i\alpha} \sin \pi(\nu_i + \mu_\alpha)| = 0 \qquad (65)$$

Equations (61) and (65) jointly determine all the discrete levels. Namely, all the discrete levels should lie at the intersections of the curve represented by Eq. (61) with that represented by Eq. (65). For each energy level, Eq. (60) determines a pair of values (ν_1, ν_2) on a two-dimensional plot ν_1 vs. ν_2. The curve determined by Eq. (65), $F(\nu_1, \nu_2) = 0$, will pass through and connect all the discrete levels belonging to these two-channels. The parameters $U_{i\alpha}$ and μ_α are introduced through the diagonalization of the reaction matrix R and are determined by matching the wave functions describing the two different configuration spaces, i.e., the wave function for the dissociation channel i appropriate at large distances and that for the close-coupled channel α appropriate at small distances. If these parameters, i.e., $U_{i\alpha}$ and μ_α, are *energy independent*, it is clear that the curve represented by Eq. (65) is periodic on the two-dimensional plot ν_1 vs. ν_2. Indeed, the analysis of perturbed noble gas Rydberg spectra demonstrates convincingly the periodicity of channel interaction.[24] The interaction of Rydberg channels with doubly excited channels leads to the introduction of energy-dependent parameters and therefore the above periodicity is broken. As we have discussed, energy-dependent parameters are then needed since the doubly excited channels require an additional label to characterize the effect of radial correlations. This effect has been singled out as one of the reasons to study double-excitation using hyperspherical coordinates. Alkaline earth spectra exemplify this situation. Extensive studies have been carried out experimentally[121,122] as well as theoretically[123] for these spectra.

The normalized wave function for $r > r_0$ can be represented as a superposition of the dissociation channels i or the close-coupled channels α in the form

$$\psi_n = \sum_i \Phi_i P_i^{(n)}(r) Z_i^{(n)} \tag{66}$$

where Φ_i is the ion-core wave function, $P_i^{(n)}(r)$ is the excited electron wave function, and $Z_i^{(n)}$ represents the set of expansion coefficients in the i-channel representation. Φ_α, $F_\alpha^{(n)}(r)$, and $A_\alpha^{(n)}$ have similar meanings in the α-channel representation. The oscillator strengths of these two mutually interacting channels are represented by[124]

$$f_n = \frac{E_n - E_0}{\cos^2 \theta \sin^2 \pi(\mu_1 - \mu_2)}$$

$$\times \left\{ D_1 \sin \pi(\nu_1 + \mu_2) + D_2 \sin \pi(\nu_2 + \mu_1) \right.$$

$$\left. \times \left[\frac{d(-\nu_1)}{d\nu_2} \right]^{1/2} \right\}^2 \bigg/ \left\{ \nu_1^3 + \nu_2^3 \left[\frac{d(-\nu_1)}{d\nu_2} \right] \right\} \tag{67}$$

where D_α, $\alpha = 1, 2$, is the energy-independent dipole moment parameter.

The expression $d(-\nu_1)/d\nu_2$ is the slope of the two-dimensional quantum defect plot of ν_1 vs. ν_2. Its explicit expression in terms of MQDT parameters is

$$\frac{d(-\nu_1)}{d\nu_2} = \frac{1}{\cos^2 \pi\nu_2} \frac{\Gamma}{(\tan \pi\nu_2 - \alpha)^2 + \Gamma^2} \tag{68}$$

It has a maximum at $\tan \pi\nu_2 = \alpha$ and a width at half-maximum of Γ, which measures the interaction strength between these two channels. α and Γ depend on the quantum defect parameters μ_α and $U_{i\alpha}$. The expression in Eq. (68) has a symmetrical Lorentz shape, for constant quantum defect parameters, modified by the factor $1/\cos^2 \pi\nu_2$. When there is no interaction, $\theta = 0$, and $\Gamma \equiv 0$. For the energy level E_n, having $\nu_{1,n} = n - \mu$, we see clearly from Eqs. (67) and (68) that

$$f_n \to (E_n - E_0)|D_1|^2/\nu_{1,n}^3$$

Therefore, for unperturbed Rydberg levels, the oscillator strength scales like $\nu_{1,n}^{-3}$, with $\nu_{1,n} = n - \mu_1$. However, when the interaction is strong, the slope $d(-\nu_1)/d\nu_2$ can be large and the mixing angle θ need not be small. In this case the oscillator strength does not follow the $\nu_{1,n}^{-3}$ law.

4.3. Paramagnetism: Channel Mixing Effects on Magnetic Shifts

We first consider the situation for magnetic fields which are weak in the sense that the paramagnetic potential is smaller than the Coulomb potential but larger than the diamagnetic interaction. We also assume that the effect due to nuclear spin is negligible. The paramagnetic potential will produce the linear Zeeman shift, $(e\hbar/2m)gBM$. The Landé g-factor for the nth discrete level, g_n, is given by the expectation value of the operator $g = 1 + \mathbf{s} \cdot \mathbf{j}/\mathbf{j} \cdot \mathbf{j}$ calculated with the nth level's wave function ψ_n. If we use a MQDT wave function, such as in Eq. (66), the Landé g-factor takes the following form[24,125]:

$$g_n = \sum_{ij} g_{ij} Z_i^{(n)} Z_j^{(n)} = \sum_{\alpha\beta} g_{\alpha\beta} A_\alpha^{(n)} A_\beta^{(n)} \tag{69}$$

The g factor may be expressed in terms of two types of channel representation, the i- and α-channel representations. The dissociative channel i corresponds to the excited electron at large distance from the ion core, where spin–orbit coupling is likely to be important. Thus it is a good approximation to label the i-channel according to a jj coupling scheme. The close-coupled channel α, on the other hand, is more suitably characterized by an LS coupling scheme, since the interaction is dominant at distances close to the ion core. Because of channel interactions, the coupling scheme of a perturbed Rydberg series will depend on the degree of mixing among the relevant channels. This information is imbedded in the mixing coefficients A_α or Z_i. The Landé g-factor depends only on the angular momentum coupling scheme and it is diagonal in an LS coupling scheme. Values of g-factors, measured by linear Zeeman effect experiments, have been utilized as a sensitive probe of the angular momentum coupling scheme of a given spectral line. One therefore expects that the measured g-values along a perturbed Rydberg series would reflect the degree of channel mixing.

A recent set of measurements of the g-factor along a series of $J = 2$ Rydberg states having the configurations $5snd$ in Sr in the region of strong mixing between the $^1D_2$ and $^3D_2$ Rydberg series, shows a variation which is in agreement with the predictions of MQDT.[125] Figure 19 shows the results. Esherick[125] uses a five-channel MQDT for the channels, $5snd$ $^1D_2$, $5snd$ $^3D_2$, $4d5s$ $^1D_2$, $4d5s$ $^3D_2$, and $4p^2$ $^1D_2$ to fit his data. The quantum defect parameters, $U_{i\alpha}$ and μ_α, thus obtained are used to compute the mixing coefficients $A_\alpha^{(n)}$. A pure LS coupled scheme is used to evaluate g_α. Equation (69) is used to compute the g_n along the series. From Figure 19, one notes that the strongest mixing occurs for $n = 16$, and the g-factors vary between 1 (that for a pure $^1D_2$ state) and 7/6 (that for a pure $^3D_2$ state). This success in treating the variation of g-factors over the whole perturbed series as a single problem rather than dealing with one state at a time, demonstrates the power of MQDT.

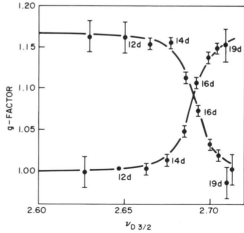

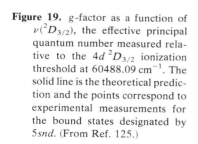

Figure 19. g-factor as a function of $\nu(^2D_{3/2})$, the effective principal quantum number measured relative to the $4d\,^2D_{3/2}$ ionization threshold at 60488.09 cm^{-1}. The solid line is the theoretical prediction and the points correspond to experimental measurements for the bound states designated by $5snd$. (From Ref. 125.)

4.4. Diamagnetism: Magnetic Contribution to Channel Mixing

In this section, we will discuss in detail how to disentangle complex perturbed Rydberg spectra by measuring the diamagnetic shifts.[120] We choose the alkaline earth atoms as examples. The alkaline earth spectra are typified by singly excited Rydberg series which overlap series of double-excited states. This circumstance makes the identification of the spectral lines difficult. For a given excitation energy, a singly excited Rydberg state has a larger radial extension from the ion core than a doubly excited state. However, channel interaction between these two types of states diminishes the distinction. In fact, the mixings in alkaline earth atoms are so strong as to "hybridize" the spectra in the sense that the oscillator strengths are redistributed between these two types of channels throughout the whole channel including both discrete and continuum portions. The MQDT has been rather successful in analyzing these spectra by introducing energy-dependent parameters for both μ_α and θ.[121,126]

The Ba even-parity spectrum is ideal for our study, since the doubly excited channels having configurations $5dnd$ and $5dns$ are embedded among the singly excited Rydberg channels having configurations $6snd$ and $6sns$. The spectrum has been analyzed by MQDT,[122] and the diamagnetic shifts have been measured.[8] The upper part of Figure 20 shows the two-dimensional quantum defect plot of ν_1 vs. ν_2 for the even-parity Ba spectrum. The pair of parabolic curves representing the channel interaction of the Rydberg channel $6sns$ $^1S_0$ with the doubly excited channel $5dnd$ $^1S_0$ are indicated in the figure. Note that the quantum defects of the Rydberg levels belonging to the $6sns$ series are nearly constant (having $\mu_1 \sim 0.2$) except near the doubly excited level $5d7d$ $^1S_0$, which is marked by a cross in the figure. The Rydberg level $18s$ (with $\mu_1 \sim 0.1$) is nearly degenerate with $5d7d$ (with $\mu_1 \sim 0.4$) and both are mixed strongly with each other as indicated by the big change in quantum defect relative to those of their neighbors. The channel interactions within $J = 2$ are more complex. The solid curves representing the situation for five interacting channels, $6snd$ $^{1,3}D_2$, $5dnd$ $^1D_2$, $^3F_2$, and $5dns$ $^1D_2$ are also indicated in the same figure. The mixing coefficients $Z_i^{(n)}$ and the quantum defect parameters, μ_α and $U_{i\alpha}$ for $J = 0$ and $J = 2$ channels have been obtained by a MQDT fitting to the data.[122] They will be used to evaluate the diamagnetic shifts. The spacing between neighboring Rydberg levels for $n = 30$ is $\Delta E_{30,29} \cong 10$ cm^{-1}, whereas the diamagnetic shift for the $n = 30$ level at a magnetic field strength of $B = 40$ kG is $\Delta\sigma \cong 5$ cm^{-1}. We thus have $\Delta E_{30,29} > \Delta\sigma$, and hence we can use perturbation theory to evaluate the diamagnetic shifts.

We use the MQDT wave functions as the zero-field unperturbed wave functions. The diamagnetic potential $H_D = (e^2/8m_ec^2)B^2r^2 \sin\theta$ mixes

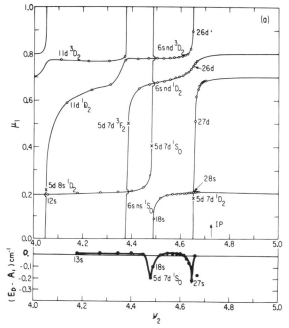

Figure 20. Diamagnetic shifts of the perturbed spectrum of Ba atoms. The upper figure shows the two-dimensional quantum defect plot μ_1 vs. ν_2 of even parity, $J = 0$ and $J = 2$ channels. The relevant channels and levels are marked in the figure. The lower figure shows the difference of the diamagnetic shift from the hydrogenic value, $E_D - A_1$. The solid dots are experimental data from Ref. 8 and the solid curve results from calculations using Eqs. (70) and (76). (From Ref. 120.)

levels with orbital angular mementum l such that $\Delta l = |l' - l| = 0, 2, \ldots$, and having the same magnetic quantum number m and parity. Because of its r^2 dependence, the diamagnetic potential is negligible near the ion core. This situation is well suited for the MQDT wave functions in Eq. (66), because in evaluating the matrix elements, it is only important to know the radial function for large r, outside of the inner ion core. We expect to obtain a good approximation for the radial integrals by using the Coulomb wave functions with dissociation channel indices i, n, and l, and putting $n = \nu$ after the integrals are evaluated. For Ba even-parity states, we choose channel 1 as $6sns\ ^1S_0$ and channel 2 as $5dnd\ ^1S_0$. The first-order diamagnetic energy for a two-channel perturbed Rydberg spectrum is

$$\frac{e^2}{8m_e c^2} B^2 \langle r^2 \sin^2 \theta \rangle_n = A_1 \left[1 + \frac{A_2}{A_1} \left(\frac{\nu_2}{\nu_1} \right)^3 \frac{d(-\nu_1)}{d\nu_2} \right]$$

$$\times \left[1 + \left(\frac{\nu_2}{\nu_1} \right)^3 \frac{d(-\nu_1)}{d\nu_2} \right]^{-1} \qquad (70)$$

where A_1 and A_2 represent hydrogenic diamagnetic matrix elements for channels 1 and 2, respectively. The cross term between $6sns$ and $5dnd$ vanishes since the core states $6s$ and $5d$ are orthogonal, i.e., $\langle 6s | 5d \rangle \equiv 0$. The hydrogenic diamagnetic matrix element for an n, l, $m = 0$ state is[69]

$$A_{l,l}(n,0) = +4.97 \times 10^{-15} B^2 \frac{n^2 [5n^2 + 1 - 3l(l+1)](l^2 + l - 1)}{(2l+3)(2l-1)} \qquad (71)$$

and the off-diagonal matrix element is

$$B_{l,l-2}(n,0) = -4.97 \times 10^{-15} B^2 \frac{5n^2 l(l-1)}{2(2l-1)} \left\{ \frac{(n^2 - l^2)[n^2 - (l-1)^2]}{(2l+1)(2l-3)} \right\}^{1/2} \qquad (72)$$

The explicit forms of A_1 and A_2 can be obtained easily by replacing $n \to \nu_1$, $l = 0$ for channel 1 and $n \to \nu_2$, $l = 2$ for channel 2 in Eq. (71), respectively:

$$A_1 = A_s(\nu_1) = 4.97 \times 10^{-15} B^2 \frac{\nu_1^2 (5\nu_1^2 + 1)}{3}$$

$$A_2 = A_d(\nu_2) = 4.97 \times 10^{-15} B^2 \frac{5\nu_2^2 (5\nu_2^2 - 17)}{21}$$

The energy matrix elements are expressed in units of cm^{-1} and B is in gauss.

For negligible channel interaction, the slope of the ν_1 vs. ν_2 plot in Figure 20 is $d(-\nu_1)/d\nu_2 \approx 0$ at an eigenvalue belonging to channel 1, and $d(-\nu_1)/d\nu_2 \gg 1$ for $\varepsilon = \nu_{2n}$ at an eigenvalue belonging to channel 2. Thus, the shift expressed by Eq. (70) becomes $E_D^{(1)} \approx A_1 \approx \nu_1^4 B^2$ at an eigenvalue belonging to channel 1 and $E_D^{(1)} \approx A_1 \approx \nu_2^4 B^2$ at an eigenvalue belonging to channel 2. Therefore the larger the radial excitation the bigger the diamagnetic shift. Since the diamagnetic potential is always positive, the shift is to the blue. For the Ba atom the doubly excited level $5d7d\ {}^1S_0$ lies between the levels $6s17s$ and $6s18s$ so that all three are nearly degenerate in energy. Without interaction, the doubly excited level $5d7d$ is much more compact in radial size than the neighboring Rydberg levels, therefore it is expected that the doubly excited level $5d7d$ should have a smaller diamagnetic blue shift than the $17s$ and $18s$ levels. Also the $18s$ should have a larger blue shift than that of $17s$. Channel interaction changes the above simple picture and the shift is described by Eq. (70). The quantum defect parameters needed to evaluate the slope $d(-\nu_1)/d\nu_2$ have been obtained by fitting to the energy levels.[120] For a field of $B = 35.7$ kG, calculations based on Eq. (70) give diamagnetic shifts for $17s$, $5d7d$, and $18s$ of 0.29, 0.17, and 0.28 cm^{-1}, respectively. It is interesting to note that the $18s$ level actually has a shift which is smaller than that of the $17s$ level because of the channel mixing with the more compact doubly excited $5d7d$ level. The lower portion of Figure 20 shows the difference between the diamagnetic shifts for Ba and those for the corresponding levels of hydrogenlike atoms, $E_D - A_1$, along the $6sns$ $^1S_0$ series. Note that the ns levels ($n = 13$ to 17) are relatively unperturbed Rydberg levels whereas $5d7d$ and $18s$ are perturbed.

When the Rydberg excitation reaches $\nu_n \approx 30$, diamagnetic l-mixing becomes important for field strengths $B \approx 40$ kG. A second-order perturbation calculation of the energy shifts is required[120]:

$$E^{(2)}_{\nu_n l} = \sum_{\nu'_n l'} \frac{|\langle \nu_n l | H_D | \nu'_n l' \rangle|^2}{E_{\nu_n l} - E_{\nu'_n l'}} \tag{73}$$

We consider only states with $\nu_n = \nu'_n$ and $l' = l + 2$. We have

$$E_{\nu_n l} - E_{\nu'_n l'} \cong \frac{\text{Ry}\,(\mu'_l - \mu_l)}{\nu_n^3} \tag{74}$$

where $\nu_n = n - \mu_l$ and $\nu'_n = n - \mu'_l$. Again we use MQDT wave functions to evaluate the matrix element in Eq. (73):

$$\langle \nu_n l | H_D | \nu'_n l' \rangle = \sum_i Z_i(\nu'_i, l) Z_i(\nu'_i, l') B'_{l,l}(\nu_i, 0) \delta_{l,l-2} \tag{75}$$

Here $B_{l,l'}(\nu_i, 0)$ is the off-diagonal matrix element of Eq. (72) in which we have made the replacement $n \to \nu_i$ and $\nu_i \cong \nu'_i$; ν_1 and ν_2 are the effective quantum numbers defined in Eq. (60) for different ionization limits; and Z_i is the mixing coefficient, defined in Eq. (66), which describes the extent of channel interactions. In order to see the diamagnetic l-mixing, we consider matrix elements connecting Ba $6sns$ levels with Ba $6snd$ levels, all having $m = 0$. We make use of Eqs. (73), (74), and (75) to obtain the second-order diamagnetic energy shift:

$$E_D^{(2)} = \frac{(4.97 \times 10^{-15})^2}{\text{Ry}} \frac{5}{9} \cdot \frac{B^4 \nu_{1l}^{11}}{\mu_l' - \mu_l} \cdot \left[1 + \left(\frac{\nu_2}{\nu_1}\right)^3 \frac{d(-\nu_1)}{d\nu_2}\right]_l^{-1}$$

$$\times \left[1 + \left(\frac{\nu_2}{\nu_1}\right)^3 \frac{d(-\nu_1)}{d\nu_2}\right]_{l'}^{-1} \tag{76}$$

Here we have put $\nu_l = \nu_l'$, $l = 0$ and $l' = 2$. The zero-field channel interactions have been taken into account through the slopes, $[d(-\nu_1)/d\nu_2]_l$, for $6sns$, $J = 0$ and $6snd$, $J = 2$ channels.

For noninteracting Rydberg spectra, $[d(-\nu_1)/d\nu_2] = 0$ for $l = 0, J = 0$ and $l = 2, J = 2$. The second-order shift is proportional to $B_{\nu_1}^4$ [11] and inversely proportional to the difference in quantum defects, $\Delta = \mu_l - \mu_{l'}$, of neighboring l-mixed levels. For field strengths $B = 35.7$ kG, the l-mixing will not be important until $\nu_1 \approx 30$. A MQDT calculation of l-mixing based on Eqs. (70) and (76) has been carried out along $6sns$ series and the result is shown in Figure 20. The anomalous diamagnetic shifts around $27s$ are due to the l-mixing between the $6sns$ channel and the $6snd$ channels. The $6snd$ channels are perturbed by the interloping level $5d7d$ $^1D_2$, as demonstrated by the zero-field quantum defect plot in Figure 20. Once again the compactness of the $5d7d$ $^1D_2$ state is reflected in the smaller diamagnetic shift relative to its neighboring levels. Therefore, the diamagnetic interaction can be used as a probe to measure the extent of radial excitation and to disentangle complex perturbed Rydberg spectra.

The Quasi-Landau Resonances and the Role of the Quantum Defect

We turn now to the discussion of the role of the quantum defect on the shapes of quasi-Landau resonances. [11] The energy spacings between quasi-Landau resonances are determined by the effective potential, comprising the Coulomb and diamagnetic potentials, over a very large range of the coordinate perpendicular to the magnetic field direction z, as shown by the WKB results. [16,35,36] Thus the spacings do not depend on the detailed short-range interaction between the excited electron and the ion core. Essentially, the $3/2\hbar\omega$ spacing calculated in WKB approximation is *universal* for all atoms, in agreement with all the experimental findings, as discussed in Section 2. However, the electron–core short-range interaction represented by the quantum defect, will affect the shapes of these resonances. This is seen in Figure 1, where the shapes of the resonances for Sr are markedly different from those for Ba. In the quasi-Landau region, the electron is moving along the potential ridge formed by the Coulomb and diamagnetic potentials. The motion of the electron is stationary when it is in the direction ρ perpendicular to the magnetic field direction. The spherically symmetric Coulomb potential couples the motion in ρ with that in z and eventually directs the electron's motion along the z axis and escapes, with a characteristic time τ. Since the Coulomb potential is strongest near the origin, this coupling strength will be sensitive near the origin as well. Since the short-range electron–ion core interactions vary from atom to atom so will the shapes of the quasi-Landau resonances. These resonances can be characterized, in addition to their energies, by two parameters, the phase shift ϕ and the width Γ. The phase shift ϕ, due to electron–core interaction, is defined relative to the hydrogenic value, which has a phase shift ϕ_0. The width Γ measures the lifetime, $\tau = \hbar/\Gamma$, of these resonances. ϕ and Γ are related near the resonance by $\tan(\phi - \phi_0) = \Gamma/2(E - E_0)$, where E_0 is the resonance position. With the above physical picture, we can now interpret the observations shown in Figure 1.

For the Sr principal series, the zero-field spectrum near the threshold is dominated by $5s\varepsilon p\ ^1P_1^0$ and perturbed by a strongly bound $4d5p\ ^1P_1^0$ state.[124] The effect of $4d5p\ ^1P_1^0$ on the quasi-Landau resonances is to produce a small phase shift ϕ, and thus produce a finite width Γ. On the other hand, the zero-field spectrum near the ionization threshold in Ba is dominated by $5d8p\ ^1P_1^0$. The interaction of $5d8p\ ^1P_1^0$ with the continuum background $6sp\ ^1P_1^0$ produces a large phase shift ϕ and thus a greater width Γ.

4.5. The Stark Effect: Coupling of Parabolic Channels by Scattering from the Ion Core

Recently, Fano[119] has succeeded in formulating a nonperturbative theory of the Stark effect of nonhydrogenic Rydberg atoms in terms of the quantum defect and local frame transformation parameters. The extension of the theory of the Stark effect to atoms other than hydrogen requires the addition of a short-range non-Coulombic potential $v(r)$ representing the effect of the ionic core on the excited electron.

The eigenfunctions $|n'm\rangle$ for the potential $-e^2/r + e\mathbf{F} \cdot \mathbf{r} + v$ are obtained by a linear superposition of the eigenfunctions in Eq. (53) with different n, but same m. The mixing coefficients are presented by a reaction matrix K in parabolic coordinates.[119] The short-range electron–core interaction is dominated by the non-Coulombic potential which produces a phase shift δ_l or a quantum defect $\mu = \delta_l/\pi$ in the electron's wave function. The phase shift δ_l is the eigenvalue of the reaction matrix K represented in the space of orbital angular momentum l

$$N_{l\varepsilon}^* K_{ll}(\varepsilon) N_{l\varepsilon} = -\tan \delta_l(\varepsilon) \tag{77}$$

Fano[119] has succeeded in transforming the K-matrix from the spherically symmetric frame having orbital quantum number l into the cylindrically symmetric frame of parabolic coordinates by a local frame transformation

$$K_{nn'}(\varepsilon) = \sum_l \alpha^*_{lm}(\beta_n, \nu)K_{ll}(\varepsilon)\alpha_{lm}(\beta_{n'}, \nu) \qquad (78)$$

where n represents the parabolic quantum number. The local frame transformation is performed in the region where the field strength F is not important, namely, as $(\xi, \eta) \to 0$ and $r \to 0$.

The photoionization cross section is proportional to

$$\sum_{n'n} D^*_{n'm}(\varepsilon)\{1 + [N^*_\varepsilon K(\varepsilon)N_\varepsilon]^2\}^{-1}_{n'n}D_{nm}(\varepsilon) \qquad (79)$$

where $D_{nm}(\varepsilon) = N^*_{n\varepsilon}\alpha^*_{lm}(\beta_n, \nu)d_{lm}(\varepsilon)$. The photoionization cross section for atoms other than hydrogen is expressed in terms of the quantum defect in Eq. (77) and the frame transformation parameters of (78). Theoretical calculation has been carried out by Harmin,[118,119] and compared favorably with experiments.[107]

For the hydrogen atom, the quantum defect is zero, i.e., $\delta_l \equiv 0$ and Eq. (79) reduces to

$$\sum_{n'n} D^*_{n'm}(\varepsilon)\delta_{n'n}D_{nm}(\varepsilon) = \sum_n D^2_{nm} \qquad (80)$$

and it reproduces the photoionization cross section for hydrogen given by Eq. (56).

5. Competition of Magnetic and Electric Forces

5.1. Introduction

In the last section, we have discussed various magnetoelectric effects, paramagnetic, diamagnetic, Stark, and motional Stark, on excited atoms separately. These potentials (other than paramagnetic) have one thing in common, namely, they are all long-range potentials, tending asymptotically as $\sim r^n$, where $n = 1$ and 2 for electric and magnetic fields, respectively. For a given atom and a fixed field strength, the effect scales according to principal quantum number n as n^α, where $\alpha \geqslant 2$, but with different magnitude. The effects are enriched by the different symmetry and/or constants of motion imposed by the external fields on the otherwise isotropic atoms.

We shall discuss here conditions under which the spectroscopic observables, i.e., level shifts, splittings, and intensities, undergo changes along a

Rydberg series in terms of external field strength, symmetries, and atomic species. These changes occur, for example, in the case of the diamagnetic effect[5,11] for a field strength $B \approx 5 \times 10^4$ G, from the l-mixing region for principal quantum number $n \lesssim 30$ through the n and l-mixing region and up to the quasi-Landau region with $\frac{3}{2}\hbar\omega_c$ spacing around the zero-field ionization limit (see Figure 1). In the l-mixing region, the diamagnetic potential acts as a perturbation on the Coulomb dominated potential between the excited electron and the ion core and breaks the isotropy of the unperturbed atom by intermixing the l-components. It preserves, however, the parity and cylindrical symmetry of the state.

For the same field strength, as n increases, the relative magnitude of the Coulomb versus diamagnetic potential also changes. The Rydberg electron's Coulomb binding energy is reduced whereas the diamagnetic energy is enhanced. The magnetically induced l-mixing manifolds belonging to different principal quantum numbers n would interact for $n \gtrsim 30$. This is the n and l-mixing region.[5,11] The n and l-mixing becomes complete for electron excitation near the zero field ionization limit where the joint action of Coulomb and diamagnetic potentials forms a "potential ridge".[44,45] The quasi-Landau resonances correspond to the electron's motion propagating along the potential ridge and its orbit being confined in the direction perpendicular to the magnetic field. As a result of the competition between Coulomb and diamagnetic potentials, the spectral lines regroup themselves from the n and l-mixing region into the $\frac{3}{2}\hbar\omega_c$ spacing of the quasi-Landau region. It is now clear that there is a geometric symmetry as far as the electron's motion in relation to the potential ridge is concerned. However, what mechanism controls the transition of the electron's motion from the n and l-mixing region into the quasi-Landau region is not yet clear. Such competition of potentials in changing the excited electron's motion seems to be a general phenomena. We shall discuss specific examples involving both magnetic and electric fields. We shall discuss first the role of motional Stark effect on diamagnetism and then the effects due to crossed external electric and magnetic fields.

5.2. The Induced Stark Effect: Coupling of Magnetic Sublevels by Nuclear Motion

It has been known for some time that for an excited atomic system in an external uniform magnetic field, there is an induced electric field due to the motion of the whole atom, i.e., the motional Stark effect. The strength of this term is rather weak compared with the magnetic effect and it is customary to neglect it. However, the conspicuous effects of this term were not appreciated until quite recently.[13,127] Briefly, for a given atom, say

Li, with mass M, temperature $T \approx 10^{3\circ}$ K, and magnetic field strength $B \approx 5 \times 10^4$ G, the induced motional Stark field is $E \approx 70$ V/cm. The effect can be perturbative or violent depending on the spectral range. Figure 21 shows the Li ground-state photoabsorption spectrum in this magnetic field. For $n \leqslant 20$, the effect is basically perturbative, namely, the motional Stark effect induces only the weak parity violating l-mixing spectra, i.e., even l components, $l = 0$ and 2, and the m-mixing components, i.e., $m = 0, +2$, in addition to the $l = 1$ and $m = +1$ spectra for right-hand circularly polarized light for normal allowed diamagnetic spectra. These weak forbidden components are documented in Figure 22 by a straightforward diagonalization calculation.[13] In the second region, $30 \leqslant n \leqslant 20$, the

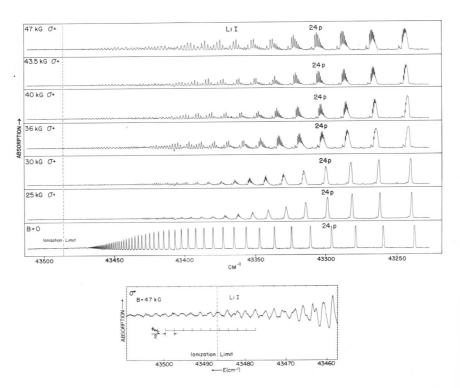

Figure 21. (a) Microdensitometer recordings of the plate transmission of the Li principal series with magnetic field strengths $B = 47$, 43.5, 40, 36, 30, and 25 kG (σ^+ polarization) and $B = 0$. The Li vapor pressure is ≈ 0.1 Torr. (b) Enlarged microdensitometer tracing of Li in the region across the ionization limit with $B = 47$ kG and σ^+ polarization. The Li vapor pressure is ≈ 0.7 Torr. (From Ref. 13.)

strength of the forbidden components induced by the motional Stark effect begins to be comparable to that of those due to the diamagnetic effect. In the third region, $n \geqslant 30$, the spectra are dominated by the motional Stark effect and altered by the quasi-Landau spacings from $\frac{3}{2}\hbar\omega_c$ into $\frac{1}{2}\hbar\omega_c$ (cf. Figure 21b).

A perturbative estimation[15] of the amplitude ratio of optical transitions to states with $m \neq 1$ and $m = 1$ serves as a criterion to illustrate the results discussed above. This ratio is given by the ratio of the matrix element of the perturbation energy, $m_e V_\perp W_c \chi$, to the energy difference of successive Zeeman levels with $|\Delta m| = 1$, i.e., $\mu_0 B = \frac{1}{2}\hbar\omega_c$:

$$P = \left[\frac{m_e V_\perp W_c \langle n, l, m | \chi | n, l + 1, m \pm 1 \rangle (1 + D_n)}{\frac{1}{2}\hbar\omega_c + R(2/n^3)\Delta\mu} \right]^2 \sim \frac{m_e}{M} \cdot \frac{kT}{R} n^4 \quad (81)$$

Here $V_\perp = |\mathbf{V}_{th} \times \hat{B}|$, $\frac{1}{2}MV_\perp^2 \approx kT$ and $R = \hbar^2/2m_e a_0^2 = 13.6 \text{ eV}$. The effect of the quantum defect is taken care of in D_n and $\Delta\mu$ as discussed in Section 4. For Li, this ratio is about 0.1 for $n \approx 20$ but increases to 1.0 for $n > 30$. At this point the m-mixing due to the motional Stark effect is complete. The diamagnetic interaction creates a series of quasi-Landau levels with different quantum number n_L having a spacing $(\frac{3}{2})\hbar\omega_c$ for a

Figure 22. Li absorption spectrum for the $n = 21$ manifold in a magnetic field $B = 47.8 \text{ kG}$. The curves are densitometer traces from photoabsorption measurements taken at a vapor pressure of 0.1 mm. The calculated values of positions corresponding to different nominal m components, $m = 1$, 2, or 3 are marked. The vertical lines represent calculated values of the square of the eigenvector component belonging to $l = 1$, $m = 1$ allowed transitions. Experimental line centers, the $B = 0$ position ($\sigma_0 = 43237.31 \text{ cm}^{-1}$) of $n = 21$, and the magnitude of the linear Zeeman shift ($\mu_0 B$) for $B = 47.8 \text{ kG}$ are indicated. The absorption features with $m = 3$ and 2 become far more prominent in spectra at higher pressures. (From Ref. 13.)

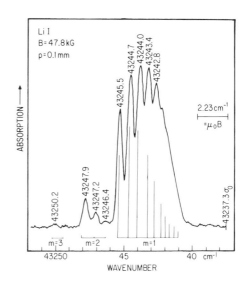

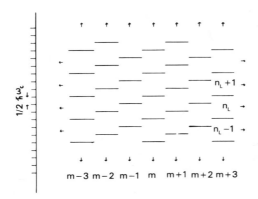

Figure 23. Array of levels $E_{n_L m} = (m + 3n_L)\mu_0 B$. (From Ref. 15.)

specific value of m. The motional Stark effect generates different m components due to strong m-mixing. New series of quasi-Landau levels belonging to different m components are formed and are displaced by the linear Zeeman shift of orbital levels by $\mu_0 B = \frac{1}{2}\hbar\omega_c$. This picture produces a two-dimensional array of levels, $E_{n_L m} = (m + 3n_L)\mu_0 B$, as shown in Figure 23. This interpretation explains the $\frac{1}{2}\hbar\omega_c$ spacing in the photoabsorption spectra of Li vapor in high magnetic fields.[15]

The two-dimensional array of levels can be described by an eigenvalue problem in terms of a two-dimensional finite difference equation:

$$(H_{n_L m} - E_{n_L m})U_m^{n_L} = \sum_{n_L'}\left[A_{n_L n_L'}^{m,m+1}U_{m+1}^{n_L'} + A_{n_L n_L'}^{m,m-1}U_{m-1}^{n_L'}\right] \quad (82)$$

This expression preserves the translational invariance in both n_L and m variables where $A_{n_L n_L'}^{mm'} \doteq A_{n_L n_L'}^{mm'}\delta_{m',m+1}$ represents a two-dimensional tridiagonal matrix. If the values of n_L and m extend to infinity, the coupling $A_{n_L n_L'}^{mm'}$ will be *uniform*. One can show by induction that the eigenvalues of Eq. (82) are equally spaced, $E_{n_L m} = (m + 3n_L)\mu_0 B$, and the eigenvectors are uniform. This indicates that the motional Stark resonances are not only equally spaced with spacing $\frac{1}{2}\hbar\omega_c$, but each resonance is of equal intensity.

5.3. Crossed External Electric and Magnetic Fields: Transitions between $\frac{3}{2}\hbar\omega_c$ and $\frac{1}{2}\hbar\omega_c$ Spacing

It is only natural now to study the competition of forces of an excited electron under the combined influence of Coulomb, magnetic, and electric fields.[120] The effective potential is

$$V = -\frac{e^2}{r} + a(x^2 + y^2) - bx = -\frac{e^2}{r} + a\left[\left(x - \frac{1}{2}\frac{b}{a}\right)^2 + y^2\right] - \frac{1}{4}\frac{b^2}{a} \quad (83)$$

where $a = \frac{1}{8}m_e\omega_c^2$, and where $b = m_e V_\perp \omega_c$ for a motional Stark field and $b = eF$ for a dc external Stark field. In the above potential the electric field is perpendicular to the magnetic field. Note on the right how the linear Stark term results in a shift of the electronic distribution off-center whereas the constant term renormalizes the system's energy. When the electric field becomes increasingly large, a potential "bow" develops in the outer region. The motion of an excited electron in such a double valley potential has not been studied much.[16]

We now turn our attention to the remarkable occurrence of either $\frac{1}{2}\hbar\omega_c$ or $\frac{3}{2}\hbar\omega_c$ spacings as a result of competing effects of different forces. We discuss the conditions under which one observes either one of the two spacings in terms of the atomic mass, the quantum defect, and the external magnetic and electric fields.[120] There are two relevant quantum numbers which govern the transition between these two modes. The first is the effective quantum number for the quasi-Landau resonance around the threshold where the Coulomb field is comparable to the diamagnetic potential. As given for instance by WKB [cf. Eq. (18)] this is

$$n_L = (41.4)B^{-1/3} \qquad (84)$$

where B is in units of 5×10^4 G. This number represents the onset where "n and l mixing" is complete, giving rise to the $\frac{3}{2}\hbar\omega_c$ spacing. The quasi-Landau resonances appear for final states having $l_f - m_f =$ even.[11,14] Since we are in the strong magnetic field domain, the linear Zeeman effect reduces to the Paschen–Back limit where orbit and spin become uncoupled. We can thus deal with orbital motion only. Atoms with isotropic initial states, i.e., $l_i = 0$, would show the resonances only in σ-polarization, one-photon transitions. If the lower state is anisotropic, i.e., $l_i \geq 1$, these resonances can be seen in both σ and π polarizations. This is because a one-photon transition from the lower state with orbital angular momentum l leads to final states with $l' = l \pm 1$, whereas σ and π polarizations induce transitions in the magnetic field to upper states with $m' = m$ and $m' = m \pm 1$, respectively. Thus, both polarizations, σ and π can lead to final states satisfying the $l' - m' =$ even rule.

On the other hand, the presence of a transverse electric field (due to either an external dc field or a motional Stark field) mixes both m and l. However, it still confines the electron's motion in the same plane perpendicular to the magnetic field. If this "m-mixing" is not appreciable by the time the above n, l-mixing is complete, one will see a $\frac{3}{2}\hbar\omega_c$ mode. If, instead, this m-mixing is complete by that stage (i.e., before n_L is reached), one should observe the $\frac{1}{2}\hbar\omega_c$ mode. The index of Stark m-mixing is the relative probability of transitions to the m-forbidden components. Replacing the motional Stark potential in Eq. (81) by $e\mathbf{r} \cdot \mathbf{F}$ for an external Stark field, the value n_S at which m-mixing is complete is defined as the point where the above ratio equals unity [120]

$$R = \left[\frac{eFn_s^2 a_0 (1 + D_n)}{eB\hbar/2m_e c + \text{Ry}\,(2/n_s^3)\Delta\mu} \right]^2 = 1 \tag{85}$$

Here $\Delta\mu$ is the difference in quantum defect between the p and f orbitals, and D_n is the quantum defect correction discussed in Section 4.

We now compare these two indices, n_L and n_S. Setting them equal will give the minimum F field necessary for any B to go from the 3/2 to the 1/2 mode with all lines having equal intensity. To get a compact expression for this, replace n_S in the denominator of Eq. (85) by n_L from (84) to obtain

$$n_S = \left[\frac{a\,\Delta\mu + b}{0.529 \times 10^{-5}}\, \frac{B}{F} \right]^{1/2} \tag{86}$$

where F is in V/cm, $a = 0.385$, and $b = 0.29$ for hydrogenlike atoms. Equating Eqs. (86) and (84) we have the condition for the critical electric field, F_c, below which one observes the 3/2 and above which the 1/2 mode. For Li, $\Delta\mu = 0.05$, $F_c = 34B^{5/3}$. For Ba, $\Delta\mu = 0.15$, $F_c = 38B^{5/3}$, where B is in 5×10^4 G and F in V/cm. For experiments performed in a cell, the motional Stark field F_{th} is always present. Assuming a temperature $T \approx 10^{3\circ}$ K for both Li and Ba, the induced Stark field for a B field in units of 5×10^4 G is $F_{th} = 70B$ for Li and $F_{th} = 18B$ for Ba. For the case of Li, it is noted that the motional Stark field is more than enough to give the full m-mixing and, thereby, the $\frac{1}{2}\hbar\omega_c$ mode. It is not until one reaches $B > 17 \times 10^7$ G that one can have $F_c > F_{th}$ and thus observe $\frac{3}{2}\hbar\omega_c$ in Li. On the other hand, in Ba at 5×10^4 G $F_c > F_{th}$ so that the m-mixing is

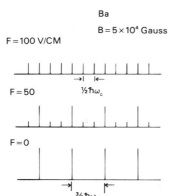

Ba

$B = 5 \times 10^4$ Gauss

F = 100 V/CM

F = 50 $\frac{1}{2}\hbar\omega_c$

F = 0

$\frac{3}{2}\hbar\omega_c$

Figure 24. Schematic spectrum of Ba near threshold in crossed electric and magnetic fields.

not appreciable and one observes the $\frac{3}{2}\hbar\omega_c$ mode. In fact, this is true for all fields larger than 7.5×10^3 G. Only at small fields is $F_{th} > F_c$. One dramatic effect of this study of competition of forces is that one can change the spacing from $\frac{3}{2}\hbar\omega_c$ into $\frac{1}{2}\hbar\omega_c$ by applying an external transverse electric field. For example, for Ba in the vapor cell at $B = 5 \times 10^4$ G, one will observe the $\frac{3}{2}\hbar\omega_c$ mode up to an F_{ext} value of about 50 V/cm. After that one should see the $\frac{1}{2}\hbar\omega_c$ mode and the $\frac{1}{2}\hbar\omega_c$ resonances become uniform in intensity for $F_{ext} > 100$ V/cm. Figure 24 shows the schematic spectra of this phenomenon for Ba. The ratio in Eq. (85) measures the ratio of intensity of peak height of the spectral resonance with its neighbors in Figure 24. Other theoretical work on atoms in joint external magnetic and electric fields has been reviewed by Bayfield.[19]

6. General Properties of Atoms in Magnetic Fields of Astrophysical Strength

The possibility of magnetic fields of order 10^7 G on white dwarf stars and of 10^{12} G on neutron stars has stimulated theoretical interest in the nature of atoms and of atomic processes under such high-field conditions. Magnetic fields of 10^9 G or greater are capable of significantly compressing even the motion of ground-state electrons in the direction perpendicular to the magnetic field. The theory for the lowest levels of atomic hydrogen under strong field conditions has been discussed in Section 2.2.3 above. Atoms heavier than hydrogen have not been as well studied in the high-magnetic-field domain. Some of their general properties are, however, qualitatively understood. Furthermore, the presence of a high magnetic field makes possible new states of matter and also introduces characteristic resonance behavior in atomic scattering processes. We discuss these features briefly below. Note that the astrophysical applications of the theory have been reviewed by Garstang,[17] and thus we do not discuss them here.

6.1. Atomic Shell Structure

For magnetic fields larger than 10^9 G the magnetic field confines electronic motion in the direction perpendicular to the field to within a cylinder of radius smaller than a_0, the Bohr radius. In the axial direction the electron is still primarily influenced by the attractive Coulomb field. In fact, for the lowest state of motion in z the electron's wave function is of even parity and hence has a large amplitude near the nucleus. Physically, the binding energy of this lowest state increases with increasing magnetic field strength. Higher states of motion in the z-direction are not nearly so tightly bound.

It is instructive to consider the limit of infinitely strong magnetic field strength.[77,128,129] This case will have energy levels similar to those for free electrons in a uniform magnetic field [128]

$$E = \hbar \left(\frac{eB}{\mu c}\right) (n_\rho + \tfrac{1}{2}|m| - \tfrac{1}{2}m + \tfrac{1}{2}) + E_z \qquad (87)$$

Here n_ρ is the quantum number for motion in the ρ direction, m is the magnetic quantum number, and E_z is the energy for motion along the z axis. One sees that for a neutral atom of atomic number Z, the lowest energy state would be that in which all electrons were in the lowest Landau level $n_\rho = 0$ with magnetic quantum numbers $m = 0, 1, \ldots, Z - 1$, since each Landau level is infinitely degenerate. All electrons would also be in the strongly bound ground state for motion along z (i.e., the lowest bound energy for E_z) and all would have their spins antialigned with the magnetic field. For successively larger m values, the mean radius ρ of the electrons would become slightly larger. Thus the atomic shell structure would resemble a set of concentric cylinders of finite length, as shown in Figure 25.

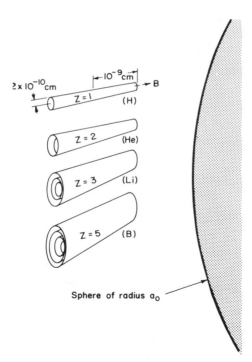

Figure 25. Shapes and sizes of light atoms in a 2×10^{12} G field. (From Ref. 129.)

For any finite strength magnetic field, the atomic structure for a particular atom will lie between the extremes of normal atomic structure and the structure described above for an extremely large magnetic field. Because of the nonseparability of the combined Coulomb and magnetic potentials it is not possible to theoretically trace the development of a particular atomic level with increasing magnetic field, although several authors have made approximate attempts to do so.[63] However, Cohen *et al.*,[77] Ruderman,[129] and Rau *et al.*[130] have given detailed quantitative discussions of atomic structure in fields of the order of 10^{12} G.

6.2. Magnetic-Field-Induced Binding

The magnetic field confinement of electronic motion in the ρ direction has important implications for the binding of the electron. Whereas attractive three-dimensional potentials do not necessarily have bound states, all attractive *one*-dimensional potentials have at least one bound state. Thus, for magnetic fields strong enough to confine electronic motion to one dimension (i.e., along the field) the electron will always have at least one bound state if the potential it moves in is at all attractive. This point was made by Avron *et al.*[131] in a short report which states that H^- can be shown to have an infinity of bound states in a large magnetic field and that He^- has at least one bound state in a large magnetic field. Note that polarization effects of the outer electron on the neutral atom lead to an attractive potential between the two. Also, Larsen[132] has performed variational calculations on H^- in a magnetic field. He then used his calculations to interpret certain features in experimental magneto absorption data on CdS as due to photodetachment of negative donor ions.

Related to this work on negative ions are theoretical calculations of Ozaki and Tomishima on the H_2^+ molecule in a uniform magnetic field.[133] In the absence of a magnetic field, the $1\pi g$ state of H_2^+ is an antibonding state. For sufficiently strong magnetic fields, however, Ozaki and Tomishima find that this state changes to a bonding state. Clearly, then, the phenomenon of magnetic-field-induced binding is quite a general one.

6.3. Landau Level Resonances

As discussed earlier in this article, the existence of resonances above the zero-field ionization limit in studies of atomic photoabsorption in the presence of laboratory-sized magnetic fields (i.e., $B \approx 10^5$ G) was a novelty. These resonances were explained theoretically as due to quasibound motion of the photoelectrons in the direction perpendicular to the magnetic field. Eventually, due to Coulomb interactions with the ionic core, the electron escapes from the ion along the magnetic field direction.

For very high magnetic fields similar behavior is predicted theoretically for a number of ionization processes. Thus sharp resonances are predicted in atomic photoionization,[134] in photodetachment of negative ions,[135] and in electron scattering.[136,137] Such resonance behavior is thus a very general feature of continuum electron motion in combined Coulomb and uniform magnetic fields.

Acknowledgments

We wish to thank Ugo Fano, David Harmin, A. R. P. Rau, and Ken Taylor for many stimulating discussions. Alan Edmonds, David Harmin, and Richard Pullen have kindly provided results prior to publication.

References and Notes

1. P. Zeeman, *Phil. Mag.* **5**, 43 (1897).
2. J. Stark, *Berl. Akad. Wiss.* **40**, 932 (1913).
3. F. Paschen and E. Back, *Ann. Phys. (Leipzig)* **39**, 897 (1912).
4. J. H. Van Vleck, *The Theory of Elastic and Magnetic Susceptibilities*, Oxford University Press, Oxford (1932).
5. F. A. Jenkins and E. Segré, *Phys. Rev.* **55**, 52 (1939); E. Segré, *Nuovo Cimento* **11**, 304 (1934).
6. L. D. Landau, *Z. Phys.* **64**, 629 (1930).
7. W. R. S. Garton and F. S. Tomkins, *Astrophys. J.* **158**, 839 (1969).
8. R. J. Fonck, F. L. Roesler, D. H. Tracy, K. T. Lu, F. S. Tomkins, and W. R. S. Garton, *Phys. Rev. Lett.* **39**, 1513 (1977).
9. P. Jacquinot, S. Liberman, and J. Pinard, Centre National de la Recherche Scientifique, Laboratoire Aimé Cotton, Orsay, France, report No. 273 (1977); S. Feneuille, S. Liberman, J. Pinard, and P. Jacquinot, *C.R. Acad. Sci. Paris, Ser. B* **284**, 291 (1977); J. C. Gay, D. Delande, and F. Biraben, *J. Phys. B* **13**, L729 (1980).
10. M. G. Littman, M. L. Zimmerman, and D. Kleppner, *Phys. Rev. Lett.* **37**, 46 (1976).
11. K. T. Lu, F. S. Tomkins, and W. R. S. Garton, *Proc. R. Soc. London, Ser. A* **364**, 421 (1978).
12. R. R. Freeman, N. P. Economou, G. C. Bjorklund, and K. T. Lu, *Phys. Rev. Lett.* **41**, 1463 (1978).
13. K. T. Lu, F. S. Tomkins, H. M. Crosswhite, and H. Crosswhite, *Phys. Rev. Lett.* **41**, 1034 (1978).
14. U. Fano, *Colloq. Int. C.N.R.S.* **273**, 127 (1977).
15. H. Crosswhite, U. Fano, K. T. Lu, and A. R. P. Rau, *Phys. Rev. Lett.* **42**, 963 (1979).
16. A. R. P. Rau, *J. Phys. B* **12**, L193 (1979); and *Comments At. Mol. Phys.* **10**, 19 (1980).
17. R. H. Garstang, *Rep. Prog. Phys.* **40**, 105 (1977).
18. K. J. Kollath and M. C. Standage, *Progress in Atomic Spectroscopy*, Part B, p. 955, Eds. W. Hanle and H. Kleinpoppen, Plenum Press, New York (1978).
19. J. E. Bayfield, *Phys. Rep.* **51**, 319 (1979).
20. D. Kleppner, *Les Houches Summer School*, Session 28, Eds. J. C. Adam and R. Ballian, Gordon Breach, New York (1981).

21. F. S. Ham, *Solid State Phys.* **1**, 127 (1955).
22. M. J. Seaton, *Proc. Phys. Soc.* **88**, 801 (1966); *J. Phys. B: Atom. Molec. Phys.* **11**, 4067 (1978).
23. U. Fano, *Phys. Rev. A* **2**, 353 (1970); *J. Opt. Soc. Am.* **65**, 979 (1975); *Phys. Rev. A* **15**, 817 (1977).
24. K. T. Lu and U. Fano, *Phys. Rev. A* **2**, 81 (1970); K. T. Lu, *Phys. Rev. A* **4**, 579 (1971); C. M. Lee and K. T. Lu, *Phys. Rev. A* **8**, 1241 (1973).
25. W. E. Lamb, *Phys. Rev.* **85**, 259 (1952).
26. H. A. Bethe and E. E. Salpeter, *Quantum Mechanics of One- and Two-Electron Atoms*, Springer-Verlag, Berlin (1957).
27. B. P. Carter, *J. Math. Phys.* **10**, 788 (1969).
28. H. Grotch and R. A. Hegstrom, *Phys. Rev. A* **4**, 59 (1971).
29. M. Sh. Ryvkin, *Dokl. Akad. Nauk SSSR* **221**, 67 (1975). [Eng. Transl.: *Sov. Phys. Dokl.* **20**, 192 (1975).]
30. J. E. Avron, I. W. Herbst, and B. Simon, *Ann. Phys. (N.Y.)* **114**, 431 (1978).
31. R. F. O'Connell, *Phys. Lett.* **70A**, 389 (1979).
32. G. Wunner and H. Herold, *Astrophys. Space Sci.* **63**, 503 (1979).
33. G. Wunner, H. Ruder, and H. Herold, *Phys. Lett.* **79A**, 159 (1980).
34. H. Herold, H. Ruder, and G. Wunner, *J. Phys. B* **14**, 751 (1981).
35. A. R. Edmonds, *J. Phys. (Paris)* **31**, C4, 71 (1970).
36. A. F. Starace, *J. Phys. B* **6**, 585–590 (1973).
37. R. Gajewski, *Physica* **47**, 575 (1970).
38. W. R. S. Garton, F. S. Tomkins, and H. M. Crosswhite, *Proc. R. Soc. London, Ser. A* **373**, 189 (1980).
39. A. R. Edmonds and R. A. Pullen, Imperial College Preprints ICTP/79–80/28, 29, 30; and R. A. Pullen, thesis, Imperial College, London (1981).
40. V. I. Arnold and A. Avez, *Ergodic Problems of Classical Mechanics*, Benjamin, New York (1968). See also D. Delande and J. C. Gay, *Phys. Lett.* **82A**, 393 (1981).
41. I. Percival, *J. Phys. B* **6**, L229 (1973).
42. C. J. Goebel and T. W. Kirkman (to be published).
43. C. Lanczos, *The Variational Principles of Mechanics*, University of Toronto Press, Toronto (1949).
44. U. Fano, *Phys. Rev. A* **22**, 2660 (1980).
45. U. Fano, *J. Phys. B* **13**, L519 (1980).
46. G. H. Wannier, *Phys. Rev.* **90**, 817 (1953).
47. C. W. Clark and K. T. Taylor, 1st Europe. Conf. Atomic Phys. Abstracts, p. 144 (1981); *J. Phys. B* **15**, 1175 (1982).
48. J. Killingbeck, *J. Phys. B* **12**, 25 (1979).
49. J. E. Avron, B. G. Adams, J. Cizek, M. Clay, M. L. Glasser, P. Otto, J. Paldus, and E. Vrscay, *Phys. Rev. Lett.* **43**, 691 (1979).
50. D. R. Bates and A. Damgaard, *Phil. Trans. R. Soc. London, Ser. A* **242**, 101 (1949).
51. J. Picart, A. R. Edmonds, and N. Tran Minh, *J. Phys. B* **11**, L651 (1978).
52. G. Risberg, *Ark. Fys.* **28**, 381 (1965).
53. C. M. Brown, R. H. Naber, S. G. Tilford, and M. L. Ginter, *Appl. Opt.* **12**, 1858 (1973).
54. C. W. Clark, *Phys. Rev. A* **24**, 605 (1981).
55. Ref. 26, p. 29.
56. G. Szegö, *Orthogonal Polynomials*, American Mathematical Society, Providence, Rhode Island (1939).
57. The substance of this paragraph has been largely developed from remarks by U. Fano (private communication).
58. M. L. Zimmerman, M. M. Kash, and D. Kleppner, *Phys. Rev. Lett.* **45**, 1092 (1980).

59. C. W. Clark and K. T. Taylor, *J. Phys. B* **13**, L737 (1980).
60. C. W. Clark and K. T. Taylor, *Nature* **292**, 437 (1981).
61. A. R. Edmonds, *J. Phys. B* **6**, 1603 (1973).
62. C. B. Crawford, *Comm. ACM* **16**, 41 (1973).
63. H. Hasegawa, in *Physics of Solids in Intense Magnetic Fields*, Ed. E. D. Haidemenakis, Plenum Press, New York (1969), Chap. 10.
64. A. F. Starace and G. L. Webster, *Phys. Rev. A* **19**, 1629 (1979).
65. G. L. Webster, Doctoral Dissertation, The University of Nebraska-Lincoln (1981).
66. H. Hellmann, *Einführung in die Quantenchemie*, Deuticke, Leipzig, Germany (1937); R. P. Feynmann, *Phys. Rev.* **56**, 340 (1939).
67. See Ref. 65, Tables III–V for the best variational results in each instance.
68. A. Baldereschi and F. Bassani, in *Proceedings of the 10th International Conference on the Physics of Semiconductors, Cambridge, Massachusetts,* AEC Oak Ridge CONF-700801, Eds. S. P. Keller, J. C. Hensel, and F. Stern, U.S. AEC., Washington, D.C. (1970), pp. 191–196.
69. L. I. Schiff and H. Snyder, *Phys. Rev.* **55**, 59 (1939).
70. R. J. Elliott and R. Loudon, *J. Phys. Chem. Solids* **15**, 196 (1960).
71. H. Hasegawa and R. E. Howard, *J. Phys. Chem. Solids* **21**, 179 (1961).
72. A. G. Zhilich and B. S. Monozon, *Fiz. Tverd. Tela (Leningrad)* **8**, 3559 (1966) [*Sov. Phys. Solid State* **8**, 2846 (1967)].
73. V. Canuto and D. C. Kelly, *Astrophys. Space Sci.* **17**, 277 (1972).
74. Y. Yafet, R. W. Keyes, and E. N. Adams, *J. Phys. Chem. Solids* **1**, 137 (1956).
75. D. M. Larsen, *J. Phys. Chem. Solids* **29**, 271 (1968).
76. E. P. Polkatilov and M. M. Rusanov, *Fiz. Tverd. Tela (Leningrad)* **10**, 3117 (1968) [*Sov. Phys. Solid State* **10**, 2458 (1969)].
77. R. Cohen, J. Lodenquai, and M. Ruderman, *Phys. Rev. Lett.* **25**, 467 (1970).
78. J. Callaway, *Phys. Lett. A* **40**, 331 (1972).
79. A. K. Rajagopal, G. Chanmugan, R. F. O'Connell, and G. L. Surmelian, *Astrophys. J.* **177**, 713 (1972).
80. (a) E. R. Smith, R. J. W. Henry, G. L. Surmelian, R. F. O'Connell, and A. K. Rajagopal, *Phys. Rev. D* **6**, 3700 (1972); (b) G. L. Surmelian and R. F. O'Connell, *Astrophys. J.* **190**, 741 (1974).
81. L. W. Wilson, *Astrophys. J.* **188**, 349 (1974).
82. H. S. Brandi, *Phys. Rev. A* **11**, 1835 (1975).
83. A. R. P. Rau and L. Spruch, *Astrophys. J.* **207**, 671 (1976).
84. R. R. dos Santos and H. S. Brandi, *Phys. Rev. A* **13**, 1970 (1976).
85. R. K. Bhaduri, Y. Nogami, and C. S. Warke, *Astrophys. J.* **217**, 324 (1977).
86. J. M. Wadehra, *Astrophys. J.* **226**, 372 (1978).
87. D. J. Hylton and A. R. P. Rau, *Phys. Rev. A* **22**, 321 (1980).
88. D. Cabib, E. Fabri, and G. Fiorio, *Solid State Commun.* **9**, 1517 (1971).
89. H. C. Praddaude, *Phys. Rev. A* **6**, 1321 (1972).
90. R. H. Garstang and S. B. Kemic, *Astrophys. Space Sci.* **31**, 103 (1974).
91. W. Ekardt, *Solid State Commun.* **16**, 233 (1975).
92. A. Galindo and P. Pascual, *Nuovo Cimento B* **34**, 155 (1976).
93. E. A. Hylleraas and B. Undheim, *Z. Phys.* **65**, 759 (1930); E. C. Kemble, *The Fundamental Principles of Quantum Mechanics*, McGraw-Hill, New York (1937), Section 51C.
94. J. E. Avron, I. W. Herbst, and B. Simon, *Phys. Lett.* **62A**, 214 (1977); *Phys. Rev. A* **20**, 2287 (1980).
95. J. E. Avron, *Ann. Phys. (N.Y.)* **131**, 73 (1981).
96. L. Haines and D. Roberts, *Am. J. Phys.* **37**, 1145 (1969).

97. L. D. Landau and E. M. Lifshitz, *Quantum Mechanics* (*Non-Relativistic Theory*), 3rd ed., Pergamon, Oxford (1976).
98. C. Lanczos, *Z. Phys.* **65**, 431 (1930).
99. E. Segré and G. C. Wick, *Proc. R. Soc. Amsterdam* **36**, 534 (1933).
100. R. E. Langer, *Phys. Rev.* **51**, 669 (1937).
101. M. H. Rice and R. H. Good, Jr., *J. Opt. Soc. Am.* **52**, 239 (1962).
102. D. S. Bailey, J. R. Hiskes, and A. C. Riviere, *Nucl. Fus.* **5**, 41 (1965).
103. R. J. Damburg and V. V. Kolosov, *J. Phys. B* **9**, 3149 (1976); *Phys. Lett.* **61A**, 233 (1977); *J. Phys. B* **11**, 1921 (1978).
104. H. M. Silverstone, *Phys. Rev. A* **18**, 1853 (1978); and P. M. Koch, *Phys. Rev. Lett.* **41**, 99 (1978).
105. B. E. Cole, J. W. Cooper, and E. B. Saloman, *Phys. Rev. Lett.* **45**, 887 (1980); and J. W. Cooper and E. B. Saloman, *Phys. Rev. A* **26**, 1452 (1982).
106. W. Sandner, K. Safinya, and T. Gallagher, *Phys. Rev. A* **23**, 2488 (1981).
107. T. S. Luk, L. DiMauro, T. Bergeman, and H. Metcalf, *Phys. Rev. Lett.* **47**, 83 (1981).
108. A. R. P. Rau and K. T. Lu, *Phys. Rev. A* **21**, 1057 (1980); K. T. Lu, *J. Opt. Soc. Am.* **68**, 1446 (1978).
109. E. Luc-Koenig and A. Bachelier, *Phys. Rev. Lett.* **43**, 921 (1979); *J. Phys. B* **13**, 1743 (1980); *J. Phys. B* **13**, 1769 (1980).
110. D. F. Blossey, *Phys. Rev. B* **2**, 3976 (1970).
111. V. D. Kondratovich and V. N. Ostrovskii, *Sov. Phys. JETP* **52**, 198 (1980).
112. D. A. Harmin, *Phys. Rev. A* **24**, 2491 (1981).
113. C. Bender and T. T. Wu, *Phys. Rev.* **184**, 1231 (1969); *Phys. Rev. Lett.* **16**, 461 (1971); *Phys. Rev. D* **7**, 1620 (1973).
114. L. Benassi, V. Grecchi, E. Harrell, and B. Simon, *Phys. Rev. Lett.* **42**, 704 (1979).
115. H. J. Silverstone, B. G. Adams, Jiri Cizek, and P. Otto, *Phys. Rev. Lett.* **43**, 1498 (1979).
116. D. Park, *Z. Phys.* **159**, 155 (1960).
117. A. R. P. Rau, unpublished.
118. D. A. Harmin, *Phys. Rev. A* **26**, 2656 (1982).
119. U. Fano, *Phys. Rev. A* **24**, 619 (1981); and D. A. Harmin, *Phys. Rev. Lett.* **49**, 128 (1982).
120. K. T. Lu and A. R. P. Rau, unpublished; and 7th International Conference on Atomic Physics Abstracts, p. 21, MIT, 1980; and J. C. Gay, L. R. Pendrill, and B. Cagnac, *Phys. Lett.* **72A**, 315 (1979).
121. J. A. Armstrong, J. J. Wynne, and P. Esherick, *J. Opt. Soc. Am.* **69**, 211 (1979).
122. M. Aymar, P. Camus, M. Dieulin, and C. Marillon, *Phys. Rev. A* **18**, 2173 (1978); M. Aymar and O. Robaux, *J. Phys. B: Atom. Molec. Phys.* **12**, 531 (1979).
123. C. H. Greene, *Phys. Rev. A* **23**, 661 (1981).
124. K. T. Lu, *Proc. R. Soc. London, Ser. A* **353**, 431 (1977).
125. J. J. Wynne, J. A. Armstrong, and P. Esherick, *Phys. Rev. Lett.* **39**, 1520 (1977).
126. J. Geiger, *J. Phys. B* **12**, 2277 (1979).
127. M. Rosenbluh, T. A. Miller, D. M. Larson, and B. Lax, *Phys. Rev. Lett.* **39**, 874 (1977); M. Rosenbluh, R. Panock, B. Lax, and T. A. Miller, *Phys. Rev. A* **18**, 1103 (1978).
128. L. D. Landau and E. M. Lifshitz, *Quantum Mechanics*, 2nd Ed., Addison-Wesley, Reading, Massachusetts (1965), Section 111.
129. M. Ruderman, in *Physics of Dense Matter* (I.A.U. Symposium No. 53), Ed. by C. J. Hansen, D. Reidel, Boston (1974), pp. 117–131.
130. A. R. P. Rau, R. O. Mueller, and L. Spruch, *Phys. Rev. A* **11**, 1865 (1975).
131. J. Avron, I. Herbst, and B. Simon, *Phys. Rev. Lett.* **39**, 1068 (1977).
132. D. M. Larsen, *Phys. Rev. Lett.* **42**, 742 (1979).
133. J. Ozaki and Y. Tomishima, *J. Phys. Soc. Jpn.* **49**, 1497 (1980).
134. S. M. Kara and M. R. C. McDowell, *J. Phys. B* **14**, 1719 (1981).

135. W. A. M. Blumberg, W. M. Itano, and D. J. Larsen, *Phys. Rev. A* **19**, 139 (1979).

136. K. Onda, *J. Phys. Soc. Jpn.* **45**, 216 (1978).

137. G. Ferrante, S. Nuzzo, M. Zarcone, and S. Bivona, *J. Phys. B* **13**, 731 (1980).

138. M. Robnik, *J. Phys. A* **14**, 3198 (1981).

139. E. A. Soloviev, *Pis'ma Zh. Eksp. Teor. Fiz.* **34**, 278 (1981) [*JETP Lett.* **34**, 265 (1981)]; *Zh-Eksp. Teor. Fiz.* **82**, 1762 (1982).

140. D. R. Herrick, *Phys. Rev. A* **26**, 323 (1982).

141. J. J. Labarthe, *J. Phys. B* **14**, L467 (1981).

8

X Rays from Superheavy Collision Systems

P. H. Mokler and D. Liesen

1. Introduction

A new and quite fascinating field of atomic physics became accessible in the year 1976 when the heavy ion accelerator UNILAC of the GSI Darmstadt went into operation: the investigation of atomic processes in extended electromagnetic fields with a coupling strength $Z\alpha \gtrsim 1$ (with $1/\alpha = 137$).

Such strong fields are produced in close collisions between a projectile Z_1 and a target nucleus Z_2 with $Z = Z_1 + Z_2 \gtrsim 137$ (for general review of the theoretical aspects see, e.g., Ref. 1 and references therein). If the collision is adiabatic with respect to the innermost bound electrons, then these electrons are exposed to the field of the nearly unscreened summed nuclear charge Z which far exceeds that of all available elements—provided, both nuclei come sufficiently close to each other. The quasi-stationary states of the innermost bound electrons formed during the collision are usually called "quasimolecular" states.[2] The quasimolecular states may become even "quasiatomic" states in the case where the internuclear distance $\mathbf{R}(t)$ becomes smaller than the K-shell radius of the united atom with charge Z. The Compton wavelength $\lambdabar_C = \hbar/mc \approx 386\,\mathrm{f}$ may serve as a typical value for this radius.[3]

Indeed, electrons bound to extremely high nuclear charges exhibit features which qualitatively and quantitatively differ from the usual

P. H. Mokler and D. Liesen • Gesellschaft für Schwerionenforschung MBH, Darmstadt, Federal Republic of Germany.

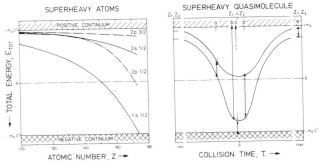

Figure 1. (a) Total energy for bound electrons in superheavy atoms
as a function of the atomic number Z from Ref. 5, (b)
Schematic behavior of the total energy of bound electrons in
a superheavy quasimolecule as a function of collision time
(for the indicated transitions see text).

conceptions. For simplicity, we consider in more detail the case of electrons
bound to an atom and not a quasimolecule with a nuclear charge $Z \geqslant 100$
(for a detailed discussion of the theoretical treatment of such systems see,
e.g., Ref. 4). In Figure 1a the total energy of some electronic levels are
plotted as a function of Z.[5] Three predicted properties of inner-shell
electrons in superheavy elements should be noted[6]:

(i)   A tremendous increase in the binding energies with increasing
      Z. In particular, the K level approaches the negative Dirac sea
      around $Z \approx 173$, which means that the binding $E_B \approx 2\,mc^2$,
      where m is the electron rest mass.

(ii) A drastic increase in the spin-orbit splitting for p states with
      increasing Z. For $Z \geqslant 137$ this splitting causes a swapping in the
      ordinary level order of $p_{1/2}$ and $s_{1/2}$ states. For example, near
      $Z = 173$ the splitting for the L shell amounts to about mc^2.

(iii) A considerable shrinking of the spatial electron density distribu-
      tions. For instance, the K shell radius, r_K, shrinks from roughly
      700 f for the Pb atom to only 100 f for an atom with a charge
      $Z = 164$: This is shown in Figure 2, which gives the radial
      probability distributions for the two cases.[7,8]

As already mentioned, these effects can only be studied experimentally
in close adiabatic collisions between very high Z_1, Z_2 partners, because up
to now all experimental attempts to produce superheavy elements or to
detect them in nature have been unsuccessful.[9] In order to realize a close
and adiabatic collision the following condition has to be fulfilled[10]:

$$0.02 \leqslant v/u = \eta \leqslant 0.16 \qquad (1)$$

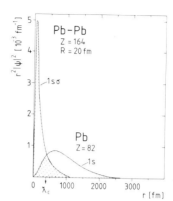

Figure 2. Spatial electron density distribution for a $1s$ electron in a Pb atom and a $1s\sigma$ electron in a Pb–Pb quasiatom (at an internuclear distance of $20\,f$) according to Ref. 8.

The ratio $v/u = \eta$ is called the adiabaticity parameter; v is the projectile velocity, and u the orbital velocity of the considered electrons bound to the quasimolecule. For a review of this subject see, e.g., Ref. 11.

According to the adiabatic condition, the strongest bound electrons in a superheavy quasimolecule rearrange continuously to the time-varying two-center potential of both nuclei. Thus, we expect in close collisions, binding energies in the quasimolecule up to the ones in the corresponding superheavy atoms. For instance, in Pb–Pb collisions, $E_B \approx 780\,\text{keV}$ (see Figure 1a), and hence $u \approx c$ where c is the velocity of light. For the production of a superheavy quasimolecule Eq. (1) will give an upper limit for the reduced projectile energy $E_1/A_1 \lesssim 11\,\text{MeV/N}$, where A_1 is the number of the nucleons in the projectile.

A lower limit follows from the condition that the distance of closest approach should be smaller than the K-shell radius of the united atom; assuming a Rutherford trajectory, we get as a lower limit for the reduced projectile energy $E_1/A_1 \gtrsim 0.5\,\text{MeV/N}$.

Additionally, in order to avoid complications with nuclear reactions, we should stay below the Coulomb barrier. Hence, we investigate in this article superheavy quasimolecular systems at reduced projectile energies typically between 1.4 and 6 MeV/N.

At the corresponding projectile velocities the trajectories of the nuclei can be described using classical mechanics; only the motion of the electrons has to be treated within quantum mechanics.[12] In Figure 1b the time evolution of the binding energy, $E_B(t)$, for some of the innermost electrons are sketched for a superheavy quasimolecular collision. Before and after the collision ($t = -\infty$ and $t = +\infty$, respectively) we have the atomic binding energies of both the separated collision partners and, in between, the binding energies of the quasimolecular orbitals (MO). If the distance of closest approach, $R_0 = |\mathbf{R}(t = 0)|$, becomes sufficiently small, the electronic configuration of the quasimolecule corresponds to about that of the respective superheavy atom with $Z = Z_1 + Z_2$.

As the time evolution of $E_B(t)$ is different for different trajectories, a more unique way of representing a quasimolecule is a correlation diagram, where the binding energies are plotted as a function of the internuclear distance, $E_B(R)$. Several attempts to calculate static values of $E_B(R)$ [that is $E_B(R)$ with both nuclei fixed in space at an internuclear distance R] for superheavy quasimolecules have been made using different approximations for the two-center potential.[13-16] In Figure 3, three correlation diagrams for a symmetric and two asymmetric superheavy quasimolecules (with $Z_1 + Z_2 = 132$, 164, 178), are shown according to Fricke and co-workers.[15-17]

In addition to the relativistic effects for superheavy atoms discussed above, two specific features are found in superheavy quasimolecules which are not observed in lighter, nonrelativistic systems:

(i) The deepest bound levels—in particular the $1s\sigma$ level—do not get flat in the R representation even for the smallest internuclear distances (no "run-way" effect[13]). For very heavy systems, such an effect seems to be observable for all $ns\sigma$ and $np_{1/2}\sigma$ states.

(ii) New level crossings (or pseudocrossings) appear and common ones may disappear due to the large spin-orbit splitting and the strong binding increase for the $ns\sigma$ and $np_{1/2}\sigma$ levels at small R. Consequently, the mechanisms of inner-shell excitation in superheavy quasimolecules or quasiatoms may differ from the well-known processes in lighter collision systems.

Because the collision system is a dynamical system where the internuclear distance and, thereby, the two-center Coulomb field depend on time, energy and momentum can be transferred to electrons bound to the quasimolecule. Therefore these electrons may be excited to higher-lying, empty orbitals or to the continuum (ionization). The investigation of these excitation processes—either by an analysis of the decay channels of the excited states or by a direct observation of the emitted particles—provides a tool for studying the behavior of electrons in very strong fields. The investigation of very strong bound systems can help one to get an understanding of the excitation mechanisms and to get an access to the peculiarities of the wave functions and the binding energies of highly relativistic, extended, bound states which cannot be studied otherwise.

Because of nonadiabatic effects during the collision—i.e., the two-center potential varies with time (see Figure 1b)—the molecular levels may couple with each other (see process a in Figure 1b) or with the empty

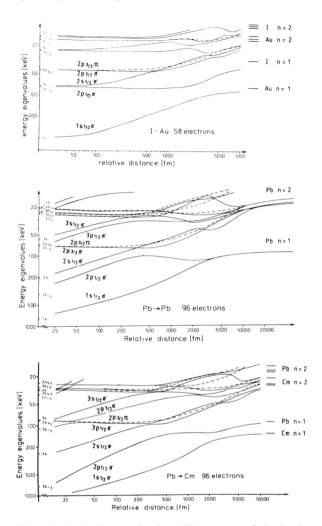

Figure 3. Level diagrams for the collision systems I–Au (top), Pb–Pb (middle), and Pb–Cm (bottom). The diagrams are based on relativistic self-consistent Dirac–Fock calculations (Refs. 15–17).

positive continuum (process b). Both processes can lead to an inner-shell vacancy. On the outgoing part of the collision, the vacancy may be transferred via its MO (e.g., $1s\sigma$, $2p_{1/2}\sigma$) to an inner shell (e.g., K) of the separated atoms. This inner-shell excitation can be observed via the characteristic x-ray emission of both the collision partners (see process e, Figure

1b). The decay via Auger electron emission is of minor importance for these high-Z partners[18] and can be neglected.

Due to the short lifetime of an inner MO vacancy—roughly 10^{-18} sec for $1s\sigma$ around $Z \approx 164$ which has to be compared with a typical collision time of roughly 10^{-20} sec—the inner MO vacancy has also a small chance of decaying radiatively during the collision itself. By such processes (see d in Figure 1b) a quasimolecular x-ray continuum is emitted.[19,20]

Processes leading to vacancies in the negative continuum and the joint positron emission (see c in Figure 1b) are not considered in this article; the same is true for the δ-electron emission (see b in Figure 1b).[1]

The extremely strong increase of the binding energy with increasing Z raises the question of the absolute value of the total inner-shell excitation cross section and of its dependence on Z and on the projectile velocity v. As will be seen in Section 3, the Z dependence of the total inner-shell excitation clearly demonstrates the quasimolecular character of the excitation mechanism. Measurements of the inner-shell excitation probability $P(b)$ as a function of the impact parameter b (see Section 4) should reveal the excitation mechanisms acting in these highly relativistic systems. Namely, it is well known from experiments with lighter collision partners[21-25] that the possible two excitation mechanisms (radial and rotational coupling[12]) exhibit a quite different dependence on the impact parameter. Thus, an analysis of experimental data on inner-shell excitation obtained in impact parameter controlled collisions can help to distinguish the different processes. The absolute value of the excitation probability $P(b)$ and the range of those impact parameters b which contribute predominantly to the excitation process provides information about the spatial shrinking of the innermost wave functions and about the corresponding binding energies (Section 4).

For the superheavy collision systems, the quasimolecular radiation is of special interest (Section 5), since, owing to the high vacancy production probability and the large radiation transition rates, appreciable photon yields are expected. The spectral shape as well as the impact parameter dependence gives valuable information on the binding of the innermost quasimolecular levels.

In a final section, we will discuss in more detail several methods of obtaining from the experimental data spectroscopic information about the innermost quasimolecular orbitals. Clearly, these methods form only a first step towards a spectroscopy of orbitals in superheavy quasimolecules, and their limitation will be demonstrated. However, the results of these attempts turn out to be encouraging enough for a further systematic investigation to be made along the way.

Before discussing the experimental results, we will first give a survey of the experimental methods used so far.

2. Experimental Procedures

2.1. The X-Ray Information

The emitted X rays are usually detected with solid state detectors [Si(Li) for low- and Ge(I) for high-energy X rays, respectively] which have sufficiently good energy resolution. In Figure 4 a typical experimental setup is sketched. This setup can be used to determine the normalized total x-ray emission as well as the coincidence rate between scattered particles and emitted X rays. At first we concentrate only on the emitted single spectra (that is without the coincidence condition). Typical x-ray spectra measured at 90° to the beam direction are plotted in Figures 5–9. Figure 5 displays the impact energy dependence of the characteristic K-radiation of both collision partners for 1.4–5.9 MeV/N Pb → Au collisions.[26] Figure 6 shows the Z_2 dependence of the projectile K-radiation for 3.6 MeV/N U → Z_2 collisions.[27] Figure 7 gives the equivalent dependence for the L-radiation in 1.4 MeV/N Pb → Z_2 collisions.[28] Figures 8 and 9 depict continua for quasimolecular radiation. The x-ray continuum around 8 keV presented in Figure 8 for low-energy I-Au collisions is attributed to the quasimolecular

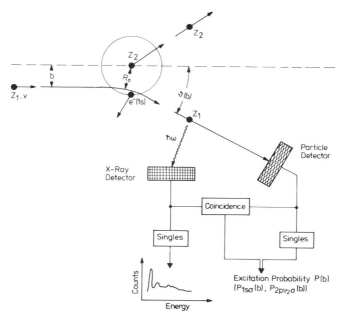

Figure 4. Typical experimental setup to measure the inner-shell excitation in heavy ion–atom collisions by an observation of emitted X rays and scattered particles.

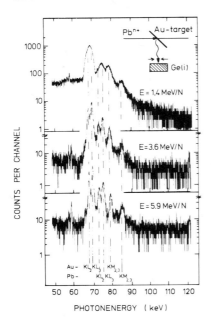

Figure 5. K x-ray spectra at 90° observation for 1.4, 3.6, and 5.9 MeV/N Pb on Au collisions demonstrating the kinematical line broadening (Ref. 26).

M-radiation for $Z = 132$.[29] In Figure 9 the x-ray continuum beyond the Pb-K radiation—the quasimolecular $1s\sigma$ spectrum—is shown for 4.2 MeV/N ^{208}Pb–^{208}Pb collisions.[20]

We find that the heavy ion–atom induced x-ray spectra depend significantly on the atomic numbers of the collision parameters Z_1, Z_2, on the projectile energy E_1/A_1, and that they deviate considerably from the well-known photon or electron induced spectra. From the characteristic spectra shown in Figures 5–7, we may deduce the typical features for heavy ion–atom induced characteristic x-ray emission:

(i) The lines are shifted to higher x-ray energies compared to the transitions in singly ionized atoms (diagram lines). This early observed feature[30] is caused by a simultaneous multiple excitation of higher-lying shells. This multiple excitation increases with increasing E_1/A_1 and varies with Z_1 and Z_2.[31]

(ii) The relative intensities for lines belonging to one shell deviate from those for singly excited atoms. For the K shell in heavy collision partners, deviations of a factor of 2 are found.[31] For the L shell pronounced intensity fluctuations as a function of Z_1/Z_2 are reported.[28,32] Comparing line shifts and intensity ratios with configuration calculations the average multiple excitation in higher shells during x-ray emission can be estimated.[28]

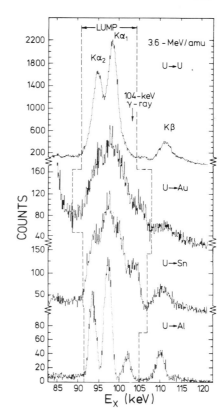

Figure 6. Projectile K x-ray spectra for 3.6 MeV/N U $\rightarrow$ Z_2 collisions for $Z_2 = 13$, 50, 79, and 92 (Ref. 27).

(iii)   The x-ray lines are broadened. On the one hand there is an inherent line broadening caused by the distribution of various configurations which contribute to the x-ray emission; this broadening is comparable to the energy resolution of the x-ray detector (typically below 500 eV for the K-radiation[26]). However, this effect alone can not explain the drastic broadening seen for instance in Figure 5 for the K-radiation at small impact energies. As will be shown in Section 2.3.2 in more detail, this effect is determined by the kinematics of the collision and caused by the Doppler effect. Since at low impact energies close collisions are necessary for an inner-shell excitation—which results in a strong deflection of the primary ion and a high recoil velocity for the target atom—the linewidth observed at 90° to the beam increases at lower impact energies; see Figure 5. The dependence of the kinematics on the masses of both collision partners can be read from Figure 6.

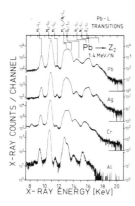

Figure 7. Projectile L x-ray spectra for 1.4 MeV/N Pb $\rightarrow Z_2$ collisions for $Z_2 = 13, 24, 47, 82$ (Ref. 28).

Background effects may also contribute to the x-ray emission. For instance, in the U-K spectra shown in Figure 6 we find a line at 104 keV, which is caused by an E2 γ-ray transition of the Coulomb excited U nucleus. Nuclei with higher-lying excited Coulomb states may not only decay via γ-ray emission but also by conversion creating an atomic inner-shell vacancy. In such cases the nuclear contribution has to be carefully subtracted in order to get the "atomic" excitation.

For the continuous x-ray emission, see Figures 8 and 9, additional background effects have to be considered carefully. First, target contaminations may effectively be Coulomb excited, see Figure 9. And, the Compton scattered γ rays may contribute to the continuum. Furthermore, background contributions, such as bremsstrahlung from the emitted δ-electrons,[33–35] nucleus–nucleus bremsstrahlung,[36–38] and even room background have to be taken into account. Also, the detector response function may change the observed spectra.

Summarizing, from measured line shifts and intensity ratios information on the configurations contributing to the x-ray emission of the multi-excited collision partners can be extracted. In particular, realistic transition rates and fluorescence yields may be extracted.[28] By this means true excitation cross sections for the inner-shell excitation can be gained from the measured x-ray yields. From the line shapes, especially from the linewidth at 90° observation, information on the impact parameters relevant for the excitation can be extracted. The x-ray continua give information on the x-ray transitions within the quasimolecule after a proper subtraction of background effects.

2.2. Total X-Ray Yields

Total excitation cross sections are determined via the absolute x-ray emission of the transition under consideration normalized to the numbers

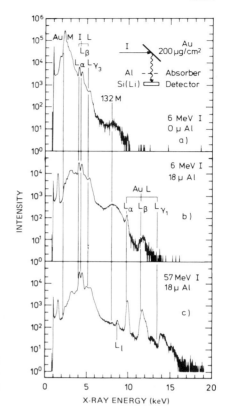

Figure 8. X-ray spectra for low-energy I–Au collisions (6 to 57 MeV). The x-ray continuum around 8 keV is attributed to quasimolecular M radiation (Ref. 29).

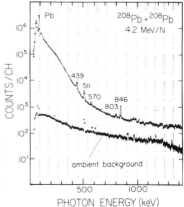

Figure 9. The x-ray continuum for 4.2 MeV/N Pb–Pb collisions beyond the characteristic Pb–K radiation. After subtraction of an ambient background and spurious γ-ray lines from contaminants most of the radiation is caused by the quasimolecular $1s\sigma$ radiation (Ref. 20).

of incoming particles and target atoms. The number of incoming particles can be determined either by a Faraday cup or by means of the Rutherford scattering cross section.[27,31] The latter method has the advantage of canceling out the thickness of the target in the ratio of observed x-rays to scattered primary particles. The x-ray detectors are normally placed at an angle of 90° to the beam direction. Suitable absorbers serve to reduce the count-rate of low-energy photons which are produced with very high cross sections in close collisions.

The x-ray production cross sections measured in this way can be converted into vacancy production cross sections using single-vacancy fluorescence yields ω_i for the ith shell from Ref. 18. For the K shell it is found that $\omega_K \gtrsim 50\%$ for atoms with atomic number $Z \gtrsim 30$. Due to these very high values of the single-vacancy fluorescence yields, possible influences from simultaneous excitation of electrons in higher-lying orbitals are rather small. This fact makes the conversion of experimental x-ray data into vacancy-production data for high-Z collision systems much more reliable than for low-Z partners.

The experimental uncertainties in the total inner-shell vacancy production cross sections are typically 25% to 40%[27,28,31,39] and usually come from the following sources: Efficiency, solid angle, and absorber corrections for the x-ray detection, angle and solid angle of the particle detector, counting statistics, background subtractions, and dead-time corrections from electronic setup and the data-handling system.

For the MO x-ray radiation, in particular, the experimental arrangements have to be designed very carefully in order to reduce optimally disturbing contributions to the continuous spectrum. Consequently, only an extremely attentive analysis of the data will give the true MO x-ray continua (Refs. 19, 20, 38, 40 and references therein).

2.3. Impact Parameter Dependences

In order to determine the impact-parameter dependence of the excitation probability $P(b)$ two experimental techniques have been used so far: The conventional scattered particle–x-ray coincidence method[41,42] and the novel method of the analysis of the "Doppler-shifted" shape of the x-ray spectra emitted at an angle of 0° relative to the beam direction.[43,44]

2.3.1. The Coincidence Method

In this procedure one measures coincidences between X rays emitted by one or both of the collision partners and primary particles scattered through a certain angle θ in the lab system. In order to achieve a high coincidence count rate the solid angle of the x-ray detector is kept as large as possible (typically 5% to 10% of 4π) and the particle detector is made to cover the whole azimuth. Position-sensitive particle detectors offer the big advantage of considerably reducing the time required for the measurements, because several scattering angles can be measured simultaneously. For this purpose position-sensitive gas detectors[45,46] and specially prepared surface barrier detectors[47] have been used.

By standard fast-slow coincidence techniques, coincidences between X rays and particles are observed and mostly an on-line computer is used to store the data in an event-recording mode. Thus, an effective way of processing all available data is achieved; the final data analysis proceeds off-line by setting "software windows" on the time spectrum and on the interesting parts of x-ray and particle spectra.

The number of true coincidences divided by the number of all particles scattered through θ gives—after correction for efficiency and solid angle of the x-ray detector and for possible dead-time effects—the x-ray production probability per scattered particle as a function of θ. This angular dependence is transformed into an impact parameter dependence using an appropriate classical description of the nuclear trajectory. In most cases the Rutherford trajectory in a pure Coulomb potential is appropriate; therefore, we get the simple connection between impact parameter b and center-of-mass scattering angle θ_{CM}

$$b = a_0 \cdot \cot (\theta_{CM}/2) \tag{2}$$

Here

$$a_0 = \frac{Z_1 Z_2 e^2}{2E_{CM}} = \frac{Z_1 Z_2 e^2}{2E_1} \cdot \frac{A_1 + A_2}{A_2} \tag{3}$$

equals half the distance of closest approach in a head-on collision between the heavy ion (with charge Z_1, mass A_1, and lab-energy E_1) and the target atom (with charge Z_2 and mass A_2). Especially at large impact parameters (typically larger than the corresponding shell radii of the separated atoms) the effect of electronic shielding has to be taken into account by the introduction of a suitable screening parameter into the Coulomb potential.[48]

At scattering angles $\theta \geq 25°$ in the lab system (the actual value depends mainly on the mass ratio A_1/A_2) recoiling target atoms coming from small impact parameter collisions may contribute significantly to the scattering intensity at the angle θ. Taking proper energy spectra of the scattering intensity[49] or looking for kinematical coincidences by means of a special particle detection system[50] recoils and scattered primary particles can be separated and thus $P(b)$ for very small b (typically below $15\,f^{(39,49)}$) can be determined.

The usual experimental setup is schematically shown in Figure 4. Special care has to be taken to overcome the problem of slit-scattering which occurs at those slits which determine the angular divergence of the primary beam; typically, this divergence leads to an angular uncertainty $\Delta\theta/\theta \approx 5\%-10\%$. The thickness of the target should be so small that contributions to x-ray and scattering intensity from multiple collision processes are negligible.

2.3.2. The Line-Shape Technique

The second method mentioned above relies on the fact that in energetic heavy ion–atom collisions the measured x-ray lines are broader than the corresponding x-ray lines emitted after excitation by photons or electrons. In heavy ion–atom encounters the width of the lines is mainly determined by the satellite contribution to the radiation due to simultaneous excitation of many higher-lying levels (see Section 2.1) and by the kinematics of the collision.[26,51] The second contribution depends via the well-known Doppler formula on the x-ray observation angle and on the collision velocity.[26,43,44,51–53] Thereby, the energy E_x of X rays observed in the lab system is given by[51]

$$E_x = E_{x0} \frac{[1 - (v_r/c)^2]^{1/2}}{1 - (v_r/c)\cos\alpha} \tag{4}$$

with

$$\cos\alpha = \cos\psi\cos\Theta + \sin\psi\sin\Theta\cos\phi \tag{5}$$

In Eqs. (4) and (5) E_{x0} is the photon energy in the rest frame of the emitting particle, $v_r = |\mathbf{v}_r(E_1, b)|$ is the velocity of the emitting particle, α is the observation angle with respect to $\mathbf{v}_r/v_r$, ψ is the angle of the emitting particle in the lab system with respect to the direction of the incoming projectile, ϕ is the azimuthal angle of the emitting particle, and Θ is the x-ray detector angle with respect to the direction of the incoming projectile. The geometry is plotted in Figure 10 for the case that the recoiling target atom is the x-ray emitter; $\mathbf{v}'$ gives velocity and direction of the scattered primary projectile.

Equation (5) shows that only for $\Theta = 0°$ (observation of the X rays emitted parallel to the direction of the incoming beam) the observed x-ray

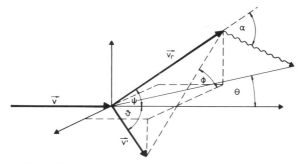

Figure 10. Collision geometry corresponding to Eq. (5).

energy is independent of the azimuth and given by

$$E_x(\Theta = 0°) = E_{x0} \frac{[1 - (v_r/c)^2]^{1/2}}{1 - (v_r/c)\cos\psi} \tag{6}$$

Rutherford scattering provides an unambiguous relation between the angle of the emitting particle and the impact parameter b. Thus, by means of Eq. (6) there is a simple correspondence between photon energy E_x and impact parameter b. With the aid of the differential Rutherford scattering cross section the shape of $P(b)$ can be deduced directly from the shape of the observed x-ray spectrum (for a detailed description see Ref. 43).

Since up to now predominantly x-ray emission from recoiling target atoms has been studied, we will consider that case in some detail. Besides trivial normalization factors, the line shape $dn(E_x)/dE_x$ of x rays emitted by target atoms under 0° observation angle is given by

$$\frac{dn(E_x)}{dE_x} = 2\pi b P(b) \cdot \frac{db}{dE_x} = 2\pi P(b) \frac{b}{\sin\theta_{\rm CM}} \cdot \frac{db}{d\theta_{\rm CM}} \cdot \frac{d\theta_{\rm CM}}{dE_x} \cdot \sin\theta_{\rm CM}$$

$$= 2\pi P(b) \frac{a_0^2}{4} \frac{1}{\sin^4(\theta_{\rm CM}/2)} \frac{d\theta_{\rm CM}}{dE_x} \sin\theta_{\rm CM} \tag{7}$$

Using Eq. (6) and the relation

$$\psi = \frac{\pi}{2} - \frac{\theta_{\rm CM}}{2}, \qquad \frac{v_r}{2} = 2\frac{A_1}{A_1 + A_2}\frac{v}{c}\cos\psi \tag{8}$$

one can demonstrate directly the unambiguous correspondence between the line shape $dn(E_x)/dE_x$ and the impact parameter b. To first order in

v/c, where from Eq. (6)

$$E_x(\Theta = 0°) \approx E_{x0}\left(1 + 2\,\frac{A_1}{A_1 + A_2}\,\frac{v}{c}\cos^2\psi\right) \tag{9a}$$

this connection is explicitly given by[43,44]   (9b)

$$\frac{dn(E_x)}{dE_x} = \text{const} \cdot \frac{a_0^2}{v}P(b)\left[1 + \left(\frac{b}{a_0}\right)^2\right]^2\left(\frac{E_x}{E_{x0}}\right)$$

Thus, $P(b)$ can directly be inferred from $dn(E_x)/dE_x$. The absolute scale in $P(b)$ can be fixed, for instance, via the relation

$$\sigma = 2\pi \int_0^\infty bP(b)\,db \tag{10}$$

if the total excitation cross section is known. The details of the experimental setup and of the final analysis of the data are extensively described in Ref. 44.

It can be seen from Eqs. (3) and (4) that at an observation angle of $\Theta = 90°$ the x-ray lines are strongly Doppler-broadened (cf. Figure 5). The width of this broadening can be related to a quantity which is characteristic for the excitation process[26,51]: The mean impact parameter $\bar{b}$ defined by

$$\bar{b} = \int_0^\infty b[bP(b)\,db] \Big/ \int_0^\infty [bP(b)]\,db \tag{11}$$

In order to determine $\bar{b}$ one has to calculate from Eqs. (3) and (4) with $\Theta = 90°$ the energy distribution E_x for a given set of values (b, E_1, ϕ). Then all events with energies between E_x and $E_x + dE_x$ have to summed up, weighted by the product of a reasonable trial dependence for $P(b)$ and the differential Rutherford scattering cross section, and normalized by the total excitation cross section.

Suggested by many experimental[39,44,47,49,53–59] and theoretical[14,60–65] data mostly, an exponential form for $P(b)$, such as

$$P(b) = P(0)\,e^{-b/a} \tag{12}$$

has been used to fit the measured line shape and to extract $\bar{b}$. Actually, $\bar{b}$ does not depend in a critical manner (within 25%) on the chosen $P(b)$ distribution.[51] For a pure exponential as given by Eq. (12) one finds from

Eqs. (10) and (11)

$$\bar{b} = 2a \tag{13a}$$

$$\sigma = 2\pi P(0)a^2 \tag{13b}$$

and for a rectangular distribution[12]

$$P_r(b) = \begin{cases} P_r(0) & \text{for } b \le a_r \\ 0 & \text{for } b > a_r \end{cases} \tag{14}$$

with an appropriate cutoff radius a_r

$$\bar{b} = \tfrac{2}{3}a_r \tag{15a}$$

and

$$\sigma_1 = \pi P_r(0)a_r^2 \tag{15b}$$

Since Eqs. (13) and (15) can be regarded as two limiting cases for the actual $P(b)$ distribution, the values of $\bar{b}$ calculated from Eq. (13a) and (15a) provide two extreme cases for $\bar{b}$. They are combined by the requirement for equal total cross section

$$2\pi P(0)a^2 = \pi P_r(0)a_r^2 \tag{16}$$

Finally, we mention that the analysis of observed x-ray lines at $0°$ observation angle is especially sensitive to collisions with very small impact parameters and to collision systems with very steep $P(b)$ distributions. At $90°$ observation angle one investigates predominantly $\bar{b}$, thus somewhat larger impact parameters. $\bar{b}$ characterizes that region of impact parameters which contribute mostly to the total excitation cross section.[66]

2.4. Data Reduction

2.4.1. Background Effects

Some words of caution about background effects are in order. Mainly two effects have to be considered: contributions to inner-shell vacancy production by nuclear Coulomb excitation—see Figure 9—and subsequent K conversion and contributions from double excitation of the inner shells followed by hypersatellite transitions. Nuclear Coulomb excitation can be important (depending on the actual collision system) for all the measurements, singles, and coincidence ones. This background effect is usually corrected for by setting windows on the simultaneously measured γ-ray spectrum. The corresponding intensities are then subtracted from the K x-ray data with the aid of the conversion coefficients from Refs. 67 and 68 and the angular dependence of the γ rays calculated from Ref. 69. Contributions from hypersatellite transitions are of great importance especially for the line-shape techniques. These transitions can disturb only the high-energy tail of the investigated x-ray line to an appreciable amount—especially if the "single-electron excitation" of the orbital under consideration is large, since to first order the double excitation probability is proportional to $[P(b)]^2$.[70-72] Moreover, all continuous radiation such as quasimolecular radiation etc. has to be considered carefully for the line-shape techniques. The problems in extracting real continuous spectra have been discussed already in connection with the MO radiation; see Section 2.1.

An extremely careful analysis of the time spectrum in a coincidence experiment is necessary, if one deals with a ratio of true to random coincidences which is in the order of unity or less. It has been found[73,74] that the shape of the time spectrum changes drastically with possible time structures in the ion beam. Such structures are mainly caused by the actual operation conditions of the ion source and are very sensitive to the gas pressure in the ion source. Since the number of random coincidences depends quadratically on the intensity of the ion beam, these spurious time structures influence predominantly the random spectrum. Therefore, a determination of the number of true coincidences may become very difficult—especially if $P(b)$ and, thus, the ratio of true to random coincidences is small.

2.4.2. K-Vacancy Sharing

Since we will mainly discuss the excitation of the K shells of one or both of the collision partners, we will consider the data reduction for the case of K x-ray emission in more detail. An inspection of the correlation diagrams given in Figure 3 shows that vacancies created in the $2p_{1/2}\sigma$ and in the $1s\sigma$ orbital during the collision end up in the K shells of the collision partners. Hence, the sum of the impact parameter dependence of the K X rays from the heavier $[P_H(b)]$ and the lighter $[P_L(b)]$ of both the partners gives the sum of $2p_{1/2}\sigma$ and $1s\sigma$ excitation probability. In general, it is not possible to infer directly from the K x-ray data the $2p_{1/2}\sigma$ and $1s\sigma$ excitation probabilities separately, because $2p_{1/2}\sigma$ vacancies can couple down to the $1s\sigma$ orbital and vice versa during the collision due to radial coupling.[75-77] This process is often referred to as "$2p\sigma$–$1s\sigma$ vacancy sharing."

If $w(b)$ gives the $2p_{1/2}\sigma-1s\sigma$ coupling probability as function of impact parameter b, then $P_H(b)$ and $P_L(b)$ can be related to the impact parameter dependences $P_{2p_{1/2}\sigma}(b)$ and $P_{1s\sigma}(b)$ by

$$P_H(b) = [1 - w(b)]P_{1s\sigma}(b) + w(b)P_{2p_{1/2}\sigma}(b) \tag{17}$$

and

$$P_L(b) = w(b)P_{1s\sigma}(b) + [1 - w(b)]P_{2p_{1/2}\sigma}(b) \tag{18}$$

Thus, from Eqs. (17) and (18) follows that only if

$$w(b)[P_L(b) + P_H(b)] \ll P_H(b) \tag{19}$$

then

$$P_{1s\sigma}(b) \approx P_H(b) \tag{20}$$

and from Eq. (10) correspondingly

$$\sigma_{1s\sigma} \approx \sigma_H \tag{21}$$

Under the condition given in Eq. (19), the $1s\sigma$ excitation can directly be inferred from the K vacancy production in the heavier partner; similarly, the K vacancy production in the lighter partner gives the $2p_{1/2}\sigma$ excitation.

Concerning the transition probability $w(b)$ up to now only a theoretical model for light, nonrelativistic collision systems is available.[78] This model has been tested experimentally for light ($Z_1 + Z_2 = 39$ and 50)[23] and medium-heavy ($Z_1 + Z_2 = 68$)[24] collision systems. The two—in this context–central points of the theoretical analysis and the experimental findings in both cases are as follows:

(1) The $2p_{1/2}\sigma-1s\sigma$ transitions take place at internuclear distances which are comparable to the K shell radii of the separated atoms; for collisions with impact parameters smaller than these radii $w(b)$ can be replaced by $w(0)$ to a good approximation.

(2) The transition probability $w(0)$ may be calculated from the nonrelativistic Demkov–Meyerhof model,[75,76] which gives

$$w(0) = (1 + e^{2x})^{-1} \tag{22}$$

with

$$2x = (2/m)^{1/2}(\sqrt{I_H} - \sqrt{I_L})/v \approx \pi|Z_H - Z_L|/v \tag{23}$$

In Eq. (23), m is the electron mass and I_H and I_L are the K-shell binding energies of the heavier and the lighter collision partner, respectively.

It should be kept in mind that the determination of the $1s\sigma$ excitation from the K-shell excitation of the heavier partner and of the $2p_{1/2}\sigma$ excitation from the K-shell excitation of the lighter partner, can depend critically on the value of the $1s\sigma$–$2p_{1/2}\sigma$ transition probability also for asymmetric systems, where $w(0)$ is small. For rather symmetric systems, the condition implied by Eq. (19) is no longer fulfilled and for symmetric systems with $w(0) = 1/2$ it is not possible to distinguish between $1s\sigma$ and $2p_{1/2}\sigma$ excitation only from characteristic K x-ray data. However, it has been found[26,39,47,54,56] that in such systems the $2p_{1/2}\sigma$ excitation strongly dominates—perhaps with the exception at the smallest impact parameters ($b \lesssim 50$ f). Thus, we will denote in symmetric and nearly symmetric systems the sum of $P_H(b)$ and $P_L(b)$ just as $P_{2p_{1/2}\sigma}(b)$ and correspondingly the sum of σ_H and σ_L just as $\sigma_{2p_{1/2}\sigma}$.

Finally, we mention that it seems to be indispensible to work out a relativistic model of the $2p_{1/2}\sigma$–$1s\sigma$ vacancy-sharing in order to perform a more reliable analysis of experimental data.

3. Total Excitation Cross Sections

3.1. Experimental Results

In this chapter we will first give a survey on the excitation in all shells after a heavy ion–atom collision. Thereafter, we summarize the K-shell excitation results in more detail.

3.1.1. Survey at 1.4 MeV/N

In order to get a general survey on the excitation of the collision partners after a heavy ion–atom collision, excitation cross sections for all shells have been extracted from x-ray spectra for 1.4 MeV/N Pb → Z_2 using normal fluorescence yields.[18] These cross sections are plotted in Figure 11 for both the collision partners as a function of Z_2.[28,39] Such a Z_2 representation of cross sections in heavy ion–atom collisions was first chosen by Specht.[79] Since for $Z \gtrsim 30$ the fluorescence yields $\omega_K \gtrsim 0.5$ the multiple excitation of outer shells will not significantly change the results for the K-shell excitation presented in Figure 11 (except possibly those for the lightest target atoms). Within the experimental accuracy of roughly 50%, the same will hold true for the projectile L-excitation cross sections. For this case a more accurate data analysis will give reliable excitation cross sections, even for the various L subshells.[28] The projectile

M shell as well as most of the target L shell cross sections will be altered by changes in ω. Nevertheless, the given order of magnitude is certainly correct.

The various excitation cross sections shown in Figure 11 as a function of Z_2 cover a range of 11 orders of magnitude. The most striking features are the Z_2 resonances[79] in the cross sections. These resonances are typical for a quasimolecular excitation process (see for instance Ref. 11). As is indicated in the figure, they are centered around collision systems with matching atomic levels of both partners. At these points we have a level swapping in the corresponding quasimolecule. This is elucidated for instance by Pb (or Au) $\rightarrow Z_2$ correlation diagrams shown in Figure 3. The Pb (Au)-K shell correlates with the $1s\sigma$ level for lighter Z_2 partners and with the $2p_{1/2}\sigma$ level for heavier Z_2 partners. For the symmetric case, $Z_1 = Z_2$, both levels correlate to the Pb-K shell. For near symmetric collision systems, $Z_1 \approx Z_2$, the $1s\sigma$ and $2p_{1/2}\sigma$ level will approach each other at large internuclear distances and the $2p_{1/2}\sigma$–$1s\sigma$ vacancy sharing near the level matching causes Z_2 resonances observed in Figure 11.

From the considerations in Section 2.4.2 we associate the Pb-K excitation for $Z_2 < Z_1$ and $Z_2 \neq Z_1$ with the $1s\sigma$ excitation and the target K excitation with the $2p_{1/2}\sigma$ excitation if the $2p_{1/2}\sigma$–$1s\sigma$ vacancy sharing is small. For the K–K level matching region, $Z_1 \approx Z_2$, the $2p_{1/2}\sigma$ excitation is given by the sum of K excitation in both the collision partners as the $1s\sigma$ cross section seems to be orders of magnitude smaller.[80] Around the L–K sharing region the $2p_{1/2}\sigma$ excitation is covered by the $3d\sigma$ excitation.[81] Correspondingly, the L-cross sections around symmetry, $Z_1 \approx Z_2$, are determined via $4f\sigma$ promotion.

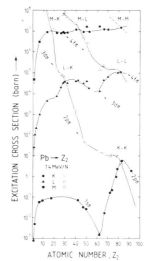

Figure 11. Survey on projectile and target atom excitation after 1.4 MeV/N Pb $\rightarrow z_2$ collisions (Refs. 28, 39). The cross sections for the different shells are plotted as a function of Z_2. The curves are to guide the eye.

As can be seen from Figure 11 the Z_2 oscillations get weaker for the higher shells, almost diminishing for the Pb-M excitation. This possibly shows the limits of the applicability of the MO model, because for Pb-M electrons, the adiabaticity parameter, $\eta = v/u = 0.2$; this value is outside the limits given by Eq. (1). M electrons can already be promoted at large internuclear distances, R_{pr}, to orbitals with such low binding energies that they will be ionized with a high probability, $P \approx 1$. Assuming such a rectangular $P(b)$ distribution as given in Eq. (14), we extract for the Pb-M shell excitation cross section from Eq. (15b), a promotion radius of 2×10^{-9} cm which is in the order of the Pb-M shell radius ($\sim 10^{-9}$). Hence, the Pb-M excitation cross section is roughly determined by the geometrical dimension of the Pb-M shell.

Using such a picture also for the Pb-L excitation near symmetry, the order of magnitude of the cross section can be explained, too. From the L cross section of roughly 10^{-19} cm^2 we deduce a promotion radius of roughly 2×10^{-10} cm. This value is about a factor of 2 smaller than the $4f\sigma$ promotion radius extracted from the correlation diagram shown in Figure 3. The difference can be easily explained by a reduced excitation probability, $P_r < 1$.

The inner shells—i.e., the K and partly the L shell of the projectile and the K shell of most of the target atoms—correlate to more tightly bound levels in the united atom (u.a.) system so that the promoted electrons cannot be ionized effectively. Consequently, similar rough estimates as approximately possible for the outer shells are not possible for the inner shells.

We would like to point out that for $Z_2 \leqslant 10$ all the Pb cross sections can be well described by direct excitation models applied for inverted systems $Z_2 \rightarrow$ Pb.[66,82-87] In the following, we will concentrate on K-shell excitation, i.e., $1s\sigma$ and $2p_{1/2}\sigma$ excitation, which seems to be governed by a direct excitation out of the corresponding quasimolecular levels, too[39]; but also couplings to higher-lying MOs seems to contribute to the inner-shell excitation.[39]

3.1.2. The K-Shell Excitation

Systematic measurements of K-shell excitation cross sections for superheavy collision systems have been reported for various projectiles in the range $53 \leqslant Z_1 \leqslant 92$: For fission products ($Z_1 \approx 54$) as projectiles,[79] for I ions[88,89] (cf. also Ref. 11), for Xe projectiles,[31,90] for Pb projectiles,[31] and for U ions.[27] The data for 3.6–5.9 MeV/N Xe and Pb $\rightarrow Z_2$ collisions from Ref. 31 are given in Tables 1 and 2, respectively, for both the collision partners. In Figure 12 the excitation cross sections for 3.6 MeV/N Xe,[90] Pb,[31] and U[27] projectiles are plotted as an example. It should be pointed

Table 1. K-Vacancy Production Cross Sections for Xe → Z_2 Collisions[a]

Xe → Z_2	Projectile cross sections			Target atom cross sections		
	3.6 MeV/amu	4.7 MeV/amu	5.9 MeV/amu	3.6 MeV/amu	4.7 MeV/amu	5.9 MeV/amu
$_6$C	14	77.5 ± 10	160 ± 20			
$_{13}$Al		165 ± 30	335 ± 40			
$_{24}$Cr	85.6	163 ± 30				
$_{32}$Ge	73.6	216 ± 42				
$_{42}$Mo	369.1	1615 ± 158	1790 ± 173			37812 ± 5130
$_{47}$Ag	887.6	1897 ± 191	3086 ± 243	8498	7117 ± 1260	14386 ± 1002
$_{50}$Sn	1261	2503 ± 187	3844 ± 272		4642 ± 423	8233 ± 502
$_{52}$Te		3783 ± 292	4095 ± 263	2360	3667 ± 274	6162 ± 400
$_{57}$La	1617	3025 ± 100	6044 ± 392		1650 ± 152	2915 ± 216
$_{62}$Sm	858		5432 ± 770	98	446 ± 40	987 ± 151
$_{70}$Yb	803		3081 ± 187	17.3		99.5 ± 13
$_{74}$W	480	1597 ± 82	2820 ± 154	20.2	21.2 ± 2	44.7 ± 7.4
$_{79}$Au	390	1113 ± 86	2207 ± 239	8.8	11.8 ± 2.7	22.7 ± 5.6
$_{82}$Pb	360	941.5 ± 97	1861 ± 65	4.1	9.4 ± 2	16.3 ± 7.1
$_{83}$Bi	(1450)	853 ± 171	1744 ± 101	3.8	6.7 ± 2	12.8 ± 3.8
$_{90}$Th				(14.4)		
$_{92}$U	880	1400 ± 600	3429 ± 232	1.5	9.8 ± 6.6	

[a] The K cross sections for target atoms and projectiles are given in barns for 3.6, 4.7, and 5.9 MeV/N impact energies (Ref. 31).

Table 2. K-Vacancy Production Cross Sections for $Pb \rightarrow Z_2$ Collisions[a]

$Pb \rightarrow Z_2$	Projectile cross sections			Target atom cross sections		
	3.6 MeV/amu	4.7 MeV/amu	5.9 MeV/amu	3.6 MeV/amu	4.7 MeV/amu	5.9 MeV/amu
$_6$C	2.05 ± 0.7	1.85 ± 0.6	7.46 ± 2.6			
$_{13}$Al		3.07 ± 0.6				
$_{24}$Cr						
$_{32}$Ge		2.1 ± 0.8				
$_{42}$Mo	2.33 ± 0.6	4.70 ± 1.7	9.47 ± 2.9	15084 ± 3080	16134 ± 6300	17036 ± 5000
$_{47}$Ag	3.1 ± 0.9	3.4 ± 1.3	6.5 ± 1.5	1789 ± 380	1991 ± 835	3794 ± 950
$_{50}$Sn	3.3 ± 1		49.5	1434 ± 265		2387
$_{52}$Te	3.5 ± 0.9	4.8 ± 1.8	13.8 ± 5	409.2 ± 71	540 ± 227	2240 ± 420
$_{57}$La	4.2 ± 0.8		14.7 ± 5	269 ± 29		877 ± 250
$_{62}$Sm	3.9 ± 0.9	6.34 ± 2.3	21.2 ± 7.6	244.5 ± 33	144 ± 59	869 ± 254
$_{70}$Yb	6.1 ± 1.4	11.9 ± 4.4		151.9 ± 28	155 ± 60	
$_{74}$W	15.2 ± 3	30.6 ± 11	78 ± 23	150.4 ± 22.5	177 ± 68	408 ± 122
$_{79}$Au	39.7 ± 6.3	68.2 ± 18	82.0 ± 36	92.7 ± 14	135.5 ± 49	116.5 ± 52
$_{82}$Pb	60.9 ± 12.2	79.4 ± 26	112.6 ± 50	60.9 ± 12.2	79.4 ± 26	112.6 ± 50
$_{83}$Bi	87.5 ± 7.1	110 ± 38	130 ± 43	53.4 ± 5	79.6 ± 30	89.1 ± 34
$_{90}$Th	142.7 ± 28	172 ± 57	181 ± 80	6.6 ± 5	22.2 ± 9.4	/
$_{92}$U	120.7 ± 18	169.0 ± 53	155 ± 70	6.1 ± 2.9	6.2 ± 2.6	15.7 ± 6

[a] The K cross sections for target atoms and projectile are given in barns for 3.6, 4.7, and 5.9 MeV/N impact energies (Ref. 31).

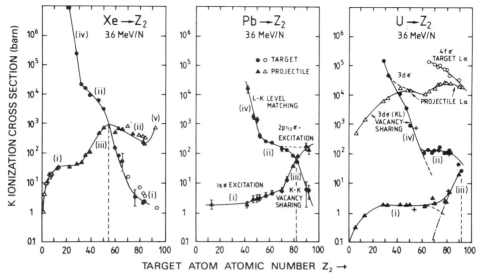

Figure 12. Excitation cross sections for the K-shell of projectile and target atom in 3.6 MeV/N Xe, Pb, and U → Z_2 collisions (from Refs. 27, 31, 90). For the U → Z_2 system cross sections for L-shell excitation are also given. All curves are to guide the eye.

out that the various data for comparable regions agree with each other within the experimental uncertainties of typically 30%–50%. In particular, inverted collision systems—i.e., systems where the role of projectile and target atom are interchanged, $Z_2 \rightarrow Z_1$—yield to the same excitation cross sections.

In Figure 12 the various regions associated with the different excitation processes are indicated: (i) $1s\sigma$ excitation, (ii) $2p_{1/2}\sigma$ excitation, (iii) $1s\sigma$–$2p_{1/2}\sigma$ (K–K) vacancy sharing, (iv) $3d\sigma$–$2p_{1/2}\sigma$ (L–K) vacancy sharing, and (v) $2p_{1/2}\sigma$–$3d\sigma$ (K–L) vacancy sharing in the case of the lighter Xe projectile. For the heavier projectiles (Pb and U) all the regions seen for the Xe projectile are shifted to correspondingly higher Z_2 values. As there are no targets available considerably heavier than U the regions beyond symmetry cannot be measured for U projectiles.

Dependences on impact energy representative for $1s\sigma$ and $2p_{1/2}\sigma$ excitation for Xe and Pb projectiles are given in Figure 13. For the $2p_{1/2}\sigma$ excitation, cross sections from symmetric collision systems are selected and the summed cross sections for both the collision partners are plotted. For low impact energies we have a rather strong energy dependence flattening out at higher energies around 3–6 MeV/N. This effect seems to be more pronounced for the heavier collision system. To extract the $1s\sigma$ excitation, cross sections from the heavier partner in asymmetric collision systems are

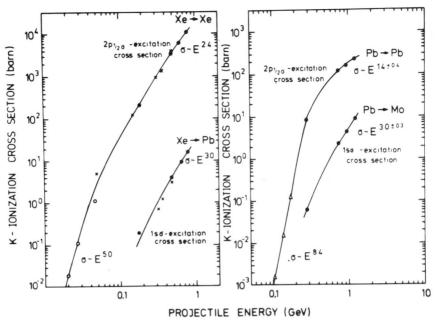

Figure 13. *K*-excitation cross sections as a function of the impact energy. For symmetric systems the summed *K* cross section is given representing the $2p_{1/2}\sigma$ excitation (Xe → Xe are extrapolated solid target values). For the asymmetric systems the cross section for the heavier collision partner is given representing the $1s\sigma$ excitation. The data are taken from Refs. 26, 31, 80, 90, 164 (partially neighboring systems are used). The curves are to guide the eye.

used. For the systems shown, Xe → Pb and Pb → Mo—about roughly inverted systems—the $1s\sigma$ cross sections and their energy dependences agree reasonably well.

3.2. Discussion

In this chapter the experimental results on inner-shell excitation are discussed within the quasimolecular picture and compared to the available theoretical models. $3d\sigma$ and $2p_{1/2}\sigma$ excitation are discussed first. After the *K*-shell vacancy sharing process, the $1s\sigma$ excitation is treated.

3.2.1. $2p_{1/2}\sigma$ Excitation and Higher MOs

(a) *3dσ Excitation.* As was shown already in Section 3.1.1 the excitation of outer shells (Pb-*M* shell for 1.4 MeV/N Pb → Z_2) can roughly be explained by geometrical considerations. For more tightly bound electrons (Pb-L shell) the excitation has to be described more specifically within the MO model. Considering the correlation diagrams, cf. Figure 3, the strongest promoted levels, in particular the $4f\sigma$ and $3d\sigma$ levels, get flat over a large range of internuclear distances (at small R) within the quasimolecule, with a binding energy almost equal to the united atom (u.a.) binding energy. This is also true for the $2p_{1/2}\sigma$ level as long as $Z_1 + Z_2 \lesssim 130$.

If the direct excitation to high-lying orbitals or to the continuum is the dominant mechanism for the vacancy creation in these MOs, the cross sections should essentially depend only on the corresponding binding energies in the u.a. system.[66] Hence, plotting the MO vacancy creation cross sections—extracted from the characteristic x-ray emission of one or both the collision partners—as a function of the u.a. atomic number $Z = Z_1 + Z_2$ a unified picture should result. In Figure 14 the $4f\sigma$, $3d\sigma$, and $2p_{1/2}\sigma$ excitation cross sections are shown using various projectiles at 1.4 MeV/N.[39] The cross sections converge to a common curve for each MO separately when considered as a function of $Z_1 + Z_2$ and seem to be

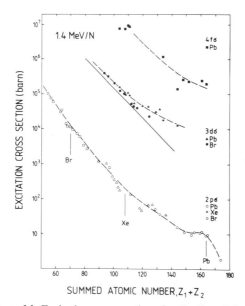

Figure 14. Excitation cross sections for the $2p\sigma$, $3d\sigma$, and $4f\sigma$ quasimolecular levels deduced from 1.4 MeV/N collisions. The data are taken from Refs. 28, 31, and 90. The dashed curves should guide the eye. For the $3d\sigma$ excitation the full curve gives the result of a calculation based on a united atom model (Refs. 81, 91).

independent of the specific Z_1/Z_2 combination. This fact was used in Ref. 39 as an argument for a dominance of direct excitation out of the promoted MOs where the u.a. binding energies are already approached. Consequently, excitation models based on a united atom model[91,92] should be used to describe the experimental data.

For the $3d\sigma$ excitation Meyerhof[81] applied the u.a. model of Briggs.[91] Here the vacancy creation is treated as a direct excitation caused by the coherent movement of both nuclei with respect to the center of the u.a. wave function. This calculation agrees reasonably with the experimental data (solid line in Figure 14). Only beyond $Z_1 + Z_2 \gtrsim 130$, the experimental cross sections deviate from an approximate exponential law to higher values, probably caused by an increase in relativistic effects.

(b) $2p_{1/2}\sigma$ *Excitation.* For the $2p_{1/2}\sigma$ excitation in not too heavy systems, u.a. excitation descriptions have been applied.[80,91,92,93] Nevertheless, all these approaches do not give a satisfying result for the high-Z region considered here, see Ref. 94. To extract actual values for the medium heavy Z region, the numerical prescription given in Ref. 93 seems to be the most reliable one.

Beyond $Z_1 + Z_2 \gtrsim 130$, the $2p_{1/2}\sigma$ excitation also seems to depend only on $Z_1 + Z_2$ despite the fact that there is no "run way" effect for the $2p_{1/2}\sigma$ level at small R for such superheavy systems (see Figures 1 and 3). That there is indeed only a $Z_1 + Z_2$ dependence of the $2p_{1/2}\sigma$ cross section for these superheavy systems is once more demonstrated in Figure 15 using Xe, Pb, and U projectiles at 3.6, 4.7, and 5.9 MeV/N.[27,31]

In order to describe the $2p_{1/2}\sigma$ excitation for the superheavy systems, relativistic approaches have to be used. Soff *et al.*[95] calculated $2p_{1/2}\sigma$ cross sections from the two-center Dirac equation, taking into account only the monopole part of the two-center potential, which is given by

$$V_{\text{MON}}(R, r) = \begin{cases} (Z_1 + Z_2)e^2/r & \text{for } r \geq R/2 \\ 2(Z_1 + Z_2)e^2/R & \text{for } r < R/2 \end{cases} \tag{24}$$

where r is the electron coordinate. Within this approximation only radial couplings can contribute to the excitation process. The results of the calculations for 4.7 MeV/N are plotted in Figure 16. The experimental $Z_1 + Z_2$ dependence is correctly reproduced by the calculations, whereas there is a factor of 4 discrepancy in the absolute values.

The fact that a monopole approximation seems to give an adequate description of the general dependence of the $2p_{1/2}\sigma$ excitation, can be understood by comparing the geometrical regions responsible for $2p_{1/2}\sigma$ excitation in the quasimolecule with the spatial dimensions of the corresponding wave functions. Considering the data of the impact parameter dependence for the $2p_{1/2}\sigma$ excitation (cf. Section 4), the vacancy production

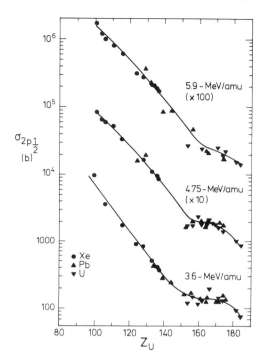

Figure 15. $2p_{1/2}\sigma$ excitation cross sections as a function of $Z_u = Z_1 + Z_2$ for Xe, Pb, and U $\to Z_2$ collisions at 3.6, 4.7, and 5.9 MeV/N. The data are from Refs. 27 and 31; the solid curves are to guide the eye.

takes place at mean internuclear distances smaller than the typical dimension of the $2p_{1/2}$ u.a. wave function. Hence, the molecular wave function is determined mainly by the total charge $Z_1 + Z_2$ and less by the charge ratio Z_1/Z_2, and consequently, the $2p_{1/2}\sigma$ cross section will depend mainly on $Z_1 + Z_2$.

The discrepancy in absolute values of the $2p_{1/2}\sigma$ excitation (Figure 16) may be caused by multicollision processes and/or multistep processes. Including into the calculations couplings to dynamically opened higher-lying MOs (multistep) the calculated cross sections may increase by the appropriate factor.[64,96] On the other hand, multicollision processes as discussed below may also significantly contribute to the $2p_{1/2}\sigma$ excitation in superheavy systems.

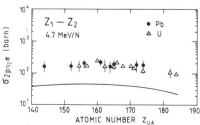

Figure 16. $2p_{1/2}\sigma$ excitation cross sections at 4.7 MeV/N. The data from Pb and U $\to Z_2$ collision (Refs. 27, 31) are compared to calculations performed in Ref. 95.

(c) *Multicollision Processes.* The importance of multicollision processes for the $2p_{1/2}\sigma$ excitation in the medium heavy Z region has been stressed in Ref. 80. At low impact velocities (i.e., for regions where the relevant $2p_{1/2}\sigma$ cross sections are small compared to the $3d\sigma$ cross sections) multicollision processes may contribute significantly to the total cross section. In such a process a projectile vacancy (mostly L shell) formed in a first collision with a high probability has a chance to survive until it undergoes a second collision where it may be transferred via MO couplings (mostly rotational) in the quasimolecule into the K-shell of the lighter collision partner. Such multicollision contributions are most important around symmetry. At higher impact energies their relative contribution decreases compared to the direct excitation because the latter increases much faster with E_1/A_1.[80]

For 1.4 MeV/N Pb projectiles we find around symmetry a $2p_{1/2}\sigma$ cross section enhancement, see Figures 11 and 14. However, this structure is also visible for U projectiles at the same $Z_1 + Z_2$ values, i.e., at more asymmetric collisions; see Figure 15. Considering the correlation diagrams (Figure 3) we find for $Z_1 + Z_2 \approx 160$ that due to the dramatic increase in binding energy of ns and $np_{1/2}$ states, preformed projectile M vacancies have a chance to couple down to the u.a. L shell.[97] on the way out such a vacancy may finally be transferred to the $2p_{1/2}\sigma$ level at larger distances. For the heaviest systems the last coupling may get less probable due to the increasing energy gap between the $2p_{1/2}\sigma$ and $2p_{3/2}\sigma$ levels, thus reducing there the multicollision contributions.

From the projectile L and M cross sections σ_L and σ_M (see Figure 11) we can estimate the fractions f_L and f_M of 1.4 MeV/N Pb ions having on the average an L or an M vacancy in the target, respectively:

$$f_{L/M} = \sigma_{L/M}(\rho \cdot L/A_2) \cdot v \cdot \tau_{L/M} \tag{25}$$

where v is the collision velocity, $\tau_{L/M}$ the lifetime of a L or M vacancy produced in the ion with the corresponding cross section $\sigma_{L/M}$, and $\rho \cdot L/A_2$ the number of target atoms per unit volume. Around symmetry ($Z_1 \approx Z_2$) we estimate for 1.4 MeV/N Pb ions fractions of roughly $f_L \approx 10^{-3}$ and $f_M \approx 0.3$. With these numbers and using appropriate statistical factors for depicting the right MO level for the preformed vacancy we can estimate the overall transfer cross sections (σ_{trans}^i) from the multicollision contribution

$$\sigma_{ML} = g_M f_M \sigma_{\text{trans}}^M + g_L f_L \sigma_{\text{trans}}^L \tag{26}$$

Estimating from Figure 11 a multicollision contribution of $\sigma_{MC} \approx 3b$ we extract transfer cross sections of $\sigma_{\text{trans}}^L \approx 5 \times 10^3 b$ and $\sigma_{\text{trans}}^M \approx 100b$ assuming for each case that the corresponding other transfer contribution vanishes. Both cross sections lead to reasonable transfer crossing radii of $r_{trans}^L \gtrsim 500$ f and $r_{trans}^M \gtrsim 50$ f. Which term is the dominant one has to be tested experimentally.

We would like to add that at higher collision velocities the enhancement in the $2p_{1/2}\sigma$ cross section observed around $Z_1 + Z_2 \approx 160$ gets broader and finally vanishes. This feature is also expected for a multicollision process.

3.2.2. $1s\sigma$–$2p_{1/2}\sigma$ Vacancy Sharing

Following the procedure outlined in Section 2.4.2 the ratio of the K shell cross sections σ_H/σ_L can be approximated as[31]

$$S = \sigma_H/\sigma_L \approx w/(1-w) + \sigma_{1s\sigma}/\sigma_{2p_{1/2}\sigma} \qquad (27)$$

where $w \equiv w(0)$ and given by Eqs. (22) and (23).

Despite the fact that atomic binding energies are used to calculate S, the measured sharing ratios are well described over orders of magnitudes.[31,81] Only for the heaviest systems and larger values of x the second term in Eq. (27), i.e., the $1s\sigma$ excitation, has to be considered. In Figure 17 the K–K sharing ratio is plotted for Xe and Pb → Z_2 collisions at 3.6–5.9 MeV/N. The straight line gives $w/(1-w)$, the first term of Eq. (27); the dashed line includes the $1s\sigma$ excitation [i.e., second term in Eq. (27)] according to Ref. 31.

Neglecting the scatter in the data the agreement between theory and measurement is fair for such very heavy systems. For small $\Delta Z/v$ values ($\lesssim 1.5$) this is surprising as no relativistic corrections are used to calculate

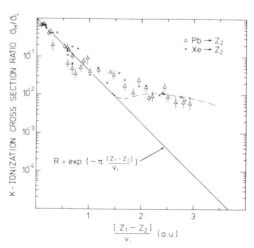

Figure 17. Sharing ratio for the K-shell excitation of the heavier and lighter collision partners as a function of $|\Delta Z|/v$ [according to Eqs. (22), (23), and (27)] (Ref. 31). The solid line gives just the first term of Eq. (27); the dashed line includes the second term, i.e., the original $1s\sigma$ excitation.

the sharing probability w. However, one should not conclude that there are no relativistic corrections necessary for such superheavy collision systems. In fact, considering the Pb–Cm correlation diagram in Figure 3, one may speculate on the existence of a second $1s\sigma$–$2p_{1/2}\sigma$ sharing region at the smallest possible internuclear distances (≈ 30 f) where both levels tend to approach again. Such an effect may cause an impact parameter dependence of w and deviations from the simple model applied so far.

Comparing the K cross sections just outside the sharing region one finds a remarkable difference between heavy and superheavy collision systems: For heavy collision systems the cross section ratio $\sigma_{2p_{1/2}\sigma}/\sigma_{1s\sigma}$ is typically in the order of 10^4 whereas for superheavy systems we deduce from Figure 17 a value of roughly 10^2. This large value indicates a strong relative increase in $1s\sigma$ cross section for superheavy collision systems.

3.2.3. $1s\sigma$ Excitation

The $1s\sigma$ excitation cross section can be inferred from the K-shell radiation of the heavier collision partner for asymmetric systems, i.e., outside the K–K sharing region. $1s\sigma$ excitation cross sections obtained by this method are shown in Figure 18 deduced from Xe, Pb, and U $\to Z_2$

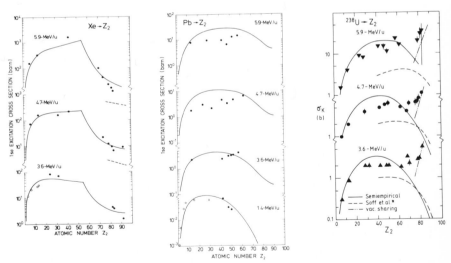

Figure 18. $1s\sigma$ excitation cross sections as a function of Z_2 at impact energies between 1.4 and 5.9 MeV/N. Left graph: Xe $\to Z_2$ collisions (Ref. 31); Middle: Pb $\to Z_2$ collisions (Ref. 31); Right side: U $\to Z_2$ collisions (Ref. 27). The full lines give the semiempirical prescription for the $1s\sigma$ excitation according to Ref. 98. The dashed lines give the monopole approximation according to Refs. 101 and 103. The dashed–dotted lines give the vacancy sharing for the U case.

collisions, respectively.[27,31] The solid lines are semiempirical predictions according to Ref. 98. There, the $1s\sigma$ excitation cross sections are deduced from a direct transfer of $1s\sigma$ electrons to the continuum using the plane wave born approximation for protons, σ_{PWBA},[99,100] and multiplying these cross sections by appropriate correction factors for the Coulomb deflection of the collision partners (C), for the increased binding of electrons in the quasimolecule (B), and for relativistic effects (R):

$$\sigma_{1s\sigma} = Z_L^2 \sigma_{PWBA} \cdot C \cdot B \cdot R \qquad (28)$$

Z_L is the atomic number of the lighter collision partner. According to Ref. 31 we have for 3.6 MeV/N Xe → Pb collision correction factors of $C = 0.64$, $B = 6.3 \times 10^{-3}$, and $R = 3.4 \times 10^3$. Thus, the $1s\sigma$ cross section reduction caused by the increased $1s\sigma$ binding (factor B) is roughly balanced by the cross section enhancement originating from relativistic effects (factor R). For heavier systems the relativistic correction gives an even stronger cross-section enhancement: For instance for 3.6 MeV/N U → Pb collisions, we have $C = 0.3$, $B = 0.6 \times 10^{-7}$, and $R = 1.6 \times 10^6$; see Ref. 27.

Considering these large correction factors only relativistic calculations including an appropriate approximation for the two-center Coulomb potential of the scattering particles can give reliable results. First reliable calculations in first-order perturbation theory have been published by Betz *et al.*[101] using the monopole approximation [cf. Eq. (24)] for the two-center potential and correspondingly for the electron wave functions (cf. also Refs. 61, 62). In Figure 18 predictions from such perturbed stationary states calculations are given. These calculations give roughly a factor of 3 to 4 to small cross sections. On the other hand, the Z and energy dependences agree reasonably with the experimental ones. Hence, the basic ideas of the calculation—i.e., a direct excitation produced by the R-dependent monopole part of the two-center potential—seems to be justified.

The discrepancy in the absolute values can possibly be explained if multistep processes (many excitations to higher-lying, bound orbitals or to the continuum during one collision) are taken into account.[64,96] In these sophisticated multichannel calculations where all possible channels (ns states) are coupled which each other the degree of ionization of the projectile (the Fermi level) at the beginning of the collision serves as a parameter.

One fundamental implication results from the monopole approximation of Betz *et al.*[101] The relevant matrix element $\langle \Psi_{E_f} | \partial/\partial R | \Psi_{1s\sigma} \rangle$ for direct excitation into the continuum, where Ψ_{E_f} is the continuum state and $\Psi_{1s\sigma}$ the $1s\sigma$ wave function, can be written as a function of just $Z = Z_1 + Z_2$[62]:

$$\langle \Psi_{E_f} | \partial/\partial R | \Psi_{1s\sigma} \rangle \propto D(Z)/R \qquad (29)$$

Hence, the $1s\sigma$ cross sections for superheavy collision systems should depend in first order only on $Z_1 + Z_2$ and not on Z_1/Z_2 or similar quantities. (Similar arguments can be used for the $2p_{1/2}\sigma$ excitation cross section of superheavy collision systems, see above.)

Figure 19 shows the measured $1s\sigma$ excitation cross sections as a function of $Z_1 + Z_2$ for the measured collision systems with Xe, Pb, and U projectiles at 3.6, 4.8, and 5.9 MeV/N. For a given impact energy all the data points tend to converge beyond $Z_1 + Z_2 \gtrsim 135$. Such a unified Z dependence, which certainly does not exist for lighter collision partners, is an essential new feature appearing for superheavy collision systems with $Z_1 + Z_2 \gtrsim 135$. It is caused by the dominance of relativistic effects for these systems.[60–64]

The results of the monopole approximation are given as dashed–dotted lines in Figure 19. The general Z dependence is well described. Inclusion of multistep processes does not change this dependence but increases the absolute values towards the experimental results.[64,96]

It will be shown in Section 4.2 that the $1s\sigma$ excitation cross section can also be calculated analytically using the coupling matrix element given in Eq. (29).[65] Only transitions from the $1s\sigma$ level to the continuum boundary are considered. The results which are given as dashed lines in Figure 19 agree somewhat better with the measured values than the numerical calculations from Refs. 14 and 101.

Recently, Berinde et al.[102] tried a further approach to calculate the $1s\sigma$ excitation based on the united atom model.[91] Details of this calculation

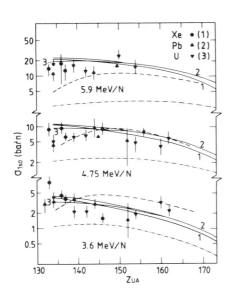

Figure 19. $1s\sigma$ excitation cross section as a function of $Z_1 + Z_2$ at 3.6, 4.7, and 5.9 MeV/N. The data points are from Refs. 27 and 31. The dashed–dotted lines gives the monopole approximation according to Refs. 101 and 103 (without multistep processes), the dashed line gives the direct excitation model (Ref. 65) and the solid lines give the SCA calculations of Ref. 102 (the curves for Xe, Pb, and U projectiles are numbered by 1, 2, and 3, respectively).

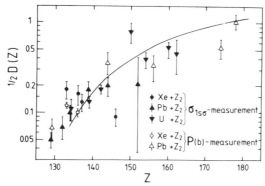

Figure 20. A comparison of theoretical $D(Z)$ values (solid line from Refs. 62 and 63) with extracted ones from total cross section and coincidence measurements (open and solid points, respectively).

are not yet published. The results are given in Figure 19 as solid lines. The lines for the various projectiles almost converge and thus indicate once more the unified $Z_1 + Z_2$ dependence of the $1s\sigma$ cross section beyond $Z_1 + Z_2 = 137$.

The unified Z dependence originates in the simple dependence of the relevant matrix element on the total charge Z via the function $D(Z)$ and on the internuclear distance R [see Eq. (29)]. The physical meaning of $D(Z)$ will become much clearer, when we discuss the impact parameter dependence of the $1s\sigma$ excitation probability. Values for $D(Z)$ can be extracted from the total cross section measurements as well as from impact parameter dependent measurements. A comparison of extracted experimental values with the calculated $D(Z)$ function[62,63] is given in Figure 20. There is a reasonable agreement between measurements and theory. Perhaps the experimental $D(Z)$ curve is slightly flatter particularly at lower Z values. This good agreement indicates a correct understanding and treatment of the relativistic effects for the $1s\sigma$ electrons in superheavy collision systems.

4. Impact Parameter Dependences

4.1. Experimental Results

We have seen that the data for systems $Z\alpha \geq 1$ can be understood within the framework of a superheavy, extremely relativistic quasiatom. However, the most convincing evidence for the formation of a short-living superheavy quasiatom comes from the impact parameter dependence of the K vacancy production probability $P(b)$. All measurements of $P(b)$ performed up to now[39,44,47,53–59] clearly show that in contrast to light collision systems[21–23] and references therein, and medium-heavy ones[24,25] the $1s\sigma$ and the $2p_{1/2}\sigma$ orbitals are excited predominantly at impact parameters small compared to the corresponding shell radii of the united atom.

In this chapter we will at first present the experimental results for nearly symmetric ($|Z_H - Z_L| \leq 10$) and for asymmetric collision partners. Then these data will be discussed in terms of the excitation of the $2p_{1/2}\sigma$ and of the $1s\sigma$ MO.

4.1.1. Nearly Symmetric Systems

In Figure 21 the results of a coincidence experiment for the system Pb + Au ($Z = 161$, $|Z_H - Z_L| = 3$) at a projectile energy of 3.6 MeV/N are shown.[56] Plotted is $P_H(b) + P_L(b) \approx P_{2p_{1/2}\sigma}(b)$. This sum gives the probability of producing a K-shell vacancy in either of the two partners per collision. Several striking features show up in the data:

1. The absolute value of the excitation probability is nearly 50% at the smallest impact parameters and is still as large as 10^{-4} at impact parameters $b \approx 1000$ f.

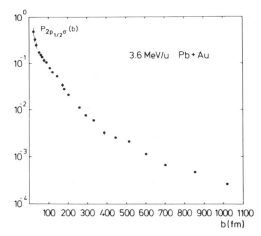

Figure 21. Sum of the Pb and Au K vacancy production probability, $P_{2p_{1/2}\sigma}(b)$, in nearly symmetric Pb + Au collisions at 3.6 MeV/N projectile energy as a function of the impact parameter b (coincidence experiment).

2. The excitation probability drops monotonically with increasing b. Because at least three regions are observed in which the $2p_{1/2}\sigma$ excitation probability decreases with a different slope, K-shell vacancies in this nearly symmetric collision system are probably produced by more than one mechanism.
3. The mean impact parameter calculated from Eq. (11) is $\bar{b} = 245$ f; for comparison we give the radius of the $2p_{1/2}$ shell of the united atom $Z = 161^{(7,8)}$: $\langle r_{2p_{1/2}}\rangle = 210$ f. Thus, in the nearly symmetric system Pb + Au the $2p_{1/2}\sigma$ orbital is predominantly excited in collisions with an impact parameter roughly equal to the $2p_{1/2}$ radius of the united atom.

Figure 22 gives a survey of all available data for the same collision partners. The measured impact parameters cover the region $0.05\langle r_{2p_{1/2}}\rangle \leqslant b \leqslant 5\langle r_{2p_{1/2}}\rangle$ at projectile energies $1.4\,\text{MeV/N} \leqslant E_1/A_1 \leqslant 6.3\,\text{MeV/N}$. For comparison, the results of theoretical calculations for the process of direct excitation of the $2p_{1/2}\sigma$ orbital to the continuum at three different projectile energies are given.[103] The general trends of the experimental data, namely, the steep projectile energy dependence between $1.4\,\text{MeV/N}$ and $3.6\,\text{MeV/N}$ and the fast decrease of $P_{2p_{1/2}\sigma}(b)$ with b for the lowest energy, are reproduced by the calculations. However, they do not give the correct absolute value, probably due to the neglecting of multistep processes. These features were already found for the total excitation cross sections (Section 3.2).

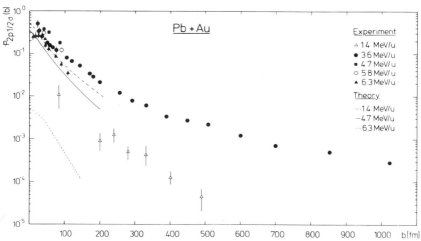

Figure 22. Sum of the Pb and Au K vacancy production probability, $P_{2p_{1/2}\sigma}(b)$, as a function of the impact parameter b at different projectile energies (coincidence experiment). The theoretical curves which give the $2p_{1/2}\sigma$ excitation probability for the process of direct excitation to the continuum are taken from Ref. 103.

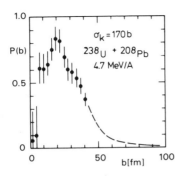

Figure 23. Impact parameter dependence of the Pb
K vacancy production probability in 4.7 MeV/N
U + Pb collisions (Ref. 44). $P(b)$ has been derived
from a line-shape analysis of Pb K X rays under
0° observation angle.

Experimental data from Ref. 44 for the system U + Pb at 4.7 MeV/N
projectile energy at extremely small impact parameters $b \leqslant 40$ f are plotted
in Figure 23. These data are obtained from a line-shape analysis of Pb K
X rays, and thus, from excited target atoms at 0° observation angle and
are normalized using the total Pb K excitation cross section from Ref. 31.
The most striking features of these data are a peak in the Pb K vacancy
production probability at $b \approx 25$ f and an almost zero probability at impact
parameters $b \leqslant 5$ f. Considering the relatively small error bars, these data
convincingly demonstrate the advantage of the line-shape techniques for
collisions with small impact parameters compared to the conventional
coincidence techniques. The latter suffer from the strong fall-off of the
Rutherford cross section with scattering angle which makes collisions with
small impact parameters very improbable.

The experimental data presented in Figures 21–23 are up to now the
only data on the impact parameter dependence of the K vacancy production
probability obtained from measurements of characteristic K X rays in
nearly symmetric systems. Many more experiments have been performed
using asymmetric collision partners. This is due to the fact that in those
cases the interpretation of the data is considerably simplified, because the
$2p_{1/2}\sigma$–$1s\sigma$ vacancy sharing is negligible [cf. Eqs. (19)–(23)]. As a consequence, the $1s\sigma$ excitation probability can directly be inferred from the
K X rays of the heavier partner.

4.1.2. Asymmetric Systems

Results of coincidence measurements of the K vacancy production
probability for the heavier partner as a function of the impact parameter
for the two-collision systems Xe + Au ($Z = 54 + 79 = 133$) and Pb + Cm
($Z = 82 + 96 = 178$) at different projectile energies are given in Figure
24.[55-57] While the $2p_{1/2}\sigma$–$1s\sigma$ vacancy sharing is smaller than 4×10^{-3}
and, therefore, negligible for Xe + Au collisions, the sharing has to be

considered for Pb + Cm. However, due to the small transition probability $[w(0) = 3.3 \times 10^{-2}$ at 5.9 MeV/N assuming Eqs. (17)–(23) are applicable] this correction has practically no influence on the data at impact parameters below 80 f. At projectile energies of 4.7 MeV/N and 5.9 MeV/N contributions due to Coulomb excitation of the Cm nuclei and subsequent K conversion in collisions with $b \leqslant 30$ f had to be corrected for. In the worst case, namely, at the smallest impact parameter $b = 20$ f at 5.9 MeV/N these contributions amounted to up to 38% of the total Cm X-ray yield. For the projectile energies 3.6 MeV/N and 4.2 MeV/N, this nuclear background was also negligible for the smallest impact parameters or at least smaller than the statistical error bars shown in Figure 24. The Pb + Cm system is the highest Z-system which could be used up to now for an investigation of the $1s\sigma$ excitation, because on the one hand the relatively weak radioactive Cm target could be handled quite easily (in contrast to heavier transuranium elements). On the other hand the Pb + Cm system is sufficiently asymmetric that the $2p_{1/2}\sigma - 1s\sigma$ vacancy sharing is small and, thus, can be taken into account properly using Eqs. (17)–(23).

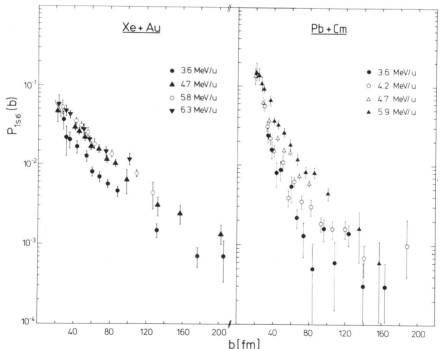

Figure 24. Impact parameter dependence of the $1s\sigma$ orbital, derived from the K vacancy production probability of the heavier partner in the asymmetric systems Xe + Au and Pb + Cm at several projectile energies (coincidence experiments) (Refs. 56, 57).

Experimental data (obtained also from coincidence measurements) for two projectile–target combinations with a total charge Z between Xe + Au ($Z = 133$) and Pb + Cm ($Z = 178$) are given in Figures 25 and 26.[59] Plotted are K vacancy production probabilities in the heavier partner for the systems Sm + U ($Z = 62 + 92 = 154$) (see Figure 25) and Sm + Pb ($Z = 62 + 82 = 144$) (see left side of Figure 26) at three projectile energies. While for Sm + U collisions the vacancy sharing is negligible for all measured impact parameters and, thus, $P_H(b) = P_{1s\sigma}(b)$, this is not the case for Sm + Pb collisions. The right side of Figure 26 shows $P_{1s\sigma}(b)$ as determined from the data on the left side and from the simultaneously measured Sm K vacancy production probability using Eqs. (17)–(23). One clearly sees that at the lowest projectile energy of 3.6 MeV/N $P_H(b) = P_{1s\sigma}(b)$ for all $b \leqslant 250$ f [$w(0) = 3.335 \times 10^{-3}$], while at the highest energy of 5.9 MeV/N $P_H(b) = P_{1s\sigma}(b)$ only for $b \leqslant 120$ f [$w(0) = 1.16 \times 10^{-2}$]. It is obvious from a comparison between the data in Figures 26a and 26b that the $2p_{1/2}\sigma-1s\sigma$ vacancy-sharing limits the extension to which a proper experimental determination of the $1s\sigma$ excitation probability is possible.

Three general conclusions can be drawn from the data in Figures 24–26:

(1) The absolute value of the $1s\sigma$ excitation probability at small impact parameters also is very high: typically 10^{-3}–10^{-1} for $b \leqslant 100$ f. At the smallest impact parameters, $P_{1s\sigma}$ is even larger for the high-Z than for the low-Z systems. This finding directly reflects the enhancement of the spatial density of the relativistic $1s\sigma$ wave functions in the vicinity of the nuclei with increasing Z.

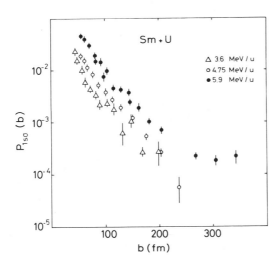

Figure 25. $1s\sigma$ excitation probability as a function of the impact parameter b, inferred from the U K-vacancy production probability in Sm + U collisions at different projectile energies (coincidence experiments) (Ref. 59).

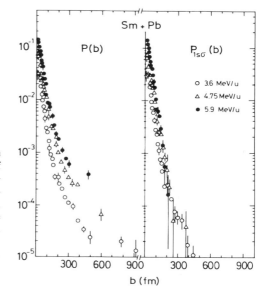

Figure 26. Pb K-vacancy production probability, $P(b)$, as a function of the impact parameter b in Sm + Pb collisions at different projectile energies (left side). Excitation probability $P_{1s\sigma}(b)$ in the same collision system after correction for the $2p_{1/2}\sigma$–$1s\sigma$ vacancy sharing (right side). The data are obtained from coincidence experiments (Ref. 59).

(2) $1s\sigma$ excitation is confined to collisions with very small impact parameters. For a projectile energy of 4.7 MeV/N one gets from the Xe + Au data in Figure 24 a mean impact parameter $\bar{b} = 85$ f and from the Pb + Cm data $\bar{b} < 55$ f. Thus, the dominant contribution to the total $1s\sigma$ excitation cross section comes from collisions with impact parameters which are considerably smaller than the K-shell radius of the united atom (≈ 350 f and 100 f, respectively[7,8]).

(3) As a consequence of the smaller mean impact parameter $\bar{b}$ for the higher-Z compared to the lower-Z systems, the $1s\sigma$ excitation probability drops more steeply as a function of impact parameter the higher the total nuclear charge.

These three features demonstrate the dominance of the relativistic effects in very-high-Z collision systems: The spatial shrinking of the $1s\sigma$ wave functions combined with a strong increase of the components with very high momenta in the corresponding momentum wave function. The dominance of these effects prevents an extrapolation of experimental data obtained for nonrelativistic systems into the superheavy region.[92,104]

The results of systematic investigations of the shape of $P_{1s\sigma}(b)$ at very small impact parameters are plotted in Figure 27 for the systems Xe + Au ($\bullet$) and Xe + Pb ($\blacktriangle$) at 4.7 MeV/N projectile energy from Refs. 55 and 39, 105, respectively. The insert in the lower left part of Figure 27 gives a survey of all available data at $b \leq 80$ f for Xe + Pb collisions: At 4.7 MeV/N from coincidence experiments ($\blacktriangle$)[39,105] and from a line-shape

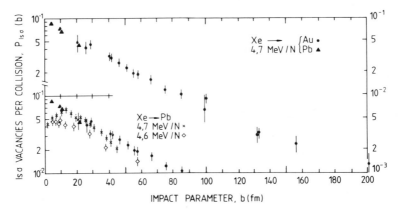

Figure 27. Excitation probability of the $1s\sigma$ orbital in 4.7 MeV/N Xe + Au and Xe + Pb collisions for impact parameters $b < 200$ f. The inset gives a compilation of all experimental data on $1s\sigma$ excitation in Xe + Pb collisions at impact parameters $b \lesssim 60$ f: 4.7 MeV/N data from a line-shape analysis ($\times$) (Ref. 44) and from coincidence experiment ($\blacktriangle$) (Refs. 39, 105) and 4.6 MeV/N data from coincidence experiment ($\blacklozenge$) (Ref. 49).

analysis ($\times$),[44] and at 4.6 MeV/N ($\blacklozenge$) from coincidence experiments.[49] The data obtained from the line-shape analysis show a peak at $b \approx 15$ f which is not observed in the coincidence data. It cannot be excluded that the peak is caused by a "background" effect [for example a contribution from hypersatellite transitions (cf. Section 2.4)] which were not considered in the data reduction. On the other side, the two sets of coincidence data also disagree at $b \lesssim 20$ f, although they may not be compared directly, because they have been taken at slightly different projectile energies. However, it seems improbable that between 4.6 and 4.7 MeV/N $P_{1s\sigma}(0)$ increases by about a factor of 2 (this results from an extrapolation of the data in the insert to $b = 0$). It should be mentioned that the data from Refs. 39 and 105 exclude any azimuthal dependence as a reason for the observed discrepancy. Additionally, these data show a negligible statistical error.

4.2. Discussion

4.2.1. $2p_{1/2}\sigma$ Excitation Probability

Suggested by the extremely high absolute values of the summed K-shell excitation probability $P_H(b) + P_L(b)$ and by the much weaker falloff with the impact parameter b in nearly symmetric collision systems (Figures 21 and 22) compared to asymmetric systems (Figures 24–26) we have denoted $P_H(b) + P_L(b)$ as $P_{2p_{1/2}\sigma}(b)$ in the system Pb + Au. This is in analogy with the interpretation of experimental data in much lighter, symmetric or nearly symmetric collision systems. However, it should be kept in mind that in our extremely high Z systems contributions from the excitation of the $1s\sigma$ orbital at the smallest impact parameter due to the relativistic effects may become important.

Looking on the Pb–Pb correlation diagram (Figure 3) which should be appropriate for Pb + Au, one finds only a weak R dependence of the binding energy of the $2p_{1/2}\sigma$ orbital for $500 \, \mathrm{f} \leqslant R \leqslant 1000 \, \mathrm{f}$. The data in Figure 21 show that in collisions with the corresponding large impact parameters where $(b \approx R_0)$ the excitation probability depends only slightly on b. However, for $b < 500 \, \mathrm{f} \; P_{2p_{1/2}\sigma}(b)$ increases rapidly with decreasing impact parameter. In the corresponding range of internuclear distances R, the $2p_{1/2}\sigma$ electron is drastically stronger bound with decreasing R (cf. Figure 3). This increase in binding which is combined with a strong shrinkage of the relativistic $2p_{1/2}\sigma$ wave function in space, favors the excitation. Namely, it is easier to transfer the large momentum needed for an excitation of the tightly bound $2p_{1/2}\sigma$ electron, because corresponding to the spatial shrinkage, the distribution of the $2p_{1/2}\sigma$ wave function in momentum space contains higher and higher components. As a consequence, the mean momentum $\bar{q} \approx \bar{h}/\bar{b}$ also becomes larger and this typical relativistic effect[101] will cause an increase of $P_{2p_{1/2}\sigma}(b)$ at $b \leqslant 500 \, \mathrm{f}$.

However, the situation is complicated by the fact that at internuclear distances $R \leqslant 400 \, \mathrm{f}$ several molecular orbitals come relatively close to each other (see Figure 3). Thus, $2p_{1/2}\sigma$ vacancies can also be produced by couplings between these orbitals, either by rotational or radial couplings. Vacancies in the corresponding orbitals may be created in an earlier state of the same collision or by a multiple-collision mechanism.[80] These processes compare to the $2p\sigma$ excitation mechanism in lighter collision systems which have been investigated experimentally (Refs. 21–25 and references therein) and theoretically[106] in great detail; for a general survey see, e.g., Refs. 11, 84, 107, and 108. First estimations have shown[17] that in very heavy systems the radial couplings are not small compared to the well-known $2p\pi$–$2p\sigma$ rotational coupling, as is the case in light, nonrelativistic systems.[109,110]

Additional support for the assumption that in nearly symmetric collision systems the $2p_{1/2}\sigma$ excitation probability is strongly influenced by couplings between molecular orbitals, comes from the data for U + Pb presented in Figure 23. The shape of the curve strongly reminds one of the "kinematic peak" in the $2p\pi$–$2p\sigma$ rotational coupling.[110] Indeed, the data in Figure 23 can be explained reasonably well by a calculation which considers a $2p_{3/2}\pi$–$2p_{3/2}\sigma$ rotational and a subsequent $2p_{3/2}\sigma$–$2p_{1/2}\sigma$ radial coupling.[111] The results are presented in Figure 28. The calculation also predicts a prominent adiabatic maximum at larger impact parameter

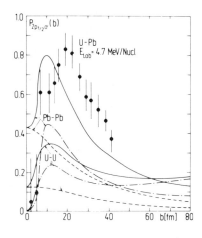

Figure 28. Excitation probability of the $2p_{1/2}\sigma$ orbital for different collision systems as a function of the impact parameter. The experimental data points for U + Pb are taken from Ref. 44 and the theoretical results for Pb + Pb and U + U from Ref. 111. The solid lines give the summed predictions for direct $2p_{1/2}\sigma$ ionization (dashed lines) and for $2p_{1/2}\sigma$ excitation from a coupled channel calculation with initial vacancies in the $2p_{3/2}\pi$ state.

(for example in 4.7 MeV/N U + U collisions at $b \sim 175$ f). However, such a maximum does not show up in the data for the quite similar system Pb + Au (cf. Figure 21).

Figure 29 shows the $2p_{1/2}\sigma$ excitation probability in Pb + Au collisions for impact parameters below 100 f at different projectile energies, while Figure 30 gives the dependence on the projectile energy at fixed impact parameters. In principle, contributions from $1s\sigma$ excitation to the data in Figure 29 cannot be excluded, but it seems unlikely that such contributions lead to a lower total excitation probability at 6.3 MeV/N than at

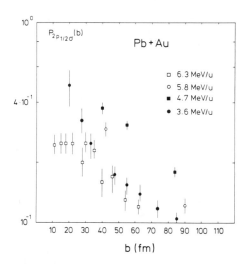

Figure 29. Sum of the Pb and Au K vacancy production probability $P_{2p_{1/2}\sigma}(b)$, in collisions with impact parameters $b < 100$ f at different projectile energies.

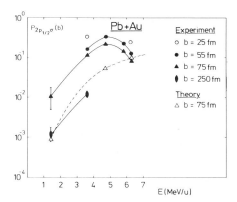

Figure 30. $P_{2p_{1/2}\sigma}$ at fixed impact para-meters b as a function of the projectile energy in the nearly symmetric col-lision system Pb + Au. The full lines are drawn to guide the eye. The dashed line give the theoretical result for the process of direct $2p_{1/2}\sigma$ excitation to the continuum (Ref. 103).

3.6 MeV/N. It may be that a superposition of projectile-energy-dependent couplings between several molecular orbitals finally results in the different shapes of $P_{2p_{1/2}\sigma}(b)$ at the small b and in the peculiar energy dependence in Figure 30. The process of the direct excitation of the $2p_{1/2}\sigma$ electron to the continuum gives, namely, a monotonic dependence on the projectile energy. This is shown in Figure 30 by the dashed line for $b = 75$.[103] Multistep processes which might shift the absolute value of the dashed line in Figure 30 towards the experimental points also increase monotonically with increasing energy.[64,96] Thus, the latter two processes alone cannot explain the data in Figure 30.

As has been shown in Section 2.3 the analysis of the Doppler-broad-ening of x-ray lines at 90° observation angle allows for the determination of mean impact parameters using Eq. (11). With the total $2p_{1/2}\sigma$ excitation cross sections for Pb + Au from Ref. 31 and for U + Th from Ref. 27, average $2p_{1/2}\sigma$ excitation probabilities $\bar{P}$ defined by[26]

$$\sigma_{2p_{1/2}\sigma} = \bar{P} \cdot \pi \cdot (\bar{b})^2 \qquad (30)$$

can be obtained. Figure 31 gives the results as a function of the projectile velocity. For Pb + Au a projectile-energy dependence very similar to the coincidence data in Figure 30 is observed which flattens out for the still heavier system U + Th ($Z = 182$) between 3.6 and 5.9 MeV/N. $\bar{P}$ for the process of direct $2p_{1/2}\sigma$ ionization strongly increases with energy for both collision systems (dashed lines in Figure 31).[26,60] These data demonstrate that first, preliminary information on $2p_{1/2}\sigma$ excitation probability can easily be obtained from a measurement of single x-ray spectra without any coincidence conditions.

At present, the process of $2p_{1/2}\sigma$ excitation in nearly symmetric systems is not well understood. This appears not to be the case in the rather asymmetric Pb + Cm system. Figure 32 gives the results of coincidence

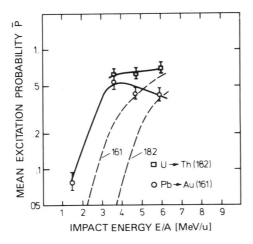

Figure 31. Average $2p_{1/2}\sigma$ excitation probability as a function of the projectile energy in the nearly symmetric collision systems U + Th and Pb + Au (Ref. 26). The solid lines are to guide the eye; the dashed lines represent the contributions from direct $2p_{1/2}\sigma$ excitation, according to Refs. 26, 60.

measurements of the $2p_{1/2}\sigma$ excitation probability which is inferred from the Pb K vacancy production.[56] At all four different projectile energies, the data can be reasonably well described by the simple experimental expression given in Eq. (12) with a projectile energy independent decay constant a $\approx$ 50 f. The absolute value of the $2p_{1/2}\sigma$ excitation probability increases monotonically with energy. Thus, the dominant $2p_{1/2}\sigma$ excitation

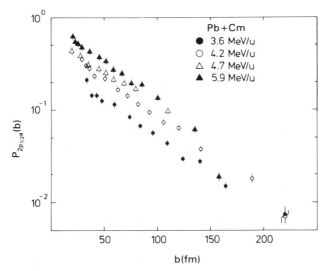

Figure 32. Excitation probability of the $2p_{1/2}\sigma$ orbital as a function of the impact parameter in asymmetric Pb + Cm collisions (coincidence experiments) (Ref. 56).

mechanism in this rather asymmetric system at impact parameters below 200 f is probably a superposition of the direct $2p_{1/2}\sigma$ excitation to the continuum or to higher-lying orbitals and of the process of multistep excitation.

It should be added that the total $2p_{1/2}\sigma$ excitation cross section as a function of Z shows a peculiar cross-section enhancement around $Z \approx 164$ (Figures 14 and 15); this was interpreted by essential changes in the correlation diagrams (see Section 2.2 and Figure 3). Hence, the difference in the $2P_{1/2}\sigma(b)$ data for Pb + Au and for Pb + Cm (Figure 32) may just reflect this effect.

We want to emphasize that the importance of the theoretically postulated multistep[64,96] can be verified from a comparison between experimental results (presented in Figures 24 and 30) and corresponding theoretical calculations for the Pb + Cm system. Such calculations should reproduce both the experimental $1s\sigma$ and the $2p_{1/2}\sigma$ data with the same initial state of ionization of the projectile (Fermi level). However, to our knowledge only calculations for the excitation of the $1s\sigma$ orbital have been available up to now.

Summarizing, we state that apart from the question about the dominant $2p_{1/2}\sigma$ excitation mechanism in nearly symmetric systems, the data clearly show that the $2p_{1/2}\sigma$ orbital is excited mainly in very close collisions. From the 3.6 MeV/N data for Pb + Au in Figure 21 one infers that about 70% of the total $2p_{1/2}\sigma$ excitation cross section of $\sigma = 120b$[31] results from collisions with impact parameters below the mean impact parameter $\bar{b} = 245$ f and more than 50% from collisions with $b < 150$ f. For the same projectile energy in the Pb + Cm system one finds from Figure 32 that 60% of the total excitation cross section of $\sigma = 56b$ comes from collisions with impact parameters below $\bar{b} = 100$ f. The values for $\bar{b}$ in both cases are comparable to the radii of the united atom $2p_{1/2}\sigma$ shell ($\langle r_{2p_{1/2}} \rangle \approx 210$ f and 100 f for $Z = 161$ and 178, respectively.[7,8] Thus, electrons in the $2p_{1/2}\sigma$ orbital are excited predominantly in the field of a "superheavy quasiatom" with the combined nuclear charge $Z = Z_1 + Z_2$. This central finding, which comes out even clearer in the case of $1s\sigma$ excitation, allows for a first experimental step towards the study of the electronic properties of superheavy atoms, such as excitation mechanisms and strengths, wave functions, and binding energies.

4.2.2. $1s\sigma$ Excitation Probability

It is a general finding (see Figures 25–27) that the excitation probability of the $1s\sigma$ orbital increases monotonically with increasing projectile energy. This dependence is plotted for the system Xe + Au in Figure 33 at three fixed impact parameters. The theoretical results which are based on coupled

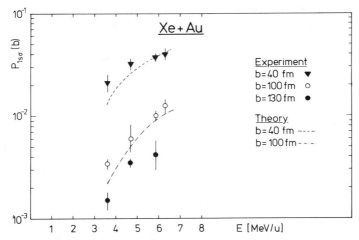

Figure 33. $1s\sigma$ excitation probability $P_{1s\sigma}$ at fixed impact parameters in the asymmetric collision system Xe + Au as a function of the projectile energy (Ref. 56). The theoretical data are taken from Refs. 96 and 112.

channel calculations including multistep contributions (64,112) are shown by the dashed and the dashed–dotted lines. They agree quite well with the experimental data, both in shape and in absolute magnitude. The same good agreement is found for $P_{1s\sigma}(b)$ in the same collision system at a fixed projectile energy of 4.7 MeV/N (see Figure 34).

A comparison between experiment and theory for the much heavier system Pb + Cm is given in Figures 35 and 36. The difference between the

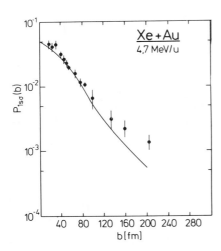

Figure 34. $1s\sigma$ excitation probability $P_{1s\sigma}$ at a projectile energy of 4.7 MeV/N for the asymmetric system Xe + Au as a function of the impact parameter (Ref. 56). The full line gives the theoretical calculation (Refs. 96 and 112).

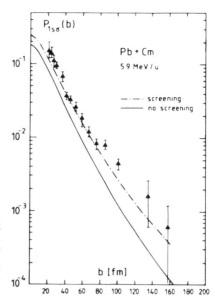

Figure 35. Comparison between experimental (Ref. 57) and theoretical (Ref. 112) results for the $1s\sigma$ excitation probability in the system Pb + Cm at a projectile energy of 5.9 MeV/N. In the dashed-dotted line the screening of the $1s\sigma$ electron is approximately taken into account.

two theoretical curves—both including multistep contributions—is that in the curve denoted as "screening," a heuristic, reduced $1s\sigma$ binding due to the influence of electron screening was taken into account.[112] For the higher projectile energy of 5.9 MeV/N (see Figure 35) the agreement between experimental and theoretical results including electron screening is good for $b \geqslant 50$ f. However, for collisions with impact parameters below 50 f the experimental data are systematically larger than the theoretical ones. The steeper b dependence of the experimental compared to the theoretical data comes out much clearer in the 3.6 MeV/N data plotted in Figure 36. At the smallest b, the experimental results are up to a factor of 3 larger than the theoretical ones. In order to get some more systematics about the Z dependence of these deviations, which were not yet observed for Xe + Au collisions, detailed calculations for the systems Sm + Pb ($Z = 144$) and Sm + U ($Z = 154$) (see Figures 25 and 26) would be helpful.

It has already been stated in Section 4.1.2 that the $1s\sigma$ excitation probability drops more steeply the higher Z is. As for $2p_{1/2}\sigma$ excitation this is again a consequence of the much stronger binding of the $1s\sigma$ electron. It is easy to show[66,113] that, in order to transfer an energy ΔE in a collision with velocity v, a minimum momentum transfer given by

$$q_0(\Delta E) = \Delta E/\hbar v \qquad (31a)$$

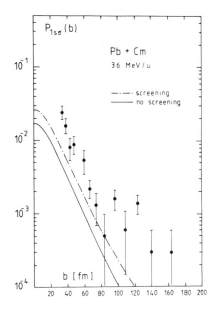

Figure 36. Same as in Figure 35 at a projectile energy of 3.6 MeV/N.

is needed. For an ionization process

$$\Delta E = E_f + E_B \tag{31b}$$

where E_f is the kinetic energy of the ejected electron ("δ electron") and $|E_B|$ is the absolute value of its binding energy which may depend on the internuclear distance R. It has been found experimentally[114–116] that the average kinetic energy of the δ electron is small compared to its rest mass. Thus, $E_f + E_B \approx E_B$ and one gets

$$q_0(\Delta E) \approx q_0(E_B) = \frac{E_B(R)}{\hbar v} \tag{32}$$

Based on the idea of the united atom approximation[91] a "Bang–Hansteen scaling rule" can be established,[51,66,117–119] which combines the mean impact parameter $\bar{b}$ and the mean minimum momentum transfer $\bar{q}_0 \equiv q_0(\bar{R}_0)$ via

$$\bar{b}\bar{q}_0 = c \tag{33}$$

with $\bar{R}_0 = R_0\,(\bar{b})$ and c a constant of the order of unity. Because for not too small b the experimental data in Figures 24–26 can be described by an exponential function as given in Eq. (12), we get from Eqs. (13a) and (27)

$$a = \frac{1}{2}\bar{b} = \frac{c}{2\bar{q}_0} \approx \frac{1}{2\bar{q}_0} \tag{34}$$

and thus

$$P_{1s\sigma}(b) \approx P(0)\, e^{-2b\bar{q}_0} \tag{35}$$

From Eqs. (32)–(35) it is immediately seen that at a given velocity a higher binding $E_B(\bar{R}_0)$ results in a smaller $\bar{b}$ and, thus, in a steeper falloff of the $1s\sigma$ excitation probability.

A somewhat unpleasant feature of Eq. (35) is that it contains $\bar{R}_0$, the distance of closest approach at $\bar{b}$ which depends by Eq. (11) on $P_{1s\sigma}(b)$. This is overcome if the R dependence of the $1s\sigma$ binding energy is taken into account. Following the suggestions given in Ref. 62 an analytic expression for $P_{1s\sigma}(b)$ very similar to Eq. (29) has been derived.[65] Using the following three assumptions:

1. Calculation in first-order perturbation theory;
2. The radial coupling matrix element between the $1s\sigma$ state and the final state of the electron with energy E_f is given by Eq. (29):

$$\langle \Psi_{E_f} | \partial/\partial R | \Psi_{1s\sigma} \rangle \propto D(Z)/R$$

3. Substitution of E_f for all final states by $E_f = mc^2$;

$P_{1s\sigma}(b)$ is given for not too small b, and $Z \geqslant 137$ by

$$P_{1s\sigma}(b) = \tfrac{1}{2} D(Z)\, e^{-2R_0 q_0(R_0)} \tag{36}$$

Compared to Eq. (35), b has been replaced by $R_0 = a_0 + (a_0^2 + b^2)^{1/2}$ [cf. Eq. (2)], the distance of closest approach in a collision with impact parameter b, and $P(0)$ by the function $D(Z)$. The physical interpretation of $q_0(R_0)$ is given by the third assumption: Fixing $E_f = mc^2$, the excitation of the $1s\sigma$ electron into continuum states $(E_f > mc^2)$ is considered to be compensated to a large extent by the excitation into weakly bound states $(E_f < mc^2)$. Thus, the energy transfer $\Delta E(R_0)$ is approximately the $1s\sigma$ binding energy at R_0 and from Eqs. (31a) and (32) follows

$$q_0(R_0) = \frac{E_{1s\sigma}(R_0)}{\hbar v} \tag{37}$$

It should be emphasized that the assumptions leading to Eqs. (36) and (37) are by no means self-evident. The use of perturbation theory may be questionable at probabilities as large as 10^{-2}–10^{-1} (cf. Figures 24–26). The validity of the second and the third assumption has to be investigated very thoroughly experimentally and theoretically especially in view of recent experimental[58,120] and theoretical[64,96,112] data.

Starting from a somewhat different point of view, namely, from the similarity between the $1s\sigma$ excitation in very high-Z systems and the process of electron[121] and pair[122] conversion by nuclear monopole transitions, an expression for $P_{1s\sigma}(b)$ has been derived[123] which corresponds to Eq. (36). In the calculation in Ref. 123 a characteristic time $\hat{t}$ enters which is determined by the dynamics of the collisional excitation process and which in first order $\sim R_0/v$. Most important, it has been shown in Ref. 123 that multistep processes do not essentially change the functional dependence of $P_{1s\sigma}$ on b; only the absolute value is influenced. This had already been stated earlier empirically.[56] An important consequence of this finding is the independence of the mean impact parameter $\bar{b}$ on multistep processes, since according to the definition given, Eq. (11) $\bar{b}$ depends only on the shape and not on the absolute value of $P_{1s\sigma}(b)$.

The validity of identifying the energy transfer ΔE with the binding energy $E_{1s\sigma}(R_0)$ in Eqs. (36) and (37) can only be justified by a comparison to experimental results. This can be done for different collision systems most conveniently by plotting the experimental $P_{1s\sigma}(b)$ data as $\ln P_{1s\sigma}(b)/\frac{1}{2}D(Z)$ vs. $R_0 q_0(R_0)$ using theoretical values of $E_{1s\sigma}(R_0)$ from Ref. 124. According to Eq. (36) this normalized plot gives one projectile velocity independent straight line with a slope -2. This graph is shown in Figure 37. The values of $D(Z)$ for the different systems have been determined by a fit of Eqs. (36) and (37) to the $P_{1s\sigma}(b)$ data for each collision system separately.

Obviously, the data fall with some reasonable scatter on the predicted straight line. This agreement strongly supports the identification of the energy transfer $\Delta E(R_0)$ with the $1s\sigma$ binding energy. Thus, one is able to describe the excitation probability $P_{1s\sigma}(b)$ for highly relativistic systems by the surprisingly simple relation given in Eqs. (36) and (37). Note that for larger impact parameters, the functional dependence of $P_{1s\sigma}$ on b from Eq. (36) agrees with that from Eq. (35), because $b \approx R_0$ and $E_{1s\sigma}$ is, at least for not too small internuclear distances, only weakly dependent on R_0. A direct presentation of the experimental $P_{1s\sigma}(b)$ data vs. $R_0 q_0(R_0)$ is made in Figure 38. The full lines which give the predictions of Eqs. (36) and (37) with $E_{1s\sigma}(R_0)$ taken from Ref. 124 agree reasonably well with the data—except in the Pb + Cm system, where a systematic deviation of the experimental data towards higher $P_{1s\sigma}(b)$ values is found. These deviations, which have already been observed in the comparison between the data and the extended coupled channel calculation (see Figures 35 and 36), will be discussed in more detail in Section 6.

The $1s\sigma$ data for the system Sm + Pb are plotted vs. R_0, the distance of closest approach, in Figure 39. This graph allows for a comparison with the values calculated from Eqs. (36) and (37) for different projectile velocities. These values are given by the dashed–dotted, solid, and dashed

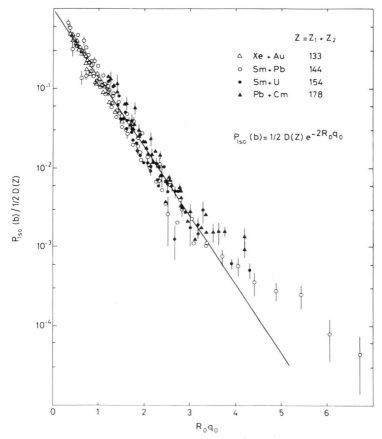

Figure 37. Normalized excitation probability $P_{1s\sigma}/\frac{1}{2}D(Z)$ as a function of $R_0 \cdot q_0$ for four different collision systems with $133 \leqslant Z_1 + Z_2 \leqslant 178$ at projectile energies between 3.6 MeV/N and 6.3 MeV/N (Refs. 56, 57, 59). The full line gives the $R_0 q_0$ dependence of Eq. (36) which is also explicitly given in the figure. $\frac{1}{2}D(Z)$ has been determined by a least-squares fit of Eqs. (36) and (37) to the experimental data for every collision system separately.

lines, respectively. At $R_0 \gtrsim 250$, the experimental data are disturbed by the $2p_{1/2}\sigma - 1s\sigma$ vacancy-sharing (see Figure 26), and thereby a proper determination of $P_{1s\sigma}$ is no longer possible. At $R_0 \leqslant 40$ f, the impact parameter b is no longer large compared to a_0 and then Eq. (36) has to be replaced by a somewhat complicated expression.[65]

From Eq. (36) follows that the function $\frac{1}{2}D(Z)$ gives the $1s$ excitation probability in the united atom limit ($R_0 \to 0$) in a superheavy atom; according to Refs. 62, 63 this "strength function" is independent of the projectile

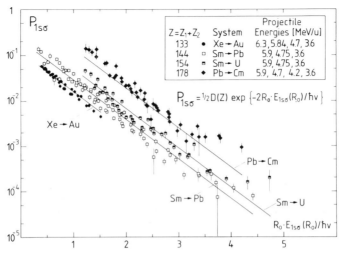

Figure 38. $1s\sigma$ excitation probability as a function of $R_0 q_0 = R_0 \cdot E_{1s\sigma}(R_0)/\hbar v$ for four collision systems with $133 - Z_1 + Z_2 - 178$ at different projectile energies as given in the inset (Refs. 56, 57, 59). The solid line gives the $R_0 q_0$ dependence of Eq. (36) (also given in the figure) for every collision system separately.

velocity. $D(Z)$ is plotted vs. Z in Figure 20; the open symbols are extracted from $P_{1s\sigma}(b)$ data as described above. The solid symbols are derived from a comparison between measured[27,31] and calculated[65] total $1s\sigma$ excitation cross sections. $\sigma_{1s\sigma}$ can easily be calculated from Eqs. (10), (36), and (37) using the parametrization[124]

$$E_{1s\sigma}(R_0) = A(Z)/R_0^{\chi(Z)} \tag{38}$$

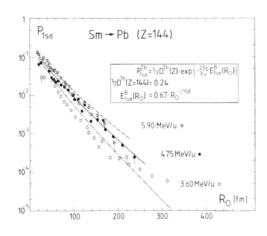

Figure 39. $1s\sigma$ excitation probability for Sm + Pb collisions at three different projectile energies as a function of the distance of closest approach in a collision with an impact parameter b (Ref. 59). The dashed–dotted, solid, and dashed lines are the predictions of Eq. (36). The relevant theoretical values from Refs. 62, 63, and 124 are given in the figure.

where A and χ are functions only of Z, with the result[65]

$$\sigma_{1s\sigma}(\theta, Z) = \frac{D(Z)}{1 - \chi} 4\pi a_0^2 \left[\frac{\Gamma(\delta, y)}{y^\delta} - \frac{1}{2} \frac{\Gamma(\delta/2, y)}{y^{\delta/2}} \right] \qquad (39)$$

$\Gamma(x, y)$ is the incomplete Γ-function[125] and

$$\delta = \frac{2}{1 - \chi}, \qquad y = 4a_0 \frac{E_{1s\sigma}(2a_0)}{\hbar v} = 4a_0 q_0(2a_0) \qquad (40)$$

Using Eqs. (39) and (40) $D(Z)$ can be determined from the experimental data in Refs. 27, 31. Note that Eqs. (36) and (39) explicitly show that $1s\sigma$ excitation in very high Z systems is just a function of v and of $Z = Z_1 + Z_2$ and not of Z_1/Z_2. Such a dependence on Z is a typical consequence of the fact that in highly relativistic systems, the two-center potential can be very well approximated by the monopole part of a multipole expansion given by Eq. (24).

The full curve plotted in Figure 20 gives the theoretical value of the $1s$ excitation probability in the united atom limit.[62,63] Despite the broad scatter of the data points, the theoretical prediction that the relativistic effects cancel the decrease of the $1s\sigma$ excitation due to the increased binding of the $1s\sigma$ electron, is confirmed by the data for the whole Z region.

5. Quasimolecular Radiation

5.1. Spectra and Emission Characteristics

If in a quasimolecular collision a vacancy is brought into or created in an inner molecular level (double or single collision mechanism, respectively), this vacancy has a certain chance to decay radiatively during the time of existence of the quasimolecule itself. In such cases quasimolecular radiation can be observed; for a review of this field see for instance Refs. 11 and 108 and the references cited therein. The (radiative) decay rate for an inner MO vacancy depends on the transition energy between the relevant MOs, which varies with the internuclear distance. Integrating over all the possible transition energies an x-ray continuum for the MO radiation is expected.

A first hint on such a radiation continuum was already found by Coates in 1934[126] studying Hg → Sn collisions. The real discovery of the quasimolecular radiation was made by Saris and co-workers[127,128] in 1971 for light collision systems like Ar → KCl. In the same year similar investigations had been expanded into the region of superheavy quasimolecules

(I → Au) by Mokler *et al.*[129,130] A large variety of investigations followed since that time covering the whole $Z_1 + Z_2$ range.

5.1.1. Spectral Shapes

Shape and structure of the x-ray continuum depends on various parameters:

    i. radiative transition rates,
    ii. vacancy creation mechanisms,
    iii. collision dynamics, and
    iv. electron binding energies of the relevant MOs and their R dependences.

All the factors together determine the spectral shape and x-ray intensity of the MO radiation. Transition rates (i) and vacancy probabilities (ii) are mainly responsible for the x-ray yield. As both are large for superheavy collision systems high MO intensities are expected here. On the other hand, both factors can influence the spectral shape. For instance, as transition rates increase strongly with transition energy the high-energy parts in the x-ray spectra may be accentuated. The size of the collision time (iii) in comparison to the transition rate (radiative lifetime) is not the only factor determining the MO x-ray yield (for superheavy collision systems the time ratio is in the region of percent). Moreover, possible structures in the MO spectra will be smeared out by the finite collision time via the uncertainty principle, point (iii).[131,132] Nevertheless, the dominant influence on the MO x-ray spectra is point (iv), the dependence of the MOs on the internuclear distance. Hence, the first information on binding energies in superheavy quasimolecules resulted from quasimolecular x-ray spectra.[129,130,133]

The influence of quasimolecular binding energies and their R dependences on the spectral shape is demonstrated by the spectra shown in Figures 8 and 9. For the I–Au collision system (Figure 8) we observe a pronounced, peaklike structure in the spectrum at about 8 keV.[129,130,133] Measurements of the Doppler shift of these X rays for different observation angles indicate an emitter velocity equal to the velocity of the center of mass of the collision system.[134,135] Hence, the structure at 8 keV is attributed to quasimolecular—or even to quasiatomic—radiation, or more precisely to M-MO radiation, i.e., to transitions to MOs correlating to the united atom $3d$ levels. As these levels are relatively flat over a large region of internuclear distances (compare the corresponding correlation diagram in Figure 3) the peaklike structure in the spectra can be understood as an integration over all possible trajectories. For higher projectile energies this structure is smeared out by the collision dynamics (see Figure 8).

Similar structures were found for neighboring collision systems. In Figure 40 the Z dependence of the position of the structure is plotted. Also shown are the (mean) x-ray energies for structures observed in other collision systems (cf. Ref. 11). The relevant MO levels attributed to the different structures can be read from the inserted correlation diagram. Unique to all the observed structures is that the relevant MO levels show a flat minimum as a function of R. Moreover, the observed Z dependence is roughly in agreement with a Z^2 law as is expected for the corresponding binding energies.[11,136] Hence, the band structures and their shift with Z give information on the binding energy of the corresponding quasimolecular level around the shallow minimum and its dependence on Z. Such investigations gave a first opportunity of penetrating into the region of superheavy electronic systems.

For MO radiation from the $1s\sigma$ level we cannot expect such structures in the spectrum as the $1s\sigma$ level is always strongly decreasing with R, especially for small R. This behavior, in connection with the collisional broadening, causes a strongly decreasing x-ray continuum beyond the

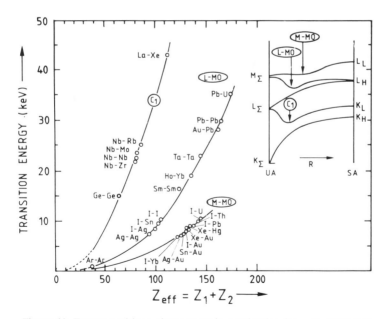

Figure 40. Energy positions of structures in quasimolecular x-ray spectra as a function of $Z_1 + Z_2$ according to Ref. 11. In addition to the data summarized in Ref. 11, values from Ref. 136 are also given.

characteristic K lines, cf. Refs. 11, 108. For Pb–Pb collisions the MO
continuum extends up to x-ray energies around 1000 keV (see Figure 9),
indicating correspondingly high $1s\sigma$ binding energies. Also in this case the
velocity of the continuum emitter was shown to be the velocity of the mass
center proving the quasimolecular character of the radiation.[20]

In Figure 41 the measured MO photon yield for Pb–Pb collisions[20]
is compared with a calculated MO spectrum involving transitions to the
$1s\sigma$ level only.[137] Between 400 and 600 keV the calculation agrees reasonably well with the data indicating that the used $1s\sigma$ binding energies from
the monopole approximation indeed describe the actual binding energies
during the collision.[138] For lower x-ray energies transitions to higher MO
(e.g., $2p_{1/2}\sigma$) increase the measured x-ray yield. How far the deviation
beyond 600 keV is caused by deviations from the binding energies given
in Ref. 124, by experimental effects, or by approximations made in the
calculations is not yet clear. Nucleus–nucleus bremsstrahlung contributions
appear not to be responsible here for the observed photon yield,[137] neither
is secondary electron bremsstrahlung.[33]

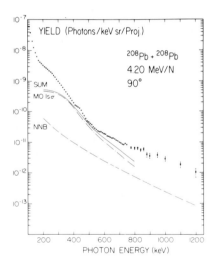

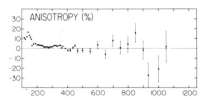

Figure 41. Thick target photon yield for 4.2
MeV/N ^{208}Pb–^{208}Pb collisions using highly
enriched ^{208}Pb targets (Ref. 20). The theoretical curves for the $1s\sigma$ molecular radiation, the nucleus–nucleus bremsstrahlung,
and the sum of both are from Ref. 137. The
lower part of the figure shows the measured
anisotropy also beyond 100 keV (Ref. 20).

5.1.2. Anisotropies

As it is difficult to extract precise information on the MO binding energies from structureless x-ray spectra there was a search for other, more unique features of the MO x-ray emission. In 1973, an anisotropic emission characteristic—preferring an MO x-ray emission perpendicular to the beam direction—was discovered for the quasimolecular M-radiation from I–Au collisions.[135,139,140] The anistropy has its maximum at x-ray energies around the u.a. transition energy.

Equivalently an anisotropy peak was found for the quasimolecular K radiation also nearby the expected united atom $K\alpha$ transition energy.[141,142] Since that time a lot of investigators[143–148] have studied the anisotropy for the MO x-ray emission. In particular, Wölfli and co-workers performed systematic studies of the MO anistropy.[145–148] Following the presentation given in Ref. 11, in Figure 42 most of the anisotropy peak positions reported in the literature are plotted as a function of the u.a. atomic number $Z_1 + Z_2$. The data coalesce into various groups which can be associated with the

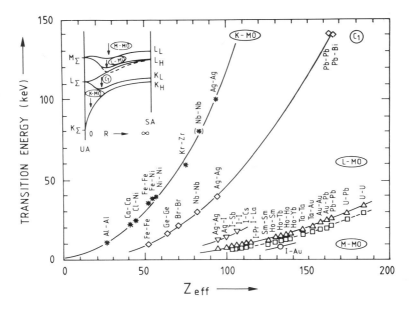

Figure 42. Positions of maxima in the anisotropy spectra for the quasimolecular radiation. The representation comprises the data summarized in Ref. 11 and a selection of more recent values given in the literature (Refs. 20, 146–148). For more detailed data see Ref. 148.

quasimolecular levels indicated in the figure. Within each group we find once more roughly a Z^2 dependence for the peak position indicating the connection between the peak positions and the binding energies of the corresponding MO levels.

Unfortunately, for the superheavy collision systems the measured peak positions do not coincide with calculated transition energies in the quasimolecule as is the case for the heavy collision molecules.[146,148] How far an inclusion of a correct screening can alter this point for L- and M-MO transitions is not clear. For K-MO radiation ($1s\sigma$) from superheavy collision systems no anisotropy is observed at all, as is seen from the lower part of Figure 41.[20]

The absence of an anisotropy for the $1s\sigma$ radiation for superheavy collision systems such as Pb–Pb was predicted by Anholt.[149] In these calculations–as in most of the other ones[150–154]—the anisotropies are supposed to be caused by an alignment of the feeding MO levels. The supposition of outer shell alignment as a reason for the anisotropies was confirmed by measurements of the linear polarization of the MO radiation for light collision systems.[155] In this case, the alignment of the feeding levels could be extracted by comparing these data with calculations from Briggs et al.[154] Hence, the absence of a $1s\sigma$ anisotropy for Pb–Pb collisions seems to indicate a suppression of alignment in the feeding MO levels[20] due to relativistic effects.

Here, we would like to point out that using the same proposition of an alignment as reason for an anisotropy, Kirsch[156] calculates a positive anistropy for the $1s\sigma$ radiation from Pb–Pb collisions. From the kinematic dipole model of Hartung and Fricke[153] no predictions are available for this case. On the other hand most of the available theories[149,153,154] seem to predict the $2p\sigma$ anisotropy in Pb–Pb collisions roughly at the right position. There, transitions into the $2p_{1/2}\sigma$ level beyond about 1000 f seem to be responsible for the anisotropy[20] (compare the appropriate correlation diagram in Figure 3).

5.2. Coincidence Measurements

The resulting inability to uniquely correlate structures in MO radiation spectra or even structures in MO anisotropy spectra to well-defined features in the correlation diagrams, necessitates more specific experiments for the investigation of the MO radiation of superheavy systems. Such investigations are for, instance, coincidence measurements studying the time correlations between the MO x-rays and characteristic x-ray[157,158] and/or scattered particles.[159–161]

5.2.1. X-Ray/X-Ray Coincidences

The disadvantage of singles, or noncoincident MO x-ray spectra is that the various possible transitions from different MOs feeding the vacant MO level can not be disentangled. By specifying the feeding level, the MO x-ray spectra contains specific information on both MO levels: If a special MO transition occurs the vacancy is transferred from the parent level (e.g., $1s\sigma$) to the feeding level (e.g., $2p_{1/2}\sigma$)—see Figure 1b. After this transition the vacancy will be transferred via—originally— the feeding level (e.g., $2p_{1/2}\sigma$) to the separated atom with the emission finally of a characteristic X ray. Hence, measuring characteristic x-ray/MO x-ray coincidences, transitions between specific MOs may be selected.[157]

For instance, measuring MO-K/characteristic-K x-ray coincidences, for Pb–Pb collisions the $1s_{1/2}\sigma \rightarrow 2p_{1/2}\sigma$ transition may be selected; compare the correlation diagram in Figure 3. In Figure 43 singles and coincidence spectra are compared for this case at 4.8 MeV/N.[158] The discrepancy between theory[137,138] and experiment is not yet clear. Possibly large crossing probabilities with levels adjacent to the $2p_{1/2}\sigma$ level may alter the above stated correlation. As the excitation probabilities in inner shells are tremendously large for superheavy quasimolecules (see Section 4), also the possibility of a double vacancy creation has to be considered. Multiple vacancy creation seems to determine mainly the ratio between singles and coincidence data, shown in Figure 43.[158,159]

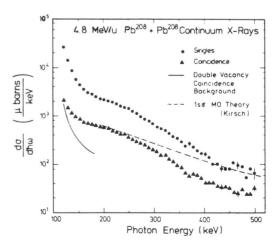

Figure 43. The x-ray continuum beyond the characteristic K-radiation for 4.8 MeV/N ^{208}Pb–^{208}Pb collisions measured directly (●) and in coincidence to the characteristic Pb-K radiation (▲) (Ref. 158). The calculated $1s\sigma$-MO spectrum from Refs. 137 and 138 is shown, too.

5.2.2. X-Ray/Particle Coincidences

For x-ray/x-ray coincidences the measurement still integrates over all possible trajectories. Hence, measuring the impact parameter dependence of the MO x-ray emission, more differential information is expected as was also the case for the characteristic radiation. Fortunately, for inner-shell MO x-ray emission from superheavy collision systems a single collision mechanism prevails.[20] That means an inner-shell vacancy formed during a collision will decay radiatively during the same collision. This is in contrast to lighter collision systems, where, due to the smaller transition rates a double collision mechanism dominates, that is, a vacancy formed in a first collision has a chance to survive to a second collision, where it can be transferred into an inner MO level and there causes a MO x-ray transition (see, e.g., Ref. 11).

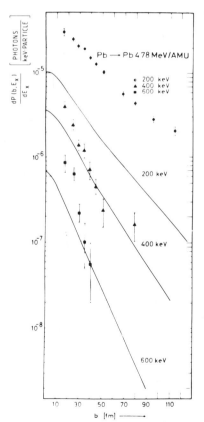

Figure 44. Impact parameter dependence for the quasimolecular radiation from 4.78 MeV/N Pb collisions at 200, 400, and 600 keV x-ray energy (Ref. 160). The corresponding calculations are taken from Ref. 138.

Due to the large $1s\sigma$ vacancy creation probabilities for superheavy collision systems at small R_1 (see Section 4) and due to the short (radiative) lifetime of the $1s\sigma$ vacancy, (on the average roughly a percent of $1s\sigma$ vacancies decay radiatively during the collision for superheavy systems) the MO x-ray impact parameter dependence is a measurable quantity. In Figure 44 first preliminary results for the K-MO radiation from 4.78 MeV/N Pb–Pb collisions are given.[159] The impact parameter dependence for MO x rays having about 200, 400, and 600 keV energy is plotted; compare, e.g., the MO singles spectrum in Figures 41 and 43 for the given x-ray energies. The higher the MO x-ray energy, the steeper the impact parameter dependence. From the arguments given in Section 4 immediately follows that in order to get higher MO x-ray energies, closer collisions have to be involved, and that the R dependence of the $1s\sigma$ binding energy govern the falloff width.

In Figure 44 the preliminary results are compared with available predictions.[138] A more recent experiment[161] gave better results, i.e., better statistics also at large impact parameters. Now, a reasonably good agreement between experiment and theory seems to result, indicating correctly predicted large binding energies in the quasimolecule. A separation of radiative transitions to the $1s\sigma$ and $2p_{1/2}\sigma$ levels seems to be possible as both give a different impact parameter dependence. In the same experiment also triple coincidences have been measured, i.e., MO X ray/characteristic X ray/scattered particles coincidences. Consequently, the full MO x-ray information can be extracted for Pb–Pb collisions now and will hopefully give new insights into the superheavy quasimolecule with united Z of 164.

6. Spectroscopy of Superheavy Quasimolecules

One of the main aims in all the studies presented so far is to get access to a "spectroscopy" of the innermost bound electrons in superheavy quasimolecules or quasiatoms. Talking about spectroscopy, one should always keep in mind that one does not talk about spectroscopy in a classical sense on an object which is more or less at rest in space during the time of investigation. Spectroscopy on quasiatoms or quasimolecules implies the handling of extremely short-lived objects whose features are strongly determined by the dynamics of the collision. It follows from the uncertainty principle that for typical collision times of $(10^{-19}-10^{-20})$ sec, the uncertainty in energy is of the order of (6.6–66) keV and, thus, not always small compared to the adiabatic binding energy.

Accepting this inerent, systematic uncertainty the steps towards a spectroscopy of superheavy quasimolecules are mainly based upon the surprisingly good agreement between the experimental data for the impact parameter dependence of the $1s\sigma$ excitation and the results of Eqs. (36) and (37). Here, three approaches have been proposed: [51,56,57,60,62,65]

1. Extraction of $E_{1s\sigma}(R)$ from measured $P_{1s\sigma}(b)$ data; [56,57,60,62]
2. Fixing lower bounds to $E_{1s\sigma}(R)$ from the slope of $P_{1s\sigma}(b)$; [65]
3. Determination of $E_{1s\sigma}(R)$ from experimental data on the mean impact parameter $\bar{b}$. [51]

(a) *ad 1.* Since the value of the function $D(Z)$ which determines the absolute value of the $1s\sigma$ excitation is not known up to now with the necessary high accuracy of a few percent, a direct extraction of $E_{1s\sigma}(R_0)$ from $P_{1s\sigma}(b)$ date via Eqs. (36) and (37) is not possible. This problem can be overcome by a measurement of $P_{1s\sigma}(b)$ at different projectile velocities v_i, v_j, and at different impact parameter b_i, b_j, such that

$$a_i + (a_i^2 + b_i^2)^{1/2} = a_j + (a_j^2 + b_j^2)^{1/2} = VR_0 \tag{41}$$

where a_j and a_j are given by Eq. (2) for v_i and v_j, correspondingly. Under the reasonable assumption that $E_{1s\sigma}(R_0)$ does not depend on v, one obtains from Eqs. (36) and (37) for the $1s\sigma$ binding

$$E_{1s\sigma}(R_0) = \frac{\hbar v_i v_j \ln\left[P_{1s\sigma}(b_i)/P_{1s\sigma}(b_j)\right]}{2R_0(v_i - v_j)} \tag{42}$$

The errors $\Delta E_{1s\sigma}(R_0)$ introduced by this method may become considerably large, because they depend mainly on $v_i - v_j$:

$$\Delta E_{1s\sigma}(R_0) = \left\{\left[\frac{\Delta P_{1s\sigma}(b_i)}{P_{V1s\sigma}(b_i)}\right]^2 + \left[\frac{\Delta P_{1s\sigma}(b_j)}{P_{1s\sigma}(b_j)}\right]^2\right\}^{1/2} \cdot \frac{\hbar}{2R_0(1/v_j - 1/v_i)} \tag{43}$$

where $\Delta P_{1s\sigma}(b_i)$ and $\Delta P_{1s\sigma}(b_j)$ are the experimental errors in $P_{1s\sigma}(b_i)$ and $P_{1s\sigma}(b_j)$, respectively. Now typically $\Delta P_{1s\sigma}/P_{1s\sigma} \approx 0.1$, and one obtains for the typical projectile energies of 3.6 MeV/N and 5.9 MeV/N

$$\Delta E_{1s\sigma}(R_0) = 0.07 \frac{\hbar c}{R_0(c/v_j - c/v_i)} \approx \frac{5.5\,[\text{MeV f}]}{R_0[\text{f}]} \tag{44}$$

From Eq. (43) follows that $\Delta E_{1s\sigma}(R_0)$ is independent of $E_{1s\sigma}(R_0)$. Thus, the relative error $\Delta E_{1s\sigma}(R_0)/E_{1s\sigma}(R_0)$ becomes large at small values of the binding energy.

As an example, we consider in more detail the highest-Z system Pb + Cm, where the strongest $1s\sigma$ binding is excepted. In Figure 45, $P_{1s\sigma}$

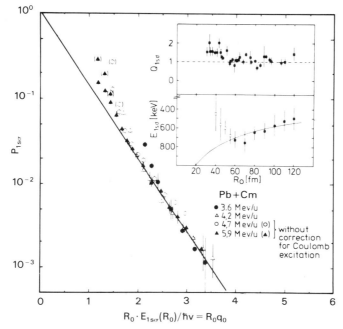

Figure 45. $1s\sigma$ excitation probability for Pb + Cm collisions as a function of $R_0 \cdot q_0$ at different projectile energies (Ref. 57). The data in brackets compared to those without brackets at the same values of $R_0 q_0$ give the contribution of nuclear Coulomb excitation to the atomic vacancy production. The solid line is the prediction of Eq. (36). The inset is explained in the text.

from Figure 24 is plotted vs. $R_0 q_0$ for four different projectile energies. The data in brackets indicate the contribution of the nuclear Coulomb excitation to the Cm K vacancy production probability determined as described in Section 2.4. For large values of $R_0 q_0$, all the data follow the prediction of Eqs. (36) and (37), which is given by the solid line. However, the experimental data exceed the calculated ones for all projectile velocities below $R_0 \approx 50$ f. This is shown in the upper part of the insert of Figure 45 which gives the ratio Q of the measured to the calculated $P_{1s\sigma}$ data as a function of R_0. From Eq. (42) calculated values of $E_{1s\sigma}(R_0)$ are plotted in the lower part of the insert. Clearly, for $R_0 > 50$ f, the agreement between the full points and the theoretical calculation (full curve from Ref. 124) is good; this fact demonstrates convincingly the extreme strong binding ($\geqslant 500$ keV) of the $1s\sigma$ electron in a system with $Z = 178$. For $R_0 \leqslant 50$ f the "experimental" binding energies (open symbols) indicate a much weaker binding than calculated. It is not clear, whether this discrepancy is caused by dynamical effects (e.g., due to the extremely strong dependence of $dE_{1s\sigma}/dR_0$ on R_0 at small internuclear distances) or by other contributions not considered in detail up to now (for example the unknown self-energy of the $1s\sigma$ electron in these strong fields may weaken the binding.[162] It must be emphasized that the simple prescription which has been used to derive "experimental" binding energies, cannot give an unambiguous conclusion about $E_{1s\sigma}$ at small R_0. An agreement between the experimental results and the predictions of Eqs. (36) and (37) confirms only the consistency of the model, because the calculation of the coupling matrix element [Eq. (29)] and of the binding energy [Eq. (38)] are based on the same theoretical treatment of superheavy collision systems.[1,63] Figure 46 gives $1s\sigma$ binding energies from Eq. (42) as a function of Z for the two internuclear distances 50 and 100 f, respectively, in comparison to theoretical results from Ref. 124 (dashed–dotted and dashed line); the solid line shows the Z dependence of an atomic $1s$ state.[5] Obviously, the experimental results are in very good agreement with the theoretical calculations for not too small values of R_0, and again confirm the extremely strong binding in superheavy quasimolecules.

(b) *ad 2.* The second approach towards a spectroscopy of the innermost levels gives a lower bound to $E_{1s\sigma}(R_0)$. From Eqs. (36) and (37) it follows that

$$-\frac{d}{dR_0}[\ln P_{1s\sigma}(b)] = \frac{2}{\hbar v}E_{1s\sigma}(R_0) + \frac{2R_0}{\hbar v}\frac{d}{dR_0}[E_{1s\sigma}(R_0)] \qquad (45)$$

If both $P_{1s\sigma}(b)$ and $E_{1s\sigma}(R_0)$ monotonically decrease with increasing R_0, then

$$-\frac{d}{dR_0}[\ln P_{1s\sigma}(b)] > 0 \quad \text{and} \quad \frac{d}{dR_0}[E_{1s\sigma}(R_0)] < 0 \text{ always}$$

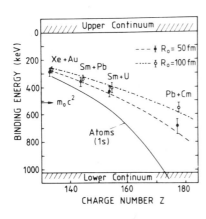

Figure 46. $1s\sigma$ binding energies as a function of Z calculated using Eq. (42) from the data in Figures 24–26; the dashed–dotted and the dashed line are theoretical results from Ref. 124. The solid line gives the Z dependence of the $1s$ state in a superheavy atom (Ref. 5).

and, thus, a lower bound to $E_{1s\sigma}(R_0)$ is implied by

$$\left| \frac{d}{dR_0} [\ln P_{1s\sigma}(b)] \right| < \frac{2}{\hbar v} E_{1s\sigma}(R_0)$$

or   (46)

$$E_{1s\sigma}(R_0) > \frac{\hbar v}{2} \left| \frac{d}{dR_0} \ln [P_{1s\sigma}(b)] \right|$$

For the exponential function given in Eq. (12) one gets for not too small R_0 where $R_0 \approx b$

$$E_{1s\sigma}(R_0) > \frac{\hbar v}{2a} \qquad (47)$$

Equations (45)–(47) show that a direct conclusion from the slope of the $P_{1s\sigma}(b)$ curve to the $1s\sigma$ binding energy[163] only holds, if $E_{1s\sigma}(R_0)$ is a monotonically decreasing function of R_0. Figure 47 shows lower bounds to the $1s\sigma$ binding from Eq. (46) for the system Xe + Au at four different projectile velocities; the solid lines are theoretical values from Ref. 124.

(c) *ad 3*. The third method starts from the mean impact parameter $\bar{b}$ and the corresponding mean distance of closest approach

$$\bar{R}_0 = R_0(\bar{b}) = a_0 + (a_0^2 + \bar{b}^2)^{1/2} \qquad (48)$$

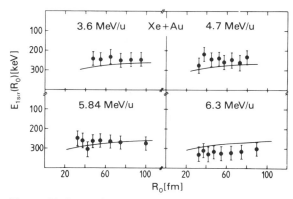

Figure 47. Lower bound to the $1s\sigma$ binding energy as a function of R_0 as calculated from Eq. (46) for different projectile energies (Ref. 65). The solid lines give static values of the binding energies and are taken from Ref. 124.

Using the mean momentum transfer q_0 [cf. Eqs. (32)–(34)]

$$\bar{q}_0 = q_0(\bar{R}_0) = \frac{E_{1s\sigma}(\bar{R}_0)}{\hbar v} \tag{49}$$

one finds from Eqs. (11), (36), and (37) after some algebra

$$\bar{b} \cdot q_0 = \xi^2 \frac{K_2(\xi) \exp(\xi)}{2(1+\xi)} \tag{50}$$

with $\xi = 2a_0 \cdot \bar{q}_0$ and $K_2(\xi)$ a modified Bessel function.[125] Since in most experimental cases

$$\xi = \frac{2a_0}{\hbar v} E_{1s\sigma}(R_0) \approx 1[\text{MeV}^{-1}] \cdot E_{1s\sigma}(\bar{R}_0) < 1 \tag{51}$$

one gets by a series expansion of the right-hand side in Eq. (50)

$$\bar{b} \cdot \bar{q}_0 \approx 1 + \tfrac{1}{2}\xi^2 - \tfrac{1}{3}\xi^3 \pm \cdots \tag{52}$$

Equations (50) and (51) give reasons for the intuitive suggestion made in Ref. 51, namely, to use the Bang–Hansteen scaling rule [Eq. (33)] for a spectroscopy of quasimolecular orbitals by setting

$$\bar{b} \cdot \bar{q}_0 = c \tag{33}$$

with c being a constant in the order of unity. This constant has been calibrated experimentally from the impact parameter dependence of atomic positron production at a fixed final kinetic energy E_f.[117] Then the momentum transfer from Eqs. (31a) and (31b)

$$q_0 = \frac{\Delta E}{\hbar v} = \frac{2mc^2 + E_f}{\hbar v} \tag{53}$$

is known and $\bar{b}$ is obtained from the experimental data. The result given in Ref. 117 is

$$c = 1.12 \pm 0.15$$

for the three systems Pb + Pb, Pb + U, U + U at 5.9 MeV/N projectile energy. For these three systems one gets from Eq. (50) with q_0 known from Eq. (53)

$$c = \begin{cases} 1.12 & \text{for U + U} \quad (\xi = 1.19) \\ 1.10 & \text{for U + Pb} \quad (\xi = 1.07) \\ 1.10 & \text{for Pb + Pb} \; (\xi = 1.01) \end{cases}$$

A simple expression for $E_{1s\sigma}(\bar{R}_0)$ independent of the actual value of c can be obtained from Eq. (51). Setting on the right side of Eq. (51) in zeroth order, $\bar{q}_0 = 1/\bar{b}$ and, thus, $\xi = 2a_0/\bar{b}$, one obtains

$$E_{1s\sigma}(\bar{R}_0) \approx \frac{\hbar v}{\bar{b}} \left(1 + \frac{2a_0^2}{\bar{b}^2} \right) \tag{54}$$

Higher orders can be obtained by an iterative procedure in which, from Eqs. (54) and (49), new values for $\bar{q}_0$ are successively calculated.

Table 3 gives the results calculated from Eq. (54) for the systems Xe + Au, Sm + Pb, and Sm + U. For comparison, the theoretical values $E_{1s\sigma}(\bar{R}_0)_{th}$ from Ref. 124 are also given. The experimental values for $E_{1s\sigma}(\bar{R}_0)$ clearly demonstrate the increase of the $1s\sigma$ binding with increasing Z. Unfortunately, it is not possible to derive reliable $\bar{b}$ values for the heaviest system Pb + Cm. An analysis of the corresponding data in Figure 24 shows that due to the very steep b dependence, a considerable part of the $1s\sigma$ excitation occurs at impact parameters smaller than the measured ones.

Table 3. $1s\sigma$ Binding Energies $E_{1s\sigma}(\bar{R}_0)$ Extracted from Mean Impact Parameters $\bar{b}^a$

System	Energy (MeV/N)	$\bar{b}$ (f)	$\bar{R}_0$ (f)	$E_{1s\sigma}(\bar{R}_0)$ (keV)	$E_{1s\sigma}(\bar{R}_0)_{th}$ (keV)
Xe + Au	3.6	74 ± 4	86 ± 4	243 ± 16	264
	4.7	72 ± 4	81 ± 4	281 ± 17	266
	5.84	71 ± 4	78 ± 4	315 ± 18	267
	6.3	71 ± 4	77 ± 4	325 ± 20	267
Sm + Pb	3.6	56 ± 6	69 ± 6	335 ± 40	343
	4.75	74 ± 4	83 ± 5	276 ± 25	333
	5.9	68 ± 4	75 ± 6	332 ± 25	339
Sm + U	3.6	54 ± 5	68 ± 5	353 ± 40	420
	4.75	60 ± 6	70 ± 6	350 ± 45	417
	5.9	60 ± 6	68 ± 7	380 ± 45	420

^{a} $\bar{R}_0$ is the corresponding mean distance of closest approach. The theoretical values $E_{1s\sigma}(\bar{R}_0)_{th}$ are taken from Ref. 124.

Certainly, at the moment, one is only dealing with the first steps towards a spectroscopy of innermost levels in superheavy collision systems, where the (surprisingly successful) approaches are based on quite simple physical pictures of the excitation processes. The reasons for the success of these first attempts and also the limits of the underlying models are not well understood and have still to be revealed more clearly.

The methods summarized in this chapter for extracting binding energies in superheavy quasimolecules are based on the characteristic x-ray emission. The original hope of extracting such values directly from the quasimolecular radiation—either from the spectral shape or from anisotropy spectra—did not yield a satisfying result for superheavy quasimolecules, despite the fact that the overall picture of transient high binding energies was confirmed. The most recent approach[161] to measuring the impact parameter dependence of MO radiation may change this fact drastically. Hopefully, detailed spectroscopic information on superheavy quasimolecules will also result here. Additional information on the strongly bound electrons in the quasimolecule results from an investigation of the electrons and positrons emitted during the atomic collision process; see Figure 1. The results are described in Chapter 10 by H. Backe and Ch. Kozhuharov in this volume.

Acknowledgments

One of us, D.L., gratefully acknowledges the hospitality of his colleagues during a one year's stay at the FOM Institute, Amsterdam, where several parts of this article were written. The many critical discussions on the subject and the comments on the text by Frans Saris are especially appreciated.

References

1. J. Reinhardt and W. Greiner, *Rep. Progr. Phys.* **40**, 219 (1977).
2. U. Fano and W. Lichten, *Phys. Rev. Lett.* **14**, 627 (1965).
3. E. Kankeleit, private communication.
4. B. Fricke, in *Progress in Atomic Spectroscopy*, Part A, p. 183, Plenum Press (1978).
5. B. Fricke and G. Soff, *At. Data Nucl. Data Tables* **19**, 83 (1977).
6. S. J. Brodsky and P. J. Mohr, in *Structure and Collisions of Ions and Atoms* ed. J. A. Sellin, *Topics in Current Physics V*, Springer-Verlag, New York (1975), p. 3.
7. G. Soff and B. Müller, *Z. Phys.* **A280**, 243 (1977).
8. G. Soff, thesis, University of Frankfurt (1976).
9. G. Herrmann, *Nature* **280**, 543 (1979), and references therein.
10. P. Armbruster, G. Kraft, P. H. Mokler, B. Fricke and H. J. Stein, *Phys. Scr.* **10A**, 175 (1974).

11. P. H. Mokler and F. Folkmann, in *Structure and Collisions of Ions and Atoms*, ed. I. A. Sellin, *Topics in Current Physics V*, p. 201, Springer-Verlag, New York (1978).

12. Q. Kessel and B. Fastrup, *Case Stud. At. Phys.* **3**, 137 (1973).

13. B. Müller and W. Greiner, *Z. Naturforsch.* **31a**, 1 (1976).

14. G. Soff, W. Betz, B. Müller, W. Greiner and E. Merzbacher, *Phys. Lett.* **65A**, 19 (1978).

15. B. Fricke, T. Morović, W.-D. Sepp, A. Rosen and D. E. Ellis, *Phys. Lett.* **59A**, 375 (1976).

16. W.-Sepp, B. Fricke, and T. Morović, *Phys. Lett.* **81A**, 258 (1981).

17. B. Fricke, private communication; see also Ref. 97.

18. W. Bambynek, B. Craseman, Z. W. Fink, H.-H. Freund, H. Mack, C. D. Swift, R. E. Price, and P. Rao, *Rev. Mod. Phys.* **44**, 716 (1972); M. O. Krause, *J. Phys. Chem. Ref. Data* **8** 307 (1979).

19. W. E. Meyerhof, D. L. Clark, Ch. Stoller, E. Morenzoni, W. Wölfli, F. Folkmann, P. Vincent, P. H. Mokler, and P. Armbruster, *Phys. Lett.* **70A**, 303 (1979).

20. Ch. Stoller, E. Morenzoni, W. Wölfli, W. E. Meyerhof, F. Folkmann, P. Vincent, P. H. Mokler, and P. Armbruster, *Z. Phys.* **A297**, 93 (1980).

21. N. Luz, S. Sackmann, and H. O. Lutz, *J. Phys.* **B12**, 1973 (1979).

22. R. Schuch, G. Nolte, and H. Schmidt-Böcking, *Phys. Rev. A* **22**, 1447 (1980), and references therein.

23. R. Schuch, H. Schmidt-Böcking, R. Schulé, and J. Tserruya, *Phys. Rev. Lett.* **39**, 79 (1977).

24. D. Liesen, J. R. Macdonald, P. H. Mokler, and A. Warczak, *Phys. Rev. A* **17**, 897 (1978).

25. R. Anholt, Ch. Stoller, and W. E. Meyerhof, *J. Phys.* **B13**, 3807 (1980).

26. S. Hagmann, P. Armbruster, H.-H. Behncke, D. Liesen, and P. H. Mokler, *Z. Phys.* **A298**, 1 (1980).

27. R. Anholt, H.-H. Behncke, S. Hagmann, P. Armbruster, F. Folkmann, and P. H. Mokler, *Z. Phys.* **A289**, 349 (1979).

28. W. A. Schönfeldt, Thesis, University of Köln (1981), and report GSI-81-7 (Darmstadt, 1981), unpublished.

29. P. H. Mokler, H. J. Stein, and P. Armbruster, *Phys. Rev. Lett.* **29**, 827 (1972).

30. P. H. Mokler, *Phys. Rev. Lett.* **26**, 811 (1971).

31. H.-H. Behncke, P. Armbruster, F. Folkmann, S. Hagmann, J. R. Macdonald, and P. H. Mokler, *Z. Phys.* **A289**, 333 (1979).

32. W. A. Schönfeldt and P. H. Mokler, Abstracts to 11th Int. Conf. on the Physics of Electronic and Atomic Collisions, Kyoto (1979), p. 768, and references cited therein.

33. F. Folkmann, C. Gaerde, T. Huus, and K. Kemp, *Nucl. Instrum. Methods* **116**, 487 (1974).

34. F. Folkmann, in *Ion Beam Surface Layer Analysis*, eds. O. Meyer, G. Lieker, F. Käppler, Plenum Press, New York (1976), p. 695.

35. K. Ishii, S. Morita, and H. Tawara, *Phys. Rev. A* **13**, 131 (1976).

36. D. H. Jakubassa and M. Kleber, *Z. Phys.* **A273**, 29 (1975).

37. J. Reinhardt, G. Soff, and W. Greiner, *Z. Phys.* **A276**, 285 (1976).

38. H. P. Trautvetter, J. S. Greenberg, and P. Vincent, *Phys. Rev. Lett.* **37**, 202 (1976).

39. P. H. Mokler, P. Armbruster, F. Bosch, D. Liesen, D. Maor, W. A. Schönfeldt, H. Schmidt-Böcking, and R. Schuch, in *Inner-Shell and X-Ray Physics of Atoms and Solids*, Eds. D. J. Fabieau, H. Kleinpoppen, and L. M. Watson, pp. 49–62, Plenum Press, New York (1981).

40. P. Vincent, C. K. Davis, and J. S. Greenberg, *Phys. Rev. A* **18**, 1878 (1978).

41. H. J. Stein, H. O. Lutz, P. H. Mokler, and P. Armbruster, *Phys. Rev. A* **5**, 2126 (1972).

42. H. Schmidt-Böcking, I. Tserruya, H. Zeckl, and K. Bethge, *Nucl. Instrum. Methods* **120**, 329 (1974).

43. D. Schwalm, A. Bamberger, P. G. Bizetti, B. Povh, G. A. P. Engelbertink, J. W. Olness, and E. K. Warburton, *Nucl. Phys.* **A192**, 449 (1972).

44. J. S. Greenberg, H. Bokemeyer, H. Emling, E. Grosse, D. Schwalm, and F. Bosch, *Phys. Rev. Lett.* **39**, 1404 (1977).

45. G. Gaukler, H. Schmidt-Böcking, R. Schuch, R. Schulé, H. J. Specht, and I. Tserruya, *Nucl. Instrum. Methods* **141**, 115 (1977).

46. J. Stähler and G. Presser, *Nucl. Instrum. Methods* **176**, 79 (1980); **177**, 427 (1980).

47. H.-H. Behncke, D. Liesen, S. Hagmann, P. H. Mokler, and P. Armbruster, *Z. Phys.* **A288**, 35 (1978).

48. E. Everhart, G. Stone, and R. J. Carbone, *Phys. Rev.* **99**, 1287 (1955); J. Lindhard, V. Nielsen, and M. Scharff, *Mat. Fys. Medd. Dan. Vid. Selsk*, **36**, 10 (1968); V. K. Nikulin, *Sov. Phys. Tech. Phys.* **16**, 21 (1971); F. W. Bingham, U.S. NBS Report No. SC-RR-66-506 (unpublished).

49. R. Anholt, W. E. Meyerhof, and Ch. Stoller, *Z. Phys.* **A291**, 287 (1979).

50. P. Fuchs *et al.*, GSI Jahresbericht 1977, J-1-78.

51. P. Armbruster, H.-H. Behncke, S. Hagmann, D. Liesen, F. Folkmann, and P. H. Mokler, *Z. Phys.* **A288**, 277 (1978).

52. Hagmann, P. Armbruster, F. Folkmann, G. Kraft, and P. H. Mokler, *Phys. Lett.* **A61**, 451 (1977).

53. R. Anholt, *Phys. Rev. A* **17**, 834 (1978).

54. J. R. Macdonald, P. Armbruster, H.-H. Behncke, F. Folkmann, S. Hagmann, D. Liesen, P. H. Mokler, and A. Warczak, *Z. Phys.* **A284**, 57 (1978).

55. D. Liesen, P. Armbruster, H.-H. Behncke, and S. Hagmann, *Z. Phys.* **A288**, 417 (1978).

56. D. Liesen, P. Armbruster, H.-H. Behncke, F. Bosch, S. Hagmann, P. H. Mokler, H. Schmidt-Böcking, and R. Schuch, Progress Report at the XI Conference on Electronic and Atomic Collisions, Kyoto 1979, N. Oda and K. Takayanagi, ed., pp. 337 (1979).

57. D. Liesen, P. Armbruster, F. Bosch, S. Hagmann, P. H. Mokler, H. Schmidt-Böcking, R. Schuch, H. J. Wollersheim, and J. B. Wilhelmy, *Phys. Rev. Lett.* **44**, 983 (1980).

58. R. Anholt, W. E. Meyerhof, Ch. Stoller, and J. F. Chemin, Book of Contributions to the 7th International Conference on Atomic Physics, p. 92, Cambridge, Massachusetts (1980).

59. D. Liesen, P. Armbruster, F. Bosch, D. Maor, P. H. Mokler, H. Schmidt-Böcking, R. Schuch, and A. Warczak, Abstracts to the XIIth Int. Conf. on the Physics of Electronic and Atomic Collisions, p. 890, Gatlinburg (1981).

60. G. Soff, B. Müller, and W. Greiner, *Phys. Rev. Lett.* **40**, 540 (1978).

61. G. Soff, W. Betz, B. Müller, W. Greiner, and E. Merzbacher, *Phys. Lett.* **65A**, 19 (1978).

62. B. Müller, G. Soff, W. Greiner, and V. Ceausescu, *Z. Phys.* **A285**, 27 (1978).

63. G. Soff, W. Greiner, W. Betz, and B. Müller, *Phys. Rev. A* **20**, 169 (1979), and references therein.

64. J. Reinhardt, B. Müller, W. Greiner, and G. Soff, *Phys. Rev. Lett.* **43**, 1307 (1979).

65. F. Bosch, D. Liesen, P. Armbruster, D. Maor, P. H. Mokler, H. Schmidt-Böcking, and R. Schuch, *Z. Phys.* **A296**, 11 (1980).

66. J. Bang and J. M. Hansteen, *Mat. Fys. Medd. Dan. Vid. Selsk.* **31**, No. 13 (1959).

67. R. S. Hager and E. C. Seltzer, *At. Data Nucl. Data Tables* **A4**, 1 (1968).

68. F. Rösel, H. M. Fries, K. Alder, and H. C. Pauli, *At. Data Nucl. Data Tables* **21**, 292 (1978).

69. A. Winther and J. de Boer, in *Coulomb Excitation*, ed. by K. Alder and A. Winther, Academic Press, New York (1966), pp. 53.

70. J. R. Macdonald, R. Schulé, R. Schuch, H. Schmidt-Böcking, and D. Liesen, *Phys. Rev. Lett.* **40**, 1330 (1978).

71. J. S. Greenberg, P. Vincent, and W. Lichten, *Phys. Rev. A* **16**, 964 (1977).

72. P. Vincent, private communication.
73. J. U. Andersen, L. Kocbach, E. Laegsgaard, M. Lund, and C. D. Mack, *J. Phys. B* **9**, 3247 (1976).
74. D. Liesen, A. N. Zinoviev, and F. W. Saris, *Phys. Rev. Lett.* **47**, 1392 (1981).
75. Y. N. Demkov, *Sov. Phys. JETP* **18**, 138 (1969).
76. W. E. Meyerhof, *Phys. Rev. Lett.* **31**, 1341 (1973).
77. E. E. Nikitin, *Adv. Quant. Chem.* **5**, 135 (1970).
78. J. S. Briggs, Technical Report No. 594, Harwell, Oxfordshire (1974) (unpublished).
79. H. J. Specht, *Z. Phys.* **185**, 301 (1965).
80. W. E. Meyerhof, R. Anholt, and T. K. Saylor, *Phys. Rev. A* **16**, 169 (1978).
81. W. E. Meyerhof, *Phys. Rev. A* **18**, 414 (1978).
82. E. Merzbacher and H. W. Lewis, in *Handbuch der Physik*, Vol. 34, ed. S. Flügge, Springer-Verlag, Berlin (1958), pp. 166–192.
83. J. D. Garcia, R. J. Fortner, and T. M. Kavanagh, *Rev. Mod. Phys.* **45**, 111 (1973).
84. D. H. Madison and E. Merzbacher, in *Atomic Inner Shell Processes*, ed. B. Crasemann, Academic Press, New York (1975), p. 1.
85. J. M. Hansteen, O. M. Johnsen, and L. Kocbach, *At. Data Nucl. Data Tables* **15**, 305 (1975).
86. W. Brandt and G. Lapicki, *Phys. Rev. A* **20**, 465 (1979).
87. H. Paul, *At. Data Nucl. Data Tables* **24**, 243 (1979).
88. W. E. Meyerhof, T. K. Saylor, S. M. Lazarus, A. Little, R. Anholt, and L. F. Chase, *Phys. Rev. A* **14**, 1653 (1976).
89. S. Hagmann, P. Armbruster, G. Kraft, P. H. Mokler, and H.-J. Stein, *Z. Phys.* **A288**, 353 (1978).
90. R. Anholt and W. E. Meyerhof, *Phys. Rev. A* **16**, 190 (1977).
91. J. S. Briggs, *J. Phys.* **B8**, L485 (1975).
92. C. Foster, Th. Hoogkamer, P. Woerlee, and F. W. Saris, *J. Phys.* **B9**, 1943 (1976).
93. W. N. Lennard, I. V. Mitchell, and J. S. Forster, *Phys. Rev. A* **18**, 1949 (1978).
94. V. Sethu Raman, W. R. Thorson, and C. F. Lebeda, *Phys. Rev. A* **8**, 1316 (1973).
95. G. Soff, W. Betz, G. Heiligenthal, and W. Greiner, *Fizika* **9**, Suppl. 4, 721 (1977).
96. G. Soff, J. Reinhardt, B. Müller, and W. Greiner, *Z. Phys.* **A294**, 137 (1980).
97. M. Mann, P. H. Mokler, B. Fricke, W.-D. Sepp, W. A. Schönfeldt, and H. Hartung, *J. Phys. B* **15**, 4199 (1982).
98. R. Anholt, *Phys. Rev. A* **17**, 983 (1978).
99. B. H. Choi, E. Merzbacher, and G. S. Khandelwal, *At. Data Nucl. Data Tables* **5**, 291 (1973).
100. G. S. Khandelwal, B. H. Choi, and E. Merzbacher, *At. Data nucl. Data Tables* **1**, 103 (1969).
101. W. Betz, G. Soff, B. Müller, and W. Greiner, *Phys. Rev. Lett.* **37**, 1046 (1976).
102. V. Zoran, A. Beriude, and D. Fluerasu, *J. Phys. B* **15**, 2027 (1982).
103. G. Soff, private communication; the details of the calculations are described in Ref. 14.
104. D. Burch, W. B. Ingells, H. Wiemann, and R. Vandenbosch, *Phys. Rev. A* **10**, 1254 (1974).
105. P. H. Mokler, F. Bosch, D. Liesen, W. A. Schönfeldt, and R. Schuch, Abstracts to 7th Int. Conf. on Atomic Physics, Mass. Inst. Technology (1980), p. 94.
106. K. Taulbjerg, J. S. Briggs, and J. Vaaben, *J. Phys.* **B9**, 1351, and references therein.
107. J. S. Briggs, *Rep. Prog. Phys.* **39**, 217 (1976).
108. W. E. Meyerhof and K. Taulbjerg, *Ann. Rev. Nucl. Sci.* **27**, 279 (1977).
109. J. S. Briggs and K. Taulbjerg, *J. Phys.* **B8**, 1909 (1975).
110. J. S. Briggs and J. H. Macek, *J. Phys.* **35**, 579 (1972).
111. G. Heiligenthal, W. Betz, G. Soff, B. Müller, and W. Greiner, *Z. Phys.* **A285**, 105 (1978).

112. G. Soff, B. Müller, and W. Greiner, *Z. Phys.* **A299**, 189 (1981).
113. L. D. Landau and E. M. Lifschitz, *Quantenmechanik*, Akademie-Verlag, Berlin (1974), p. 593 ff.
114. C. Kozhuharov, P. Kienle, D. Jakubassa, and M. Kleber, *Phys. Rev. Lett.* **39**, 540 (1977).
115. F. Bosch, H. Krumm, B. Martin, B. Povh, K. Traxel, and Th. Walcher, *Phys. Lett.* **B78**, 568 (1978).
116. E. Berdermann, F. Bosch, H. Bokemeyer, M. Clemente, F. Güttner, P. Kienle, W. Koenig, C. Kozhuharov, B. Martin, B. Povh, H. Tsertos, and Th. Walcher, GSI Scientific Report 1979, ISSN 0174-0814.
117. P. Armbruster and P. Kienle, *Z. Phys.* **A291**, 399 (1979).
118. J. Bang and J. M. Hansteen, *Phys. Lett.* **72A**, 217 (1979).
119. J. Bang and J. M. Hansteen, Nordita Preprint (1980), ISSN 0106-2646.
120. I. Tserruya, B. M. Johansson, and K. W. Jones, *Phys. Rev. Lett.* **45**, 894 (1980).
121. E. L. Church and J. Weneser, *Phys. Rev.* **103**, 1035 (1956).
122. R. J. Lombard, C. F. Perdrisat, and J. H. Brunner, *Nucl. Phys.* **A110**, 41 (1968).
123. E. Kankeleit, "Collisional Quasiatoms, Facts and Speculations," Invited Lecture at the XII Summer School on Nuclear Physics, Mikolajki, Poland 1979.
124. G. Soff, J. Reinhardt, W. Betz, and J. Rafelski, *Phys. Scr.* **17**, 417 (1978).
125. M. Abramowitz and J. A. Stegun, *Handbook of Mathematical Functions*, Dover Publ., New York (1965).
126. W. M. Coates, *Phys. Rev.* **46**, 542 (1934).
127. F. W. Saris, W. F. van der Weg, H. Tawara, and R. Laubert, *Phys. Rev. Lett.* **28**, 717 (1972).
128. F. W. Saris, I. V. Mitchell, D. C. Santry, J. A. Davies, and R. Laubert, in *Proc. of Int. Conf. on Inner Shell Ionization Phenomena and Future Applications* (eds. R. W. Fink et al.), Atlanta (1972), p. 1255.
129. P. H. Mokler, H.-J. Stein, and P. Armbruster, *Phys. Rev. Lett.* **29**, 827 (1972).
130. P. H. Mokler, H.-J. Stein, and P. Armbruster, in *Proceedings of the International Conferences on Inner Shell Ionization Phenomena and Future Applications*, (eds. R. W. Fink et al.), Atlanta (1972), p. 1283.
131. J. S. Briggs and J. Macek, *J. Phys.* **B7**, 1312 (1974).
132. W. Lichten, *Phys. Rev. A* **9**, 1458 (1974).
133. P. H. Mokler, S. Hagmann, P. Armbruster, G. Kraft, H.-J. Stein, K. Rashid, and B. Fricke, in *Atomic Physics* 4 (eds. G. zu Putlitz, E. W. Weber, and A. Winnacker), Plenum Press, New York (1975), p. 301.
134. P. H. Mokler, P. Armbruster, F. Folkmann, S. Hagmann, G. Kraft, and H.-J. Stein, in *The Physics of Electronic and Atomic Collisions IX* (eds.: J. S. Risley and R. Geballe), University of Washington Press (1976), p. 501.
135. F. Folkmann, P. Armbruster, S. Hagmann, G. Kraft, P. H. Mokler, and H.-J. Stein, *Z. Phys.* **A276**, 15 (1976).
136. M. P. Stöckli, thesis, Eidgen. Techn. Hochschule Zürich (1979), Diss ETH 6299.
137. J. Kirsch, W. Betz, J. Reinhardt, G. Soff, B. Müller, and W. Greiner, *Phys. Lett.* **B72**, 298 (1978).
138. J. Kirsch, W. Betz, J. Reinhardt, B. Müller, W. Greiner, and G. Soff, *Z. Phys.* **A292**, 227 (1979).
139. P. H. Mokler, G. Kraft, and H. J. Stein, in *Proc. of the Int. Conf. on Reactions between Complex Nuclei* (eds. R. L. Robinson, F. K. McGowan, J. B. Ball, and J. H. Hamilton), North-Holland, Amsterdam (1974), p. 134; see also G. Kraft, H.-J. Stein, P. H. Mokler, and S. Hagmann, *Verh. Dtsch. Phys. Ges.* (*VI*) **9**, 77 (1974).
140. G. Kraft, P. H. Mokler, and H.-J. Stein, *Phys. Rev. Lett.* **33**, 476 (1974).
141. J. S. Greenberg, C. K. Davis, and P. Vincent, *Proc. of the Int. Conf. on Reactions between*

Complex Nuclei (eds. R. L. Robinson, F. K. McGowan, J. B. Ball, and J. H. Hamilton), North-Holland, Amsterdam (1974), p. 135.

142. J. S. Greenberg, C. K. Davis, and P. Vincent, *Phys. Rev. Lett.* **33**, 473 (1974).

143. W. E. Meyerhof, T. K. Saylor, and R. Anholt, *Phys. Rev. A* **12**, 2641 (1975).

144. W. Frank, H.-H. Kaun, P. Manfrass, N. V. Pronin, and Y. P. Tretyakov, *Z. Phys.* **A279**, 213 (1976).

145. Ch. Stoller, W. Wölfli, G. Monani, M. Stöckli, and M. Suter, *J. Phys.* **B10**, L347 (1977).

146. W. Wölfli, E. Morenzoni, Ch. Stoller, G. Bonani, M. Stöckli, *Phys. Lett.* **A68**, 217 (1978).

147. Ch. Stoller, W. Wölfli, G. Bonani, E. Morenzoni, and M. Stöckli, *Z. Phys.* **A287**, 33 (1978).

148. E. Morenzoni, thesis, Eidgen. Techn. Hochschule Zürich (1981), Diss. ETH 6 777.

149. R. Anholt, *Z. Phys.* **A288**, 257 (1978).

150. B. Müller and W. Greiner, *Phys. Rev. Lett.* **33**, 469 (1974).

151. M. Gros, P. T. Greenland, and W. Greiner, *Z. Phys.* **A280**, 31 (1977).

152. J. S. Briggs and K. Dettmann, *J. Phys.* **B10**, 1113 (1977).

153. H. Hartung and B. Fricke, *Z. Phys.* **A288**, 345 (1978).

154. J. S. Briggs, J. H. Macek, and K. Taulbjerg, *J. Phys.* **B12**, 1457 (1979).

155. P. H. Mokler, W. N. Lennard, and I. V. Mitchell, *J. Phys.* **B13**, 4607 (1980).

156. J. Kirsch, thesis, University of Frankfurt (1980).

157. J. J. O'Brien, E. Liarokapis, and J. S. Greenberg, *Phys. Rev. Lett.* **44**, 386 (1980).

158. P. Vincent, in *Inner-Shell and X-Ray Physics of Atoms and Solids*, Eds. D. J. Fabieau, H. Kleinpoppen, and L. M. Watson, p. 117, Plenum Press, New York (1981).

159. K. E. Stiebing, K. Bethge, F. Bosch, S. Hagmann, D. Liesen, P. H. Mokler, W. Schadt, H. Schmidt-Böcking, R. Schuch, and P. Vincent, in *Book of Abstracts to the European Conference on Atomic Physics*, Heidelberg (1981), p. 799; and Report GSI-82-14 (Darmstadt, 1983).

160. R. Schuch, G. Nolte, H. Schmidt-Böcking, K. E. Stiebing, K. Jones, B. Johnson, P. H. Mokler, D. Liesen, S. Hagmann, and P. Armbruster, in *Report Energiereiche Atomare Stöße, EAS-1* (eds. J. Eichler, B. Fricke, R. Hippler, D. Kolb, H. O. Lutz, and P. H. Mokler), Kassel (1980), p. 61; see also K. E. Stiebing, thesis, University of Frankfurt (1983) and Ref. 159.

161. K. Bethge, F. Bosch, S. Hagmann, D. Liesen, D. Maor, P. H. Mokler, W. Schadt, H. Schmidt-Böcking, R. Schuch, K. E. Stiebing, and P. Vincent, in Report EAS-2 (eds. see Ref. 160), Kassel (1981), p. 68.

162. K. T. Cheng and W. R. Johnson, *Phys. Rev. A* **14**, 1943 (1976).

163. P. Kienle, Atomic Physics of High Z-Systems, Invited lecture presented at the *VII International Conference on Atomic Physics*, Cambridge (eds. D. Kleppner and F. M. Pipkin), Plenum Press, New York (1980), p. 1.

164. P. Gippner, K.-H. Kaun, W. Neubert, F. Stary, and W. Schulze, *Nucl. Phys.* **A245**, 336 (1975).

NOTE: Paper submitted Summer 1981.

9

Recoil Ion Spectroscopy
with Heavy Ions

H. F. BEYER AND R. MANN

1. Introduction

Violent ion–atom collisions resulting in the excitation of inner-shell electrons have been studied theoretically and experimentally for many years. Some limiting cases have been studied extensively, specifying collision parameters such as the collision energy and the atomic numbers of the projectile (Z_1) and the target (Z_2). Most of the early work was related to light-particle impact at moderately low collision energies.

Perturbative techniques have been applied to account theoretically for the ionization-excitation and electron capture by a point charge. Molecular orbital theory was most successful in describing the inner-shell vacancy production in heavy-ion–atom collisions at low impact velocities.

Inner-shell vacancy production in ion–atom collisions may be detected by several experimental techniques such as observation of characteristic photons or electrons emitted by the excited atom, measurement of the inelastic energy loss of the projectile,[1] or direct observation of ejected electrons.[2-4] In some experiments the gain or loss of charge from the projectile beam[5] is observed or coincidences between the scattered projectiles and emitted X rays or electrons[6-8] are detected. Other experiments are related to the processes of radiative electron capture[9-11] or the observation of quasimolecular X rays.[12-17]

H. F. BEYER AND R. MANN • Gesellschaft für Schwerionenforschung MBH, Darmstadt, Federal Republic of Germany.

K-vacancy production in light projectile asymmetric systems ($Z_1 \ll Z_2$) has been thoroughly investigated. Total K-shell ionization cross sections have been found to be in general agreement with direct Coulomb ionization theories such as the plane-wave Born approximation, PWBA,[18] the semiclassical approximation, SCA,[19] and the binary encounter approximation, BEA.[20]

There were also a number of investigations on light-target asymmetric systems where the traditional role of projectile and target was interchanged.[21-23] Total K x-ray production has been studied as a function of the incoming charge of the projectile. Projectiles carrying an initial K vacancy are of particular interest as the K–K electron transfer is a major channel to be studied.[24-31]

In recent years these limiting cases were left accumulating data on the more general case of collisions $A^{q+} + B$. With the advent of large heavy-ion accelerators the studies were extended to higher bombarding energies, higher charge states, and large atomic numbers of the incident ions.

It was possible to isolate some aspects of these systems rather than account for the physical state of all the many electrons involved in the collision. Most studies on K-vacancy production did not discriminate on all the final ionic states but only identified which collision partner had the K vacancy. A large number of experiments have been carried out on these collision systems and compared to some theoretical models. The magnitude and energy dependence of the measured cross sections have been compared with Coulomb ionization theories. Perturbative corrections to these theories have extended the range of validity of the models.[32,33]

The molecular-orbital model of Fano and Lichten[34] has been used to identify K-vacancy production via rotational coupling between $2p\sigma$ and $2p\pi$ molecular orbitals in symmetric[35] and asymmetric[36,37] systems. Superheavy systems ($Z_1 + Z_2 > 173$) have gained particular interest as the K shell ionization probability is governed by QED effects (See Chapter 8 in this volume by P. H. Mokler and D. Liesen).

Restriction on the limiting case of light-particle impact or identifying only single innermost vacancies in heavy-ion–atom collisions obscures the fact that in most of the inelastic collision regimes studied experimentally multiple ionization occurs. Additionally, when using solid targets, successive collisions may give rise to a high degree of multiple ionization-excitation. Extensive use of this fact has been made in the beam-foil spectroscopy where one is more interested in preparing an ion beam of any degree of ionization for deducing atomic lifetime data rather than understand the excitation processes themselves.

Similar high excitation states, however, may also result from a *single* collision when highly stripped heavy projectiles available from large accelerators are used. The kinetic energy imparted to the highly stripped target atom can be very small in comparison to the fast heavy ion of the same state of excitation produced by beam-foil interaction. Such experiments are within the scope of the present article.

We will review on the status of understanding of the production of very high charge state ions in single ion–atom collisions. We will restrict ourselves to the most recent developments in this field. The experimental methods which have been applied to investigate these collision systems will be introduced in Section 3. Section 4 is devoted to recent experimental results on the production of high charge low velocity ions whereas Section 5 is devoted to the new experiments which use highly charged recoil ions in secondary experiments.

2. Some Aspects of Highly Ionized Atoms

In early spectroscopic studies a bulk of atomic data (note, e.g., the tabulation by C. E. Moore[38] was accumulated mostly concerning atoms excited in their valence shells. Low states of ionization prevailed which were accessible by standard light sources[39] and light-particle impact. Allowed dipole transitions was the only deexcitation channel studied extensively. Such excitation regimes are by no means representative for excited atomic systems in nature since most of the mass in nature is found in star atmospheres, where the majority of elements are in high states of ionization-excitation.

With laboratory devices such as artificial plasmas or particle accelerators it has become possible to study highly ionized atoms as well. There is, however, an intermediate state in the development of these devices (moderate plasma temperatures and bombarding energies) where one has to face the problem of simultaneous production of many-charge states and new previously unobserved spectral lines causing complicated overlapping spectral features. These difficulties in the interpretation of many-electron systems are one reason for the necessity to excite few-electron–ion stages where the number of excited states is drastically reduced. This results in a relatively simple excitation pattern comparable to that of the few-hole states produced under the light-particle impact regimes.

It has been recognized that few-electron heavy ions produced, e.g., in the beam-foil source, provide a testing field for basic physical concepts, in particular the theory of quantum electrodynamics[40] [see also the contribution of H. J. Andrä (Chapter 20 of Part B of this work) and that of H. J. Beyer (Chapter 12 of Part A)]. A large fraction of the few-electron ions can be in states of high excitation energy and high electronic angular momentum. The relaxation of such excited states very often occurs by violation of the selection rules valid for allowed radiative and Auger processes. Higher-order radiative processes like spin–orbit and spin–spin interactions then govern the deexcitation, the strength of which rapidly increases with the nuclear charge yielding fairly large transition probabilities at high Z.

It is not only the possibility for studying atomic structure which makes the highly ionized atoms so attractive. Fundamental collisional and radiative processes important for the interpretation of emission spectra from natural and artificial plasma sources may be studied in accelerator based experiments. Fast-ion stripping in the beam-foil interaction has been recognized to be reciprocal to slow-ion stripping by fast electrons in a hot plasma.[41] A few-electron ion may, however, also be created by stripping off a large number of electrons in a single encounter with an energetic heavy ion. The kinetic energy of such a highly stripped recoil target ion in the laboratory may be small compared to its total electronic excitation energy, thus resembling much more the thermodynamic state of an ion in a hot plasma. Stripping of slow target atoms in single collisions and subsequent recapturing of electrons can thus be studied in a cold gaseous environment.

3. Experimental Approaches

The experimental techniques which have been used to study the recoil target ions produced by impact of fast highly stripped heavy projectiles, are described in this section. The heavy-ion beams have been provided by accelerators such as the University of Aarhus tandem Van de Graaff facility and the UNILAC at Darmstadt. Experiments have been reported employing projectiles ranging from hydrogenlike or bare swift heavy ions like F^{9+} [42] up to very heavy ions in high charge states like U^{40+}.[43,44] The specific energy of these ions has been in the range of 1 MeV/amu. Monatomic targets such as He, Ne, and Ar and some molecular targets for example N_2, O_2, and CH_4 have been investigated.

In general the design of a collision experiment suitable for investigation of recoil ions will be similar to those already depicted in previous chapters (Part B, Chapters 20 and 30) of this series. Few specific examples will be given below. Most of the data on the distribution of the electronic states produced by the collision are related to the x-ray and Auger decay of atoms carrying an initial K vacancy. The significance of this spectral range is due to the fact that multiple target ionization drastically increases in collisions close enough to ionize the K shell. Thus, high charge-state recoil ions may be studied most effectively by x-ray and Auger-electron spectroscopy. Another approach is to analyze the charge to mass ratio q/m of the recoiling target ions. This method disregards the electronic states of the various charge state ions but may efficiently yield complementary information on the recoil ion production mechanism.

3.1. Experimental Techniques

3.1.1. Auger-Electron Spectroscopy

Figure 1 shows a scheme of a typical gas collision experiment suitable for measuring electrons originating from single collisions of heavy ions with target gas atoms. A momentum and charge-state analyzed heavy ion beam enters the target vacuum chamber through a system of collimators, traverses the collision center, and gets finally stopped in a Faraday cup. The gas target is provided by a gas-jet device consisting of a tandem gas-inlet system, a nozzle, a gas reflector plate, and a gas exhaust. Such an arrangement, providing a high-pressure gradient near the collision center, maintains high vacuum in the target chamber when operating the gas jet near the 0.1 mbar pressure range. The use of a gas jet also allows the observation angle of an electron-energy analyzer to be continuously varied. Alternatively, an open gas cell may be used, having the advantage of a homogeneous gas density in the observed collision region but a fixed observation angle in most cases.

A spherical deflector analyzer operated in the retarding field mode as used by Mann et al.[45] has proved very useful in the analysis of Auger electrons emitted from light target ions. By using this device, electrons emitted into a certain acceptance cone are decelerated in a retarding field and focused by electrostatic lenses to the entrance slit of the deflector analyzer. The voltage between the hemispheres of the spherical deflector analyzer is kept constant whereas the voltage applied to the retarding grid is scanned to obtain a spectrum. By this method, the instrumental linewidth is independent of the primary electron energy in the range between 10 eV and 3 keV and may be as small as 1 eV or less.

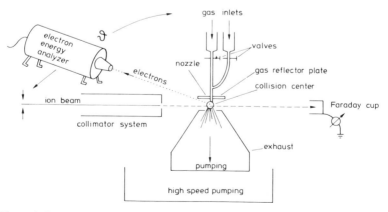

Figure 1. Experimental setup for measuring Auger electrons emitted from gas targets after excitation with heavy ions.

Electrons emitted by the fast-moving projectiles generally do not affect the analysis of the light-target Auger electrons because they are kinematically shifted by a rather large amount of energy depending on the observation angle. Although the laboratory kinetic energies of the recoil ions are small compared to those of the projectiles, kinematic effects in most cases are not negligibly small. The influence of the kinematics has been analyzed by Rudd et al.[46] and by Gordeev and Ogurtsov[47] and more recently by Dahl et al.[48] Most important is the Doppler effect caused by the motion of the particle from which the electron is emitted. It comprises energy shift, line broadening, and change of observed line intensity.

Let the electrons be emitted by a source particle of mass M which is scattered as a result of the collision, through a fixed angle β, and which has a finite kinetic energy T. The electron-energy analyzer is placed at an observation angle δ and let E_0 be the Auger electron energy in the emitter coordinate frame. For that condition Dahl et al.[48] derive an energy shift

$$E_{av} - E_0 = 2\left(\frac{m_e}{M} TE_{av}\right)^{1/2} \cos\delta \cos\beta - \frac{m}{M} T \cos^2\beta \qquad (1)$$

and a Doppler width

$$E_D = 4\left(\frac{m_e}{T} TE_0\right)^{1/2} \sin\beta \sin\delta \qquad (2)$$

The Doppler profile $D(E)$ is given by

$$D(E) \propto (E/E_0)[(\tfrac{1}{2}\Delta E_D)^2 - (E - E_{av})^2]^{1/2} \qquad (3)$$

E_{av} is the mean position of the Auger electron line and m_e the electron mass. The treatment by Gordeev and Ogurtsov[47] and that by Rudd and Macek[49] results in a $D(E)$ proportional to $[1/2(\Delta E_D)^2 - (E - E_{av})^2]^{-1/2}$ which essentially determines the shape of the profile. If the Auger electron line is a Gaussian contour of width γ in the emitter coordinate frame, the convolution with the Doppler profile will yield a distribution as illustrated in Figure 2 for different values of $(\tfrac{1}{2}\Delta E_D)/\gamma$. At sufficiently large values of the broadening ΔE_D the distribution has nothing in common with the shape of the initial Gaussian contour.

In a real scattering experiment, however, Auger electrons with energy E_0 may be emitted from particles scattered through different angles β having different kinetic energies T. Thus, the observed electron energy distribution $f(E)$ will be a convolution of several functions[47]: The distribution $f_0(E_0)$, describing the Auger-electron line shape in the source frame

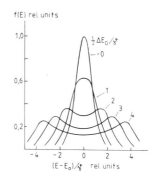

Figure 2. Deformation of a Gaussian contour of width γ under the action of Doppler broadening. (From Ref. 47.)

of reference, the Doppler distribution function $D(E - E', \beta, T)$, the apparatus function of the energy analyzer $G(E' - E_0)$, and the function $\Psi(\beta, T)$, describing the scattering angle and kinetic energy distribution:

$$f(E) = \iiiint f_0(E_0) \cdot D(E - E', \beta, T)$$
$$\cdot G(E' - E_0) \cdot \Psi(\beta, T) \cdot dE_0 \, dE' \, dT \sin \beta \, d\beta \qquad (4)$$

If the functions f_0 and G are known sufficiently well and are narrow enough the scattering angle and kinetic-energy distribution function $\Psi(\beta, T)$ may be deconvoluted from the measured line shape $f(E)$. Determination of, e.g., the recoil energy distribution by deconvolution of Ψ thus in principle may allow access to physical parameters such as the ionization probability as a function of impact parameter.

3.1.2. X-Ray Spectroscopy

For measuring the target X rays emitted after heavy ion bombardment an experimental configuration similar to that used for electron spectroscopy may be suitable with the electron analyzer exchanged by an x-ray spectrometer. As an example we reproduce a schematic diagram of the experimental arrangement used by Demarest and Watson[50] (see Figure 3). They have employed a 5-in., Johansson-type, curved crystal spectrometer designed by Applied Research Laboratories.[51] In this device the Bragg angle is scanned by a remotely controlled linear wavelength drive. The x-ray detection element is a flow mode proportional counter operated at atmospheric pressure. The counter gas is isolated from the vacuum by a thin foil of low-Z material to keep window-absorption losses small. The Bragg-angle range of approximately $15° < \theta < 70°$, accessible by the spectrometer together with the $2d$ spacing of the crystal, determines the useful wavelength range. As the wavelength resolution is proportional to $\tan \theta$

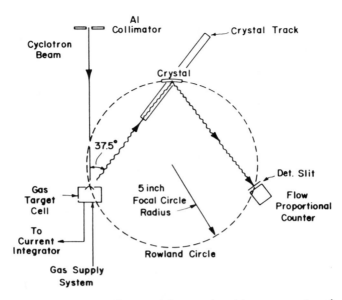

Figure 3. Schematic diagram of the experimental arrangement used by Demarest and Watson (Ref. 50).

it is desirable to select a crystal which enables one to work in the higher θ range for the x rays of interest. For measuring K x rays from elements near Ar, a NaCl crystal, oriented in the 200 plane, or from elements near Ne a RbAP (Rubidium acid phthlate) crystal, oriented in the 100 plane, has frequently been used.

In the setup of Figure 3 the crystal spectrometer is oriented in the horizontal plane to view X rays emitted at a mean angle of 142.5° with respect to the ion-beam direction. The target gas is contained in a high-pressure gas cell closed by a foil window through which the ion beam enters and the X rays are viewed. Other experiments employ two different windows, i.e., one beam-entrance window and one exit window for the X rays. The entrance window has to be of high mechanical strength and heat resistance whereas for the x-ray window high transmission efficiency is required. Using closed target cells one has to take care of effects such as energy loss and charge-state distribution of the beam after foil transmission, x-ray production at the entrance window, molecular-target decomposition, and x-ray absorption. These effects may be circumvented using a differentially pumped gas cell which has an open entrance as well as an open x-ray exit.

As the spectra are taken in the multiscaler mode they have to be normalized to the collected beam charge. The beam stop, however, does not serve as an accurate beam monitor when it is located in the gas. In that case a low-resolution x-ray detector, e.g., a proportional counter or a Si(Li) detector provides a good indication of the beam intensity.

3.1.3. Delayed Coincidence Technique

For several reasons it is interesting to study the time distribution of photons or electrons resulting from the deexcitation of the excited ions. The delayed coincidence technique[52,53] makes use of a pulsed beam-excitation source. The time difference between an excitation pulse and detection of a relaxation photon or electron is registered using a standard time to amplitude converter (TAC). Good timing resolution is required in both the excitation and photon–electron branch, because the time jitter between both pulse trains essentially determines the time resolution of a delayed coincidence spectrum. For starting the TAC the train with the lower counting statistics (usually the photon train) is used whereas the stop pulse is provided by the train with higher counting statistics. The delayed coincidence technique has been widely used in a large number of atomic and molecular lifetime measurements with pulsed electron excitation. Until recently the method has not been applied to atomic physics using a fast heavy ion beam.

We mention the experiments recently conducted by Sellin *et al.*[54,55] who measured the lifetime of metastable autoionizing states of Ne and O employing a 1.4 MeV/amu Cu beam from the UNILAC at GSI. The experimental setup is essentially that of Figure 1 and the method for deducing lifetimes by the delayed coincidence technique is illustrated in Figure 4. The accelerator beam has a micro-time structure with a period of 37 nsec whereas the beam-pulse duration is about 1 nsec. A specially designed phase probe[56] serves to derive a suitable excitation pulse which is used to stop the TAC. The start pulse is derived from the channeltron of the electron analyzer which is set to an analyzing voltage corresponding to the desired Auger-electron line. As the phase probe provides a periodic signal several maxima appear in the pulse-height distribution of the TAC output. From the slope between to maxima in a lin-vs.-log plot (see Figure 4) the lifetime of the excited metastable target ion can be derived.

Several processes affect the extraction of lifetime data; among those are cascade repopulation, recoil drifts out of the viewing range of the spectrometer, and collisional quenching. The latter comprises all processes which change the state of primary excitation by means of a second collision in the target gas. In return the method can be applied to study those effects (see Section 5). For that purpose it is imperative to have the whole time information for the complete electron spectrum. Using an on-line computer to process the timing pulses it is possible to set several time windows on

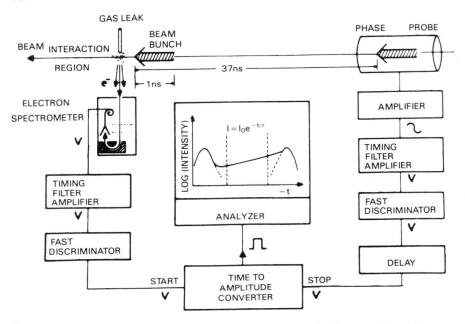

Figure 4. Lifetime measurements of target ions by the delayed coincidence technique. (From Ref. 54.)

the Auger-electron spectrum, thus allowing for discrimination between prompt and delayed processes.

3.1.4. Time-of-Flight Spectroscopy

In a recent experiment Cocke[57] has measured the charge state distribution of low-velocity recoil ions produced by 25–45-MeV Cl ions impinging on thin targets of helium, neon, and argon. He used a time-of-flight apparatus shown schematically in Figure 5. The method is based on the fact that for the collision systems studied the recoil kinetic energies are very small, typically less than 30 eV. The beam-target interaction region is situated in an approximately uniform electric field which is obtained by setting the flat bottom of the gas cell to ground potential and applying a negative voltage V_g (in the kV range) to a high-transmission Ni mesh 3 mm above the beam axis. The postacceleration velocity of the recoil is measured by the time of flight necessary to reach the channeltron detector located approximately 4 cm apart. As the velocity is proportional to $(q/m)^{1/2}$, where q and m are recoil charge and mass, respectively, the flight-time spectrum gives a direct measure of the charge-state spectrum. Time zero is defined by using a pulsed beam with a pulse width of less than 3 nsec

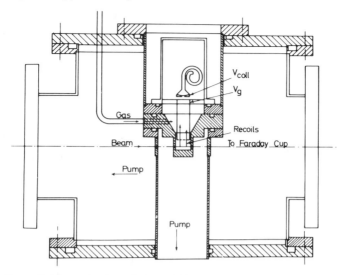

Figure 5. Schematic of recoil time-of-flight apparatus used by Cocke
(Ref. 57).

and a repetition time of 4 μ sec. The electronics are similar to those used
by the delayed coincidence technique. A TAC is started by the channeltron
pulses and stopped by a signal correlated to the beam bunches. From the
peak width of the time-of-flight spectra one may obtain an estimate of the
average initial recoil velocity provided the geometry of the time-of-flight
apparatus is known. Absolute recoil production cross sections can be
derived by normalizing the charge-state spectra to the known cross sections
for singly charged recoil production by proton bombardment.

3.2. Characteristics of Experimental Techniques

The experimental techniques described above bear common and complementary properties the main characteristics of which are schematically
summarized in Table 1. By Auger and x-ray spectroscopy information is
extracted about the distribution of excited states bearing an initial K
vacancy. The time-of-flight technique yields charge state distributions of
all recoil ions including the ground-state ions. Because many multiplet
states are populated simultaneously, a complicated line structure is observed
in the Auger-electron and x-ray spectra. When many different charge states
are simultaneously observed a tremendous number of transitions govern
the Auger-electron spectra resulting in overlapping-line features which
often are difficult to interpret. If additionally the recoil velocities are not
very small the kinematic line broadening represents a further limiting factor

Table 1. Experimental Techniques Used to Study Highly Charged Recoil Ions

Property	Auger spectroscopy	X-ray spectroscopy	TOF spectroscopy
Extractable information	Excited states distribution		Charge state distribution
Accessible electronic states	K-vacancy bearing states		All
Spectral features	Complicated line structure		Simple
Resolution limited by	Kinematics	Instrumental	Instrumental
Counting rate	Low	Very low	Very high

for the interpretability of the data. These effects are less severe in the case of the x-ray spectra because the total number of transitions is reduced as a consequence of the dipole selection rule governing the K x-ray transitions. Furthermore, the photons are less sensitive to the Doppler effect than the electrons. In that context we should recall that the kinematic line broadening observed in the Auger electron spectra in return may be used to measure the emitter velocity, provided a suitable line can be isolated in the spectrum. As opposed to the recoil charge state analysis, which is a high-counting-rate experiment, the other two techniques suffer from the necessity of working with low-efficiency instruments. In particular the low integrated crystal reflectivities cause very low counting rates in experiments employing crystal spectrometers and thin gas targets.

4. Production of Highly Charged Target Ions

4.1. Monatomic Targets

4.1.1. X-Ray Emission

In 1969 an experiment was reported by Richard et al.[58] finding a shift to higher energy of target K X rays produced by heavy ion bombardment. The effect could quickly be attributed to a removal of screening outer-shell electrons which accompanies K-vacancy production in heavy-ion–atom collisions, and the subject has been thoroughly studied in subsequent years.[59] With the higher resolution of crystal spectrometers the $K\alpha$ x-ray satellites were studied and could be explained by the various degrees of L-shell ionization produced in K-shell ionizing collisions.

The x-ray satellites arising from collisions in *solid* targets have been extensively studied[60–63] for several years. The measured relative cross sections $\sigma(KL^i)$ of single K- plus multiple L-shell vacancy production appear to fit a binominal distribution in the single L-shell ionization probability at small impact parameters $p_L(0)$. The probability $p_L(0)$ for light targets was found to increase with increasing projectile atomic number Z_1 and to saturate at values near $p_L \sim 0.4$.[64,65] This has been attributed to a rearrangement of the L-shell vacancies prior to the K x-ray emission.

Such rearrangement processes are no longer present when a thin atomic gas is used as a target so that the x-ray emission displays much more the pattern of the initial ionization produced by the collision. Evidence for a high level of target ionization in 80 MeV Ar on Ne collision was found by Mowat et al.,[66] who used a lithium-drifted silicon detector for measuring x-ray yields as a function of the Ar-projectile charge state. The results of these measurements are illustrated in Figures 6 and 7. As may be seen in

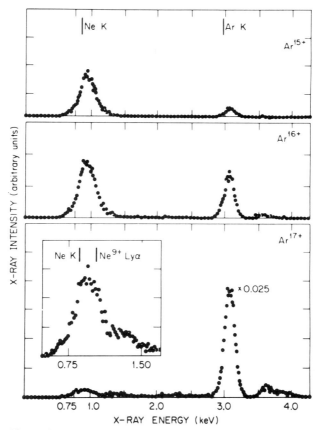

Figure 6. X-ray spectra resulting from collisions of 80-MeV Ar ions with Ne. The electronic resolution is 150 eV. The Si(Li) detector window and dead layers attenuate the neon peak by a factor of ~10 compared with the argon peak. (From Ref. 66.)

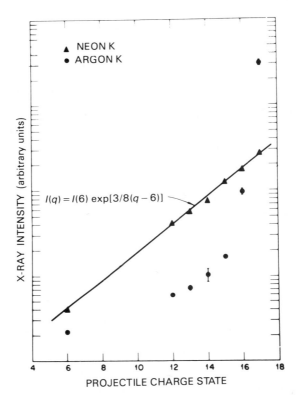

Figure 7. Neon and argon K x-ray yield versus projectile charge state. (From Ref. 66.)

the insert of Figure 6 the neon peak has a broad structure with the appearance of unresolved lines in the high-energy wing of the peak when projectiles of high charge state are used. The energies of these lines correspond to those expected for transitions in few-electron neon ions. Figure 7 shows the exponential increase of the target x-ray yield with incident charge state. This behavior can be partly understood from the rapid increase of the neon fluorescence yield[67] with increasing neon charge state.

The measurements of Mowat et al.[66] clearly indicated the need of a systematic study with instrumental resolution high enough to separate all the charge states of the target ions produced. A large number of high-resolution experiments followed concentrating on neon targets.[68–75] Figure 8 contains examples of resolved neon K x-ray lines produced by electron and heavy-ion bombardment, respectively. In contrast to the excitation by

ENERGY/keV

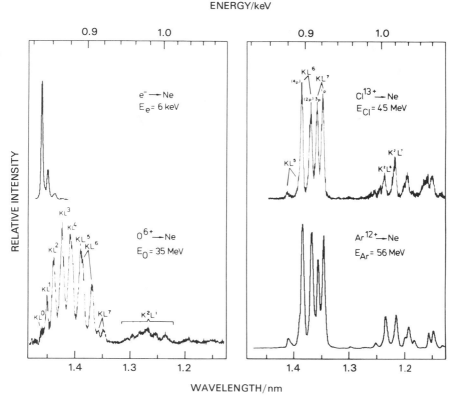

Figure 8. Neon K x-ray spectra obtained by electron and heavy ion bombardment. The oxygen- and chlorine-induced spectra are reproduced from Mathews *et al.* (Ref. 72).

electron bombardment predominantly producing the diagram line of Ne^{1+}, the heavy-ion-induced spectra reveal a many-lines structure. Using swift heavy projectiles like O^{6+} the $K\alpha$ X rays are resolved into eight components indicated by the notation KL^i with $0 \leq i \leq 7$. The identification of the $K\alpha$-satellite peaks with definite charge states is based on Hartree–Fock calculations of the corresponding transition energies. Such calculations have been performed[76–78] resulting in the conclusion that the KL^i assignment is not unique as there are spectral overlaps between different charge states.[72,69] The overlaps are due to the many multiplet states belonging to each charge state and one has to account carefully for this effect when extracting relative cross sections $\sigma(KL^i)$ for K- plus multiple L-shell ionization. We will have to return to this point in the next section. The striking feature of Figure 8 is the strong increase of the degree of multiple L-shell ionization going from O^{6+} to Cl^{13+} or Ar^{12+} bombardment. The chlorine-

and argon-induced spectra—both are quite similar—are dominated by the heliumlike and lithiumlike contributions KI^6 and KI^7 both of which split into two multiplet components. The mean number of L-shell electrons removed in these collisions amounts to 5.8.

With the increase of the projectile atomic number Z_1 not only does the satellite spectrum show a strong variation in the intensity distribution, but also there is a strong rise in intensity of higher-energy lines with wavelengths less than 1.3 nm. They are partly due to transitions of neon ions with doubly ionized K shells, i.e., hypersatellites $K^2 L^i$. Simultaneous with the increase of the total hypersatellite intensity is observed a strong rise of the contribution from very high charge states like Ne^{8+} and Ne^{9+} from which substantial amounts are produced.[72-75] The 1–2-MeV/amu projectiles we are concerned with have a velocity close to the neon K-shell orbital velocity u_K so that the scaled velocity is $v_K \equiv v/u_K \approx 1$. At such velocities the velocity dependence of the multiple ionization is rather weak.[79-81] The projectile dependence can enter through both the nuclear and ionic charge. Measurements of Kauffman et al.[82] showed that C^{6+}, N^{6+}, and O^{6+} projectiles at matched velocities produce the same extent of multiple ionization. As these projectiles have the same ionic charge but differing numbers of K-shell electrons this suggests that the K-shell electrons screen the nuclear charge during the collision. That may no longer hold true for outer-shell electrons, as demonstrated in Figure 9, where the neon satellite spectra arising from O^{4+} and Cl^{4+} bombardment, respectively,[71] are compared. The chlorine projectile bearing eight L-shell and three M-shell electrons causes much higher target ionization than does the oxygen ion with only two K-shell electrons.

It is clear that extremely high target ionization can only be reached by using very heavy ions. The extent to which one can ultimately strip neon target atoms has very recently been studied by Beyer et al.[44] using 1.4 MeV/amu Ar^{12+}, Ti^{14+}, Kr^{18+}, Xe^{24+}, Pb^{36+}, and U^{40+} ions from the UNILAC at GSI. The corresponding x-ray spectra are displayed in Figure 10 for wavelengths between 1.4 and 0.9 nm. The spectra are normalized to give an equal integral over the total spectral range displayed. Identification of the lines has been achieved by calculations of transition energies using the Dirac–Fock program of Desclaux.[83] The initial configurations are indicated in Figure 10. With only very few exceptions the observed transitions may be classified by the following three-, two-, and one-electron series:

$$(1s2l)np \rightarrow 1s^2 2l$$

$$(1s)np \rightarrow 1s^2 \tag{5}$$

$$np \rightarrow 1s$$

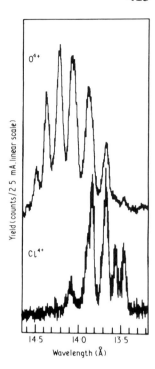

Figure 9. Neon $K\alpha$ x-ray spectra induced by 16 MeV O^{4+} and 29.4 MeV Cl^{4+} bombardment. (From Ref. 71.)

The satellite spectra observed at wavelengths greater than 1.3 nm comprise the corresponding lithiumlike and heliumlike transitions with $n = 2$. Only the small peak at 1.41 nm observed in the Ar^{12+}- and Ti^{14+}-induced spectra is due to the production of berylliumlike neon. With the use of heavier projectiles this four-electron contribution disappears whereas the ratio of the two-electron $1s2p$ intensity to the three-electron $1s2s2p$ and $1s2p^2$ intensity strongly increases. In the high-energy part of the spectra the intensity of the one-electron series increases dominating when projectiles like Pb^{36+} or U^{40+} are used. The general trend of using such very heavy projectiles is due to the distribution of the intensity among fewer spectral lines of highly stripped ions causing relatively clean spectra comparable to those of very low ionization produced by light-particle impact.

4.1.2. Relative Cross Sections $\sigma(KL^i)$

Access to the initial level populations from the measured $K\alpha$-satellite intensity distribution is often rendered difficult by the following facts: As there are Auger and $\Delta n = 0$ transitions as competing decay modes one has to know the x-ray-line fluorescence yields, which in the case of lighter

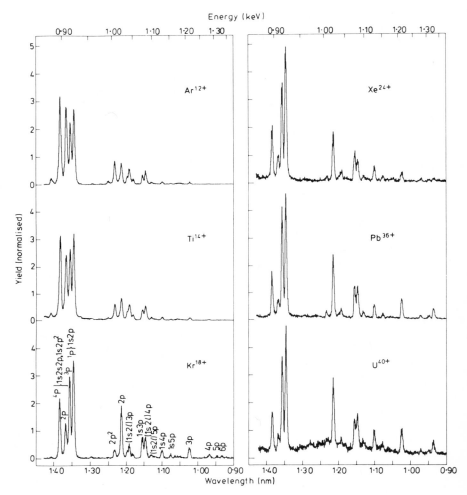

Figure 10. Ne K x-ray spectra following 1.4 MeV/amu very-heavy ion impact. The spectra are normalized to total integral intensity. (Beyer *et al.*, Ref. 44.)

atoms such as neon are sensitive functions of both the charge states and the quantum numbers of the multiplet states. The instrumental resolution is not sufficient to resolve all the multiplets produced. One, therefore, is dealing with eight distinct groups of levels KL^i belonging to eight different charge states. Because of this lack of information one has to make an assumption about the relative level population within one charge state. As long as no selective excitation mechanisms are working, the assumption of a statistical population of multiplet states seems reasonable. However,

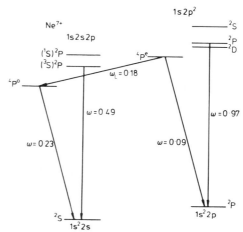

Figure 11. X-ray decay scheme of the three-electron configurations $1s2s2p$ and $1s2p^2$. (From Ref. 44.)

there may also be spectral overlaps between different charge states for which one has to correct.

In the case of neon, reliable theoretical fluorescence yields are available[84,85,77] which can be used to analyze the $K\alpha$ satellite spectra. We will now discuss in more detail the cases of high target ionization as observed in Figure 10. The KL^7 component which contains the $1s2p\ ^1P_1$ and $^3P_1$ is separated from the KL^6 component for which an x-ray decay scheme is shown in Figure 11. Only those x-ray transitions are taken into account which contribute more than 1% to the total decay rate. The fluorescence yields are extracted from Refs. 84 and 85. The two quartet transitions on the one hand and the two doublet transitions on the other hand coincide in wavelength, producing the two peaks of the KL^6 component at 1.385 and 1.368 nm, respectively (see Table 2). The quartet line has a contamination from Be-like states causing a correction which in the case of Ar^{12+} impact amounts to less than 15% and becomes zero for projectiles heavier than Kr. The relative intensity arising from a multiplet L, S may be written as

$$I_x(L, S) = g(L, S)\omega_K(L, S) \qquad (6)$$

with $g(L, S)$ denoting the statistical weight and $\omega_K(L, S)$ the multiplet fluorescence yield for the K x-ray transition. In the case of the $^4P^e$ states there is also a $\Delta n = 0$ cascade transition of appreciable intensity feeding the $^4P^0$ states. Therefore, the x-ray intensity arising from the $^4P^0$ states consists of a direct and a cascade contribution:

$$I_x(^4P_j^0) = \frac{\omega_K(J)}{56}\left[2J + 1 + \sum_{j'} (2J' + 1)\omega_L(J' \rightarrow J)\right] \qquad (7)$$

Table 2. Experimental Wavelength λ_{exp} Compared with Calculated Wavelengths λ_{DF}[a,b]

		Fluorescence yield			
		Chen and Crasemann (1975)	Bhalla (1975)	Initial state	Final state
λ_{exp}	λ_{DF}				
1.385	1.3852	0.225	0.236	$1s2s2p\ ^4P$	$1s^22s\ ^2S$
	1.3835	0.0718	0.0713	$1s2p^2\ ^4P$	$1s^22p\ ^2P$
1.368	1.3681	0.971	1.00	$1s2p^2\ ^2P$	$1s^22p\ ^2P$
	1.3661	0.488	0.468	$1s2s2p\ ^2P_-$	$1s^22s\ ^2S$
	1.3557	0.0119	0.0122	$1s2s2p\ ^2P_+$	$1s^22s\ ^2S$
	1.3706	0.0310	0.0315	$1s2p^2\ ^2D$	$1s^22p\ ^2P$
	1.3529	0.0738	0.0765	$1s2p^2\ ^2S$	$1s^22p\ ^2P$

[a] Reference 44.
[b] All wavelengths are given in nanometers.

with J and J' denoting the total angular momenta of the $^4P^0$ and $^4P^e$ states, respectively. From Eqs. (6) and (7) a statistically expected $^4P/^2P$ intensity ratio can be calculated which is shown in Table 3. The numbers in brackets are obtained by assuming $\omega_L(J' \rightarrow J) = 0$. The experimentally observed ratios strongly exceed the statistical expectation revealing a preferential population of quartet states. As we will see in Section 4.1.4 the nonstatistical population can be explained by cascade transitions $1s2lnl' \rightarrow 1s2l2l''$ which strongly feed the quartets but not the doublets. With this explanation and the calculated $^4P/^2P$ ratio of 0.52 one finds in the case of U^{40+} (where the cascade feeding is strongest) that only 20% of the quartet intensity is caused by direct population but 80% is due to a cascade population.

Once having corrected the relative intensities for cascades and spectral overlaps as described above, it is possible to convert the intensities to relative cross sections $\sigma(KL^i)$ by using statistically averaged fluorescence

Table 3. $^4P/^2P$ X-Ray Intensity Ratio of the Li-like Satellites[a,b]

Statistical expectation			Experimental observation							
$1s2s2p$	$1s2p^2$	Mean	Cl^{13+}	Ar^{12+}	Ti^{14+}	Ni^{16+}	Kr^{18+}	Xe^{24+}	Pb^{36+}	U^{40+}
1.19	0.19	0.52	0.90	0.93	1.07	1.02	1.45	1.91	1.99	2.58
(0.92)	(0.19)	(0.43)								

[a] Reference 44.
[b] The values within the parentheses have been obtained without inclusion of $\Delta n = 0$ transitions of the $^4P^e$ states.[a]

yields extracted for instance from the data of Chen and Craseman.[84] The relative cross sections $\sigma(KL^i)$ may be interpreted in a statistical model assuming the ejection of one or more K- and L-shell electrons to be statistically independent events to which probabilities can be assigned. The model was originally incorporated in Coulomb ionization theories[86,87] applying perturbative techniques in calculating the ionization probabilities which are functions of the impact parameter and the collision energy. In the statistical model the cross sections $\sigma(KL^i)$ should follow a binomial distribution

$$\sigma(KL^i) = \sigma_K \binom{8}{i} p_L^i(0)[1 - p_L(0)]^{8-i} \qquad (8)$$

where σ_K is the total cross section for K-shell ionization and $p_L(0)$ denotes the single L-shell ionization probability at zero impact parameter.

The x-ray data, some of which were shown in the previous section, were treated the way described and the binomial distribution was fitted to the measured cross sections $\sigma(KL^i)$. Reasonable fits have been obtained yielding $p_L(0)$ values as a fit parameter. Figure 12, which is reproduced from Ref. 88, is a plot of the L-shell ionization probability $p_L(0)$ as a function of the ionic charge q for incident 1.5 MeV/amu and 1.4 MeV/amu ions ranging from H^+ up to U^{40+}. $p_L(0)$ increases with q reaching a value of 0.95 with U^{40+} ions.

In the statistical model the L-shell electrons are treated as if they are all ejected simultaneously and are subject to the same average binding energy, justifying the assignment of a single probability p_L to all L-shell electrons. Another treatment[89] is to view the electrons as being ejected sequentially. Then the ejection of each electron is accompanied by an increase of the binding energy. Using the correct binding energies in calculating the ionization probabilities for instance in the SCA or BEA

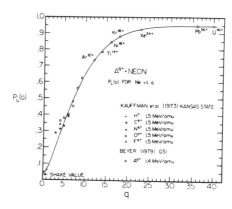

Figure 12. $P_L(0)$ of neon. (From Ref. 88.)

approximation leads to a decreasing probability with increasing number of ejected electrons. The calculated distribution of the cross sections $\sigma(KL^i)$ will, therefore, be somewhat narrower in the overall width than in the case of the simultaneous ejection model.

4.1.3. Auger-Electron Emission

For low-atomic-number elements the main decay of K-shell vacancies occurs by the Auger process. A similar development in the spectral features as discussed for the neon x-ray decay can be found in the corresponding Auger electron spectra going from light- to heavy-ion impact. Figure 13 shows K-Auger spectra[90–93,43] of neon excited by various projectiles whose atomic number varies in the range between 1 and 92. The spectrum obtained by Stolterfoht et al.[90] with 4.2-MeV proton impact reveals the typical spectral pattern of predominant single K-shell ionization. Nearly identical spectra are also obtained by electron impact and x-ray fluorescence.[94] The predominant peaks observed at the top of Figure 13 are the diagram lines due to the KL_1L_1, the KL_1L_{23}, and the $KL_{23}L_{23}$ Auger process in singly ionized neon.

Below the proton-induced spectrum a spectrum is plotted obtained by 2 MeV/amu O^{5+} impact measured by Burch et al.[91] Similar results were obtained by Matthews et al.[95] The main diagram line $KL_{23}L_{23}(^1D)$ at 804 eV is still observed in the oxygen-induced spectrum, but the spectrum is now overcrowded with a bulk of satellite lines which appear at lower energies. This shift to lower electron energy corresponds to a multiple ionization of the L-shell. The complexity of the Auger-electron spectrum produced by O^{5+} impact demonstrates that assignment of Auger satellite lines according to the charge state of the ion (i.e., the multiple L-shell vacancy configuration KL^i) is not as simple as in the corresponding x-ray spectrum.[68,70] The complex line structure is caused since Auger selection rules allow many transitions from the same initial configurations. This makes it difficult to identify observed peaks unambiguously even when high instrumental energy resolution is applied.

Therefore, attempts have been made to use average quantities such as the mean number of L-shell vacancies $\bar{n}_L$ to interpret the Auger electron spectra rather than to study individual lines. For instance it could be shown for a large range of projectiles and collision velocities that the centroid energy $\bar{E}_c$ of the Auger electron satellite distribution depends linearly on the mean number of L-shell vacancies $\bar{n}_L$.[96] The latter quantity is related to the L-shell ionization probability $p_L(0)$ through $\bar{n}_L = 8p_L(0)$, provided binomial statistics are used. Although it is not possible to resolve the seven satellites $KL^i(0 \le i \le 6)$, access to the KL^0, KL^1, and KL^6 components is possible by isolating lines observed at the low- and high-energy ends of

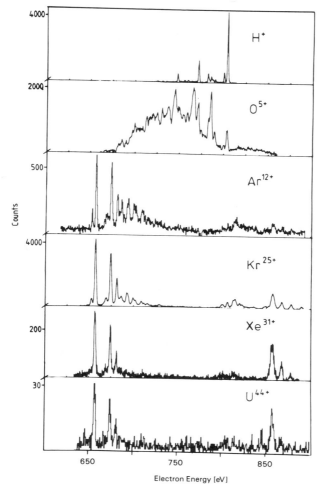

Figure 13. Auger electron spectra from neon induced by
4.2 MeV H^+ (Stolterfoht *et al.*, Ref. 90), 1.9 MeV/amu O^{5+}
(Burch *et al.*, Ref. 91), and 1.4 MeV/amu Ar^{12+}, Kr^{25+},
Xe^{31+}, and U^{44+} (Mann, Ref. 92, and Stolterfoht *et al.*, Ref.
43).

the Auger spectra. Comparison with the corresponding x-ray satellite
spectra thus allows extraction of semiempirical fluorescence yields $\bar{\omega}_i$
as a function of the charge state.[72,97,98] The semiempirical fluorescence
yields deduced this way are in agreement with statistically weighted mean
fluorescence yields obtained from calculated multiplet fluorescence
yields.[67]

The situation changes dramatically when heavier projectile ions are used achieving high degrees of target ionization. This is demonstrated in Figure 13 for 1.4 MeV/amu Ar^{12+}, Kr^{25+}, Xe^{31+}, and U^{44+} ions which have been provided by the UNILAC at GSI. Already with Ar^{12+} most of the intensity is found at very low energies, indicating a high degree of L-shell ionization. A similar spectrum was obtained with 1.3 MeV/amu Cl^{12+} ion impact.[99] As the total number of observed spectral lines is drastically reduced it is possible to attribute multiplet states to individual lines. Line identification has been achieved by comparison with calculated Auger transition energies.[76,100] The Ar^{12+} induced spectrum is composed of lines mainly originating from lithium- and berylliumlike initial states. The lines observed at energies lower than about 750 eV arise from configurations with one K-shell and mainly two or three L-shell electrons. The reader is referred to Refs. 76, 100–102 for the Auger transition energies. No extensive tabulations will be given here; rather, we restrict ourselves to the three-electron system, because it dominates at very heavy-ion impact. Table 4 contains the experimental energies of Auger lines arising from the $1s2s^2$, the $1s2s2p$, and the $1s2p^2$ configuration. These configurations become dominant when the projectile nuclear charge is further increased because the contribution from four-electron states decreases.

For krypton impact the Auger spectrum is nearly a pure three-electron spectrum. The two dominant lines at 657 and 674 eV originate from the $1s2s2p\,^4P^0$ and the $1s2p^2\,^4P^e$ state, respectively. There is very little change in the satellite spectrum when the nuclear charge is raised to the ultimate going to Xe^{31+} and U^{44+} impact, besides the fact that contributions from berylliumlike states vanish completely. Interesting is the appearance of lines at energies larger than 800 eV which could be identified as being due to the three-electron configurations $1s2lnl'$ with $3 \leq n \leq 6$.[101] They are most pronounced using projectiles heavier than argon. The distribution of these high-energy lines grows narrower with increasing nuclear charge of

Table 4. Experimental Auger Transition Energies
for $Ne^{7+\,a}$

Energy (eV)	Initial		Final	
			State	
651.7	$1s2s^2$	$^2S^e$	$(1s^2)\,^1S^e$	
656.2	$1s2s2p$	$^4P^o$	$(1s^2)\,^1S^e$	
668.9	$1s2s2p$	$^2P^o_-$	$(1s^2)\,^1S^e$	
673.6	$1s2s2p$	$^2P^o_+$	$(1s^2)\,^1S^e$	
	$1s2p^2$	$^4P^e$	$(1s^2)\,^1S^e$	
680.6	$1s2p^2$	$^2D^e$	$(1s^2)\,^1S^e$	

^{a} Reference 101.

the projectile stressing the line at 857 eV which is due to the Auger decay of the configuration $1s2p4l$. The "saturation" of the spectra with increasing nuclear charge observed in the very heavy projectile regimes Xe^{31+} and U^{44+} can be explained by the fact that no Auger processes occur for the one-electron ions and that only small contributions from autoionizing two-electron ions are produced. This leads to pure three-electron Auger spectra as only one-, two-, and three-electron configurations are produced. The same can, therefore be stated for the Auger decay channel as for the x-ray channel: By using very heavy projectiles, impressively clean spectra of few-electron configurations are produced which have nothing in common with the many blending lines produced by intermediate heavy-ion impact. It is also noteworthy that the kinematic line broadening due to target recoil is no severe limitation to the spectral resolution (see also Section 4.2).

4.1.4. Nonstatistical Population and Cascade Transitions

In Section 4.1.2 we have already noted a severe deviation from statistical population of multiplet states within the three-electron $K\alpha$ satellite. Figure 14 is an illustration of both the Auger and x-ray spectra of the Ne^{7+}

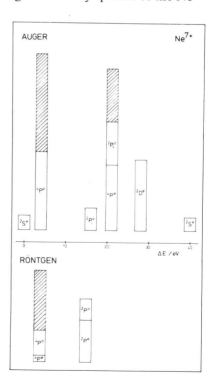

Figure 14. Intensity distribution of the three-electron lines in the Auger electron- and x-ray spectra calculated under the assumption of a statistical population (open bars). The relative energy scale applies for both the electron and the x-ray energy. The cross-hatched bars indicate the boosting of the quartet intensity observed with 1.4 MeV/amu krypton impact.

satellite. The intensity distribution is calculated assuming the multiplets of the configurations $1s2s^2$, $1s2s2p$, and $1s2p^2$ to be populated according to their statistical weights $(2S + 1) \cdot (2L + 1)$. Theoretical fluorescence yields[84,85] are used to obtain the relative decay strengths. As opposed to the simple pattern of two lines obtained in the x-ray case a structure of six lines occurs in the Auger branch for this particular charge state. The first five of these lines are readily resolved in the spectra shown in Figure 13 in the case of projectiles heavier than argon. For argon impact the $1s2p^2\,^2D^e$ line has admixtures[43] from lower charge states whereas the weak $1s2p^2\,^2S^e$ component is completely covered by those states. Comparing the experimental line intensities with the statistical expectations, one finds a strong boosting of the experimental quartet intensity which is indicated by the cross-hatched bars in Figure 14. The latter are the experimental quartet intensities for 1.4 MeV/amu Kr impact notable in Figures 10 and 13.

The selective population mechanism responsible for the strong enhancement of the quartets is a cascade feeding[44,75] from outer-shell configurations $(1s2l)nl'$ with $n \geq 3$. Transitions from such configurations are observed as high-energy lines in both the x-ray and Auger electron spectra (see Figures 10 and 13). The situation is illustrated in Figure 15 where the various Auger and x-ray decay branches are shown. Whenever the three electrons of the $(1s2l)nl'$ configuration are coupled to a doublet term $(S = 1/2)$ the Auger and x-ray decay to the K-shell are the dominant deexcitation channels as indicated in Figure 15 by the two left-most arrows. These are the transitions mentioned above observed as high energy lines. Transitions to the L shell are very weak $[\omega_L(S = 1/2) \approx 0]$. If, however, the electrons are coupled to a quartet term $(S = 3/2)$, direct decay to the K shell can only occur via second-order spin–spin interaction, hence it is negligible against the L x-ray decay which occurs via fully allowed dipole transitions with a probability of $\omega_L(S = 3/2) \approx 1$. The latter decay branch acts as a strong cascade for the quartet states of the satellite configurations $(1s2l)2l''$. Because in the case of the doublets a corresponding cascade

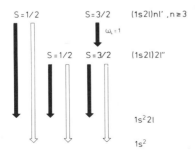

Figure 15. Decay scheme of the three-electron system. The full (open) arrows indicate the radiative (Auger) decay.

branch is missing, a severe boosting of the $^4P/^2P$ intensity ratio will result, explaining qualitatively the experimental data.

The quartet cascade transitions have not yet been measured directly for the collision systems under discussion. However, their relative intensity may be estimated from the measured K-x-ray and K-Auger intensity using a theoretical branching ratio. The problems one has to face here are twofold: (i) Each decay branch, which in Figure 15 is indicated by only one arrow, comprises quite a number of multiplet transitions. Therefore, one has to make an assumption about the relative level population relevant for the branch. (ii) Theoretical fluorescence yields have to be available for all the many configurations and multiplets.

For an estimation, statistical population of the outer-shell states will be assumed and the fluorescence yields calculated by Chen[103] for the $(1s2l)3l'$ configurations will be used. The same dependence of the fluorescence yield upon the multiplet state will then be used for the configurations $(1s2l)nl'$ with $n \geq 3$. With this procedure, for instance 40% and 50% of the high-energy Auger intensity is expected to be the cascade portion of the $^4P^e$ and $^4P^0$ lines, respectively. This gives rough agreement between estimated cascade contribution and surplus quartet intensity over the statistical expectation as indicated in Figure 14. The agreement can be improved further when the direct population is experimentally suppressed and only the cascade populates the low-lying quartets. Such investigations will be discussed in more detail in Section 5 where the origin of the outer-shell population is studied.

4.1.5. Scaling Behavior of Multiple Ionization

The degree of multiple ionization depends on the projectile atomic number Z_1, the target atomic number Z_2, and the collision velocity v_1. It is desirable to have a universal function relating these quantities over the wide range studied experimentally.

The data base for testing such functions should concern atomic gas targets. Solid and molecular-gas targets can only be used in the restricted range of light projectiles where the degree of multiple ionization is low and outer-shell rearrangement effects are not severe.[89] Attempts have been made to relate the mean number of L-shell vacancies $\bar{n}_L$ to Coulomb ionization cross sections.[87] The immediate difficulty arising is the failure of the L-shell ionization probability $p_L(0)$ to scale as Z_1^2 as predicted by Coulomb ionization theories. In the case of light target atoms bombarded with heavy ions there is also the projectile charge state effect which renders more difficult the comparison between theory and experiment.

Schneider et al.[96] suggested using the simple v_1/Z_1 dependence of the mean number of L-shell vacancies $\bar{n}_L$. They determined the $\bar{n}_L$ values from the centroid energies of the Auger satellite spectra of neon. The projectile atomic number Z_1 varied between 6 and 54 and the velocity ranged from 1 to 10 a.u. The result is shown in Figure 16 where an approximately linear dependence on v_1/Z_1 is observed. We have added the results obtained from the x-ray spectra using the relation $\bar{n}_L = 8 \cdot p_L(0)$. The x-ray data apply for bare lighter ions C^{6+} up to F^{9+},[69] and for very heavy partially stripped ions Ar^{12+} up to U^{40+}.[44] For the latter, the ionic charge q instead of the nuclear charge Z_1 is used for plotting the data. The data points closely follow a straight line. The discrepancy between the Auger-data on one hand and the x-ray data on the other hand may be partly due to the fact that fluorescence-yield effects have not been taken into account in extracting the $\bar{n}_L$ from the Auger spectra. A rather large scatter of the data would be noted if all the x-ray data available for partially stripped lighter ions were included. In particular the projectile charge-state dependence at a fixed velocity may result in a spread of $\bar{n}_L$ reaching as much as 30%.

Recently, Schmiedekamp et al.[104] studied the argon K x-rays by varying the projectiles from H to Cl and the impact energy between 1 and 5 MeV/amu. For such projectiles with atomic numbers smaller than the target atomic number, the effect of the ionic charge is expected to be less pronounced and the change of fluorescence yield with increasing ionization is smaller in argon than in neon. A universal scaling of the mean number of L vacancies was possible by taking into account the effect of increased binding as incorporated in the PWBA.[32,33] It was found that the BEA scaling law

$$\bar{n}_L = KZ_1^2 f(V_L \cdot \varepsilon^{-1/2})\varepsilon^{-2} \tag{9}$$

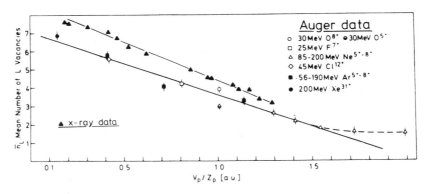

Figure 16. The scaling of the mean number of L vacancies $\bar{n}_L$ in neon. (From Ref. 96.)

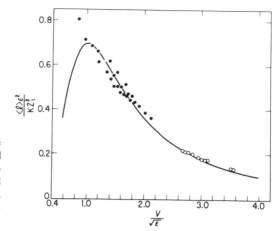

Figure 17. Scaled average satellite number measured for Ar and theoretical universal ionization curve plotted vs. scaled projectile velocity ($\bigcirc$, from Ref. 106, and $\bullet$, from Ref. 104).

worked much better than the PWBA scaling law

$$\bar{n}_L = KZ_1 f(V_L \cdot \varepsilon)\varepsilon^{-1} \tag{10}$$

where K is a constant and ε is the binding energy correction factor. f is the universal ionization function of the BEA or PWBA and V_L is the impact velocity in units of the mean L-shell orbital velocity. The form of ε used was

$$\varepsilon = 1 + (aZ_1 + bZ_1^2)g(V_L) \tag{11}$$

where a and b are fitting parameters and $g(V_L)$ is taken from Brandt and Lapicki.[105] Applying the BEA scaling, good fits were obtained to both the BEA and the PWBA universal function with a slightly better fit in case of the PWBA function. Figure 17 shows the data fitted to the scaled BEA curve using the BEA scaling. The figure is taken from Tonuma *et al.*,[106] who extended the measurements by using 4.7–7.8 MeV/amu N ions.

4.1.6. Charge State Analysis of Recoil Ions

Using the time-of-flight technique described in Section 3.1.4 Cocke[57] was able to analyze the charge-to-mass ratio of the recoil ions produced by fast heavy ion impact. He measured charge state distributions of multiply ionized recoils of He, Ne, and Ar produced in collisions with 25 to 45 MeV Cl beams in charge states 6+ to 13+. Very recently an extension of this work has been reported by Gray *et al.*,[107] who measured the neon recoil ions in coincidence with the charge state of the projectiles after the collision.

These measurements are similar to earlier coincidence experiments done at low bombarding energies in which the charge states of both projectile and recoil were analyzed. The impact parameter range was selected by requiring a certain projectile scattering angle. As shown by Kessel and Everhart,[108] the charge-state distributions of projectile and recoil were largely uncorrelated as long as inner-shell excitation was avoided. This, however, may no longer hold true for the asymmetric situation where a highly stripped projectile collides with a neutral target atom at energies which are orders of magnitude larger.

In the single spectra reported by Cocke[57] recoil charge states of 1+ and 2+, 1+ to 8+, and 1+ to 11+ were observed for He, Ne, and Ar targets, respectively. Large cross sections are found for recoil production being larger than 10^{-15} cm^2 for single ionization of argon by Cl^{12+} and the cross sections remain at least of the order of 10^{-16} cm^2 through charge state 8. The cross sections vary only slightly with impact energy between 25 and 45 MeV. They could to some extent be interpreted in terms of two models[57] which both regard the projectile as a point charge. The first of these models, suggested by Olson,[109] treats the target electrons as being totally independent. It overestimates the cross sections for the higher charge states. A better description of the experimental data is obtained with the energy deposition model which uses the statistical treatment of electron emission proposed by Russek et al.[110-112] Within this model a pronounced dependence of the recoil charge q upon the impact parameter b is obtained.

As opposed to the low-charge–low-energy collision experiments a strong correlation of the charge state spectra was found in collisions of bare and hydrogenlike ions incident on neon targets.[107] Figure 18 shows the coincident time-of-flight spectra of neon recoil ions obtained with 1 MeV/amu F^{9+} ions. A strong shift of the distribution to higher recoil charges is found when the projectile undergoes single- or double-electron capture. The shift is much more than one or two units as originating from the capture event itself. It is the impact parameter dependence of the projectile and target charge changing processes which leads to the strong correlation. In collisions where the projectile charge is changed a narrower impact parameter range is preselected causing a higher degree of target ionization. This explanation is supported by the fact that the maximum of the recoil-charge-state spectrum occurs at nearly the same position in both cases when the projectile captures and when it loses one electron. This is demonstrated in Figure 19 where the model calculations are compared to the experimental cross sections. In these calculations based on the energy deposition model the probability for producing recoil charge q and simultaneously changing the projectile charge from i to j was factorized by

$$P_q^{ij}(b) = P_q^i(b) \cdot P_{ij}(b) \qquad (12)$$

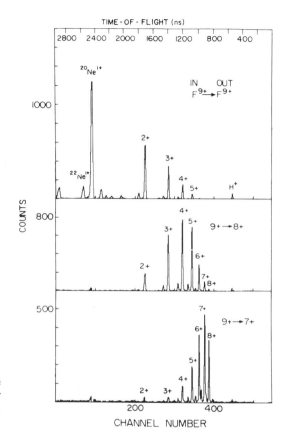

Figure 18. Neon recoil charge state time-of-flight spectra for 1 MeV/amu $F^{9+} \to$ Ne. (From Ref. 107.)

where $P_q^i(b)$ is the probability of generating recoil charge q, independent of what happens to the projectile, and $P_{ij}(b)$ is the probability of the projectile going from charge state i to j independent of what happens to the recoil. $P_{ij}(b)$ was assumed to have a Gaussian form, the width of which was used as a fit parameter. The fits shown in Figure 19 resulted in nearly the same width in both the one-electron capture and the one-electron-loss collisions of the nitrogen ions.

4.2. Recoil Energy Distribution

In earlier collision experiments performed at bombarding energies below a few hundred keV, ionization of the collision partners occurred in close collisions resulting in large scattering angles. Thus a complete determination of the scattering event was easily obtained by selecting collision parameters such as the scattering angles of the projectile and the target

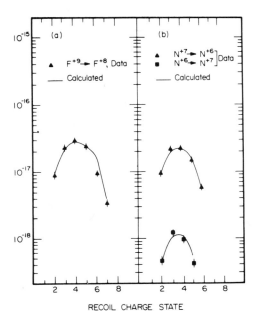

Figure 19. Neon recoil production cross sections for incident 1 MeV/amu ions which have undergone electron capture or loss in the collision. (From Ref. 107.)

recoil, and the kinetic energies of the particles (see for instance Ref. 108).

We address quite different systems where large impact parameter collisions contribute significantly to the total cross sections. For very heavy multiply charged ions colliding with lighter atoms recoil-ion production is already effective at impact parameters larger than the target K-shell radius. In such collisions the projectiles remain practically undeflected whereas the recoils are scattered at almost exactly 90°. For example, a 1.4 MeV/amu Kr ion passing a Ne atom at an impact parameter of 0.1 a.u. (approximately one K-shell radius) is scattered through a laboratory angle of only 0.048° which means a 0.8 mm deflection after traveling a path of 1 m. The corresponding recoil angle amounts to 89.88°. Therefore, it is evident that a direct measurement of the ionization probability through coincidences with scattered particles becomes unfeasible. A mean impact parameter, however, may be estimated from total production cross sections as determined from the recoil charge state spectra or from x-ray production cross sections. Cocke[57] reported cross sections of larger than 10^{-17} cm^2 for high charge state neon recoil ions produced by 0.7–1.3 MeV/amu Cl^{12+} ions. These results fit the vacancy production cross sections which were extracted from x-ray measurements reported by Mowat *et al.*[74,113] Such large cross sections correspond to a mean impact parameter of greater than 0.3 a.u. and the kinetic energy transferred to the neon target atom must be less than

10 eV. A value of 30 eV has been estimated as an upper limit for Ne^{7+} recoils from the peak width of the time-of-flight spectra.[57]

As already shown in Section 3.1.1 Auger-electron emission is very sensitive to the Doppler effect, so that the emitter velocity may be determined from the observed width of an appropriate Auger line. The decay of the $1s2s2p\,^4P^0$ state in Ne^{7+} for instance is suitable for this method. By means of a line-shape analysis of this peak, the emitter-velocity distribution can be deconvoluted provided other line broadening effects are small enough. As the natural linewidth[84,114] is negligible, only the instrumental resolution limits the applicability of the method. From Eq. (2) one obtains a lower limit of kinetic energies which are measurable with a given instrumental resolution

$$T_{min} = \frac{(\Delta E_{app})^2 M}{16 E_0 m_e} \qquad (13)$$

where ΔE_{app} is the FWHM of the apparatus resolution. For the line under discussion and a width of $\Delta E_{app} = 1$ eV the kinetic energy has to exceed at least 3 eV. Other precautions to be taken are to keep the target gas pressure low enough to prevent secondary collisions and to avoid space charges affecting the transmission of the electrons from the scattering region to the energy analyzer.

Investigations of the neon $^4P^0$ line using 1.4 MeV/amu very heavy ions have shown that the recoil kinetic energies are so small that we are just at the limit where the method starts to work. With the energy set to a width of slightly below 1 eV, preliminary results of a deconvolution procedure indicate a mean recoil energy located below 5 eV for the collision systems mentioned. Furthermore, recoils of more than 10 eV kinetic energy contribute only little to the whole distribution.

4.3. Molecular Fragmentation

From the previous sections it became evident that highly charged target atoms can be produced which receive only very low momenta from the fast heavy projectiles. When molecular targets are used instead of the monatomic targets discussed so far, additional phenomena are to be expected. Let us consider a molecule which suddenly loses many of its electrons in a fast interaction with a heavy ion. As the binding electrons are blown off, the remaining molecular fragments will dissociate due to the Coulomb forces (Coulomb explosion). Radiation emitted from atoms of an exploding molecule thus will experience the Doppler effect provided that the corresponding atomic lifetimes are large enough. It is, therefore, interesting to look at the time scale of the whole process. The stripping itself takes place in a time which is of the order of 10^{-17} sec for MeV/amu projectiles. If we take for simplicity a diatomic molecule with two nuclei of charges Z_1 and Z_2 initially at rest and separated by a distance R_0 the separation R between the two atoms increases as a function of the repulsion time t according to

$$\frac{m_1 m_2}{m_1 + m_2} \cdot \frac{d^2 R}{dt^2} - V(R) = 0 \tag{14}$$

$V(R)$ is the potential energy, which may be approximated by a simple Coulomb potential

$$V(R) = \frac{Z_1^* Z_2^*}{R} \tag{15}$$

where Z_1^* and Z_2^* denote the effective charges and m_1 and m_2 the masses of the two fragments. After a time of a few 10^{-15} sec most of the initial potential energy is converted into kinetic energy and the ion of mass m_2 reaches the final dissociation velocity

$$v_D = \left[\frac{2m_1}{m_2(m_1 + m_2)} V(R_0) \right]^{1/2} \tag{16}$$

If an excited atomic state to be studied has a mean lifetime which is large as compared to the dissociation time the radiation emitted is subject to the final emitter velocity v_D. The whole matter can then be treated as a three-step process comprising stripping, dissociation, and emission of photons or electrons.

Mann and collaborators[115,45] investigated the Coulomb explosion of simple molecules containing carbon, nitrogen, or oxygen. They used 1.4 MeV/amu heavy ions from the UNILAC and observed the Auger decay of these atoms. The spectra reveal a similar high degree of ionization as in the case of the monatomic neon target. In Figure 20 we record the K Auger spectra from carbon in methane. The domiant peak results from the $1s2s2p\ {}^4P^0 \rightarrow 1s^2 + e^-$ Auger transition, whereas the other structures are mainly due to lines of the K–LL and K–LM transitions from three- and four-electron configurations. The ${}^4P^0$ states of the atoms under discussion have lifetimes exceeding 10^{-9} sec. Therefore, the kinematics can be accounted for by a single velocity v_D which is already reached within a time of 10^{-14} sec.

The line shape of the ${}^4P^0$ Auger decay was studied experimentally[45] using a 1.4 MeV/amu Ar^{12+} beam and different molecules containing the

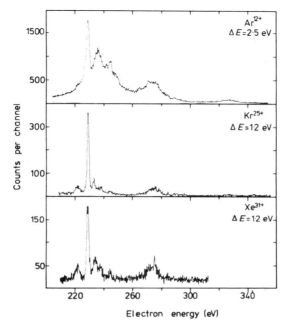

Figure 20. Auger electron energy spectrum from carbon in methane after 1.4 MeV/amu Ar^{12+}, Kr^{25+}, and Xe^{31+} impact observed under a laboratory angle of 120°. Instrumental FWHM resolution as indicated. (From Ref. 45.)

carbon, nitrogen, or oxygen atom. The experimental setup was similar to that sketched in Figure 1. The results shown in Figure 21 were obtained with a spectrometer resolution $\Delta E \leq 0.8$ eV. A great change in the peak width can be observed depending on the molecular species. When the Auger electron is emitted from an atom located in the center of a symmetrical molecule like CH_4, a small linewidth is observed. This is also to be expected as the momenta induced to the center atom are compensating each other. The other linewidths for the different molecules are in accordance with the mass ratios of the fragments assuming a two-fragment dissociation. In the case of the oxygen spectra from CO_2 the broadening is largest because the rest fragment CO is rather heavy.

The experimental data of Figure 21 have been fitted to a rectangular function convoluted with an experimentally determined spectrometer function and the background was fitted by a straight line. The rectangular shaped Auger electron intensity distribution is expected theoretically under

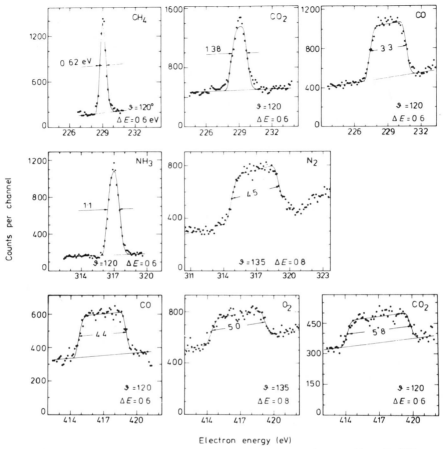

Figure 21. Experimental and fitted (full curves) lines shapes of the $(1s\,2s\,2p)^4P \rightarrow (1s^2)^1S + e^-$ Auger electron transition in carbon, oxygen, and nitrogen in different molecules after 1.4 MeV/amu Ar^{12+} impact. (From Ref. 45.)

the following assumptions which are all fulfilled to a very good approximation:

(a) The projectile-induced recoil velocity is small compared to the dissociation velocity.

(b) The Auger electrons are emitted isotropically in the frame of reference of coordinates related to the emitting ion.

(c) The dissociation velocities v_D are randomly distributed in the laboratory frame of reference.

(d) A single dissociation velocity v_D is taken into account.

The theoretical width of the rectangular distribution then simply reads[45]

$$\Delta E = 2m_e v_A v_D \qquad (17)$$

where $v_A = (2E_A/m_e)^{1/2}$ is the Auger electron velocity in the emitter frame of reference and the position of the center of the line remains unchanged. The dissociation linewidth does not depend on the electron observation angle as in the case of the primary recoil ions which reveal a $\sin \delta$ dependence. Indeed, angular distribution measurements[45] between the laboratory angles of 60° and 145° confirmed the independence of the dissociation line shape upon the observation angle within the experimental uncertainty. The linewidth ΔE depends on the molecular potential $V(R_0)$ through v_D, given by Eq. (16). Comparison with calculated molecular orbitals[116] has shown that the simple Coulomb potential (15) is a good approximation for the two-fragment-dissociation processes under discussion. Using the Coulomb potential Eq. (17) can be written as

$$\Delta E_{\text{coul}} = 4\left[\frac{m_e m_1}{m_2(m_1 + m_2)} \frac{Z_1^* Z_2^* E_A}{R_0}\right]^{1/2} \qquad (18)$$

where R_0 is the molecular bond length. Z_2— the nuclear charge of the ion from which the electron is emitted—may be assumed to be fully screened by its three electrons obtaining $Z_2^* = Z_2 - 3$. In Figure 22 the experimentally determined widths ΔE^* are compared with the theoretical Coulomb widths ΔE_{coul} given by the above equation. For Z_1^* values between 2 and 5 were inserted. From the graph it can be seen that the charges of the rest fragments are near three or four. Even though some

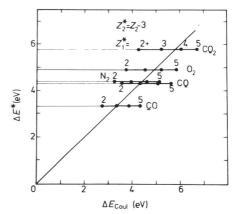

Figure 22. Experimental linewidth contributions ΔE^* (corrected for instrument resolution) plotted against calculated values for molecular Coulomb explosion. Auger electron emitting atoms of charge state Z_2^* and atomic number Z_2 are underlined and the molecular rest fragment has charge state Z_1^* shown from 2+ to 5+. (From Ref. 45.)

information about Z_1^* is obtained, several speculations about the initial projectile–molecule interaction are possible. One possibility would be a complete stripping of the atom followed by a fast recapture of electrons from the other fragment.

The molecular explosion as a consequence of a high degree of molecular electron stripping has also been observed in fast beams of molecules transmitted through thin-foil targets.[117,118] Through the kinematical transformation from the center of mass of the fast moving molecule to the laboratory frame of reference, large energy shifts and angular deviations of the break-up fragments are induced, providing a method for studying the fast-molecule-target interaction phenomena.[117–120]

4.4. Outer-Shell Rearrangement

In the K x-ray emission studies the photon energy reflects the atomic electron configuration present at the time of x-ray emission. Therefore, identification of this configuration with that initially produced by the collision is only possible if all processes are missing which occur before the decay of the K hole and which change the initial multiple vacancy configuration. In monatomic targets such fast rearrangement processes appear to the lowest extent since only atomic rearrangement has to be considered, which can be taken into account by using purely atomic decay rates. If the atom, however, is located in the environment of a molecule or even a solid, alteration of the initial vacancy configuration can occur through electron transitions from neighboring atoms.

Investigations of the $K\alpha$ satellites from solid targets[64,65] composed of low-atomic-number elements have shown that the apparent probability $p_L(0)$ for L-shell ionization increases with the atomic number of the projectile and saturates at values near $p_L(0) \approx 0.3$. The effect could be explained by postulating interatomic transitions which refill the L-vacancies prior to the x-ray emission. One interesting feature of the $K\alpha$ satellites observed from solids is the variation in the relative intensity distribution of a particular element going from one chemical compound to another.[121] This is not unexpected since the L-vacancy transfer process must involve the valence levels, i.e., the M-shell in the case of third-row elements. Therefore, the rates for L-vacancy filling will depend sensitively on the density of valence electrons, which in return is influenced by the chemical bonding. A further requirement for the L-vacancy transfer rates to influence the $K\alpha$ emission spectra is to compete with the $K\alpha$ decay rates. K-hole lifetimes of the highly ionized species under discussion are typically a few 10^{-14} sec. A measure of the response time of the valence electrons is the plasmon lifetime, which, however, may be as small as 10^{-16} sec.[122,123] Watson and collaborators[121] found that for solid compounds of Al, Si, S, and Cl the apparent degree of L-shell ionization present at the time of $K\alpha$ x-ray emission increases as the localized valence electron density of the target increases and decreases as the average total valence electron density of the compound increases. The explanation for this behavior is found by considering the availability of electrons after the collision to refill the L vacancies. When the density of valence electrons localized near the target atom is high there is also a high probability of ionizing those electrons during the collision. After the collision, therefore, they are no longer available for the fast relaxation process which leads to a higher value of the apparent degree of L-shell ionization.

Molecular gas targets are of particular importance for the study of the relaxation process since they provide relatively simple isolated systems which may be regarded as intermediate between the monatomic gases and the solids. Kauffman et al.[124] and Hopkins et al.[125] searched for the differences in the $K\alpha$ satellites from gaseous compounds of Si and S. The measurements have shown that the light gases SiH_4 and H_2S display much higher apparent degrees of L-shell ionization than do solid Si and S. A comparison between a light-ligand gas and a heavy-ligand gas is shown in Figure 23 displaying the Si $K\alpha$ x-ray spectra from SiH_4 and SiF_4 induced by 53.4 MeV Cl ions. A strong downward shift in the centroid energy of the whole satellite distribution is found for the SiF_4 along with centroid shifts of the individual satellites. Comparison of the energy positions of the satellites with Hartree–Fock calculations showed that in the case of the light-ligand gas SiH_4 the M shell is completely depleted. The behavior of SiH_4 thus appears to be closer to that of a free atom, whereas the SiF_4 is similar to a solid.

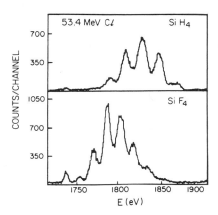

Figure 23. Si $K\alpha$ x-ray spectra induced in SiH_4 and SiF_4 by 53.4-MeV Cl ions. (From Ref. 125.)

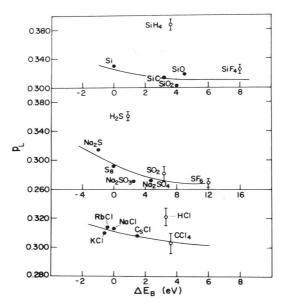

Figure 24. The variation of p_L with 2p-electron binding-energy shift as measured by ESCA. The full data points refer to measurements for solids whereas the open data points refer to measurements for gases or a liquid (CCl$_4$). (From Ref. 50.)

Demarest and Watson[50] investigated the process in some detail for a number of Si-, S-, and Cl-containing compounds bombarded with 32-MeV oxygen ions. They discussed their data in terms of the local valence-electron density which may be represented by two different quantities. The first quantity is the effective charge given by the product of the oxidation number and the Pauling bond ionicity.[126] Another measure of the local valence-electron density is the inner-shell-electron binding-energy shift as measured by electron spectroscopy for chemical analysis (ESCA). Both concepts yielded consistent results. In Figure 24 we record the apparent L-shell ionization probability as a function of the 2p-electron binding-energy shift.[50] The data from the heavy gases as well as from the liquid CCl$_4$ closely follow the same trend as do the data from the solids. At low binding energy shifts the valence electrons are mainly concentrated near the atom and, therefore, become ionized during the collision. With increasing (positive) binding-energy shifts the valence electrons are more localized near the ligands where they do not get ionized during the collision. After the collision they are readily available for the interatomic transitions leading to lower p_L values.

4.5. Lifetime Measurements

In highly ionized atoms carrying one or two K-holes, a considerable part of the excited states are metastable. These states are of particular interest since their lifetimes may give information about the decay modes, the state wave functions,[127] and relativistic and electron correlation effects,[128] by which theories[129] may be tested.

Most work in this field has been done with projectile ions ($Z \leqslant 26$) using the beam-foil time-of-flight technique.[130,131] The projectile-ion stripping and excitation by foils is very effective due to a high collision frequency inside the solid, and few electron states will be populated even in ions with relatively large atomic numbers when the projectile velocity is sufficiently high. However, precise lifetime measurements using the beam-foil technique are often disturbed by cascades from high n state populations[131,132] produced with remarkable intensities. Thus, an accurate determination of lifetimes from the decay curves needs a detailed analysis of affecting cascade channels.

Since the radiation is emitted from fast ions, strong Doppler shifts and line broadenings are disadvantages of the beam-foil excitation which cause the need for small-angular collimating detectors. Less restrictive are lifetime measurements with slow target recoil ions using pulsed projectile beams and the delayed-coincidence technique as described in Section 3.1.3.

By this method the lifetimes τ of the $(1s2s2p)\,^4P^o_{5/2}$ states in oxygen and in neon excited by a 1.4 MeV/amu Cu beam have been studied,[54,55] measuring Auger electron decay rates (Figure 4). The $^4P_{5/2}$ state is the lowest quartet state of Li-like ions which decays only through spin–spin interaction, either autoionizing to $(1s^2)\,^1S_0 + k\,^2F_{5/2}$ states or decaying to $(1s^2 2s)\,^2S_{1/2}$ by $M2$ radiation. For target atomic number $8 \leqslant Z \leqslant 10$ τ is within a proper region where it could be measured using the 37-nsec periods of the pulsed UNILAC beam.

In the experiment, a dependence of τ on the target pressure p was observed and explained[133] by "collisional quenching" through which the metastable state is destroyed by a collision of the recoiling ion with another target particle. Figure 25 exhibits $\tau(p)$ for O_2 and Ne targets. For O_2 the curve is much steeper than for Ne, reflecting the higher velocity of the oxygen ion caused by the Coulomb explosion of the O_2 molecule. The lifetime τ was obtained from an extrapolation to $p = 0$, as listed in Table 5 and compared with results from beam-foil experiments[134] and with theoretical values.[129]

In the case of Ne remarkable deviations from the theoretical lifetime are noticed for both methods. The larger lifetime in the beam-foil experiment may be explained by a cascade feeding[131] $(1s2s)\,^3S + e^- \rightarrow (1s2s2p)\,^4P^o$ from high n state populations, which was not taken into

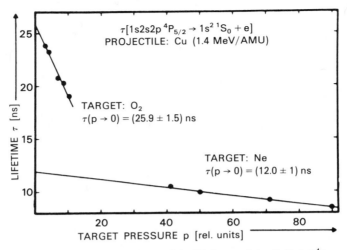

Figure 25. Pressure dependence of the lifetime τ of the $(1s2s2p)\,^4P_{5/2}$ state in O_2 and Ne. (From Ref. 133.)

account. However, for the gas-target experiment the situation is more complex. The quartet cascades which have been discussed in Section 4.1.4 will cause a delayed production of the inner-shell states $(1s2p^2)\,^4P^e$ and $(1s2s2p)\,^4P^o$ because the process arises from secondary collisions (see Section 5.1). Consequently, a lower decay rate of the $^4P^o$ term is observed. When the target pressure is high enough, i.e., the free flight time of the target recoil ion is comparable to $\tau(^4P^o)$, quenching or capture collisions producing Be-like states may occur and the apparent lifetime decreases with increasing pressure. To prevent such disturbances, lifetime measurements using the delayed coincidence technique should be performed at sufficiently low target pressures.

Table 5. Lifetime τ in Nanoseconds of the $1s2s2p\,^4P_{5/2}$ State in O and Ne Extrapolated to Pressure Zero[a]

Target	$\tau_{exp}{}^{a}$	$\tau_{exp}{}^{b}$	$\tau_{theor}{}^{c}$
O_2	25.9 ± 2	20 ± 3	23.1
Ne	12.0 ± 1	10.4 ± 1.5	8.4

[a] Reference 133.
[b] Reference 134.
[c] Reference 129.

4.6. Comparison of Projectile-, Target-, and Plasma-Ion Stripping

Similar and complementary properties of stripping fast atomic beams and atoms in a hot plasma were examined previously by Nagel.[41] In adding a third mode of high-charge-state creation, namely, the stripping of a target atom in a single collision, we are completing the comparison. The reciprocity between stripping of a fast projectile by slow electrons and stripping of a slow ion by fast electrons in a hot plasma enters through the relative velocity between ions and electrons, which is comparable for both cases. In a plasma the electron velocities v_e are characterized by the mean plasma temperature

$$T_e = \frac{2}{3}\frac{v_e}{k} \tag{19}$$

Equating the projectile velocity and the electron velocity the temperature

$$T_e(\text{eV}) = 363\,\frac{E}{A}(\text{MeV/amu}) \tag{20}$$

may be attributed to the ion beam of specific energy E/A allowing a direct comparison of plasma and projectile stripping as a function of temperature. Taking into account the gas–solid difference of projectile stripping and the high-energy wing of the Maxwellian velocity distribution of the plasma electrons it was found[41] that a reduction of the temperature given by Eq. (20) by a factor of 4 is reasonable. Doing so yields identical degrees of ionization as a function of temperature for both plasma stripping and projectile stripping in a gas target.

In contrast to the multiple electron collisions present in the beam-foil and in the plasma source, target-ion stripping proceeds via single interaction with a fast heavy ion. The charge states accessible by this technique depend on the collision parameters, primarily the nuclear charge and the velocity of the projectile.

With the beam-foil technique the limitation in the degree of projectile stripping mainly enters through the parameters set by the heavy-ion accelerator under use, i.e., the beam velocity. At present, projectiles as heavy as krypton may be stripped almost completely.

Various sources of hot plasmas have been set up including laser and plasma focus, vacuum spark, exploding wire, and large tokamak devices. Intensive x-radiation in the 1–3-keV range is emitted from these sources providing a powerful diagnostic tool (see the contribution to this series by D. D. Burgess). As copious ionization involving outer and inner shells results from all three methods the deexcitation spectra reveal quite similar features as demonstrated in Figure 26 where K x-ray spectra of lighter elements are compared.

Figure 26a displays a beam-foil spectrum[135] of 22.2 MeV fluorine ions obtained with the x-ray spectrometer viewing the foil directly. Short-lived states, therefore, contribute dominantly to the whole spectrum. The second spectrum (Figure 26b) is a K x-ray spectrum[136] of Ne recoil ions produced by 1.4 MeV/amu lead ions. Figure 26c is an Al K x-ray spectrum[137] from a plasma produced by a 1.8-J laser pulse of 0.25 nsec duration. The common feature of all three spectra is the observation of Rydberg series of the few-electron ions. Besides the line spectra shown here, continuous radiation can be observed in the plasma as well as in the beam-foil source. The same physical process, namely the pickup of free or quasifree electrons, is responsible for the free–bound continuum in a plasma and the radiative electron capture (REC) by fast projectiles. Such continua are missing in the spectra of recoil ions.

Some differences of the three light sources may be noted having practical consequences. The state of motion of the emitting ions is different. The projectiles as well as the recoils have a preferred direction in the laboratory in contrast to the thermal ions of a plasma. Very important are the differences in the kinetic energies. The plasma ions have kinetic energies of some keV and the recoil ions may be as slow as a few eV whereas the beam-foil particles may reach some tens to hundreds of MeV. With the increase of velocity there is also a strong increase in the Doppler broadening of the spectral lines due to a finite acceptance angle of the spectrometer. The deexcitation process may be disturbed by strong magnetic and electric fields present in plasma sources whereas the deexcitation in the accelerator based experiments occurs in the vacuum. In particular the recoil ions may reach similar stages of ionization as plasma ions but without the disadvantages of large kinetic energies or the disturbance from a hot and dense environment.

5. Secondary-Collision Experiments

One aspect of the high-charge low-velocity recoil ions is to use them as projectiles for secondary collisions. Thereby, a rather exotic situation is obtained where ions with a total electronic excitation energy near 1 keV collide with neutral atoms at laboratory kinetic energies of a few eV. There are two different ways of doing such experiments. The first one is to extract the recoil ions out of the primary target region by electric fields and to direct them onto a second target located some distance away. On the way between the two targets the ions might be accelerated and again decelerated or confined in and released from an ion trap. The other possibility is to

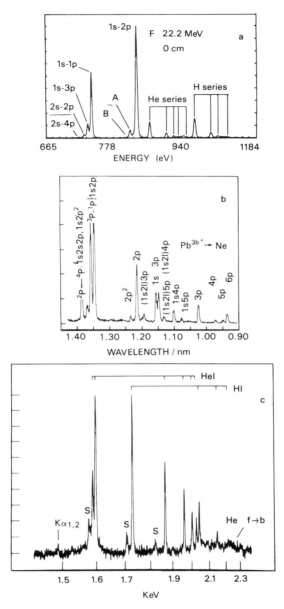

Figure 26. Comparison of K x-ray spectra from (a) beam-foil (from Ref. 135); (b) recoil-ion (from Ref. 136); and (c) plasma excitation (from Ref. 137).

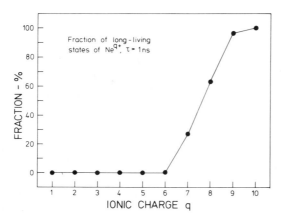

Figure 27. The fraction of states with lifetimes $\tau \geqslant 1$ nsec in neon ions bearing a K vacancy.

study primary and secondary collisions in the same target region and to distinguish between both processes through the differences in their emission spectra.

The latter method has been used[44,101,138] to investigate electron-capture collisions of K-vacancy bearing recoil ions. The possibility of observing secondary-collision effects in the deexcitation spectra of K-vacancy bearing states enters through the lifetime of the recoil-ion states. If the lifetime is sufficiently short the K x-ray or the K Auger-electron is emitted and the K hole is filled before the recoil ion can collide with a surrounding neutral target atom. In that case the x-ray and electron spectra reflect only the excited states produced directly by the heavy-ion impact. If, however, long-lived states are produced by the primary collision, the K hole may survive until it undergoes a secondary collision where the excited state is changed and the subsequently emitted x-ray or electron will bear the signature of the secondary collision. The lifetime of a K vacancy-bearing state is strongly correlated with the charge q of the ion as shown in Figure 27 for multiply ionized neon ions. For ionic charges $q \leqslant 6$ there are no states with lifetimes $\tau > 1$ nsec. If the charge, however, is raised to $q \geqslant 7$ there is a large fraction of metastable states. Therefore, secondary collisions can be readily studied by using very heavy primary ions which effectively produce highly charged metastable recoil ions. Such experiments will be discussed below.

5.1. Selective Electron Capture

When observing the K-vacancy decay of multiply ionized target atoms one generally obtains photons or electrons originating both from the direct population of promptly decaying states and from states populated through electron capture by the slowly recoiling target ions from surrounding target atoms. Therefore, one has to find ways of discriminating between both processes.

The final states resulting from electron capture by highly stripped slow ions are preferentially certain outer-shell states. Therefore, irregularities are expected in the line series observed. This selective electron capture has already been observed by Dixon and Elton[139] for ions emerging from a laser irradiated surface. They found an enhanced population of $n = 4$ excited states in C^{4+} and C^{5+} due to a selective electron capture from neutral carbon by C^{5+} and C^{6+} ions. The addition of a gaseous atmosphere in this experiment did not lead to the observation of capture from the background gas atoms. In the spectra emitted from the recoil ions,[44,101] however, charge transfer between highly stripped ions of one species and ground-state atoms of another species can be observed by using gas mixtures.

The mean free path and the corresponding mean lifetime of long-lived recoil ions is determined by the gas density. If the average time for a second collision is comparable to the atomic lifetime of the electron-collector state, nonlinearities in the pressure dependence will be observable.

A direct way of discriminating the second collision processes is to access the path or time between creation of a recoil ion and the capture event marked by the prompt emission of a photon or electron. This is possible by doing spatially or time-resolved spectroscopy.

By the capture event, the state of motion of the recoil ion is altered through the reaction kinematics which reflect the energy balance of the process. The capture-induced change of the kinetic energy can be measured by the kinematical line broadening of Auger lines.[138]

5.1.1. Collector States and Selectivity

For observation of a second-collision capture process the collector states must have lifetimes exceeding

$$\tau \geqslant (N v_{\text{rec}} \sigma_c)^{-1} \tag{21}$$

where N denotes the target-gas density, v_{rec} the recoil velocity, and σ_c the cross section for electron capture. For a gas pressure of less than 1 mbar and assuming the recoil velocity to be 10^6 cm sec^{-1} and the capture cross section to amount to less than 10^{-14} cm^2, collector lifetimes of at least a few nsec are necessary. The most likely collector configurations are bare nuclei, the $1s$, $1s2s$, $1s2p\ ^3P_{0,2}$, and the $1s2s2p\ ^4P$. The core-excited states (5) observed in the K x-ray and Auger-electron spectra, can thus be produced in two steps, i.e., the production of a metastable core in the primary collision and a subsequent capture into an outer shell.

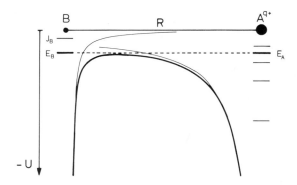

Figure 28. Electron capture in a classical one-electron model. (From Ref. 144.)

From theoretical considerations[140–143] of the capture process at high charge and low velocity it is expected that the electron is transferred to outer shells with a narrow distribution in the principal quantum number n. In order to obtain an estimate for the most probable quantum number n of the final electron state let us consider a simple classical one-electron model[144,101] (see Figure 28) for the charge transfer

$$A^{q+} + B \to A^{(q-1)+}(n) + B^+ + E_{\text{exo}} \qquad (22)$$

where E_{exo} is the exothermic energy defect. At large internuclear distances R, the electron is centered at atom B at an energy level $-J_B$ where J_B is the (positive) ionization potential of atom B. As the highly charged atom A^{q+} approaches the atom B, there are bound states centered around B and bound states centered around A. Neglecting polarization terms the energy levels are represented by

$$E_B = -J_B - q/R \qquad (23a)$$

and

$$E_A = -\frac{q^2}{2n^2} - \frac{1}{R} \qquad (23b)$$

respectively.

Regarding the charge transfer as an over-barrier transition[141] the condition for the electron to cross from B to A is the maximum of the potential barrier—which is the superposition of the two r^{-1} potentials—

should be lower than the energy E_B. The potential maximum is

$$U_{max}(R) = -\frac{(q^{1/2} + 1)^2}{R} \tag{24}$$

Equating (23a) and (23b) and the relation $U_{max} \leqslant E_B$ yields

$$-J_B - \frac{q}{R} = -\frac{q^2}{2n^2} - \frac{1}{R} \tag{25a}$$

$$-\frac{(q^{1/2} + 1)^2}{R} \leqslant -J_B - \frac{q}{R} \tag{25b}$$

Thus the principal quantum number of the hydrogenlike product state is the largest integer n not exceeding n_0 which is given by

$$n_0^2 = \frac{q^2}{2J_B[1 + (q - 1)/(1 + 2q^{1/2})]} \tag{26}$$

and the corresponding crossing distance is

$$R = \frac{q - 1}{q^2/2n^2 - J_B} \tag{27}$$

which may be regarded as an approximation for the crossing point of the respective potential-energy curves.

For the Ne^{10+} + Ne system Eq. (26) gives $n_0 = 5.3$ and for Ne^{8+} + Ne $n_0 = 4.4$. This has to be kept in mind when inspecting the experimentally observed line series. The latter do not reveal a monotonic intensity decrease with increasing principal quantum number. Depending on the degree of ionization reached, certain lines are prominent in the Auger[43] and x-ray[44] line series. As an example this is demonstrated in Figure 29 for the x-ray intensities in Ne^{9+} and Ne^{7+}. Here, the transition $6p \rightarrow 1s$ and the transition $(1s2l)4p \rightarrow 1s^2 2l$ become more and more pronounced when the target ionization is increased by heavier projectile impact. In the case of the lighter ions, i.e., Ar^{12+} for the lithiumlike and Kr^{18+} for the hydrogenlike sequence, the population of the states by the primary heavy-ion collision dominates, whereas in the case of the very heavy projectiles, a large number of collector cores (bare nuclei and $1s2l$ cores) are produced which capture electrons selectively into the 4th and 6th shell. These quantum numbers are in agreement with the values predicted above for the selective electron capture.

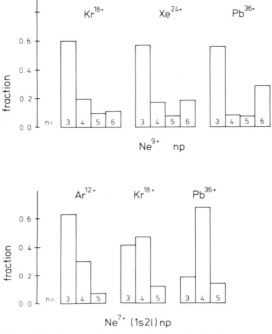

Figure 29. Observed n dependence of the hydrogen- and lithiumlike lines for various heavy projectiles. (From Ref. 44.)

5.1.2. The Dependence on the Ionization Potential

Equation (26) contains the dependence of the principal quantum number n_0 on the ionization potential J_B. This dependence was investigated experimentally[101,44] using for example Ne targets containing admixtures from other gases with different ionization potentials. In the various gas mixtures, the highly ionized Ne recoil ions can thus capture electrons not only from the neon background atoms but also from the admixed neutrals. In Figures 30 and 31 we show examples for this method. The selective electron capture by bare nuclei produced by 1.4 MeV/amu U^{40+} impact is evident in Figure 30 where the hydrogenlike series of Ne^{9+} is depicted. For comparison are added the vertical lines indicating the theoretical transition energies, SL denoting the series limit. In the case of the pure neon target we observe an intensity enhancement at $n = 6$ originating from an electron capture by Ne^{10+} ions from neutral neon atoms. When the ionization potential of the gas admixture is increased as in the Ne/He case there is an additional enhancement involving a lower shell ($n = 5$), and a

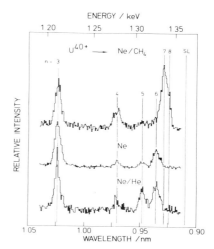

Figure 30. X-ray spectra of hydrogenlike neon excited by 1.4 MeV/amu U^{40+}. The vertical lines indicate calculated energies for the transitions $np \to 1s$ (SL = series limit). Target gases are Ne/He, Ne and Ne/CH$_4$, respectively. (From Ref. 44.)

higher shell ($n = 7$) is involved when the ionization potential is lowered going to CH$_4$.

The same trend was observed before for heliumlike recoil ions in the Auger-electron experiments, showing very clearly the selectivity of electron capture and its dependence on the target ionization potential. As an example, Figure 31 exhibits the high-energy sections of neon Auger-electron spectra from Li-like configurations induced by 1.4 MeV/amu Kr^{18+} impact. The lines originate from initial configurations which have a $1s2s$ or $1s2p$ core and an outer electron with a principal quantum number between 3 and 6. The structure observed within the KLM group reflects

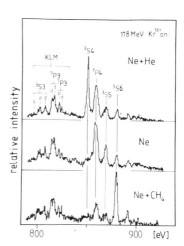

Figure 31. Ne K Auger electron spectra obtained from 1.4 MeV/amu Kr^{18+} on Ne or Ne mixtures with different gases. The notation ^{2S+1}Ln refers to the configuration $(1s2l, {}^{2S+1}L)nl'$.

the population of different l states whereas for the higher n states the multiplet splitting is too small to be resolved. The outer shell configurations with n between 4 and 6 are mainly produced secondarily by electron capture. When the ionization potential of the admixtured gas is lowered going from He to CH_4, states of higher shells, e.g., $^3S4$ to $^3S6$ having lower binding energies, are strongly populated. Simultaneously, the relative intensities from the more strongly bound states decrease, reflecting that the core states $(1s2s)\,^3S$ and $(1s2p)\,^3P$ alternatively capture electrons into different outer shells depending on the ionization potential. A residual intensity in all "capture lines" indicates a weak contribution of direct excitation by the heavy ion impact.

From both Auger-electron and x-ray spectra, consistent results are obtained for a variety of gases and other collector cores as summarized in

Table 6

Ion	Neutral	Ionization potential (a.u.)	Crossing distance R (a.u.)	Principal quantum number		Observed radiation
				Calculated	Experimental	
C^{4+}	CH_4	0.46	7.0	3.2	3	Auger
N^{5+}	NH_3	0.38	10.0	4.3	4	Auger
N^{5+}	CH_4	0.46	12.5	4.0	4	Auger
N^{5+}	H_2	0.57	12.5	3.6	3 (4)	Auger
N^{5+}	N_2	0.62	5.2	3.4	3 (4)	Auger
N^{5+}	Ne	0.79	6.7	3.0	3	Auger
N^{5+}	He	0.90	1.8	2.8	3	Auger
N^{6+}	N_2	0.62	9.9	4.0	4	Auger
O^{6+}	H_2O	0.46	7.5	4.6	4 (5)	Auger
O^{6+}	CO_2	0.50	8.0	4.4	4	Auger
O^{6+}	O_2	0.45	7.4	4.7	5 (4)	X-ray
O^{6+}	Ne	0.79	4.1	3.5	3 (4)	X-ray
O^{6+}	He	0.90	4.5	3.3	3	Auger
O^{7+}	O_2	0.45	11.3	5.3	5	X-ray
O^{7+}	Ne	0.79	8.1	4.0	4	X-ray
O^{7+}	He	0.90	3.3	3.7	4	X-ray
Ne^{8+}	NH_3	0.38	13.8	6.4	6	Auger, X-ray
Ne^{8+}	Xe	0.45	8.4	5.9	6	Auger
Ne^{8+}	CH_4	0.46	8.5	5.8	6	Auger, X-ray
Ne^{8+}	H_2	0.57	9.9	5.2	5	Auger, X-ray
Ne^{8+}	Ar	0.58	10.0	5.2	5	X-ray
Ne^{8+}	Ne	0.79	5.8	4.4	4 (5)	Auger, X-ray
Ne^{8+}	He	0.90	6.4	4.2	4	Auger, X-ray
Ne^{10+}	CH_4	0.46	16.0	7.0	7	X-ray
Ne^{10+}	H_2	0.57	11.0	6.3	6	X-ray
Ne^{10+}	Ar	0.58	11.1	6.3	6	X-ray
Ne^{10+}	Ne	0.79	7.4	5.3	6 (5)	X-ray
Ne^{10+}	He	0.90	8.2	5.0	5	X-ray

Table 6 together with the predictions for the principal quantum number n_0 obtained from Eq. (26). Good agreement is obtained between the experimental and theoretical values of the principal quantum numbers. Included in the table are the crossing distances R [Eq. (27)] corresponding to the integral values of n. From these distances, geometrical cross sections πR^2 may be derived which are in the order of some 10^{-15} cm^2. Such large cross sections are consistent with the observed nonlinear pressure dependence of capture lines resulting from a $1s2p\ ^3P_{0,2}$ core state.[101]

5.1.3. Time Delay and Kinematics

The delayed-coincidence technique provides a method for discriminating between prompt and delayed emission from target atoms. Thereby, detailed information about population processes, decay modes of excited states, and the dynamics of charge exchange reactions can be obtained. The technique has been applied for Auger-electron emission from neon excited by a pulsed heavy-ion beam from the UNILAC.[138,145] For each electron registered in the channeltron a time-to-amplitude converter (TAC) is started. The stop pulse is derived from the beam pulses which approximately have 1 nsec duration and a period of 37 nsec. For each event the time-relation relative to the excitation pulse is stored in list mode together with the channel number corresponding to the electron energy of the analyzer.

Figure 32 contains a typical time distribution for a complete (total) electron spectrum of neon where two time windows are indicated, a delayed

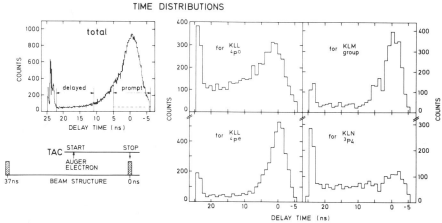

Figure 32. Time spectra for emission of Ne Auger electrons relative to a pulsed 1.4 MeV/amu Kr^{18+} beam on a 50 mTorr Ne target. The total spectrum is accumulated for all electron energies scanned. (From Ref. 145.)

one and a prompt one, which are employed for gating the electron spectra. Also shown in Figure 32 are time spectra for special lines and the *KLM* group of the electron spectra. A pronounced delayed emission is found for the *KLL* $^4P^\circ$ line and for the $^3P4$ line which was previously attributed to the capture. The $1s2s2p\ ^4P^\circ$ state is the only one having a lifetime $(8.4\ \text{nsec}^{(129)})$ greater than the instrumental time resolution of 5 nsec.

In Figure 33 a total electron spectrum from a pure neon target is shown together with delayed (10 nsec) spectra from mixtures with helium and methane using 1.4 MeV/amu Kr^{18+} projectiles in all cases. The intensities in the delayed spectra are strongly suppressed with the exception of the specific high-energy capture lines and the lines from the $^4P^\circ$ and $^4P^e$ states of the *KLL* group. The appearance of the $^4P^\circ$ line in the delayed spectrum may be explained by the lifetime effect. However, the intensity is greater than expected from the lifetime and the applied time window. This suggests that both the $^4P^\circ$ and the $^4P^e$ state which has a smaller lifetime of $0.2\ \text{nsec}^{(129)}$ are produced delayed by a second collision. Direct capture

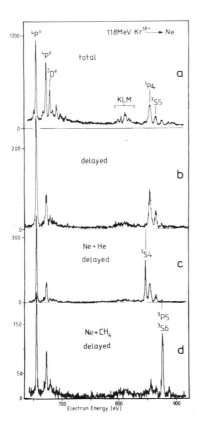

Figure 33. Total and delayed $(t \geqslant 10\ \text{nsec})$ Ne K-Auger spectra from 1.4 MeV/amu Kr^{18+} impact on pure neon and mixtures with neon. (From Ref. 138.)

into the L shell can be excluded from the theoretical model, therefore the quartet cascade as delineated in Section 4.1.4 is the origin of the delayed production of these states.[101] Whenever the electron is captured into an outer shell configuration coupled to a quartet term the direct transition to the K shell is forbidden, whereas the cascade channels feeding the low-lying quartet states are fully open. But when a doublet term is created by the capture the decay to the ground state leads to the capture lines in the high-energy part of the spectra. In Figures 33b and 33c the isolation of the capture lines is observed together with the dependence on the ionization potential producing the $^3S4$ line in the case of helium admixture and the $^3S6$ or $^3P5$ line for a mixture with methane.

For systems with a lower degree of ionization as for Ar^{12+} impact, the method seems to work even more decisively, which is demonstrated in Figure 34. There are many lines from Be-like states populated directly, and only a small fraction of He-like collector cores are produced leading to the capture lines observed in the delayed spectrum. The delayed spectra are almost the same for different projectiles. When the pressure is reduced from 30 to 1 mTorr, the secondary collisions become unlikely and only the $^4P^0$ line remains notable as a pure lifetime effect. Both the $^3P4$ and $^3S5$ capture lines and the cascade-produced $^4P^e$ line disappear at low target pressure.

When examining the line shapes of the low-lying quartets one finds systematically broader lines in the delayed spectra and narrower lines in the prompt spectra and in the delayed spectra taken at very low pressure. The explanation for this behavior is a kinematic Auger-line broadening similar to that observed in the molecular Coulomb explosion of Section

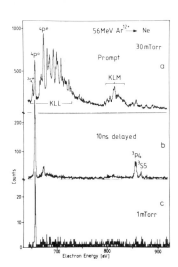

Figure 34. Prompt (± 5 nsec) and delayed ($\geqslant 10$ nsec) Ne K-Auger spectra from Ar^{12+} impact. (From Ref. 138.)

4.3. The kinetic energy of the emitting ion arises from the capture process by which the energy defect E_{exo} of the reaction is converted into the final kinetic energies of the dissociating collision partners according to their mass ratio. Using the same treatment for the linewidth as in Section 4.3 the line broadening ΔE_c due to electron capture may be obtained by

$$\Delta E_c = \{4[E_{exo} \cdot E_A m_e \cdot m_2/(m_1 + m_2)m_1]\}^{1/2} \tag{28}$$

where m_1 and m_2 are the masses of the recoil ion and of the neutral gas atom, respectively. For the investigated systems E_{exo} is about 30 eV, giving capture-induced linewidths of 1–3 eV. In Table 7 the experimental linewidths are shown for the $^4P^o$ and for the $^4P^e$ lines. The delayed $^4P^o$ may still have contributions from prompt excitation due to its lifetime. Therefore, the $^4P^e$ should reflect somewhat better the effect of the secondary capture process. The analyzer FWHM resolution for these data was 1.4 eV and the uncertainty for the widths was ±0.2 eV. In the last row of Table 7 it was attempted to extract the dissociation width ΔE_{diss} by quadratic subtraction according to

$$\Delta E_{diss} = (\Delta E_{delayed}^2 - \Delta E_{prompt}^2)^{1/2} \tag{29}$$

The data reveal the same variation with the mass m_2 as does Eq. (29), giving some confidence in the possibility of extracting the exothermic energy defect E_{exo} with improved instrumental resolution. From the exothermicity of the capture process one may extract the internuclear crossing distance $R = (q - 1)/E_{exo}$ according to the classical model. For the system $Ne^{8+} + Ne$ we obtain $R \approx 6.3$ a.u., resulting in a geometrical cross section $\pi R^2 \approx 1 \times 10^{-15}$ cm^2, which is in the order of the expected large capture cross section.

Table 7. Experimental Linewidths in eV for Ne (KLL)^{4}P Lines in the Net Prompt and in the Delayed Spectra for Bombardment with 1.4 MeV/amu Kr^{18+} ^{a}

$$\Delta E_{diss} = (\Delta E_{delayed}^2 - \Delta E_{prompt}^2)^{1/2}$$

Target (mTorr)	Ne + Xe 25 + 25	Ne 20	Ne + CH$_4$ 20 + 20	Ne + He 40 + 40	Ne + H$_2$ 25 + 25
$^4P^o_{prompt}$	1.1	1.3	1.4	1.2	1.6
$^4P^o_{delayed}$	2.7	2.1	2.0	1.9	2.2
$^4P^e_{prompt}$	1.9	1.9	1.9	1.9	2.3
$^4P^e_{delayed}$	3.9	2.8	2.4	2.4	2.5
$^4P^e$ ΔE_{diss}	3.4	2.0	1.5	1.4	1.0

^{a} The detector resolution was around 1.4 eV and the uncertainty for the widths ± 0.2 eV.

5.2. Potential Applications

Emission spectra from astrophysical and laboratory plasmas require for their interpretation the knowledge of the collisional and radiative processes occurring in the plasma. The spectral features from highly stripped ions have a particular significance since they are used to derive information on the plasma parameters such as density, temperature, and charge states.

In this context some of these processes may be studied in accelerator-based experiments under nonplasma conditions. As opposed to the beam-foil excited fast ions, highly stripped recoil ions have a thermodynamic state close to that of an ion in a hot plasma. It is, therefore, not only the term structure of the excited levels which compares to that of the plasma ions but also the charge-transfer processes at low relative velocities. Charge transfer, involving impurity ions in a controlled-fusion reactor, has received much interest as it influences the overall energy balance and plasma behavior.[146,147] For determination of the plasma cooling by charge transfer, capture cross sections have to be available whereas the distribution of excited states after electron capture might be of importance for diagnostic purposes. It was shown above that selective electron capture makes the slow highly charged ions a sensitive probe for the surrounding gas atmosphere. Conversely the injection of neutral atoms into a hot plasma[148] leads, e.g., fully stripped ions to radiate after selective electron capture, providing a diagnostic for these plasma ions.

In a number of publications[149–152] it has been stressed that pumping by selective electron capture might be a promising way towards achieving amplification by stimulated emission in the soft x-ray region. Selective electron capture has by far the largest cross sections compared to other modes of exciting highly ionized ions as for example electron collisional excitation.

References

1. J. T. Park, J. E. Aldag, J. M. George, and J. L. Peacher, *Phys. Rev. A* **14**, 608 (1976); **18**, 48 (1978).
2. M. E. Rudd and D. H. Madison, *Phys. Rev. A* **14**, 128 (1976).
3. E. S. Selov'er, R. N. Il'in, V. A. Oparin, and N. V. Dedorenko, *Sov. Phys. JETP* **15**, 459 (1962).
4. F. J. deHeer, J. Schutten, and H. Moustafa, *Physics (Utr.)* **32**, 1766 (1966).
5. T. Bratton, C. L. Cocke, and J. R. Macdonald, *J. Phys.* **B10**, L517 (1977).
6. W. Brandt, K. W. Jones, and H. W. Kramer, *Phys. Rev. Lett.* **30**, 351 (1973).
7. R. R. Randall, J. A. Bednar, B. Curnutte, and C. L. Cocke, *Phys. Rev. A* **13**, 204 (1976).
8. N. Lutz, S. Sackmann, and H. O. Lutz, *J. Phys.* **B12**, 1973 (1979).
9. H. W. Schopper, A. D. Betz, J. P. Delvaille, K. Kalata, A. R. Sohval, K. W. Jones, and H. E. Wegner, *Phys. Rev. Lett.* **29**, 898 (1972).

10. H. W. Schopper, A. D. Betz, J. P. Devaille, K. Kalata, A. R. Sohval, K. W. Jones, and H. E. Wegner, *Phys. Lett.* **47A**, 61 (1974).

11. A. R. Sohval, J. P. Delvaille, K. Kalata, K. Kirby-Docken, and H. W. Schopper, *J. Phys.* **B9**, L25 (1976).

12. F. W. Saris, W. F. van der Weg, H. Tawara, and R. Laubert, *Phys. Rev. Lett.* **28**, 717 (1972).

13. P. H. Mokler, H. J. Stein, and P. Armbruster, *Phys. Rev. Lett.* **29**, 827 (1972).

14. J. R. Macdonald and M. D. Brown, *Phys. Rev. Lett.* **29**, 4 (1972).

15. W. E. Meyerhof, T. K. Saylor, S. M. Lazarus, W. A. Little, B. B. Triplett, and L. F. Chase, Jr., *Phys. Rev. Lett.* **30**, 1279 (1973).

16. C. K. Davis and J. S. Greenberg, *Phys. Rev. Lett.* **32**, 1215 (1974).

17. W. Wölfli, Ch. Stoller, G. Bonani, M. Stuter, and M. Stöckli, *Phys. Rev. Lett.* **35**, 656 (1975).

18. E. Merzbacher and H. W. Lewis, *Handbuch der Physik*, S. Flügge, ed., Vol. 34, p. 199 (1958).

19. J. Bang and J. M. Hansteen, *K. Dan. Vidensk. Selsk. Mat.-Fys. Medd.* **31**, 13 (1959).

20. J. D. Garcia, *Phys. Rev. A* **4**, 955 (1971) and references therein.

21. F. Hopkins, A. Little, and N. Cue, *Phys. Rev. A* **14**, 1634 (1976).

22. U. Schiebel, B. L. Doyle, J. R. Macdonald, and L. D. Ellsworth, *Phys. Rev. A* **16**, 1089 (1977); **17**, 523 (1978).

23. H. Tawara, P. Richard, K. A. Jamison, T. J. Gray, J. Newcomb, and C. Schmiedekamp, *Phys. Rev. A* **19**, 1960 (1979).

24. C. L. Cocke, R. K. Gardner, B. Curnutte, T. Bratton, and R. K. Saylor, *Phys. Rev. A* **16**, 2248 (1977).

25. M. Rødbro, E. H. Pedersen, C. L. Cocke, and J. R. Macdonald, *Phys. Rev. A* **19**, 1936 (1979).

26. C. W. Woods, R. L. Kauffman, K. A. Jamison, N. Stolterfoht, and P. Richard, *Phys. Rev. A* **13**, 1358 (1976).

27. T. J. Gray, P. Richard, K. A. Jamison, J. M. Hall, and R. K. Gardner, *Phys. Rev. A* **14**, 1333 (1976).

28. R. K. Gardner, T. J. Gray, P. Richard, C. Schmiedekamp, K. A. Jamison, and J. M. Hall, *Phys. Rev. A* **15**, 2202 (1977); **19**, 1896 (1979).

29. T. J. Gray, P. Richard, G. Gealy, and J. Newcomb, *Phys. Rev. A* **19**, 1424 (1979); F. D. McDaniel, J. L. Duggan, G. Basbas, P. D. Miller, and G. Lapicki, *Phys. Rev. A* **16**, 1375 (1977).

30. H. Tawara, P. Richard, T. Gray, J. Newcomb, K. A. Jamison, C. Schmiedekamp, and J. M. Hall, *Phys. Rev. A* **18**, 1373 (1978).

31. H. Tawara, P. Richard, T. J. Gray, J. R. Macdonald, and R. Dillingham, *Phys. Rev. A* **19**, 2131 (1979).

32. G. Basbas, W. Brandt, and R. Laubert, *Phys. Rev. A* **7**, 983 (1973).

33. G. Basbas, W. Brandt, and R. Laubert, *Phys. Rev. A* **17**, 1655 (1978).

34. U. Fano and W. Lichten, *Phys. Rev. Lett.* **14**, 627 (1965).

35. J. S. Briggs and J. Macek, *J. Phys.* **B5**, 579 (1972).

36. J. S. Briggs and K. Taulbjerg, *J. Phys.* **B8**, 1909 (1975); **B9**, 1 (1976).

37. K. Taulbjerg, J. S. Briggs, and J. Vaaben, *J. Phys.* **B9**, 1351 (1976).

38. C. E. Moore, *Atomic Energy Levels*, NSRDS-NBS 35, U.S. Government Printing Office Washington, D.C. (1971).

39. P. F. A. Klinkenberg, in *Methods of Experimental Physics*, Vol. 13A, D. Williams, ed., Academic Press, New York (1976), pp. 253.

40. H. W. Kugel and D. E. Nurnick, *Rep. Prog. Phys.* **40**, 297 (1977).

41. D. J. Nagel, in *Beam-Foil Spectroscopy*, I. A. Sellin and D. J. Pegg, eds., Plenum Press, New York (1976), Vol. 2, pp. 961.

42. L. Cocke, T. J. Gray, and E. Justiniano, XIth International Conference on the Physics of Electronic and Atomic Collisions, Kyoto 1979, p. 600.

43. N. Stolterfoht, D. Schneider, R. Mann, and F. Folkmann, *J. Phys.* **B10**, L281 (1977).

44. H. F. Beyer, K.-H. Schartner, and F. Folkmann, *J. Phys.* **B13**, 2459 (1980); H. F. Beyer, F. Folkmann, and K.-H. Schartner, *J. Phys. (Paris)* **40**, C1-17 (1979).

45. R. Mann, F. Folkmann, R. S. Peterson, Gy. Szabó, and K.-O. Groeneveld, *J. Phys.* **B11**, 3045 (1978).

46. M. E. Rudd, T. Jorgensen, and D. J. Volz, *Phys. Rev. Lett.* **16**, 929 (1966).

47. Yu. S. Gordeev and G. N. Ogurtsov, *Sov. Phys.-JETP* **33**, 1105 (1971).

48. P. Dahl, M. Rødbro, B. Fastrup, and M. E. Rudd, *J. Phys.* **B9**, 1567 (1976).

49. M. E. Rudd and J. H. Macek, in *Case Studies in Atomic Physics*, Vol. 3, M. R. C. McDowell and E. W. McDaniel, eds., North-Holland, Amsterdam (1972), p. 47.

50. J. A. Demarest and R. L. Watson, *Phys. Rev. A* **17**, 1302 (1978).

51. Applied Research Laboratories, P.O. Box 129, Sunland, California 91040.

52. P. Erman, *Phys. Scr.* **11**, 65 (1975).

53. R. E. Imhof and F. H. Read, *Rep. Prog. Phys.* **40**, 1 (1977).

54. I. A. Sellin, R. Mann, H. J. Firschkorn, D. Rosich, S. Schumann, and Gy. Szabó, *Bull. Am. Phys. Soc.* **22**, 1320 (1977); and GSI-Jahresbericht GSI-J-1-78 (1977).

55. I. A. Sellin, in *Topics in Current Physics*, Vol. 1, S. Bashkin, ed., pp. 273, Springer-Verlag, Berlin (1976).

56. A. Nicklas, P. Strehl, and H. Vilhjalmsson, GSI-Pb-5-75-Mai (1976).

57. C. L. Cocke, *Phys. Rev. A* **20**, 749 (1979).

58. P. Richard, I. L. Morgan, T. Furuta, and D. Burch, *Phys. Rev. Lett.* **23**, 1009 (1969).

59. P. Richard, in *Atomic Inner-Shell Processes*, B. Craseman, ed., Academic, New York (1975), Vol. 1, p. 73.

60. A. R. Knudson, D. J. Nagel, P. G. Burkhalter, and K. L. Dunning, *Phys. Rev. Lett.* **26**, 1149 (1971).

61. F. Hopkins, D. O. Elliott, C. P. Bhalla, and P. Richard, *Phys. Rev. A* **8**, 2952 (1973).

62. J. McWherter, J. Bolger, C. F. Moore, and P. Richard, *Z. Phys.* **263**, 283 (1973).

63. C. F. Moore, D. L. Matthews, and H. H. Wolter, *Phys. Lett.* **54A**, 407 (1975).

64. R. L. Watson, T. Chiao, F. E. Jenson, and B. I. Sonobe, in *Beam-Foil Spectroscopy*, I. A. Sellin and D. J. Pegg, eds., Plenum, New York (1976), p. 567.

65. R. L. Watson, F. E. Jenson, and T. Chiao, *Phys. Rev. A* **10**, 1230 (1974).

66. J. R. Mowat, I. A. Sellin, D. J. Pegg, R. Peterson, M. D. Brown, and J. R. Macdonald, *Phys. Rev. Lett.* **30**, 1289 (1973).

67. M. H. Chen, B. Craseman, and D. L. Matthews, *Phys. Rev. Lett.* **34**, 1309 (1975).

68. R. L. Kauffman, F. F. Hopkins, C. W. Woods, and P. Richard, *Phys. Rev. Lett.* **31**, 621 (1973).

69. R. L. Kauffman, C. W. Woods, K. A. Jamison, and P. Richard, *Phys. Rev. A* **11**, 872 (1975).

70. D. L. Matthews, B. M. Johnson, and C. F. Moore, *Phys. Rev. A* **10**, 451 (1974).

71. C. F. Moore, J. Bolger, K. Roberts, D. K. Olsen, B. M. Johnson, J. J. Mackey, L. E. Smith, and D. L. Matthews, *J. Phys.* **B7**, L415 (1974).

72. D. L. Matthews, B. M. Johnson, G. W. Hoffmann, and C. F. Moore, *Phys. Lett.* **49A**, 195 (1974).

73. M. D. Brown, J. R. Macdonald, P. Richard, J. R. Mowat, and I. A. Sellin, *Phys. Rev. A* **9**, 1470 (1974).

74. J. R. Mowat, R. Laubert, I. A. Sellin, R. L. Kauffman, M. D. Brown, J. R. Macdonald, and P. Richard, *Phys. Rev. A* **10**, 1446 (1974).

75. D. L. Matthews, R. J. Fortner, D. Schneider, and C. F. Moore, *Phys. Rev. A* **14**, 1561 (1976).

76. D. L. Matthews, B. M. Johnson, and C. F. Moore, *At. Data Nucl. Data Tables* **15**, 42 (1975).

77. C. P. Bhalla, *Phys. Rev. A* **12**, 122 (1975).

78. C. P. Bhalla, *J. Phys.* **B8**, 2787 (1975).

79. A. R. Knudson, D. J. Nagel, and P. G. Burkhalter, *Phys. Lett.* **24A**, 69 (1972).

80. P. Richard, R. L. Kauffman, J. H. McGuire, C. F. Moore, and D. K. Olsen, *Phys. Rev. A* **8**, 1369 (1973).

81. D. K. Olsen and C. F. Moore, *Phys. Rev. Lett.* **33**, 194 (1974).

82. R. L. Kauffman, C. W. Woods, K. A. Jamison, and P. Richard, *J. Phys.* **B7**, 1335 (1974).

83. J. P. Desclaux, *Comput. Phys. Commun.* **9**, 31 (1975).

84. M. H. Chen and B. Crasemann, *Phys. Rev. A* **12**, 959 (1975).

85. T. W. Tunnell and C. P. Bhalla, *Phys. Lett.* **67A**, 119 (1978).

86. J. M. Hansteen and O. P. Mosebekk, *Phys. Rev. Lett.* **29**, 1361 (1972).

87. J. H. McGuire and P. Richard, *Phys. Rev. A* **8**, 1374 (1973).

88. P. Richard, in *Proceedings of the XIth International Conference on the Physics of Electronic and Atomic Collisions—Invited Papers and Progress Reports*, N. Oda and K. Takayanagi, eds., p. 125, North-Holland Publishing Company, Amsterdam (1979).

89. R. L. Watson, B. I. Sonobe, J. A. Demarest, and A. Langenberg, *Phys. Rev. A* **19**, 1529 (1979).

90. N. Stolterfoht, H. Gabler, and U. Leithäuser, *Phys. Lett.* **45A**, 351 (1973).

91. D. Burch, N. Stolterfoht, D. Schneider, H. Wiemann, and J. S. Risley, Nuclear Physics Annual Report (1974) (University of Washington, Seattle 1975) unpublished.

92. R. Mann, Ph.D. thesis (1977).

93. N. Stolterfoht, in *Proceedings of the Fourth Conference on the Scientific and Industrial Applications of Small Accelerators*, J. L. Duggan and I. L. Morgan, eds., p. 311, The Institute for Electrical and Electronic Engineers Inc., Denton, Texas (1976).

94. M. O. Krause, F. A. Stevie, L. J. Lewis, T. A. Carlson, and W. E. Moddeman, *Phys. Lett.* **31A**, 81 (1970).

95. D. L. Matthews, B. M. Johnson, J. J. Mackey, and C. F. Moore, *Phys. Rev. Lett.* **31** (1973).

96. D. Schneider, M. Prost, P. Ziem, and N. Stolterfoht, in *Abstracts of Contributed Papers of the 7th International Conference on Atomic Physics*, p. 102, Massachusetts Institute of Technology (1980).

97. D. L. Matthews, B. M. Johnson, L. E. Smith, J. J. Mackey, and C. F. Moore, *Phys. Lett.* **48A**, 93 (1974).

98. N. Stolterfoht, D. Schneider, P. Richard, and R. L. Kauffman, *Phys. Rev. Lett.* **33**, 1418 (1974).

99. D. Schneider, C. F. Moore, and B. M. Johnson, *J. Phys.* **B9**, L153 (1976).

100. S. Schumann, K.-O. Groeneveld, K. D. Sevier, and B. Fricke, *Phys. Lett.* **31**, 1331 (1973).

101. R. Mann, F. Folkmann, and H. F. Beyer, *J. Phys. B* **14**, 1161 (1981). R. Mann and F. Folkmann, *J. Phys. (Paris)* **40**, Cl-236 (1979).

102. Z. I. Kuplyauskis, *Opt. Spektrosk.* **44**, 1200 (1978).

103. M. H. Chen, *Phys. Rev. A* **15**, 2318 (1977).

104. C. Schmiedekamp, B. L. Doyle, T. J. Gray, R. K. Gardner, K. A. Jamison, and P. Richard, *Phys. Rev. A* **18**, 1892 (1978).

105. W. Brandt and G. Lapicki, *Phys. Rev. A* **10**, 474 (1974).

106. T. Tonuma, Y. Awaya, T. Kambara, H. Kumagai, I. Kohno, and S. Özkök, *Phys. Rev. A* **20**, 989 (1979).

107. T. J. Gray, C. L. Cocke, and E. Justiniano, *Phys. Rev. A* **22**, 849 (1980).

108. Q. C. Kessel and E. Everhart, *Phys. Rev.* **16**, 16 (1966).

109. R. Olson, *J. Phys.* **B12**, 1843 (1979).

110. A. Russek and M. T. Thomas, *Phys. Rev.* **109**, 2015 (1958); **114**, 1538 (1959).
111. J. B. Bulman and A. Russek, *Phys. Rev.* **122**, 506 (1961); A. Russek, *ibid.* **132**, 246 (1963).
112. A. Russek and J. Meli, *Physica* **46**, 222 (1970).
113. J. R. Mowat, I. A. Sellin, P. M. Griffin, D. J. Pegg, and R. S. Peterson, *Phys. Rev. A* **9**, 644 (1974).
114. K. T. Cheng, C. P. Lin, and W. R. Johnson, *Phys. Lett.* **48A**, 437 (1974).
115. R. Mann, F. Folkmann, and K.-O. Groeneveld, *Phys. Rev. Lett.* **37**, 1674 (1976).
116. H. Hartung, B. Fricke, T. Morović, W.-D. Sepp, and A. Rosén, *Phys. Lett.* **69A**, 87 (1978).
117. D. S. Gemmel, J. Remillieux, J.-C. Poizat, M. J. Gaillard, R. E. Holland, and Z. Vager, *Phys. Rev. Lett.* **34**, 1420 (1975).
118. H. G. Berry, A. E. Livingston, and G. Gabrielse, *Phys. Lett.* **64A**, 68 (1977).
119. Z. Vager and D. S. Gemmell, *Phys. Rev. Lett.* **37**, 1352 (1976).
120. E. P. Kanter, P. J. Cooney, D. S. Gemmell, K.-O. Groeneveld, W. J. Pietsch, A. J. Ratkowski, Z. Vager, and B. J. Zabransky, *Phys. Rev. A* **20**, 834 (1979).
121. R. L. Watson, A. K. Leeper, B. I. Sonobe, T. Chiao, and F. E. Jenson, *Phys. Rev. A* **15**, 914 (1977).
122. H. R. Philipp and H. Ehrenreich, *Phys. Rev.* **129**, 1550 (1963).
123. R. A. Pollak, L. Ley, F. R. McFeely, S. P. Kowalczyk, and D. A. Shirley, *J. Electron Spectrosc. Relat. Phenom.* **3**, 381 (1974).
124. R. L. Kauffman, K. A. Jamison, T. J. Gray, and P. Richard, *Phys. Rev. Lett.* **36**, 1074 (1976).
125. F. Hopkins, A. Little, N. Cue, and V. Dutkiewicz, *Phys. Rev. Lett.* **37**, 1100 (1976).
126. L. Pauling, *The Nature of the Chemical Bond*, Cornell U.P., Ithaca, New York (1960).
127. R. Marrus and R. W. Schmieder, *Phys. Rev. A* **5**, 1160 (1972).
128. H. H. Haselton, R. S. Thoe, J. R. Mowat, P. M. Griffin, D. J. Pegg, and I. A. Sellin, *Phys. Rev. A* **11**, 468 (1975).
129. K. T. Cheng, C. P. Lin, and W. R. Johnson, *Phys. Lett.* **48A**, 437 (1974).
130. B. Donally, W. W. Smith, D. J. Pegg, M. Brown, and A. Sellin, *Phys. Rev. A* **4**, 122 (1971).
131. H. D. Dohmann and R. Mann, *Z. Phys.* **A291**, 15 (1979).
132. F. Hopkins and P. von Brentano, *J. Phys.* **B9**, 775 (1976).
133. S. Schumann, I. A. Sellin, R. Mann, H. J. Frischkorn, D. Rosich, Gy. Szabó, K.-O. Groeneveld, *J. Phys. (Paris)* C1-221 (1979).
134. K.-O. Groeneveld, R. Mann, G. Nolte, S. Schumann, and R. Spohr, *Phys. Lett.* **54A**, 335 (1975).
135. R. L. Kauffman, C. W. Woods, F. F. Hopkins, D. O. Elliott, K. A. Jamison, and P. Richard, *J. Phys.* **B6**, 2197 (1973).
136. H. F. Beyer, R. Mann, K.-H. Schartner, and F. Folkmann, in *European Group for Atomic Spectroscopy, Summaries of Contributions*, Paris-Orsay (1979), p. 17.
137. D. J. Nagel, P. G. Burkhalter, C. M. Dozier, J. F. Holzrichter, B. M. Klein, J. M. McMahon, J. A. Stamper, and R. R. Whitlock, *Phys. Rev. Lett.* **33**, 743 (1974).
138. R. Mann, H. F. Beyer, and F. Folkmann, *Phys. Rev. Lett.* **46**, 646 (1981).
139. R. H. Dixon and R. C. Elton, *Phys. Rev. Lett.* **38**, 1072 (1977).
140. L. P. Resnyakov and A. D. Ulantsev, *Sov. J. Quant. Electron.* **4**, 1320 (1975).
141. M. I. Chibisov, *JETP Lett.* **24**, 46 (1976).
142. C. Harel and A. Salin, *J. Phys.* **B10**, 3511 (1977).
143. H. Ryufuku and T. Watanabe, *Phys. Rev. A* **20**, 1828 (1979).
144. H. F. Beyer, GSI-Report 79-6 ISSN: 0171-4546.

145. F. Folkmann, R. Mann, and H. F. Beyer, in *International Conference on X-Ray Processes and Inner-Shell Ionization*, D. J. Fabian, L. M. Watson, and H. Kleinpoppen, eds., Stirling (1980).

146. V. V. Afrosimov, Yu. S. Gordeev, A. N. Zinovev, and A. A. Korotkov, *Pis'ma Zh., Eksp. Teor. Fiz.* **28**, 540 (1978).

147. M. N. Panov, in *Proceedings of the VIIth Intern. Summer School on the Physics of Ionized Gases*, B. Navinsek, ed., p. 165, Dubrovnik (1976).

148. M. N. Panov, in *Abstracts of Contributed Papers. XIth International Conference on the Physics of Electronic and Atomic Collisions*, K. Takayanagi and N. Oda, eds., p. 437, North-Holland Publishing Company, Amsterdam (1979).

149. A. V. Vinogradov and I. I. Sobel'man, *Sov. Phys. JETP* **36**, 1115 (1973).

150. W. H. Louisell and M. O. Scully, *Phys. Rev. A* **11**, 989 (1975).

151. R. C. Elton, in *Progress in Laser and Laser Fusion*, A. Perlmutter and S. M. Widmayer, eds., p. 117, Plenum Press, New York (1975).

152. R. W. Waynant and R. C. Elton, *Proc. IEEE* **64**, 1059 (1976).

10

Investigations of Superheavy Quasiatoms via Spectroscopy of δ Rays and Positrons

HARTMUT BACKE AND CHRISTOPHOR KOZHUHAROV

1. Introduction

There exists a long-standing and very interesting problem in atomic physics, namely, the question: What is the binding energy of an electron if the strength of the Coulomb potential exceeds $Z\alpha = 1$? According to the Dirac–Sommerfeld fine-structure formula for a point charge

$$E = m_e c^2 [1 - (Z\alpha)^2]^{1/2} \tag{1}$$

the total energy of the lowest bound $1s$-state becomes imaginary for $Z\alpha > 1$. But even as early as 1945 it was realized[59] that this property of Eq. (1) is caused by the singularity of the Coulomb potential at the origin. Assuming a realistic charge distribution of the nucleus there is no restriction such as $Z\alpha < 1$ for the binding energy. Recent calculations show (cf., e.g., Ref. 81) that the binding energy exceeds $2m_e c^2 = 1.022$ MeV at a critical charge $Z_{cr} \simeq 173$.

This state "dives" into the negative energy Dirac sea and becomes a resonance embedded there. An atom with a charge number $Z > Z_{cr}$ is called overcritical. Such an atom, stripped of all electrons, would behave

HARTMUT BACKE • Johannes-Gutenberg-Universität, Mainz, Federal Republic of Germany   CHRISTOPHOR KOZHUHAROV • Gesellschaft für Schwerionenforschung mbH, Darmstadt, Federal Republic of Germany.

in a quite extraordinary way. The $1s$-shell could be spontaneously filled by two electrons from the Dirac sea, resulting in holes, which escape as positrons with kinetic energies $E_{e^+} = E_B - 2m_e c^2$, with E_B the binding energy.

The spectroscopy of these positrons could provide a new method to investigate overcritical atoms resulting in information on the Lamb shift, on finite size effects of the nuclear charge distribution in the high-Z region, etc.

Of course, atoms with charge numbers $Z > Z_{cr} \simeq 173$ do not exist in nature. The hitherto existing attempts to produce superheavy elements artificially have failed.[36] On the other hand, with the heavy ion accelerator UNILAC at GSI in Darmstadt, uranium ions can be accelerated to energies greater than 1.5 GeV. In a scattering experiment, two uranium nuclei may then approach each other to distances less than 20 f. Under these conditions, a quasi atom is formed and the electrons are exposed for the very short time of about 2×10^{-21} sec to the combined nuclear charge $Z_u e = (Z_1 + Z_2)e$. In that time, a $1s$-electron localized to the Compton wavelength covers a distance of 600 f (assuming a speed of light) and may pass this region only twice. With this in mind the justification for speaking about quasi "atoms" may be questionable. In any case, it is possible to describe the electrons quantum-mechanically in an adiabatically varying two-center potential of the scattered ions. It is assumed that at all distances, the electrons have time enough to adjust to the two-center potential. The validity of this assumption can be tested by determining if physical phenomena like δ-electron or positron emission can be described adequately by lowest-order time-dependent adiabatic perturbation theory. If this is a bad approximation, we will be able to obtain physically significant results only from coupled channel calculations.

The current theoretical (cf., e.g., Refs. 62, 60, and references cited therein) and experimental investigations of quasi atoms gives credence to the use of the adiabatic basis. In the example presented in Figure 1 the binding energy of the $1s$ state exceeds twice the rest mass of the electron for a very short time of 2×10^{-21} sec. The experimental question is: If this situation actually occurs, how can it be observed. In principle, there are four possibilities to investigate quasi atoms. All of them are based on the presumption that a very strong perturbation acts on the electrons resulting in δ-electron emission. Experimentally, either the δ-electrons or processes associated with the resultant holes can be investigated. These holes can be observed by measurement of emission of quasi-molecular X rays, emission of characteristic X rays from the separated atoms, or emission of positrons (see, e.g., Refs 3, 4, 5, 6, 20, 21, 31, 39, 40, 43, 44, 49, 55).

In this contribution we restrict ourselves to the investigation of high-energy δ rays and positrons. We shall see in Section 2 that the high-energy

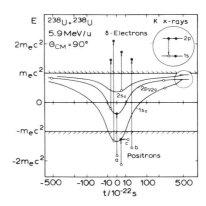

Figure 1. Binding energies of the $1s\sigma$, $2p_{1/2}\sigma$ and $2s\sigma$ levels as a function of the time for the ^{238}U $+ ^{238}$U collision system. Indicated are inner-shell ionization processes with δ-ray emission, three positron creation processes, spontaneous (c), induced (a), and direct (b), and the most probable process to fill an inner-shell vacancy—the K x-ray emission in the separated atoms (Ref. 4).

tail of the δ-ray spectrum contains detailed information about the high-energy component of the momentum distribution of the bound electrons in the quasi atom. In Section 3 we shall show that, at least in principle, the most direct way to investigate the question of the binding energy of very high Z systems is the spectroscopy of positrons, since a preionized diving level may decay by spontaneous positron emission. However, as long as the collision partners follow pure Rutherford trajectories, the diving time is very short and the positron spectra are expected to be dominated by dynamically induced positron emission from the Fourier frequency spectrum of the collision process.

We will see that "spectroscopy" is not to be understood in the usual way. What we observe experimentally are more or less structureless continua, which decrease exponentially as a function of the electron or positron energy. From these spectra, physically relevant information has to be extracted. This situation may change if the collision partners experience nuclear contact. Recent calculations show[63,73] that a sticking time of a few 10^{-21} sec would be sufficient to cause oscillations in the spectral distribution of δ rays or positrons. From these oscillations, information about the nuclear sticking time could be extracted. If the binding energy of the $1s\sigma$ level exceeds $2m_ec^2$ a nuclear sticking time of about 10^{-20} sec could produce a pronounced positron peak caused by the spontaneous decay of $1s\sigma$ vacancies. The experimental observation of such a peak could make possible a spectroscopy of superheavy collision systems in the known sense.

2. Spectroscopy of High-Energy δ Rays

2.1. General Considerations

Parallel to the fact that the observation of the ionization "messenger"— the emitted electron (δ ray, δ electron)—represents a direct way of studying both the mechanism of ionization and the behavior of the ionized bound state itself, we would like to emphasize the reasons for which high-energy δ-ray spectroscopy is well suited for investigations of strongly bound quasi-atomic states formed transiently during heavy-ion collisions. As an illustrative example let us consider a Pb–Pb encounter at relative velocity $(v_\infty/c) = 0.10$, which is much smaller than the relativistic velocities of K electrons in the Pb atom. The bombarding energy in this case runs to approximately 1 GeV and leads to a distance of closest approach for a head-on collision of 20 f, also much smaller than the K-shell radius of the colliding atoms. As already mentioned before, the wave functions of the strongly bound, fast inner-shell electrons are expected to adjust smoothly in the two-center Coulomb field of the combined charge of projectile and target nucleus for all internuclear separations. The time variation of the field may induce transitions from inner bound states to the positive energy continuum. The ionized electrons are ejected with continuous momentum distributions, the high-energy tail of which—above 100 keV up to 2 MeV—is the object of our investigations.

The reason for our interest in the high-momentum component of the spectrum is easily understood qualitatively: For pure kinematical reasons, high momentum can be transferred in a heavy ion collision only to electrons with very high initial momentum, i.e., only if the electron "stems" from those portions of the initial orbit, which are very close to the charge center. It is also evident that δ rays with extremely high energies (while of low absolute intensities) originate mostly from the sharper localized innermost shells. But even more important, a rapid increase of the high-momentum components can be expected for superheavy quasiatoms considered, since the extremely strong Coulomb field leads to enormous relativistic contraction of the electronic wave functions. Hence, the high-energy tail of the δ-ray spectrum can be attributed to the high-energy component of the momentum distribution of strongly bound quasi-atomic electrons and, thus, provide information about their wave function and energy.

The contributions from high-lying, weakly bound electronic shells to the high-energy part of the δ-ray spectrum can be neglected, due to the very weak localization of their wave functions. The latter are also not expected to vary smoothly between the atomic and the united atom limits, because of the very small velocity of the bound electrons compared to the relative motion of the ions.

It should be pointed out, however, that the observed high-energy part of the δ-ray spectrum represents a superposition of contributions from several inner shells—above all, from the quasi-atomic K and L shells. Indeed, the K shell is more sharply localized than the L shells—i.e., the value of the high-energy component is higher in the region of initial momenta considered. However, the ionization probability depends also on the energy transfer needed ($\Delta E = E_\delta + |E_B|$) to induce transition from a state with binding energy E_B to continuum state with E_δ. Thus, the stronger binding can cancel to some extent the effects of sharper localization up to δ-ray kinetic energies $E_\delta \gg |E_B|$, at which the binding can be neglected.

A preferential observation of the contributions from a particular quasi-atomic shell to the total δ-ray spectrum is possible in some cases, if only δ rays, associated with a characteristic x-ray emission of one of the collision partners, are being detected.[45] The vacancy left in a shell after δ-electron decays in a time of $\tau_{\text{hole}} \approx 10^{-17}$ sec, which is much longer than the collision time ($\tau_{\text{coll}} \approx 10^{-21}$ sec)—i.e., the decay occurs after the nuclei are well separated and the quasi-molecular picture does not hold. Thus, the hole decays predominantly via x-ray emission from a separated atom. It was shown[52,54] that a vacancy produced in the $1s\sigma$ state of the quasiatom becomes a K hole in the heavier atom, whereas a $2p_{1/2}$ vacancy occurs as a K hole in the lighter collision partner. In symmetrical or nearly symmetrical collisions ($Z_1 \approx Z_2$) vacancies produced in higher shells may be transferred to lower, stronger bound states in the outgoing phase of the collision, where at large two-center distances the energy splitting between those levels becomes small (cf. also Figure 1).

Thus, for asymmetric collisions a measurement of high-energy δ rays associated with the characteristic K X rays of the heavier collision partner leads to spectroscopy of the innermost $1s\sigma$ quasi-atomic shell.

For very low δ-ray energies and heavy collision systems, where the multiplicity of δ-ray emission per collision is higher than unity, it is possible, however, to detect a δ electron from a higher shell emitted in a collision in which a K hole has been also produced.

2.2. Physical Quantities in δ-Ray Spectroscopy

The objective of δ-ray spectroscopy in heavy-ion collisions is in principle to investigate the triple differential cross section for δ-ray emission:

$$d^3\sigma_\delta/(dE_\delta \, d\Omega_\delta \, d\Omega_p)(E_\delta, \Theta_\delta, \phi_\delta, \Theta_p, E_1, M_1, Z_1, M_2, Z_2) \qquad (2)$$

which is differential with respect to the δ-ray kinetic energy E_δ, to the solid angle element $d\Omega_\delta$ of the emitted electron, and to the solid angle element $d\Omega_p$ of the scattered ejectiles or recoils. In the experiments described below, the beam energy E_1 (or relative ion velocity at infinity v_∞) is such that the colliding nuclei interact only by the electromagnetic interaction and, therefore, follow nearly pure Rutherford trajectories. The quantity Θ_δ describes the emission angle of the electron with respect to the beam axis, ϕ_δ—with respect to the scattering plane, whereas Θ_p is the scattering angle relative to the beam direction of the detected ejectiles or recoils. All quantities are measured in the laboratory system. The cross section is futhermore a function of the charge numbers Z_1, Z_2, and masses M_1, M_2 of the projectile and target nuclei, respectively.

From an experimental point of view it is more convenient to measure the energy and solid angle differential probability for δ-electron production:

$$\frac{d^2 P_\delta}{dE_\delta \, d\Omega_\delta} = \frac{d^3 \sigma_\delta / (dE_\delta \, d\Omega_\delta \, d\Omega_p)}{(d\sigma_R / d\Omega_p)} \approx \frac{\Delta^2 N_\delta^{\text{exp}} / [\varepsilon_\delta(E_\delta) h]}{\Delta E_\delta \Delta \Omega_\delta N_p} \tag{3}$$

which is obtained by normalizing the cross section (2) with respect to the Rutherford cross section. Experimentally, the number $\Delta^2 N_\delta^{\text{exp}}$ of electrons in an energy interval ΔE_δ and solid angle element $\Delta \Omega_\delta$ is measured in coincidence with the scattered particles detected (N_p represents the total number of them). In the case of asymmetric collision systems the particle scattering trajectories can be defined unambiguously by means of kinematical coincidences if the time and/or angular resolution of the detecting system is sufficient.

This is, of course, impossible for symmetric or nearly symmetric encounters, where ejectiles and recoils cannot be distinguished. In this case a scattering angle Θ_p in the laboratory system defines two scattering angles in the c.m. system $\Theta_p^{\text{c.m.}} = 2\Theta_p$ and $(\pi - \Theta_p^{\text{c.m.}})$. Only for $\Theta_p = 45°$ is the Rutherford trajectory unambiguously defined. This angle was therefore sometimes preferred in positron experiments, to be described below.

To get $\Delta^2 P_\delta / (\Delta E_\delta \Delta \Omega_\delta) \simeq d^2 P_\delta / (dE_\delta \, d\Omega_\delta)$ the right-hand side of Eq. (2) must be calculated taking into account the detection efficiency for electrons $\varepsilon_\delta(E_\delta)$ which is close to unity and the fraction of time h the apparatus is ready for data acquisition.

If scattered particles are not observed, the triple differential cross section equation (2) must be integrated over $d\Omega_p$, yielding the double differential cross section with respect to the energy and solid angle element of the emitted electron

$$\frac{d^2 \sigma_\delta}{dE_\delta \, d\Omega_\delta} = \iint \frac{d^3 \sigma_\delta}{dE_\delta \, d\Omega_\delta \, d\Omega_p} \, d\Omega_p \approx \frac{\Delta^2 N_\delta}{\Delta E_\delta \Delta \Omega_\delta} \cdot \frac{1}{\varepsilon_\delta(E_\delta) \cdot h} \cdot \frac{1}{N_1 \cdot (N_2 / A)} \tag{4}$$

The product of the number of beam particles N_1 times the number of target atoms per unit area N_2 / A can be determined with a monitor counter of solid angle $\Delta \Omega$ which detects Rutherford scattered particles (number N_m). Thus,

$$N_1 (N_2 / A) = N_m \Big/ \iint_{\Delta \Omega} (d\sigma_R / d\Omega) \, d\Omega \tag{5}$$

It should be noted that the quantities given by Eqs. (3) and (4), i.e., cross sections and emission probabilities, have been transformed from the laboratory system to the c.m. system prior to the presentation in this article (cf., e.g., Appendix). Furthermore, cross sections or probabilities are denoted with an index "KX" if δ-ray emission was observed in coincidence to K X rays emitted from the separated atoms after the collision.

For the discussion of the δ-ray probabilities some kinematical quantities which characterize the Rutherford trajectory are of importance. These are, first of all, the distance of closest approach of the projectile and target nuclei in the scattering process

$$R_0 = a(1 + \varepsilon) \tag{6}$$

and the impact parameter

$$b = a \cot(\Theta^{c.m.}/2) \tag{7}$$

Here, $2a$ is the minimum distance of closest approach for a head-on collision (collision diameter)

$$2a = [Z_1 Z_2 e^2/(E_1 \mu/M_1)] \tag{8}$$

with $\mu = M_1 M_2/(M_1 + M_2)$ the reduced mass and ε the excentricity for the scattering angle $\Theta^{c.m.}$ in the center-of-mass system

$$\varepsilon = 1/\sin(\Theta^{c.m.}/2) \equiv \csc(\Theta^{c.m.}/2) \tag{9}$$

2.3. Experimental Arrangements

For δ-ray spectroscopy, a flexible experimental setup is needed, capable of operating in a large range of electron energies, with sufficient background suppression and reasonable detection efficiency for measurements of small cross sections and/or coincidence measurements. Two experimental setups possessing somewhat complementary features were utilized: An iron-free "orange"-type β-spectrometer[56,16,21] and an achromatic electron channel.[47,32,71,76]

Figure 2 shows a schematic view of the "orange"-type β-spectrometer, which uses a toroidal magnetic field produced by 60 current coils to momentum analyze and focus electrons emitted from the target onto a small, cone-shaped plastic scintillator counter. The momentum direction of the emitted δ rays is defined by the entrance slits. Two configurations have been used—one shown in Figure 2, where all electrons emitted between 30° and 50° (Θ_δ) relative to the beam direction and within a

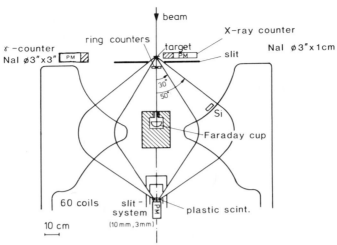

Figure 2. Experimental setup in the iron-free "orange"-type β-spectrometer. Momentum resolution $(\Delta p/p) = 0.018$, transmission $T = 0.08$. Coincidence measurements with scattered particles and/or X rays can be performed.

momentum band of $(\Delta p/p) = 0.018$ were focused, and the second configuration with $50° \leq \Theta_\delta \leq 70°$ and $(\Delta p/p) = 0.014$. The transmission efficiency in both cases ran to 0.08 of 4π ($\Omega_\delta \approx 1$ srad). The large transmission renders possible the detection of small δ-ray intensities and the performance of coincidence measurements with X rays and/or scattered particles. Characteristic K X rays were detected by two $3''\phi \times 1$ cm NaI scintillator counters, mounted 4.5 cm away from the target. All particles scattered between 9.5° and 27° (arrangement shown in Figure 2) or between 16° and 48° (second configuration, not shown here) relative to the beam axis were detected by an annular parallel-plate avalanche counter. Its active area was subdivided into eight concentric rings with individual read-out, each capable of operating up to counting rates of 10^7 sec^{-1}. Kinematical particle coincidences between particular rings were also possible, allowing an unambiguous definition of scattered particle trajectories for asymmetric collision systems. The magnetic field of the spectrometer was swept repeatedly up and down focusing δ rays with energies typically between 150 keV and 2.4 MeV. The measuring time after each field adjustment was normalized to the number of elastically scattered particles, detected by a Si-surface barrier monitor counter (BM). The multiparameter data were recorded event-by-event, together with the normalization spectra and the corresponding instantaneous value of the spectrometer magnetic field.[77] Self-supporting, ~1 mg/cm^2 metallic foils of ^{238}U, ^{232}Th, ^{208}Pb, ^{197}Au, and ^{120}Sn were bombarded with ^{208}Pb-ions, accelerated up to 4.7 MeV/u ($v_\infty/c = 0.10$) at GSI Darmstadt.

The experiment with the achromatic magnetic electron channel shown in Figure 3 was performed with C, S, and Ni beams from the Heidelberg MP Tandem Van de Graaff accelerator, with Ni, Br, and I ions from the MP Tandem in combination with the Heidelberg postaccelerator and with Pb beams from the UNILAC accelerator in Darmstadt. Zr, Pr, Pb, and U foils of thicknesses 0.5–1 mg/cm^2 were used as targets. The channel separates a momentum byte of ($\Delta p/p$) = 0.24 from the continuous distribution and focuses it through a solid angle of 4 msrad onto a cooled Si(Li) counter, whose energy resolution ran to 2.5 keV at 1 MeV. Similarly, as in the "orange" spectrometer, x-ray counters (two 2"ϕ × 1 cm NaI scintillator counters) and particle counters (annular parallel-plate avalanche counter with a Θ_p-sensitive anode but also with a ϕ_p-sensitive cathode) were mounted in the scattering chamber (not shown in the figure). Unlike the "orange" spectrometer, the relatively small solid angle allows better definition of the δ-ray emission angle relative to the beam axis or—combined with the Θ_p and ϕ_p information from the particle counter—relative to the trajectory of the scattered particles. The channel can be rotated in the plane around the target, thus allowing angular distribution measurements.[32,71,76]

With the experimental setups presented, both double-differential cross sections for δ-ray emission ($d^2\sigma_\delta/d\Omega_\delta dE_\delta$) or ($d^2\sigma_\delta/d\Omega_\delta dp_\delta$) as well as energy differential probabilities ($\Delta^2 P_\delta/\Delta E_\delta \Delta\Omega_\delta$) were investigated as a function of the δ-ray kinetic energy E_δ, or momentum p_δ, relative ion velocity v_∞, united atom charge Z_u, and projectile Θ_1 and target Θ_2 scattering angles.

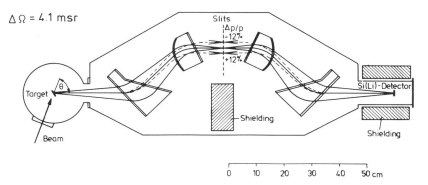

Figure 3. Achromatic magnetic electron channel. The annular parallel-plate avalanche particle counter and the NaI x-ray counters are not shown in the figure. The variable electron detection angle Θ allows measurement of the δ-ray angular distribution with respect to the beam axis and/or to the direction of the particles scattered.

Finally, the possible sources of background should be discussed, which, as a consequence of nuclear deexcitation following Coulomb excitation or other nuclear reactions, lead to the emission of electrons and, thus, to possible distortion of the δ-ray spectra under investigation. In order to keep the background intensities as low as possible the bombarding energies are mostly chosen to be far below the Coulomb barrier and ^{208}Pb ions are preferred as a projectile. In the case of Coulomb excitation of ^{208}Pb nucleus, only the decay from the first excited 3^- state (2.6145 MeV) to the ground state has to be considered and the excitation probability is rather small. In some cases also the targets consist of enriched isotopes for which, due to the nuclear structure, no significant Coulomb excitation probabilities are expected.

As for the background electrons, there are mainly electrons following internal conversion and internal pair creation. Compton electrons and electrons from external pair creation can be neglected as a background source, since in the β-spectrometers used almost no matter (or at least materials with very low Z) surrounds the target. In the case of internal conversion, one deals with monoenergetic electrons. The lines, however, are Doppler broadened and for high particle velocities and high δ-ray energies might not be easily distinguished from the continuous δ-ray spectrum. In such cases, knowing the angle between the emitted electron and the particle as well as the particle velocity (i.e., measuring Θ_δ, ϕ_δ, ϕ_p, v_p) one can correct the Doppler broadening, reproduce the initial line shape, and subtract the background peak from the δ-ray spectrum.[71]

The correction of the background due to internal pair creation is rather difficult, since one deals with continuous electron energy distributions. One should keep in mind, however, that the expected intensities are mostly lower than those from the internal conversion and that this kind of background can be successfully suppressed by a coincidence requirement with characteristic K X rays.

2.4. Test of the Basic Concept—"Light" Collision Systems with $Z_u < 107$

In order to prove the significance of the proposed method, first experiments have been carried out with "lighter" ^{16}O or ^{32}S projectiles and ^{197}Au or ^{208}Pb targets, respectively. The charge numbers of the quasiatoms studied thus remain in the region of known elements, where the theoretical description should be easier. Double-differential cross sections ($d^2\sigma_\delta/d\Omega_\delta dk_f$ or $d^2\sigma_\delta/d\Omega_\delta dE_\delta$) have been measured both for single electron spectra as well as for δ rays in coincidence with the K X rays.[45,25] The energy spectra were calculated within the framework of the Born approximation, using wave functions in the united atom limit and is given by

$$\frac{d^2\sigma_\delta}{d\Omega_\delta \, dk_f} = \left(\frac{2Z_1 e^2}{\hbar v_\infty \zeta}\right)^2 k_f^2 \int_{q_m}^{\infty} \frac{q \, dq}{q^4} |F(k_i)|^2 \tag{10}$$

with $\zeta = Z_1/(Z_1 + Z_2)$ reflecting the correlation of the quasi-atomic wave function to the center of charge, and $\hbar k_i = \mathbf{p}_i$ and $\hbar k_f = \mathbf{p}_f$ the initial and final momentum of the electron ejected, related to the momentum $\hbar q$ transferred in the collision by

$$\mathbf{k}_f = \mathbf{k}_i + \mathbf{q} \tag{11}$$

In the integral (from $q_m = q_0/\zeta$ to ∞) the factor $1/q^4$ reflects the probability for momentum transfer q in the collision, whereas the form factor given by

$$|F(k_i)|^2 = \int_0^{2\pi} d\phi_q (\langle f(\mathbf{k}_f)|e^{i\mathbf{q}\cdot\mathbf{r}}|i\rangle|^2 \tag{12}$$

(with ϕ_q the azimuthal angle of q) contains the information about the momentum k_i of the bound electron in the united atom.

The collision process can be described semiclassically within the adiabatic approximation. The transition amplitude for direct Coulomb ionization in a slow collision is given to first order by[14]

$$a_{fi} = -\int_{-\infty}^{\infty} dt \left\langle f \left| \frac{\partial}{\partial t} \right| i \right\rangle \exp\left[\frac{i}{\hbar} \int_0^t dt'(E_f - E_i)\right] \tag{13}$$

with i and f time-dependent electron states of the quasiatom. The coupling responsible for Coulomb excitation in asymmetric collisions is given by[27]

$$\frac{\partial}{\partial t} = \frac{1}{E_f - E_i} \dot{\mathbf{R}} \nabla_R \frac{Ze}{|\mathbf{r} - \zeta R|} \tag{14}$$

Calculating the transition amplitude in the united atom limit one obtains the usual Born approximation, except that the factor $1/\zeta^2$ appears as a correction for the center-of-charge transformation. As can be easily seen, Eq. (10) reduces to the Born approximation[29] for $\zeta = 1$.

It should be emphasized again that the above considerations are valid for strongly asymmetric collisions and for a light projectile and heavy targets. For very heavy and/or symmetric systems one cannot expect to describe the process within the framework of the Born approximation.

Because of its clarity, however, either first-order perturbation theory or the Born approximation can illustrate comprehensively the basic concept of these studies.

The minimum momentum transfer q_0 to an electron with a binding energy $|E_B|$ ejected to a continuum state (E_δ) can be easily obtained from the energy and momentum conservation laws

$$\Delta E = |E_B| + E_\delta = \frac{1}{2M_1}(\mathbf{P}_i^2 - \mathbf{P}_f^2) \tag{15}$$

Knowing that the initial momentum of the projectile $\mathbf{P}_i = M_1 \cdot \mathbf{v}_\infty$ changes only slightly due to ionization

$$|\mathbf{P}_i| + |\mathbf{P}_f| \simeq 2|\mathbf{P}_i| = 2M_1 v_\infty \tag{16}$$

one obtains

$$|E_B| + E_\delta = \frac{\hbar}{2M_1} q_0 2M_1 v_\infty \tag{17}$$

with

$$\hbar q_0 = |\mathbf{P}_i| - |\mathbf{P}_f| \tag{18}$$

the minimum momentum transfer to an ionized electron.

From Eq. (17) it follows that

$$\hbar q_0 = (|E_B| + E_\delta)/v_\infty \tag{19}$$

Figure 4 shows the δ-ray spectrum from ^{32}S + ^{208}Pb collisions at 120 MeV bombarding energy, corresponding to a relative velocity of $v_\infty/c = 0.09$ and to a collision diameter $2a = 18.2$ f.[25] Double differential cross section $(d^2\sigma_\delta/d\Omega_\delta\,dE_\delta)$ has been measured with good statistics over five orders of magnitude and up to very high kinetic energies. The data are in a very good agreement with calculations using Born approximation [Eq. (10)] with relativistic wave functions of the united atom ($Z_u = 98$). It should be noted that no fitting parameters have been used. Also shown are calculations in the target atom limit (dashed line) as well as results obtained using nonrelativistic wave functions (dashed–dotted line)—both underestimate the data significantly. In all three cases the cross section calculated represents a sum taken over all electrons of the K and L shell.

The data clearly show that the Cf quasiatom is formed transiently during the collision, where—at larger Z_u than for the target atom—the

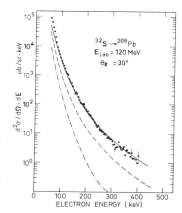

Figure 4. The δ-ray spectrum from ^{32}S + ^{208}Pb collisions at 120 MeV bombarding energy (Ref. 25). (——) *united* atom limit ($Z_u = 98$) *relativistic* Born calculation; (- - -) *target* atom limit ($Z_u = 82$) *relativistic* Born calculation; (- · · ·) *united* atom limit ($Z_u = 98$) *nonrelativistic* Born calculation.

relativistic shrinking of the wave function overcompensates the effects of stronger electronic binding. The agreement with the calculated cross sections demonstrates the direct relation between the spectral distribution of high energy δ rays and the form factor of the inner-shell electrons of the quasi atom.

Since the first measurements were carried out, considerable results have been achieved in systematic studies of high-energy δ-ray emission, covering various target projectile combinations in the region $66 \leq Z_u \leq 145$ and relative velocities $0.07 \leq (v_\infty/c) \leq 0.10$.[47,32,24,71,76,33] For purposes of systematic comparison of the data from different systems, a so-called "spectral" function $S(q_0, Z_u)$ has been introduced

$$S(q_0, Z_u) = \int_{q_m}^{\infty} \frac{dq}{q^3} |F(k_i)|^2 \qquad (20)$$

From Eq. (10) one can see that $S(q_0, Z_u)$ can be derived from the measured cross sections by a reduction of the pure kinematic effects, such as projectile velocity (v_∞) or transformation into the center of charge system (ζ). Keeping in mind that the dominant contribution to the cross section comes from $q \approx q_0$ one can replace q by q_0 in Eq. (11). Furthermore, from the fact that the relative velocity v_∞ is much smaller than that of the relativistic δ ray and from Eq. (19) one can obtain $q_0 \gg k_f$, which gives $k_i = |\mathbf{k}_f - \mathbf{q}| \approx q_0$. Thus, using the spectral function $S(Z_u, q_0)$ one can describe the experimental data as a function of the united atom charge Z_u and the minimum momentum transfer only. At given q_0 one observes a strong increase of the spectral function by increasing Z_u, thus convincingly demonstrating the quasi-atomic origin of the δ rays. The slope of the spectral function is well reproduced up to $Z_u < 137$ by a power law in $1/q_0$:

$$S(q_0, Z_u) \propto 1/q_0^{6+2\gamma} \qquad \text{with } \gamma = [1 - (Z_u \cdot \alpha)^2]^{1/2} \qquad (21)$$

extracted by expanding the form factor in power series in $1/k_i$,

$$F(k_i) \propto (1/k_i)^{4+2\gamma} + \cdots \qquad (22)$$

using hydrogenlike wave functions for a pointlike charge and integrating the first term in Eq. (20) ($k_i \approx q_0$ is assumed). The dominant role of the relativistic effects is evident, since with the much steeper falloff, $S \propto 1/q_0^{10}$, calculated in the nonrelativistic limit, no similarity with the data can be achieved.

2.5. δ-Ray Spectroscopy in the High-Z_u Region

Extending the experiments outlined before to much heavier systems, we ask first of all whether the observed δ electrons still originate from the quasiatom, as already demonstrated for lower united charge Z_u. Let us, therefore, look first at the Z_u dependence of the δ-ray emission. The principal features reflected in the data[16,17,21,18] of Figure 5 are as follows:

i. The very high kinetic energy of the δ rays observed: up to 750 keV for the Pb + Sn collision system with $Z_u = 132$ and over 1.7 keV for the Pb + Pb system ($Z_u = 164$). These are energies 8.5 or even 20 times higher than the binding energy of K electrons in a Pb-atom and imply extremely high momentum transfers.

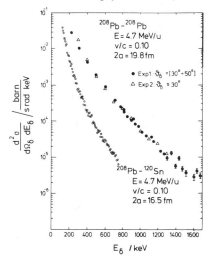

Figure 5. The δ-ray spectra from Pb + Pb ($Z_u = 164$) and Pb + Sn ($Z_n = 132$) collision systems measured at the same relative velocity $(v_\infty/c) = 0.10$. Also indicated are the distances of closest approach for a head-on collision $2a$.

ii. The large difference in the intensities of the Pb + Sn and Pb + Pb spectra of more than one order of magnitude.
iii. The much steeper falloff of the Pb + Sn spectrum reflecting the relative lack of high-momentum component of the bound state wave functions compared to the Pb + Pb distribution.

All of these imply that the electrons have been emitted from bound quasi-atomic states, where, due to the Z_u-dependent relativistic contraction of the wave function, large momentum transfers to strongly localized bound electrons become possible. This is also corroborated by the data in Figure 6, in which the double differential cross section $(d^2\sigma_\delta/d\Omega_\delta E_\delta)$ at given kinetic energy $E_\delta = 470$ keV is displayed versus the united charge Z_u for measurements at the same relative ion velocity.[18,32,47] The increase of the cross section over four orders of magnitude, while Z_u only doubles, demonstrates again the characteristic behavior of bound state wave functions in superheavy quasi atoms. For Z_u, the ionized electrons originate predominantly from the L shell with binding energies much smaller than the kinetic energy observed—i.e., at given δ-ray energy, the energy transfer required ($\Delta E = |E_B| + E_\delta$) is determined by E_δ. This does not apply to the

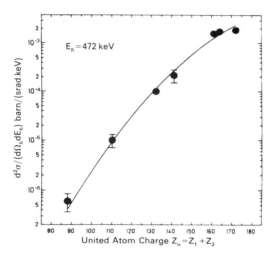

Figure 6. Double differential cross section for δ-ray emission at fixed kinetic energy $E_\delta = 427$ keV for the collision systems: C + Pb, Ni + Pb, Pb + Sn, Pb + Pr, Pb + Au, Pb + Pb, and Pb + Th as a function of the united atom charge Z_u measured at $(v_\infty/c) = 0.10$ ion relative velocity [Ni + Pb extrapolated from $(v_\infty/c) = 0.09$]. The solid line is to guide the eye.

heavier systems, where the saturation of the Z_u dependence observed is partially due to the stronger binding.

2.5.1. Asymmetric Collision Systems with $Z_u \cdot \alpha \approx 1$

For two systems with similar united charge Z_u δ rays associated with K X rays of the heavier collision partner have been measured. In Figure 7 the results for ^{208}Pb + ^{120}Sn collisions ($Z_u = 132$) are shown together with the single spectrum (Σ). The measurements were performed with the "orange"-type β-spectrometer at GSI in Darmstadt.[16-18] The relative ion velocity ran to $(v_\infty/c) = 0.10$. The corresponding spectra for the system ^{127}J + ^{208}Pb are presented in Figure 8.[76] The data were taken at MPI Heidelberg at slightly different ion velocity $(v_\infty/c) = 0.09$ with the achromatic electron channel. The solid lines in Figures 7 and 8 represent results of coupled channel calculations performed by Soff et al.[74,75] for δ-ray emission in coincidence with a $1s\sigma$-vacancy creation. Also included are multistep processes—i.e., the ejected electron can be rescattered during the collision to higher energy. These calculations reproduce very well both the absolute δ-ray yields as well as their energy distribution. It should be emphasized again that the good agreement between the theoretical calculations (which adopt the quasi-atom picture) with measurements performed with two different experimental arrangements and with inverse projectile—

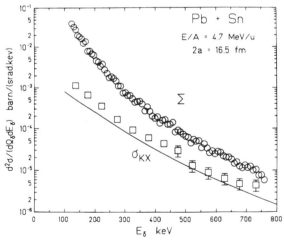

Figure 7. The δ-ray spectrum from Pb + Sn collisions at $(v_\infty/c) = 0.10$ associated with the K X rays from lead ($\square$) is compared to the results of coupled channel calculations performed by Soff et al. (Refs. 74, 75). Also shown is the single δ-ray spectrum (Σ).

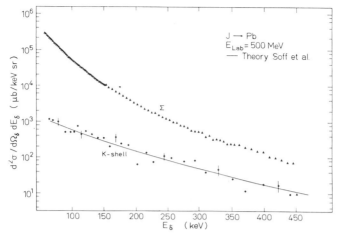

Figure 8. The δ-ray spectrum for I + Pb colliisons at $(v_\infty/c) = 0.09$ associated with the K X rays from lead (K-shell) is compared to the results of coupled channel calculations performed by Soff *et al.* (Ref. 75). Also shown is the single δ-ray spectrum (Σ). (From Ref. 76.)

target combinations strongly supports the expectations of exploiting the characteristic behavior of the innermost $1s\sigma$ bound state in an overcritical collision system, such as the Pb + Cm quasiatom with $Z_u = 178$.

As already mentioned in Section 2.1 the δ-ray spectrum associated with K X rays of the heavier collision partner in asymmetric encounters $(Z_1 \neq Z_2)$ represents the contributions from the K shell of the united atom. It should be pointed out again that this method for selecting contributions from a particular shell has its limits for cases where the probabilities (per collision) both for δ-ray emission as well as for characteristic x-ray emission reach extremely high values. For very heavy collision systems, the δ-ray multiplicity at very low kinetic energies is higher than unity—i.e., several shells are ionized in the same collision. A particular background problem occurs at those δ-ray energies when (because of the finite detection efficiencies) the K δ ray escapes but the K X ray and an electron from a higher shell are detected. Knowing the total δ-ray production probability at lower kinetic energies and the K x-ray emission probabilities, one can estimate the background mentioned above as a product of both probabilities, since at low kinetic energies the δ-ray spectrum is dominated by contributions from the weakly bound higher shells. In the cases presented in Figures 7 and 8 the single δ-ray spectrum (denoted by Σ) shows the total δ-ray production probability. The cross sections for K x-ray emission have been measured simultaneously in the (J + Pb)-collision system—in the (Pb + Sn) case those values can be taken from Ref. 55.

Comparing the single δ-ray spectra with the spectra associated with K x-ray emission one should keep in mind that a higher energy transfer is needed to ionize the K shell and emit a δ ray with certain kinetic energy compared to the energy transfer required for the ionization of weaker bound electrons to the same kinetic energy—because of the stronger binding of the K shell. It has already been mentioned that higher energy transfers can cancel to some extent the effects of stronger localization of the wave function and, thus, reduce the ionization probability. This explains only qualitatively the fact that the contributions of the K shell to the total cross section (Σ in Figures 7 and 8) are very small at low kinetic energies and do not become significant until the energy reaches very high values. On the other hand, it is also known that higher momenta can be transferred at rather small impact parameters b, due to the higher Fourier frequencies available in the Coulomb collision. This implies that the impact parameter dependence of the δ-ray emission at fixed kinetic energies is different for the emission from different shells—concentrated at smaller impact parameters for the innermost and flatter for the weakly bound states. Thus, more detailed information can be obtained comparing the cross sections as a function of both impact parameter and energy (or momentum) transfer.

Another reason for the investigation of the δ-ray production at fixed kinematical conditions—i.e., as a function of the impact parameter b or of the distance of closest approach R_0—can be extracted from the behavior of the binding energy as displayed in Figure 1: For very heavy collision systems the binding energies of the innermost shells are expected to vary strongly with the internuclear separation, especially at very small distances.[30]

Last but not least, the impact parameter dependence of the δ-ray emission represents a very sensitive test for scaling laws, as proposed for ionization probabilities determined via measurements of characteristic x-ray yields.[12,1,13,57,26] Unlike x-ray measurements the energy transfer $\Delta E = E_\delta + |E_B|$ is determined for fixed δ-ray energy, provided that the binding energy is known. And vice versa, if the validity of the scaling law is established, unknown binding energies of electrons in superheavy colliding systems can be experimentally determined to some extent.

A scaling rule for $1s\sigma$ excitation probabilities as a function of the distance of closest approach and the momentum transfer has been proposed[57,26] and successfully applied by Bosch et al.[26] considering results of x-ray measurements. The δ-ray emission probability per energy interval is given by

$$\frac{\Delta P(R_0, q_0)}{\Delta E \, \Delta \Omega} = d_0^2(Z) \left(\frac{m_e c^2}{E_\delta + m_e c^2} \right)^{C(Z)} \exp\left[\frac{-m' R_0 (E_\delta + |E_B|)}{\hbar v_\infty} \right] \quad (23)$$

with $m' = 2$ for heavy collision systems; $q_0 = (E_\delta + |E_B|)/(\hbar v_\infty)$ from Eq. (19); $E_B = E_B(R_0)$.

The exponential is based on the scaling behavior of the wave function $|\Psi|^2 \sim 1/R^2$. The factor in front of the exponential reflects the coupling strength to the continuum; $C(Z)$ is supposed to be a smooth function of Z. For x-ray measurements $\langle E_\delta \rangle = 0$ is assumed.

For the (Pb + Sn) system Figure 9 shows δ-ray emission probabilities as a function of the distance of closest approach R_0 for two fixed δ-ray kinetic energies, 540 and 720 keV, at which the K-shell contributions are more significant. The solid lines in the figure represent a result of two independent least-square fits to the data using Eq. (23) reduced to

$$\frac{\Delta P(R_0, q_0)}{\Delta E \Delta \Omega} = C_1(E_\delta) \exp\left[- \frac{2R_0(E_\delta + |E_B|)}{\hbar v_\infty} \right] \qquad (24)$$

with $C_1(E_\delta)$-parameter of the fit = const for fixed E_δ; $|E_B|$ is assumed to be constant within the region of R_0 shown. The latter assumption is supposed to lead to an error less than 10%.[30] The fitted value of $|E_B| = 310$ keV is in a good agreement with the value of 320 keV calculated by Fricke and Soff[30] for the data at $E_\delta = 720$ keV. For $E_\delta = 540$ keV one obtains a smaller value of $|E_B| = 250$ keV, which may be caused by the contributions from the L shell. No reasonable agreement with the data

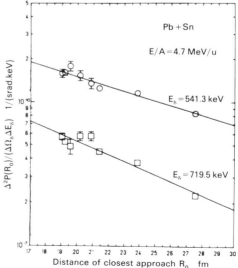

Figure 9. The δ-ray emission probability for Pb + Sn collisions as a function of the distance of closest approach R_0 for two fixed δ-ray kinetic energies: 540 keV ($\bigcirc$) and 720 keV ($\square$). The solid line represents a least-squares fit to the data $P \propto \exp(-R_0 q_0)$ with q_0 the minimum momentum transfer required.

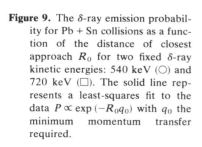

presented in Figure 9 is achieved, however, when Eq. (23) is fitted as a whole to the absolute values of all data points—i.e., with $d_0(Z)$, $C(Z)$, m', and $|E_B|$ as parameters of the least-squares fit. The disagreement is due to strong deviations from the proposed[57] kinetic energy dependence: $[(E_\delta/m_0c^2) + 1]^{-C(Z)}$. Similar behavior is observed also for lower kinetic energy δ rays and larger distances of closest approach, shown in Figure 10a for I + Pb collisions at $(v_\infty/c) = 0.09$.[33,71] The deviations from the scaling law are not unexpected, since Eq. (23) is derived for $1s\sigma$ ionization only and the spectra are dominated by δ rays from the L shell. Notwith-

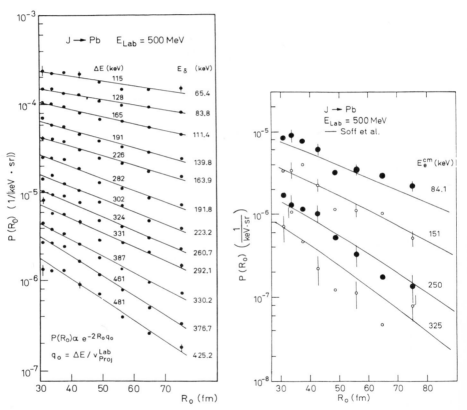

Figure 10. (a) The δ-ray emission probability for I + Pb collisions as a function of the distance of closest approach R_0 for two fixed δ-ray kinetic energies as indicated ($E^{\text{c.m.}}$). The solid line represents the results of least-squares fits to the data $P \propto \exp(-2R_0q_0)$. Also indicated in the figure are the fitted values of the energy transfer required ΔE. Data from Refs. 71, 34. (b) Emission probability for δ-rays associated with K X rays from lead (K shell, united atom) for I + Pb collisions as a function of the distance of closest approach R_0 for four fixed δ-ray kinetic energies as indicated ($E^{\text{c.m.}}$). The solid lines show results of calculations performed by Soff *et al.* (Ref. 75). (From Refs. 71, 34.)

standing, the slope of the emission probability distributions exhibits an exponential falloff with the distance of closest approach R_0, which as expected becomes steeper for higher kinetic energies, according to $P \propto \exp(-2R_0 q_0)$. The solid lines again represent a least-squares fit of Eq. (24) to the data, showing a good agreement and allowing one to extract reasonable values for the binding energy (60 keV on average).

In Figure 10b the dependence on the distance of closest approach of the emission probabilities for electrons from the K shell of the (I + Pb) united atom are displayed for four different δ-ray kinetic energies. Triple coincidences have been performed, detecting δ rays associated with particles scattered through an angle Θ_p and in additional coincidence with K X rays from lead.[34,71,76] The solid line represents coupled channel calculations, including multistep processes, performed by Soff *et al.*[75] Both slope and absolute intensities are in a good agreement within the experimental errors. This emphasizes the role of multistep processes for proper interpretation of measured absolute values. The scaling law from Eq. (24), however, again reproduces well the exponential behavior of δ-ray emission probabilities as a function of the distance of closest approach R_0 and the minimum momentum transfer required q_0; or in other words, the exponential falloff of the emission probabilities P as a function of the distance of closest approach R_0 apparently does not depend on the way energy and momentum are transferred—either in one step or in a multistep excitation. One is also led to treat incoherently the individual steps of the multistep process, e.g., for a two-step excitation from a bound state E_B to a continuum state E_δ via an intermediate state E_m the ionization probability can be reduced to $P \propto \exp[-2R_0(E_\delta + |E_B|)\hbar v_0] = P_1 \cdot P_2 = \exp[-2R_0(E_m + |E_B|)\hbar v_0] \cdot \exp[-2R_0(E_\delta - E_m)/\hbar v_0]$ (with $E_\delta > E_m > E_B$). For this, however, the sensitivity of the procedure is insufficiently known. On the other hand, the limits of application of the scaling law are demonstrated by the discrepancy in the δ-ray kinetic energy dependence. In other words, the energy distribution of the δ rays at fixed impact parameter is not described properly by Eq. (23). The origin of this disagreement is not easy to see, since the calculated energy distribution of δ rays do not differ in slope if multistep excitations are neglected.[72,37]

2.5.2. Very Heavy Collision Systems

First measurements of high-energy δ rays from very heavy quasiatoms were carried out,[16,17,21,32,18] as already mentioned, for ^{208}Pb + ^{208}Pb collisions at 4.7 MeV/u bombarding energy—far below the Coulomb barrier. For symmetrical collisions, the ion trajectory can be defined unambiguously only for scattering angles $\Theta_p^{c.m.} = \pi/2$ and the contributions from the $1s\sigma$-innermost state cannot be separated by a K x-ray coincidence requirement. The ^{208}Pb + ^{208}Pb system gives the unique opportunity to study δ-ray spectra without any distortion due to nuclear deexcitation processes. The only source of background—the decay of the Coulomb-excited 3^- state (2.614 MeV) via internal e^+e^--pair creation—is expected to be extremely weak, as shown in measurements of positron yields.[10,46,77]

In Figure 11 the measured[16,17,21,18] double-differential cross sections are compared with results of coupled channel calculations performed by Soff *et al.*,[74,75] which include multistep excitation processes. Shown are the calculated total δ-ray distribution, representing a sum over all initial bound states up to $4s\sigma$ and $4p_{1/2}\sigma$ state, as well as contributions from the $1s\sigma$ and $2p_{1/2}\sigma$ state. The experimental data exceed the calculated values by a factor of about 3. The slope of the measured spectrum, however, is reproduced well. It should be noted in this context that perturbation theory values also agree in shape with the measured spectrum, underestimating the data by more than a factor of 6.[72,37] The pronounced difference between the contributions from $1s\sigma$ and $2p_{1/2}\sigma$ states, shown in Figure 11, demonstrates again that the emission process depends strongly on the energy (or momentum) transfer $\Delta E = E_\delta + |E_B|$ required and that the $1s\sigma$-binding energy cannot be neglected even at δ-ray kinetic energies as high as 1.7 MeV.

The dependence of the emission probabilities on the ion scattering angle at several fixed δ-ray kinetic energies between 470 keV and 1.8 MeV is given in Figure 12. The experimental data represent the total emission probability—i.e., contributions from several inner shells are superimposed.[16–18] The results of the triple coincidence measurement—i.e., δ-ray

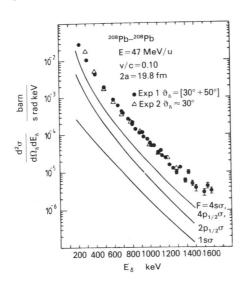

Figure 11. The δ-ray spectra from ^{208}Pb + ^{208}Pb at 4.7 MeV/u bombarding energy. (Exp. 1, "orange"-type β-spectrometer; Exp. 2, achromatic electron channel.) The solid lines show results of coupled channel calculations performed by Soff *et al.* (Ref. 74) for the total spectrum ($F = 4s\sigma$, $4p_{1/2}\sigma$ denotes the Fermi surface) as well as for the contributions from the $1s\sigma$ and $2p_{1/2}\sigma$-states.

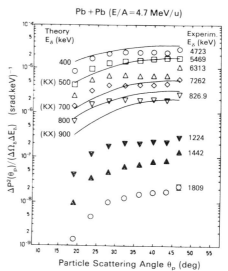

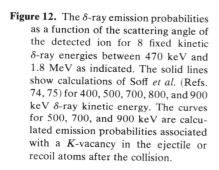

Figure 12. The δ-ray emission probabilities as a function of the scattering angle of the detected ion for 8 fixed kinetic δ-ray energies between 470 keV and 1.8 MeV as indicated. The solid lines show calculations of Soff *et al.* (Refs. 74, 75) for 400, 500, 700, 800, and 900 keV δ-ray kinetic energy. The curves for 500, 700, and 900 keV are calculated emission probabilities associated with a *K*-vacancy in the ejectile or recoil atoms after the collision.

associated with Pb-*K*-x-ray and scattered particles—do not show, within the limits of statistical errors (approximately 25%), any significant deviations (both in slope and magnitude) from the distributions presented in Figure 11. It is not surprising that the triple coincidence does not affect the slope of the distributions, since the high-energy component of the strongly relativistic wave functions of the innermost bound states ($1s\sigma$, $2p_{1/2}\sigma$, and $2s\sigma$) are expected to be very similar.[74] For symmetric scattering systems, however ($Z_1 = Z_2$), one expects that only a minor fraction of the *K* X rays is associated with vacancies produced in the $1s\sigma$ shell of the quasiatom and that the majority of the *K* X rays observed is due to $2p_{1/2}\sigma$ ionization and the following "vacancy sharing" with the $1s\sigma$ state.[52,54,1] Although the statistical errors are relatively large—from the similarity of the absolute values of the total and of the *K* x-ray coincident δ-ray production probabilities at a fixed particle scattering angle, one has to conclude that (i) either a large fraction of the *K* X rays observed is not associated with holes in the $1s\sigma$ or $2p_{1/2}\sigma$ states but results from vacancies produced in higher-lying shells, which may couple to the $1s\sigma$ orbital in the outgoing phase of the collision; or (ii) the $2p_{1/2}\sigma$-ionization probability is extremely high—of the order of unity; or (iii) both (i) and (ii) hold for the quasi-atomic system investigated at this relative velocity. On the other hand, this experimental finding allows us to compare the data in Figure 12 with calculated impact parameter dependences both of the total as well as of the *K* x-ray coincident δ-ray emission probabilities.[74,75] Since the scattering particles are identical, the theoretical emission probabilities for an impact parameter $b(\Theta_p)$ and those for the complementary $b(\pi - \Theta_p)$ have been weighted with the corresponding Rutherford cross sections σ_{Ruth} and added:

$$P = \frac{\sigma_{\text{Ruth}}(\Theta) \cdot P(\Theta) + \sigma_{\text{Ruth}}(\pi - \Theta) \cdot P(\pi - \Theta)}{\sigma_{\text{Ruth}}(\Theta) + \sigma_{\text{Ruth}}(\pi - \Theta)}$$

The agreement with the experimental data is surprisingly good, especially at larger particle scattering angles, in contrast to the factor of 3 discrepancy between the measured and the calculated total (i.e., integrated over all impact parameters) δ-ray distributions presented in Figure 11. Indeed, the calculated spectra do not include effects of the rotational coupling, however, these are not expected to play any essential role at δ-ray kinetic energies as high as those measured even at large impact parameters. The discrepancy is relatively large—i.e., it seems improbable that the main reason for it is caused by underestimating the δ-ray emission associated with particles scattered through small angles Θ_p, because these are relatively small contributions to the total yield. The experimental data in Figure 12 exhibit a steeper increase with the particle scattering angle Θ_p in the region of small Θ_p in accordance with the slope of the theoretical curves but also a relatively flat distribution at larger scattering angles Θ_p (also at high δ-ray kinetic energies), which seems to disagree with the smooth increase and slower saturation predicted by the theory. Also, the comparison of the data with the scaling law, symmetrized for Θ_p and $\pi - \Theta_p$ from Eq. (24) gives, under realistic assumptions for the energy transfer required, a fair agreement only for small particle scattering angles. Further experiments are planned to study this unexpected behavior, with additional emphasis of the dependence on the relative ion velocity, keeping in mind that for larger particle scattering angles—i.e., for smaller internuclear separations—one expects an extremely rapid change of the $1s\sigma$ and most of all the $2p_{1/2}\sigma$ binding energies (cf. Figure 1) and that the K x-ray production probabilities exhibit, at fixed impact parameter, an unexpected maximum at $(v_\infty/c) = 0.10$.[49]

3. Spectroscopy of Positrons

3.1. Remarks on Positron Creation in Strong Electric Fields

The question of pair creation in strong electric fields is a very old problem. Already in 1929, Klein,[38] within the framework of the new Dirac theory of the electron, considered the behavior of an electron wave with a total energy $E = E_T + m_e c^2$ impinging on a one-dimensional step barrier potential $V_0 > 2m_e c^2/e$ (Figure 13a). If the kinetic energy E_T of

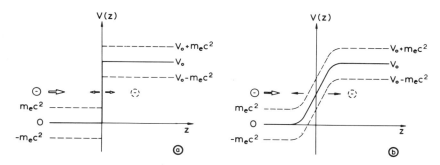

Figure 13. (a) Step barrier potential for which Klein (Ref. 38) calculated, using the Dirac equation, the reflection behavior of electron waves. (b) More realistic potential for which Sauter (Refs. 65, 66) performed corresponding calculations.

an electron traveling on the z axis from left to right is less than eV_0 for $z > 0$, an exponentially damped wave function is expected from the Schrödinger equation. According to Klein's calculations with the Dirac equation, this behavior was also found for $eV_0 - 2m_ec^2 < E_T < eV_0$ but for a kinetic energy $E_T < eV_0 - 2m_ec^2$ the wave function starts oscillating again. This means that electrons behave in a forbidden region as free particles with negative energy (and also negative momentum). This anomaly was called the Klein paradox.

To solve the Klein paradox it was necessary to understand the solutions of the Dirac equation with negative energies. This was achieved with Dirac's hole theory. According to this, all states with negative energy can be thought to be occupied with electrons. A missing electron can be then regarded as a positron. This interpretation resulted in the solution of another problem at the same time, namely, that electrons of an atom were not any longer allowed to undergo transitions to the negative energy continuum states because of the Pauli exclusion principle. Consequently, the Klein paradox, in its original form, can be interpreted in the following way: A positron wave approaching from $z > 0$ toward the step barrier at $z = 0$ results in a radiationless annihilation with an electron wave coming from $z < 0$. While the electron wave is reflected partially, the positron wave is not. With properly chosen initial conditions the continuous creation of electron–positron pairs at $z = 0$ takes place. In today's terminology, this is called spontaneous positron creation.

It is well known that potential differences having $eV_0 > 2m_ec^2$ are easily attainable. Electrostatic accelerators operate routinely with a terminal high voltage of 12 MV or greater. Nevertheless, a vacuum breakthrough which results in spontaneous electron–positron pair creation has never been observed. The reason for this was recognized by Sauter[65,66] only shortly after Klein's paper was published. Klein's assumption of a step barrier implies an infinitely high electrical field strength at $z = 0$ which, of course, is unphysical. In a more realistic potential like that presented in Figure 13b, the electrons have to tunnel through a barrier. Assuming a homogeneous electrical field, the transmission D of the electrons through this barrier is approximately given by

$$D \simeq \exp\{-\pi m_e c^2/[(\hbar/m_e c^2)e(dV/dz)]\} \tag{25}$$

This factor is unobservably small for field strengths which can be realized in the laboratory. D only gets large for critical fields

$$E_{\text{crit}} = (dV/dz)_{\text{crit}} \simeq m_e c^2/(\hbar e/m_e c) = 1.33\,\text{kV/f} \tag{26}$$

exhibiting a potential jump of $m_e c^2/e$ over a distance of the Compton wavelength of the electron $\hbar/m_e c = 386\,\text{f}$. Such fields are called strong fields.

Strong fields can in principle be realized near an atomic nucleus, but for the known elements, there are no bound states with $E_B > 2m_e c^2$ which could be occupied by electrons of the negative energy continuum accompanied by emission of positrons. This is probably the reason that for the past 50 years only a few papers have appeared dealing with the Klein paradox[15,28,78] or the binding energy in strong Coulomb fields.[67,59,80,79] With the predicted existence of superheavy elements (e.g., Ref. 58) and their feasible production by fusion or transfer reactions, interest was focused again on this field. This has resulted in a rapidly growing study of the "quantum electrodynamics of strong fields" and has been recently reviewed.[62,60]

The binding energy of the lowest bound $1s$ state as a function of the nuclear charge number Z is shown in Figure 14. Assuming a realistic charge distribution of the nucleus, calculations[81] show the binding energy exceeds $2m_e c^2 = 1.022\,\text{MeV}$ at a critical charge $Z_{\text{cr}} = 173$. Of course, atoms with charge numbers $Z > Z_{\text{cr}} = 173$ neither exist in nature nor have been produced artificially. However, in a scattering experiment at energies as high as 1.5 GeV, two uranium nuclei may approach each other to less than 20 f. Under these conditions, a quasiatom is formed and electrons are exposed for the very short time of about 2×10^{-21} sec to the combined nuclear charge $Z_u e = (Z_1 + Z_2)e$. The most direct way to investigate the question whether the binding energy of the innermost electrons exceeds $2m_e c^2$ or not should be with positron spectroscopy, since a preionized diving level may decay by spontaneous positron emission.

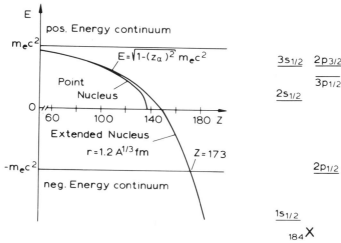

Figure 14. Binding energies of a $1s$ electron in the Coulomb potential of a point nucleus and an extended nucleus with the nuclear radius $r = 1.2\,A^{1/3}f$ as a function of the nuclear charge number Z. The right-hand side presents the level scheme for a hypothetical hydrogenlike atom with $Z = 184$.

Unfortunately, the Fourier frequency spectrum of the collision process results in a dynamically induced contribution to the positron spectrum. This is expected to be the most important positron production process for supercritical scattering systems like U + U with $Z_u = 184$. This dynamic contribution can be investigated experimentally in undercritical systems for which no level diving is expected. The experimental objective is, therefore, to try to establish level diving by comparison of the difference of positron spectra or integrated positron production probabilities for overcritical and undercritical systems.

In performing these experiments there is one important difficulty which should be mentioned. By the Coulomb excitation processes, nuclear levels of the collision partners may be excited which decay by transitions having energies greater than $2m_ec^2$. Positrons due to internal pair conversion of these transitions cannot be distinguished easily from the atomic contribution because the time delay is only on the order of 10^{-16} to 10^{-13} sec and, therefore, discrimination cannot be obtained for these events using timing or recoil techniques. The procedure for correcting nuclear positron background is based on converting the observed γ-ray spectra into positron spectra and was first done by Meyerhof.[51] This will be discussed in more detail in Section 3.4.

3.2. Physical Quantities in Positron Spectroscopy

The physical quantities in positron spectroscopy are, in principle, similarly defined as those for δ-ray spectroscopy in Section 2.2. We only have to replace "δ-ray" by "positron" and the index δ by e^+. However, there are some peculiarities which result from the fact that positron cross sections are fairly small, and, therefore, the positron spectrometers have to accept a rather large solid angle Ω_{e^+}. In this case it is in general not a simple task to transform quantities measured in the laboratory system to the c.m. system. However, we will show in the Appendix that for the spectrometer described in Section 3.3 these transformations can be performed assuming isotropy of positron emission in the c.m. system.

The quantity actually measured in the laboratory system is the energy differential probability for positron production

$$\frac{dP_{e^+}^{exp}}{dE_{e^+}} \equiv \frac{d^2\sigma_{e^+}/(d\Omega_p\,dE_{e^+})}{d\sigma_p/d\Omega_p} \simeq \frac{\Delta N_{e^+}^{exp}/[\varepsilon_{e^+}(E_{e^+})h]}{\Delta E_{e^+}N_p} \tag{27}$$

Experimentally, the number of positrons $\Delta N_{e^+}^{exp}$ in an energy interval ΔE_{e^+} (at energy E_{e^+}) must be obtained in coincidence with the number N_p of scattered particles. As in Eq. (3) the factor h is the fraction of time the apparatus is ready for data taking. Equation (27) is only approximately valid since the particle counter has a finite acceptance angle $\Theta_p \pm \Delta\Theta_p$.

If $dP_{e^+}^{exp}/dE_{e^+}$ is integrated over the positron energy the total positron production probability

$$P_{e^+}^{exp} = \frac{d\sigma_{e^+}/d\Omega_p}{d\sigma_R/d\Omega_p} \simeq \frac{N_{e^+}^{exp}/(\bar\varepsilon_{e^+}h)}{N_p} \tag{28}$$

is obtained. In this case a detector simply counting the number of positrons is sufficient. For positrons with a spectral distribution $F(E_{e^+})$ the mean detection efficiency is

$$\bar\varepsilon_{e^+} = \left[\int \varepsilon_{e^+}(E_{e^+})F(E_{e^+})dE_{e^+}\right]\Big/\left[\int F(E_{e^+})dE_{e^+}\right] \tag{29}$$

It was shown by Kankeleit[39,40] that besides the often used minimum distance of closest approach R_0 [cf. Eq. (6)] and the impact parameter $b^{(2)}$ [cf. Eq. (7)], the scattering time

$$2\hat{t} = (2a/v)[\varepsilon + 1.16 + 0.45/\varepsilon] \tag{30}$$

$[v = (2E_1/M_1)^{1/2}$, the projectile velocity at infinity] is an important quantity for understanding the positron production process. This is the time between the extrema of $\dot{R}(t)/R(t)$, whereby $R(t)$ is the relative distance and $\dot{R}(t)$ the relative radial velocity in the scattering process at time t. For undercritical systems $\hat{t}$ can be used to adequately describe the dynamically induced positron production. For a symmetrical system like U + U all these quantities are defined unambiguously only for $\theta_p = 45°$ in the laboratory system. This is, therefore, experimentally a preferred angle.

For a typical U + U experiment at $E_1 = 5.9$ MeV/u and $\theta_p = 45°$ these quantities are $2a = 17.36$ f, $\varepsilon = 1.41$, $R_0 = 20.95$ f, $2E = 1.71 \times 10^{-21}$ sec, $b = 8.68$ f. The Rutherford cross section for the projectile in the laboratory system is $d\sigma/d\Omega = 2.13$ b/sr and the integrated positron production yield is $P_{e^+} \simeq 2 \times 10^{-4}$.

3.3. Experimental Configurations for In-Beam Positron Spectrocopy

Spectrometers for use in in-beam positron spectroscopy have to have a large positron collection efficiency and a broad energy acceptance band while at the same time effectively suppressing other background radiation like electrons, γ rays, and neutrons. These requirements are fairly well satisfied by the two different types of spectrometers which will be described in this section. These are a magnetic "orange"-type β-spectrometer and positron transport systems having a Si(Li) detector for energy determination.

3.3.1. The "Orange"-Type β-Spectrometer

In Figure 15 the "orange"-type β-spectrometer[56] used for positron spectroscopy at the UNILAC accelerator in Darmstadt is schematically shown. Positrons emitted from the target between 40° and 70° with respect to the beam direction are focused by a toroidal magnetic field produced by sixty coils onto a 60-mm-diam × 100-mm-long hollow cylindrical plastic scintillator detector. Electrons cannot reach the detector. Target γ rays are strongly attenuated by a lead absorber between the target and detector. To further reduce the background, one of the positron annihilation quanta was detected by a 51-mm-diam × 100-mm-long NaI(Tl) counter placed inside of the plastic detector. Within the focused momentum band of 15% the positron detection efficiency is 2.6%.

Scattered ions are detected in coincidence with positrons using an annular parallel-plate avalanche counter with four concentric anode rings subtending scattering angles between 13.5° and 32.3°. The detector rings are capable of operating individually at counting rates up to of 10^6/sec. The energy resolution, 20% of the energy loss, is sufficient to separate

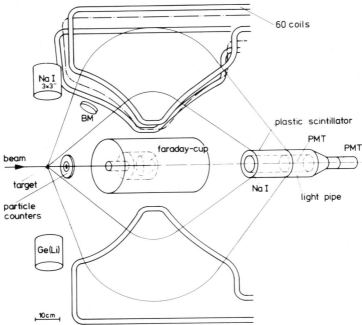

Figure 15. Experimental setup in the "orange"-type β-spectrometer. Positrons
are focused by the toroidal magnetic field onto a cylindrical plastic
scintillator containing a NaI crystal. Scattered particles are detected by
an annular parallel-plate avalanche counter with four concentric anodes.
(From Ref. 46.)

quasielastic scattered ions from fission fragments and light reaction products. This experimental setup has been used to investigate positron yields in a selected energy range between 0.44 and 0.55 MeV for the U + U, U + Pb, and Pb + Pb scattering systems at $E_1/M_1 = 5.9$ MeV/u.[46]

3.3.2. Positron Transport Systems

The other type of spectrometers in use for positron spectroscopy at GSI in Darmstadt are solenoid transport systems with normal conducting coils and a Si(Li) detector for energy determination. The improved version of the solenoid used for the pioneering positron experiments[10] is shown schematically in Figure 16. Beam and solenoid axis are perpendicular to each other. Positrons created in the target spiral in the magnetic field to a Si(Li) detector assembly 88 cm away from the target. The magnetic mirror field configuration focuses all positrons with an emission angle

$$\Theta < \Theta_{\max} = \pi/2 + \arcsin (B_T/B_{\max})^{1/2} \tag{31}$$

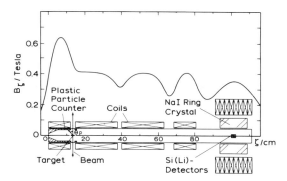

Figure 16. Schematic drawing of the solenoid spectrometer. Also shown is the magnetic field on the ζ axis. The heavy ion beam enters the vacuum chamber perpendicular to the solenoid axis (Ref. 9).

with respect to the ζ axis to the detector. B_T is the magnetic field at the target position and B_{max} its maximum value at $\zeta = 6.5$ cm. With the magnetic field configuration shown in Figure 16 the acceptance angle goes to $\Theta_{max} = 148.7°$ which corresponds to a solid angle of $\Omega/4\pi = 0.74$. Unfortunately, in the same way, all electrons resulting from Coulombic projectile target interactions or internal conversion transitions of excited nuclear levels are also focused to the detector. To exclude this very high electron background from detection without appreciably decreasing the positron detection efficiency, the detector system shown in Figure 17 has been used.[9] It consists of 2 Si(Li) diodes with a diameter of 19.5 mm and a thickness of 3 mm mounted having their surfaces parallel to the solenoid axis. One of these has an upward orientation while the other is down. The sensitive face of the upward counter looks to the left, the downward counter

Figure 17. The NaI–Si(Li) detector assembly for the detection of positrons in a view perpendicular to the solenoid axis at $\zeta = 102$ cm. The circular Si(Li) detectors have a diameter of 20 mm and a thickness of 3 mm. The fourfold segmented NaI ring crystal has the following dimensions: inner diameter 90 mm, outer diameter 204 mm, length 150 mm (Ref. 9).

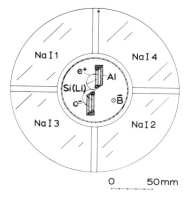

to the right. With this configuration, positrons spiraling in a right-hand manner have a good probability to enter the sensitive area of one of the Si(Li) diodes. Electrons spiraling in the opposite direction are absorbed or scattered in the 3-mm-thick aluminum back cover of the detector. Only the very few electrons with an energy greater than about 1.4 MeV can penetrate the aluminum absorber. A suppression factor for 365 keV electrons of about 150 was measured. In spite of this strong suppression of electrons the Si(Li) counting rate observed for in-beam positron experiments is caused nearly entirely by electrons. In order to select the very few positrons, the Si(Li) diodes were surrounded by a four-fold segmented NaI(Tl) ring crystal which permitted detection of the 511-keV annihilation radiation from the positron. To characterize a positron, a coincidence was required between an event in the 511-keV region in one of the NaI(Tl) segments and the total energy region in the opposite one.

Both Si(Li) diodes operate essentially independently of each other. This feature is useful in the search for peaks in the positron spectra. Such peaks are expected from a $1s\sigma$ vacancy decay in a collision having a long time delay after nuclear contact.[61] Suppose a hypothetical superheavy nucleus is formed after a $^{238}U + ^{238}U$ collision at 5.9 MeV/u and it emits positrons with a sharp energy of 300 keV. Such a superheavy nucleus moves with about 5.6% of the velocity of light. Positrons accepted by the magnetic field of the solenoid would be broadened in energy by the Doppler effect. The corresponding energy distribution is nearly rectangular with a width of 71 keV. The Si(Li) counters at the top and at the bottom are sensitive essentially to positrons emitted parallel or antiparallel relative to the beam direction, respectively. Therefore, for these counters positron distributions are expected to result in two peaks separated by about 35 keV and having reduced width of also about 35 keV. Such an observation would be a very characteristic fingerprint for a positron peak.

The positron detection efficiency obtained by unfolding the Si(Li)-detector response function is shown in Figure 18 and was measured with intensity calibrated ^{22}Na and $^{68}Ge/Ga$ sources. The calibration determination was performed in the same manner as in the in-beam experiment, namely, the 511-keV full energy peak was selected for one of the NaI counters and the total energy spectrum for the opposite NaI counter.

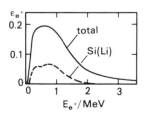

Figure 18. Positron detection efficiency of the solenoid spectrometer. The lower dashed curve, labeled "Si(Li)," corresponds to energy analyzed positrons. For the upper curve, labeled "total," the Si(Li) detector assembly serves as a passive positron catcher only (Ref. 9).

The total positron yield P_{e^+} can be obtained by integration of the energy spectrum. If this procedure is used, only the fraction of positrons which strikes the Si(Li) detector is used. Obviously, the detection efficiency can be increased by approximately a factor of 3 if the complete Si(Li) detector assembly is used as a passive positron catcher. In this case, the detection efficiency shown in Figure 18 drops off for low positron energies because of target self-absorbtion effects, and for high energies because positrons get absorbed in the vacuum chamber or miss the catcher. Nevertheless, the shape of the detection efficiency is sufficiently flat in the relevant region between 0.2 and 1 MeV. As shown below, only very few positrons have energies beyond this region.

Positrons are measured in coincidence with scattered particles or target recoils using a plastic scintillation counter with a thickness of 50 μm, shown in Figure 19. Recoils of light target contaminants like carbon or oxygen are excluded from detection electronically by pulse height discrimination. Projectile fission products resulting from fusion reactions with light target contaminants were kinematically excluded from entering the sensitive region of the plastic counter between 35° and 55°. For U-beam energies below 5.9 MeV/u a small fraction of fission fragments from target projectile interactions were also detected. However, positrons associated with these events were studied using two backward silicon surface barrier detectors (cf. Figure 19). With this configuration, elastic scattering events for nearly symmetric systems could be excluded from detection.

In the new solenoid positron detection system EPOS at GSI in Darmstadt[11] shown in Figure 20, the tube diameter of 260 mm is large

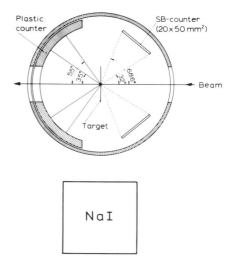

Figure 19. The particle counters as viewed in a plane perpendicular to the solenoid axis at $\zeta = 13$ cm. Scattered particles, target recoils, and/or fission products are detected in a 50-μm-thick plastic scintillator foil. Fission fragments from the projectile emitted in the backward hemisphere following nearly central collisions are detected with silicon surface barrier detectors. Target γ-radiation is measured by a 7.5 × 7.5 cm NaI detector (Ref. 6).

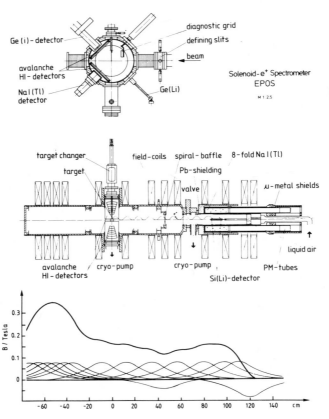

Figure 20. A schematic view of the solenoidal spectrometer EPOS
(Ref. 11). The heavy ion beam axis lies perpendicular to the
plane of the drawing. The four coils to the left are to produce a
magnetic mirror field configuration similar to that shown in Figure
16. Target electrons are excluded from detection by a pencil-like
positron detector (plastic scintillator or silicon) by an Al spiral
baffle.

enough to insert two position-sensitive parallel-plate avalanche detectors
with delay line read-out for particle detection. The angular acceptance of
each of the particle counters is $15° \leq \Theta \leq 80°$ at constant $\Delta\Phi = 36°$. The
angular resolution is better than 1°. This results in good suppression of
unwanted reaction channels. Electrons are suppressed by a spiral baffle
between target and positron detector which makes use of the different
helicity of positrons and electrons. The baffle was optimized for highest
positron transmission with minimal limitation of the momentum band
passed.[7] The probability of detecting electrons scattered by the baffle is

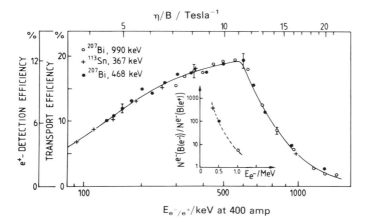

Figure 21. Transport efficiency of the solenoid-system as a function of the electron–positron kinetic energy E_{e^+/e^-} and the scaling parameter η/B, the magnetic fielding at the target and the e^+/e^- momentum in units of $m_e c$. Insert: e^--suppression factor (ratio of total e^--transmission for opposite field settings) as a function of initial kinetic e^--energy (Ref. 11).

strongly reduced by using a pencil-like positron detector of 1 cm diameter positioned along the axis of the spectrometer within an eightfold NaI(Tl) detector for detection of the positron annihilation radiation. The detection efficiency for positrons in a magnetic mirror field configuration is shown in Figure 21 as well as the suppression factor for electrons in the spiral baffle.

3.4. Evaluation of Atomic Positrons

To calculate the atomic positron spectrum $\Delta P_{e^+}^{\text{atom}}/\Delta E_{e^+}$ with Eq. (27), the number of the measured positrons $\Delta N_{e^+}^{\text{exp}}$ has to be corrected for several contributions:

$$\Delta N_{e^+}^{\text{atom}} = \Delta N_{e^+}^{\text{exp}} - \Delta N_{e^+}^{\text{int}}(Z_1, M_1) - \Delta N_{e^+}^{\text{int}}(Z_2, M_2)$$
$$- \Delta N_{e^+}^{\text{ext}}(T) - \Delta N_{e^+}(S) \tag{32}$$

These corrections include positrons produced in the internal pair decay of excited levels in the projectile and target nuclei [$\Delta N_{e^+}^{\text{int}}(Z_1, M_1)$ and $\Delta N_{e^+}^{\text{int}}(Z_2, M_2)$] as well as positrons from external pair conversion of target γ rays in the target $\Delta N_{e^+}^{\text{ext}}(T)$ and in the solenoid $\Delta N_{e^+}^{\text{ext}}(S)$. From test experiments and calculations it was concluded that

$$\Delta N_{e^+}^{\text{ext}}(S) \simeq 0 \tag{33}$$

On the other hand, the external pair creation in a 1 mg/cm^2 Pb or U target requires corrections on the order of 5% to the positron yield. Similar corrections have to be applied for the determination of the integral number $N_{e^+}^{atom}$ of positrons.

To correct for the internal pair conversion contribution, target γ-ray spectra have been measured using a 7.5 × 7.5 cm NaI(Tl) detector in the configuration indicated in Figure 19 under the same coincidence conditions with respect to scattered particles as in the positron determination. In the case of neutron emission after nuclear reactions, additional γ rays are produced in the solenoidal material by $(n, n'\gamma)$ reactions. This contribution must be subtracted to get the target γ spectrum itself.

For the ^{208}Pb + ^{208}Pb scattering system the γ-ray spectrum is very simple because only the 3^- level at 2.614 MeV and the 2^+ level at 4.09 MeV are populated by Coulomb excitation. In this case, the correction procedure for nuclear positrons is straightforward.[10] To obtain the highest possible $Z_u = Z_1 + Z_2$, ^{238}U beams have been used, which results in the excitation of a large number of nuclear levels having unknown decay characteristics. This makes the subtraction procedure of nuclear positrons rather involved. A typical γ spectrum is shown in Figure 22.

After unfolding this spectrum with the response function of the NaI(Tl) detector, the original γ distribution $dN\gamma/dE\gamma(E\gamma)$ can be transformed into

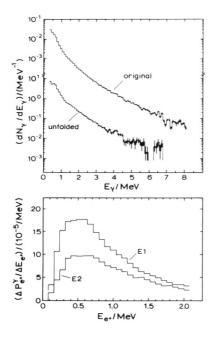

Figure 22. Target γ-ray spectrum taken with the 7.5 × 7.5 cm NaI detector shown in Figure 19. A ^{108}Pd target was bombarded with ^{238}U projectiles with an energy of 5.9 MeV/u. The γ-ray spectra were recorded in coincidence with the ^{108}Pd recoils, uniquely detected in the particle counter at $\Theta_p = 45° \pm 10°$. The spectrum labeled "unfolded" was obtained from the original spectrum by unfolding from the detector response function and by applying the efficiency correction. The lower part of this figure exhibits positron spectra calculated from the unfolded γ-ray spectrum, assuming E1 and E2 multipolarity (Ref. 6).

a positron spectrum using theoretical pair conversion coefficients$^{(68,69,70)}$ $d\beta_{M\lambda}/dE_{e^+}(E\gamma, E_{e^+}, Z)$ of multipolarity $M\lambda$

$$\frac{dN_{e^+}^{\gamma}}{dE_{e^+}}(E_{e^+}, M\lambda) = \int_{2m_ec^2}^{\infty} dE\gamma \frac{dN\gamma}{dE\gamma}(E\gamma) \frac{d\beta_{\mu\lambda}}{dE_{e^+}}(E\gamma, E_{e^+}, Z) \qquad (34)$$

In Figure 22 two positron spectra $\Delta P_{e^+}^{\gamma}/\Delta E_{e^+} = [\Delta N_{e^+}^{\gamma}/\Delta E_{e^+}(E_{e^+}, M\lambda)]/N_p$ are presented as calculated from the γ distribution assuming $E1$ and $E2$ multipolarities, respectively.

There is a remarkable difference between these which emphasizes the difficulty in correcting for nuclear background positrons when the multipolarity of the γ-ray spectrum is unknown. However, the experimentally measured positron spectrum shown in Figure 26 strongly suggests an $E1$ multipolarity if we make the plausible assumption that there are negligible amount of atomic positrons created in the U + Pd scattering system.

Further information about the γ-ray multipolarity can be gained from energy integrated positron yields which require much less beam time than positron spectra due to a higher total efficiency (cf. Figure 18). We calculate

$$N_{e^+}^{\gamma}(M\lambda) = \int_0^{\infty} dE_{e^+} \varepsilon_{e^+}^{tot}(E_{e^+}) \frac{dN_{e^+}^{\gamma}}{dE_{e^+}}(E_{e^+}, M\lambda) \qquad (35)$$

and compare this number with the experimental value $N_{e^+}^{exp}$ forming the ratio $N_{e^+}^{exp}/N_{e^+}^{\gamma}(M\lambda)$. For light scattering systems for which nuclear positron emission is expected to be dominant, this ratio is nearly 1, with the assumption of $E1$ multipolarity for the nuclear γ rays. However, it should be stressed that mixing of $E1$, $M1$, and $E2$ multipolarities with the same mean feature as pure $E1$ transitions cannot be excluded. In particular, a possible contribution due to monopole conversion ($E0$) is not taken care of by this procedure. In Figure 23 the data have been normalized to unity

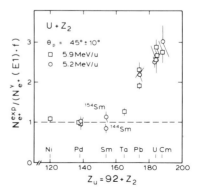

Figure 23. Ratio of measured to calculated positron yields $N_{e^+}^{exp}/[N_{e^+}^{\gamma}(E1)f]$ as a function of the united nuclear charge Z_u, assuming $E1$ multipolarity for the nuclear target γ rays, $\Theta_p = 45° \pm 10°$, U-beam ($\square$) 5.9 MeV/u, ($\bigcirc$) 5.2 MeV/u. For $Z_u > 174$ there is a clear indication for a large number of excess positrons which are attributed to atomic processes (Ref. 6).

in the region $Z_u < 160$ by the introduction of a factor f, which is close to unity.

The displayed ratio clearly exceeds unity for the U + Pb, U + U, and U + Cm scattering systems. This fraction of excess positrons is believed to originate from atomic processes for which we can write

$$N_{e^+}^{\text{atom}} = N_{e^+}^{\text{exp}} - N_{e^+}^{\text{ext}}(T) - N_{e^+}^{\gamma}(E1)f \qquad (36)$$

From comparison with Eq. (32) after consideration of Eq. (33) in the integral form, we can identify the last term of (36) as the nuclear contribution of positrons:

$$N_{e^+}^{\gamma}(E1)f = N_{e^+}^{\text{int}}(Z_1, M_1) + N_{e^+}^{\text{int}}(Z_2, M_2) \qquad (37)$$

However, it should be emphasized that this procedure is only correct if the multipolarity distribution of the γ-ray spectrum for the high $Z_u = Z_1 + Z_2$ region ($Z_u > 160$) for which atomic positrons are expected to contribute is not significantly different than in the low Z_u region ($Z_u < 160$).

3.5. Results and Discussion

Significant experimental efforts are currently in progress to investigate the question of positron creation in heavy ion collision systems with large combined nuclear charge number Z_u. Therefore, it is not possible at this time to adequately survey this dynamically developing field. For this reason we present and discuss published experimental results, in which we include information from annual reports or conference contributions which apparently are not subject to future changes.

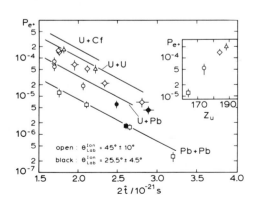

Figure 24. The total atomic positron production probabilities P_{e^+} for different systems as a function of scattering time $2\hat{t}$, Eq. (30). The following designations are used: $\triangle$, U + Cm; $\diamond$, U + U; $\bigcirc$, U + Pb; $\square$, Pb + Pb. Open points represent $\Theta_p = 45° \pm 10°$, black points $25.5° \pm 4.5°$. The full lines represent theoretical calculations (Ref. 64). The insert shows the atomic production probability P_{e^+} as a function of the united nuclear charge Z_u for a fixed scattering time $2\hat{t} = 1.75 \times 10^{-21}$. (From Refs. 8 and 6.)

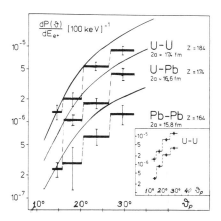

Figure 25. Energy differential atomic positron production probabilities within a positron energy interval of 100 keV centered around 490 keV as a function of the laboratory scattering angle Θ_p. The solid lines represent the results of calculations (Ref. 64). The insert shows the total number of observed positrons per scattered particle (upper curve) together with calculated background from nuclear processes (lower curve) for U + U collisions (Ref. 46).

The experimental results available up to now can be grouped into three categories: (i) measurement of total positron production probabilities P_{e^+} as a function of Z_u, scattering angle Θ_p, and projectile energy E_1 (Figure 24); (ii) measurement of the energy differential positron production probability $\Delta P_{e^+}/\Delta E_{e^+}$ at a fixed positron energy $E_{e^+} = (490 \pm 50)$ keV and a constant projectile energy $E_1/M_1 = 5.9$ MeV/u as a function of Θ_p and Z_u (Figure 25); and (iii) positron spectra $\Delta P_{e^+}/\Delta E_{e^+}$ as a function of Z_u for a scattering angle $\Theta_p = 45° \pm 10°$ and a projectile energy $E_1/M_1 = 5.9$ MeV/u (Figure 26).

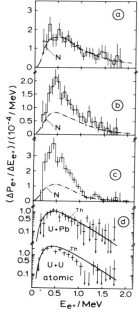

Figure 26. Positron spectra $\Delta P_{e^+}/\Delta E_{e^+}$ for (a) the nuclear U + Pd, (b) the undercritical Pb + Pb, (c) the overcritical U + U systems at a U beam energy of 5.9 MeV/u and $\Theta_p = 45° \pm 10°$. The dashed–dotted curves "N" indicate nuclear positrons as derived from γ-ray spectra, assuming $E1$ multipolarity. (d) is a semilogarithmic presentation of the spectra of atomic positrons compared to theoretical calculations indicated by "Th" (Ref. 64). (Data from Refs. 8 and 6).

The total positron production yield P_{e^+} is represented in Figure 24 as a function of the scattering time $2\hat{t}$ which was defined with Eq. (30). As already mentioned in Section 3.2, this time characterizes the dynamically induced positron creation for undercritical systems. In order to understand this, we follow a qualitative argument given by Migdal[53] where he showed that the first-order adiabatic transition probability $|a_f(t = \infty)|^2$ with

$$a_f(t = \infty) = -\int_{t'=-\infty}^{\infty} dt' \langle f|\partial/\partial t|i\rangle \exp\left[i(E_f - E_i)t'/\hbar\right] \tag{38}$$

contains a factor

$$\exp\left[-(E_f - E_i)\tau/\hbar\right] \tag{39}$$

if the condition

$$(E_f - E_i)\tau/(2\hbar) \gg 1 \tag{40}$$

is fulfilled and if the transition matrix element exhibits no singularity on the real axis. The transition amplitude $a_f(t = \infty)$ describes the excitation of an electron from the negative energy continuum with energy E_i to a bound or continuum state with energy E_f. The time τ characterizes the scattering process.

It can be shown[39,40] that in the monopole approximation the factor in front of Eq. (39) is

$$[f(E_i)g(E_f)]^{1/2}/(E_f - E_i) \tag{41}$$

The energy denominator $1/(E_f - E_i)$ originates from the application of the Hellmann–Feynman relation. The functions $f(E_i)$, $g(E_f)$ stem from the Coulomb repulsion of positrons or Coulomb attraction of electrons, respectively, in analogy to the well-known Fermi function in β decay. Then, the double differential probability for pair creation can be expressed as

$$d^2P_{e^+}/(dE_{e^+}dE_f) = f(E_i)g(E_f)\exp\left[-2\hat{t}(E_f - E_i)/\hbar\right]/(E_f - E_i)^2 \tag{42}$$

We see directly that the transition amplitude factors into an exponential which contains the dynamical aspects of positron creation and functions $f(E_i)$ and $g(E_f)$ which are connected to the electron and positron wave functions in the initial and final state. When electrons are not detected in the experiment we have to integrate over E_f and obtain the positron spectrum:

$$dP_{e^+}/dE_{e^+} = h(E_{e^+}, \hat{t})\exp\left[-2\hat{t}(E_{e^+} + 2m_ec^2)/\hbar\right] \tag{43}$$

The function $h(E_{e^+}, \hat{t})$ causes the falling off of the positron intensity at low energy, which has been observed experimentally (cf. Figure 26). If we are interested in the total positron production probability only, a further integration over E_{e^+} has to be performed, and this yields

$$P_{e^+} = i(\hat{t}) \exp\left[-2\hat{t}(2m_e c^2)/\hbar\right] \tag{44}$$

Following this simple consideration we can determine that the nearly exponential slopes observed in both dP_{e^+}/dE_{e^+} at sufficiently high positron energies E_{e^+} and in $P_{e^+}(\hat{t})$ as a function of the scattering time, are the results of the simple dynamical aspects of collisional positron creation. On the other hand, the absolute positron production probability holds detailed physical properties on the wave function in the matrix element $\langle f|\partial/\partial t|i\rangle$ of Eq. (38). Furthermore, a spontaneous positron contribution could be noticed as a deviation of the experimental findings from extrapolations into the region for which, according to theoretical calculations, the diving of the $1s\sigma$ level into the negative energy continuum is expected to occur. It was also suggested[2] in a somewhat different context that interferences of the spontaneous and direct transition amplitudes could manifest itself in deviations from a simple scaling law. However, it must be stressed at this point that second- and higher-order terms in the perturbation expansion are not negligible, as present theoretical investigations imply.[64] Keeping in mind this reservation, conclusions drawn from this model must be accepted with caution.

From Figure 24 we see that P_{e^+} can be approximated for every scattering system by an exponential function. No distinct effects of level diving are observed in P_{e^+} for the U + Cm and U + U points at lowest scattering time even though these systems are expected to be overcritical. This is also true for the Z_u dependence of P_{e^+} as shown in the insert of Figure 24 for the fixed scattering time $2\hat{t} = 1.75 \times 10^{-21}$ sec and the probability can be approximated by the relation

$$P_{e^+} \propto Z_u^{20.3} \tag{45}$$

The absence of effects of a diving level within the experimental errors of about 30% may be a consequence of insufficient experimental sensitivity in the measurement of the total positron yield P_{e^+}.

A more sensitive determination for diving characteristics would be expected to be obtained by directly measuring the low-energy portion of the positron spectra. Figure 24 represents measured spectra corrected for the detection response functions and efficiency. The spectrum of the low $Z_u = 138$ system U + Pd, for which only nuclear positrons are expected, can be reproduced by converting the target γ-ray distribution into a positron spectrum assuming $E1$ multipolarity. The nuclear background contributions for the U + Pb and U + U systems have similarly been determined. Comparing the overcritical U + U spectrum with the undercritical U + Pb spectrum, we can note small deviations in the U + U system in the low-energy part of the spectrum. Even though such an effect is in the direction that would be expected from level diving, it must be verified with better statistics.

Finally, we compare the positron production yields with theoretical calculations.[64] The slopes of these calculations as shown in Figures 24 and 25 agree quite well with the experiment. It is worth noting that from these calculations no significant effects of a diving $1s\sigma$ level are predicted. Obviously, the diving time is too short as long as the colliding nuclei follow pure Rutherford trajectories. Nevertheless, there is some indication that the theoretical calculations overestimate the positron production probability. However, it must be stressed that the absolute experimental errors are at present about 30%. But if the indicated deviations persist with improved experimental results, their consequences must be investigated. Among the possible causes can be the absence of level diving, presence of nuclear delay time effects, and, perhaps, approximations made in the theory.

4. Outlook

All positron experiments described in the preceding sections have been performed for scattering systems in which the collision partners follow nearly pure Rutherford trajectories. For the $^{238}U + ^{238}U$ systems at 5.9 MeV/u, e.g., the distance of closest approach R_0[cf. Eq. (6)] approaches the interaction radius only for very central collisions. One can speculate that only these very central collisions bring about nuclear sticking times of about 10^{-20} sec resulting in a decay of $1s\sigma$ vacancies by emission of monochromatic positrons. Principally, such a mechanism of monochromatic positron emission was first proposed by Rafelski et al.[61] Indications for structures in the positron spectra between 300 and 400 keV have been observed in recent experiments with the "orange"-type β-spectrometer[19] (cf. Figure 15) and the new positron detection system EPOS (cf. Figure 20).[23] However, it remains to be verified that such structures are not caused by internal pair decay with a multipolarity $E0$ of an excited nuclear level.

Presently, also, attempts are in progress with the solenoid spectrometer shown in Figure 16 to investigate positron spectra obtained in the presence of a perturbation of the Rutherford trajectory due to nuclear contact at a beam energy as high as 8.4 MeV/u. At such an energy, the grazing angle is about 70° in the c.m. system. Collisions at impact parameters smaller than 8.7 f result in nuclear contact with energy dissipation and consequently, most probably, fission of the very fragile ^{238}U nuclei. A preliminary background-corrected positron spectrum recorded in coincidence with fission fragments exhibits a remarkable feature. Above an energy of 700 keV the spectrum drops off as rapidly as the reference spectrum taken of 5.9 MeV/u in coincidence with Rutherford scattered particles, while from Eq. (43) and calculations based on pure Rutherford trajectories this is not expected. This effect may indicate a small time delay ($\approx 10^{-21}$ sec) during the nuclear contact in connection with changed kinematical conditions in the outgoing channel.

For second-generation experiments, the positron detection efficiency could be improved significantly. We mention the new positron detector of the "orange" spectrometer, which no longer requires the 511-keV annihilation coincidence signal and, thus, results in an improvement of the detector efficiency of about a factor of 6 in a momentum interval $\Delta p/p = 0.17$.[18] Furthermore, using the so-called Tori spectrometer[41] shown in Figure 27, a broad band acceptance of positrons with an efficiency of about 20% is expected. The main feature of this new instrument is the ability to separate electrons and positrons and to detect them simultaneously in different detectors. This is achieved by a S-shaped field configuration produced by pancake coils surrounding two quarters of a toroidal tube. Electrons and positrons leaving the target experience in the bent magnetic field of the first quarter torus a drift in their guiding center perpendicular to the spectrometer plane. According to the charges, q, this drift is opposite for electrons and positrons. It is designed to give complete separation of positrons from electrons which are absorbed in a detector while the positrons are left free to pass into the second quarter torus in which the drift is reversed.

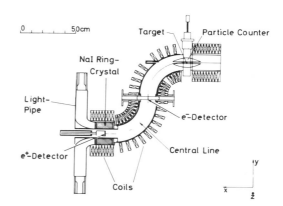

Figure 27. Tori spectrometer, beam direction, and z-drift is perpendicular to plane of drawing (Ref. 41). The e^+ detector is a Si(Li) detector with 40 mm diameter and 5 mm thickness. The NaI ring crystal is the same as shown in Figure 17.

In all second-generation experiments position-sensitive particle counters are required, with the ability to determine precisely the impact parameter in a scattering process. This could be of great importance since small physical effects at specific distances of closest approach in the collision process may be obscured with a particle counter accepting a wide Θ_p band.

With the increase of the detection efficiency for positrons, the final accuracy in the atomic positron spectra may be limited by uncertainties in the correction of nuclear background positrons. For example, if an accuracy of 5% is required in the atomic positron spectra after subtraction of a 40% nuclear background contribution, this background has to be known with an accuracy of 7.5%. Future experiments will have to establish whether the γ-ray positron conversion procedure described above can be sufficiently improved to achieve this accuracy or whether new methods will be required. The first of these to be mentioned should be a proposed channeling experiment.[42] In this experiment which uses a single U crystal, a blocking effect is expected for the atomic processes and not for the delayed nuclear positrons.

Another technique for obtaining information about the multipolarity of the nuclear γ-ray continuum is from measuring the electron spectrum in coincidence with positrons. First attempts to do this have been performed with the old solenoid spectrometer. Two Si(Li) detectors for electrons have been installed within the short coil at $\zeta = 6$ cm (cf. Figure 16) in a similar geometry as the positron counters at the other side. Electrons below an energy of about 400 keV were geometrically eliminated. In a test experiment electron spectra have been measured in coincidence with positrons for U + U at 5.9 MeV/u and $\Theta_p = 45° \pm 10°$. A preliminary analysis of this experiment indicates that the electron spectrum taken in coincidence with positrons exhibits a flatter descent as a function of energy than the corresponding electron spectrum not in coincidence with positrons. The reason for this effect lies in the distinct difference in the slope of δ-electrons and electrons from nuclear pair decay. This offers a possibility of obtaining additional information about positrons from nuclear pair decay.

Parallel to future investigations of the heaviest available system Pb + Cm leading to a quasiatom with overcritical combined charge $Z_u > 173$ it should also be emphasized that nuclear sticking time effects as mentioned above might play an important role in investigation of lighter colliding systems by means of high energy δ-ray spectroscopy.[73] The basic concept can be briefly described as follows: For deep inelastic reactions, relatively long reaction times are expected: $\sim 10^{-20}$ sec. This reaction time, treated as a delay between the incoming and outgoing phase of the collision, leads to a phase shift between the electron excitation amplitudes for the corresponding paths of the trajectory. Therefore, an interference between the incoming and outgoing path is expected which may lead to oscillations in the high-energy δ-ray spectrum with the width $\Delta E = h/\Delta T$. The experimental observation of these oscillations might allow one to determine the reaction time, i.e., to construct an "atomic clock" for the lifetime of the composite system in deep inelastic reactions.

Appendix: Transformation of the Cross Section from the Laboratory System to the C.M. System

"Orange"-Type β-Spectrometer

Albeit trivial, the transformation of the measured triple-differential cross section from the laboratory to the c.m. system is of certain importance, mainly for two reasons:

i Both δ rays and positrons are emitted with relativistic velocities—e.c. 100 keV β-kinetic energy corresponds to a velocity of $\beta = (v/c) = 0.55$ and $E_\beta = 1$ MeV to $\beta = (v/c) = 0.94$. The nonrelativistic c.m. motion of the quasiatoms with βc.m. $= (v_{\text{c.m.}}/c) = 0.05$ should not veil the fact that β rays are "always" relativistic.

ii Both the correction for the solid angle $(d\Omega^*/d\Omega)$ and for the kinetic energy (dE^*/dE) depend on the kinetic energy itself. Therefore, the *relativistic* transformation of the triple-differential cross section $[d^3\sigma/(d\Omega_\beta dE_\beta d\Omega_p)]$ should be treated carefully, since the probability of affecting the *shape* of a continuous distribution measured is latent.

In the case of the "orange"-type β-spectrometer the principal quantities and symbols needed will be as follows:

(1) $\eta = (p/m_0 c)$, momentum of the β ray in units of $m_0 c$. Since the β rays are momentum analyzed by the magnetic field of the spectrometer, their momentum is defined (in the laboratory) by the magnetic rigidity $(B\rho)$ in gauss cm and $(B\rho)(e/c) = mv = p$ or $(B\rho)/1704.5 = \eta$.

(2) $\varepsilon = E(m_0 c^2)$ is the total energy of the β ray in units of $m_0 c^2$. It is obvious that $\varepsilon^2 = \eta^2 + 1$ from $E = p^2 c^2 + (m_0 c^2)^2$.

(3) $\beta_{\text{c.m.}} = (v_{\text{c.m.}}/c)$, velocity of the center of mass in the laboratory in units of the light velocity, with $v_{\text{c.m.}} = v_\infty \cdot M_1/(M_1 + M_2)$. It should be emphasized again that the δ rays and the positrons of interest are emitted from the *combined quasi-atomic system* of the projectile and target nuclei, which moves with a velocity $v_{\text{c.m.}}$ in the beam direction. This is not the case for the background β rays, emitted after nuclear deexcitation via internal conversion of internal pair creation, which are emitted from the scattered particles and *not* from the combined system.

(4) $\bar{\Theta}$, average emission angle of the β ray with respect to the c.m. motion:

$$\bar{\Theta} = \frac{\int_{\Theta_1}^{\Theta_2} \Theta \, d\Omega}{\int_{\Theta_1}^{\Theta_2} d\Omega} = \frac{\int_{\Theta_1}^{\Theta_2} \Theta \sin \Theta \, d\Theta}{\int_{\Theta_1}^{\Theta_2} \sin \Theta \, d\Theta} \tag{A.1}$$

with Θ_1 and Θ_2 respectively, minimal, and maximal angle accepted, defined by the entrance slits of the spectrometer.

In the transformation procedure sketched below all quantities in the c.m. system are denoted by an asterisk.

The kinetic energy of a β ray in the c.m. system can be easily obtained now from the measured momentum $(B\rho)$ in the laboratory:

$$\eta = \frac{(B\rho)}{1704.54} = \frac{mv}{m_0 c} = \frac{\beta}{(1 - \beta^2)^{1/2}} = (\beta^{-2} - 1)^{-1/2} \tag{A.2}$$

$$\beta = (1 + \eta^{-2})^{-1/2} \tag{A.3}$$

$$\beta^* = \frac{\beta - \beta_{\text{c.m.}}}{1 - \beta} \frac{\cos \bar{\Theta}}{\beta_{\text{c.m.}} \cos \bar{\Theta}} \qquad \text{for } \beta_{\text{c.m.}} \ll 1 \tag{A.4}$$

The minus sign reflects the fact that β rays were measured in the forward direction:

$$\varepsilon^* = [1 - (\beta^*)^2]^{-1/2} \quad \text{and} \quad E_\beta^* = (\varepsilon^* - 1)m_0 c^2 = (\varepsilon^* - 1)511 \, [\text{keV}] \tag{A.5}$$

$$\eta^* = [(\beta^*)^{-2} - 1]^{-1/2} \tag{A.6}$$

Transforming the measured cross section $d^2\sigma/(d\Omega \, dE)$ or $d^2\sigma/(d\Omega dp)$ into the c.m. system, not only the ratios (dp^*/dp) or (dE^*/dp) are needed, but also the ratio of the solid angle elements $(d\Omega^*/d\Omega)$. The first term (dE^*/dp) or (dp^*/dp) can be easily obtained analytically from the above formulas leading to an expression of dE^* or dp^* by dp. An easier way, however, is to express $d\eta^* \, d\Omega^*$ by $d\eta d\Omega$ as follows.

The transformation (lab–c.m.) of η, Θ, and $d\eta \, d\Omega$ can be evaluated from the transformation of the parallel and normal components of the momentum (with respect to the c.m. motion):

$$\eta_\perp = \eta_\perp^* \Rightarrow \eta \sin \Theta = \eta^* \sin \Theta^* \tag{A.7}$$

$$\eta_\parallel = (1 - \beta_{\text{c.m.}}^2)^{-1/2}(\eta_\parallel^* + \beta_{\text{c.m.}}\varepsilon^*) \Rightarrow$$
$$\Rightarrow \eta \cos \Theta = (1 - \beta_{\text{c.m.}}^2)^{-1/2}(\eta^* \cos \Theta^* + \beta_{\text{c.m.}}\varepsilon^*) \tag{A.8}$$

and with the well-known relation

$$d\Omega = \sin \Theta \, d\Theta = d(\cos \Theta) \tag{A.9}$$

one obtains

$$\frac{d^2\sigma^*}{d\Omega^* dE^*} = \frac{d^2\sigma}{d\Omega \, dE} \cdot \frac{\eta^*}{\eta} \tag{A.10}$$

$$\frac{d^2\sigma^*}{d\Omega^* \, dE^*} = \frac{d^2\sigma}{d\Omega \, dp} \cdot \frac{\varepsilon}{\eta} \cdot \frac{\eta^*}{\eta} \tag{A.11}$$

It should be noted that the above expressions are valid only for a constant angle of emission in the laboratory or for isotropic emission in the c.m. system. At fixed emission angle in the laboratory the c.m. emission angle is a function of the β-ray energy:

$$\Theta^* = \arcsin\left[(\eta/\eta^*)\sin\Theta\right] \tag{A.12}$$

Solenoid Transport System

In this part we will describe some details of the correction procedures which are necessary to transform positron production probabilities from the laboratory system into the c.m. system.

Let us consider atomic positrons emitted from a "compound" system which moves with velocity $v = v_{\text{c.m.}} = \beta c$ in beam direction. From the triple differential cross section in the c.m. system

$$d^3\sigma_{e^+}^{\text{c.m.}}/(dE_{e^+} \, d\Omega_{e^+} \, d\Omega_p)(E'_{e^+}, \Theta'_{e^+}, \phi'_{e^+}, \Theta'_p, E_1, M_1, Z_1, M_2, Z_2) \tag{A.13}$$

we define the positron production probability as follows:

$$\frac{d^2 P_{e^+}^{\text{c.m.}}}{dE_{e^+} \, d\Omega_{e^+}} = \frac{d^3\sigma_{e^+}^{\text{c.m.}}/(dE'_{e^+} \, d\Omega'_{e^+} \, d\Omega'_p)}{d\sigma_R/d\Omega'_p} \tag{A.14}$$

Let us assume our positron spectrometer analyzes the positron *energy*. Spectrometers of this type were described in Section 3.3.2. Then, neglecting resolution effects by means of the apparatus, the number of positrons per energy interval dN_{e^+}/dE_{e^+} in the laboratory system is connected to the positron production probability in the c.m. system as follows:

$$\frac{dN_{e^+}}{dE_{e^+}}(E_{e^+}) = N_p \int_{\Omega_{e^+}} d\Omega_{e^+} \, \tilde{\varepsilon}(E_{e^+}, \Theta_{e^+}, \phi_{e^+}) \left(\frac{dE'_{e^+}}{dE_{e^+}}\right)$$

$$\times \left(\frac{d\Omega'_{e^+}}{d\Omega_{e^+}}\right) \frac{d^2 P_{e^+}^{\text{c.m.}}}{dE'_{e^+} \, d\Omega'_{e^+}}(E'_{e^+}) \tag{A.15}$$

with

$$E'_{e^+} = \frac{E_{e^+}}{1 + \beta[1 + 2m_e c^2/E_{e^+}]^{1/2} \cos \Theta_{e^+}} \tag{A.16}$$

The integral is taken over the solid angle Ω_{e^+} in the laboratory system which the positron spectrometer accepts. The function $\tilde{\varepsilon}$ describes the detection efficiency of positrons emitted in direction Θ_{e^+}, ϕ_{e^+} with respect to the beam axis. The angle ϕ_{e^+} is counted with respect to the solenoid axis. The functions dE'_{e^+}/dE_{e^+} and $d\Omega'_{e^+}/d\Omega_{e^+}$ result from the transformation from the c.m. system in the laboratory system. From the Doppler relation in first order in β,

$$E_{e^+} = E'_{e^+}(1 + \beta' \cos \Theta_{e^+}) \tag{A.17}$$

with

$$\beta' = \beta[1 + 2m_e c^2/E'_{e^+}]^{1/2} \tag{A.18}$$

we get

$$dE'_{e^+}/dE_{e^+} = 1 - \beta/\beta'_e \cos \Theta_{e^+} \tag{A.19}$$

with the positron velocity in the c.m. system

$$\beta'_e = (E'^2_{e^+} + 2m_e c^2 E'_{e^+})^{1/2}/(E'_{e^+} + m_e c^2) \tag{A.20}$$

From the relativistic transformation of the velocity[48] we get again in first order in β for the solid angle ratio

$$d\Omega'_{e^+}/d\Omega_{e^+} = 1 + 2\beta/\beta'_e \cos \Theta_{e^+} \tag{A.21}$$

The counting rate dN_{e^+}/dE_{e^+} at a fixed energy E_{e^+} in the laboratory system results from a superposition of positrons in an energy band in the c.m. system which is defined by the relation (A.16). However, we are interested in the positron production probability in the c.m. system. This can be achieved from equation (A.15) by the following simplifying assumptions:

i. Isotropy of positron emission in the c.m. system (monopole emission!):

$$d^2 P_{e^+}^{c.m.}/(dE'_{e^+} d\Omega'_{e^+}) = (4\pi)^{-1} dP_{e^+}^{c.m.}/dE'_{e^+} \qquad (A.22)$$

ii. $\beta \ll 1, \beta'_e \gg \beta$

iii. Linearization of the $dP_{e^+}^{c.m.}/dE'_{e^+}$ in the Doppler interval

$$E_{e^+}/(1 + \beta') \leq E'_{e^+} \leq E_{e^+}/(1 - \beta') \qquad (A.23)$$

$[dP_{e^+}^{c.m.}/dE'_{e^+}(E'_{e^+})$ is assumed to be a slowly varying function]

iv. $\tilde{\varepsilon}(E_{e^+}, \Theta_{e^+}, \phi_{e^+}) = \tilde{\varepsilon}(E_{e^+}, \pi - \Theta_{e^+}, -\phi_{e^+}) \qquad (A.24)$

After some algebra, we get these simplifications:

$$\frac{dP_{e^+}^{lab}}{dE_{e^+}}(E_{e^+}) \simeq \frac{dP_{e^+}^{c.m.}}{dE'_{e^+}}(E_{e^+}) \qquad \text{with } P_{e^+}^{lab} = \frac{N_{e^+}}{\varepsilon_{e^+}(E_{e^+})N_p} \qquad (A.25)$$

with the overall detection efficiency for positrons at energy E_{e^+}

$$\varepsilon_{e^+}(E_{e^+}) = (4\pi)^{-1} \int \varepsilon_{e^+}(E_{e^+}, \Theta_{e^+}, \phi_{e^+}) d\Omega_{e^+} \qquad (A.26)$$

For the experiments with the positron solenoidal spectrometers described in Section 3.3 the conditions (i)–(iv) are fairly well fulfilled. Although these spectrometers may accept a rather large solid angle, the transformation from the laboratory system to the c.m. system turns out to be trivial. A similar statement holds for nuclear positrons emitted from ejectiles or recoils but we will not prove it here.

References

1. P. Armbruster, H.-H. Behnke, S. Hagmann, D. Liesen, F. Folkmann, and P. H. Mokler, Z. Phys. **A288**, 277 (1978).
2. P. Armbruster and P. Kienle, Z. Phys. **A291**, 399 (1979).
3. H. Backe, Nordic Spring Symposium on Atomic Inner Shell Phenomena, Geilo, Norway, April 17–21, 1978, Lecture Notes, Vol. II, p. 191; H. Backe, in Selected Topics in Nuclear Structure, J. Styczen and R. Kulessa, Eds., Uniwersytet Jagiellonski, Krakow (1978), Vol. 2, p. 823.

4. H. Backe, in *Trends in Physics* 1978, M. M. Woolfson, Ed., Adam Hilger Ltd., Bristol (1979), p. 445.

5. H. Backe, in XVIII International Winter Meeting on Nuclear Physics, Bormio, Jan. 21–26, 1980 (University of Milan, ed.), Suppl. No. 13 (1980) 809.

6. H. Backe, *Present Status and Aims of Quantum Electrodynamics*, Lecture Notes in Physics, Vol. 143, Springer-Verlag, Berlin (1981), p. 277.

7. H. Backe, H. Bokemeyer, E. Kankeleit, E. Kuphal, Y. Nakayama, L. Richter, and R. Willwater, Laborbericht Nr. 67, Institut für Kernphysik der TH Darmstadt (1975).

8. H. Backe, W. Bonin, W. Engelhardt, E. Kankeleit, P. Senger, F. Weik, V. Metag, and J. B. Wilhelmy, GSI-Sci. Rep. 1979 80–83 (1980), p. 101.

9. H. Backe, W. Bonin, E. Kankeleit, W. Patzner, P. Senger, and F. Weik, GSI-Sci. Rep. 1979 80–83 (1980), p. 168.

10. H. Backe, L. Handschug, F. Hessberger, E. Krankeleit, L. Richter, F. Weik, R. Willwater, H. Bokemeyer, P. Vincent, Y. Nakayama, and J. S. Greenberg, *Phys. Rev. Lett.* **40**, 1443 (1978).

11. A. Balanda, H. J. Beeskow, K. Bethge, H. Bokemeyer, H. Folger, J. S. Greenberg, H. Grein, A. Gruppe, S. Ito, S. Matsuki, R. Schulte, R. Schultz, D. Schwalm, J. Schweppe, R. Steiner, P. Vincent, and M. Waldschmidt, GSI-Sci. Rep. 1979 80–83 (1980), p. 161.

12. J. Bang and J. M. Hansteen, *K. Dan. Vidensk. Selsk. Mat.-Fys. Medd.* **31**, no. 13 (1959).

13. J. Bang and J. M. Hansteen, *Phys. Lett.* **72A**, 218 (1979); *Phys. Sci.* **22**, 609 (1981).

14. D. R. Bates and R. McCarroll, *Proc. R. Soc. London, Ser A* **245**, 175 (1958).

15. F. Beck, H. Steinwedel, and G. Süssmann, *Z. Phys.* **171**, 189 (1963).

16. E. Berdermann, H. Bokemeyer, F. Bosch, M. Clemente, F. Güttner, P. Kienle, W. Koenig, C. Kozhuharov, H. Krimm, B. Martin, B. Povh, K. Traxel, and Th. Walcher, GSI-Jahresbericht (1978), p. 95.

17. E. Berdermann, H. Bokemeyer, F. Bosch, M. Clemente, F. Güttner, P. Kienle, C. Kozhuharov, H. Krimm, B. Martin, B. Povh, K. Traxel, and Th. Walcher, *Verh. Dtsch. Phys. Ges. (VI)* **14**, 492 (1979); E. Berdermann, F. Bosch, H. Bokemeyer, M. Clemente, F. Güttner, P. Kienle, W. Koenig, C. Kozhuharov, B. Martin, B. Povh, H. Tsertos, and Th. Walcher, GSI-Sci. Rep. 1978 79–11 (1979), p. 100.

18. E. Berdermann, F. Bosch, M. Clemente, F. Güttner, P. Kienle, W. Koenig, C. Kozhuharov, B. Martin, W. Potzel, B. Povh, C. Tsertos, W. Wagner, and Th. Walcher, GSI-Sci. Rep. 1979 80–3 (1980), p. 103; E. Berdermann, H. Bokemeyer, F. Bosch, M. Clemente, F. Güttner, P. Kienle, W. Koenig, C. Kozhuharov, B. Martin, B. Povh, H. Tsertos, and Th. Walcher, *Verh. Dtsch. Phys. Ges. (VI)* **15**, 1168 (1980).

19. E. Berdermann, F. Bosch, M. Clemente, F. Güttner, P. Kienle, W. Koenig, C. Kozhuharov, B. Martin, B. Povh, C. Tsertos, W. Wagner, and Th. Walcher, GSI-Sci. Rep. 1980 81–82 (1981), p. 128.

20. H. Bokemeyer, in *Selected Topics in Nuclear Structure*, J. Styczen and R. Kulessa, eds., Uniwersytet Jagiellonski, Krakow (1978), Vol. 2, p. 841.

21. H. Bokemeyer, in *Heavy Ion Physics*, A. Berinde, V. Ceausescu, and I. A. Dorobantu, eds., Proc. Predeal International School (1978), p. 489.

22. H. Bokemeyer, H. Folger, H. Grein, S. Ito, D. Schwalm, P. Vincent, K. Bethge, A. Gruppe, R. Schule, M. Waldschmidt, J. S. Greenberg, J. Schweppe, and N. Trautmann, GSI-Sci. Rep. 1980 81–82 (1981), p. 127.

23. H. Bokemeyer, H. Folger, H. Grein, S. Ito, D. Schwalm, P. Vincent, K. Bethge, A. Gruppe, R. Schule, M. Waldschmidt, J. S. Greenberg, J. Schweppe, and N. Trautmann, private communication.

24. F. Bosch, F. Güttner, B. Martin, L. Meyer-Schützmeister, P. Kienle, W. Koenig, C. Kozhuharov, H. Krimm, B. Povh, H. Scappa, J. Soltani, and Th. Walcher, in XVIII International Winter Meeting on Nuclear Physics, Bormio, Jan. 21–26, 1980 (University of Milan, ed.), Suppl. No. 13.

25. F. Bosch, H. Krimm, B. Martin, B. Povh, Th. Walcher, and K. Traxel, *Phys. Lett.* **78B**, 568 (1978).
26. F. Bosch, D. Liesen, P. Armbruster, D. Maor, P. H. Mokler, H. Schmidt-Böcking, and R. Schuch, *Z. Phys.* **A296**, 11 (1980).
27. J. S. Briggs, *J. Phys. B* **8**, L485 (1975).
28. H. G. Dosch, J. H. D. Jensen, and V. F. Müller, *Phys. Norv.* **5**, 151 (1971).
29. M. R. C. McDowell and J. P. Coleman, *Introduction to the Theory of Ion–Atom Collisions*, North-Holland, Amsterdam (1970).
30. B. Fricke and G. Soff, *At. Nucl. Data Tables* **19**, 83 (1977).
31. J. S. Greenberg, in *Electronic and Atomic Collisions*, N. Oda and T. Takayanagi, eds., North-Holland, Amsterdam (1980), p. 351
32. F. Güttner, Diplomarbeit, Universität Heidelberg (1979), unpublished.
33. F. Güttner, W. Koenig, B. Martin, B. Povh, H. Skapa, J. Soltani, Th. Walcher, F. Bosch, and C. Kozhuharov, preprint (1980), submitted to *Z. Phys. A*.
34. F. Güttner, W. Koenig, B. Martin, B. Povh, H. Skapa, J. Soltani, Th. Walcher, C. Kozhuharov, P. Kienle, and L. Meyer-Schützmeister, in *Book of Abstracts (Part II)*, European Conference on Atomic Physics (ECAP), J. Kowalski, G. zu Putlitz, and H. G. Weber, eds., Eur. Phys. Soc. C (1981), Vol. 5A, Part II, p. 899.
35. F. P. Hessberger, Diplomarbeit, Technische Hochschule Darmstadt (1977).
36. G. Hermann, *Yearbook of Science and Technology*, p. 385, McGraw-Hill (1980) and references cited therein.
37. D. H. Jakubassa, *Z. Phys.* **A293**, 281 (1979).
38. O. Klein, *Z. Phys.* **53**, 157 (1929).
39. E. Kankeleit, in *Nuclear Interactions*, B. A. Robson, ed., Lecture Notes in Physics, Vol. 92, Springer-Verlag, New York (1979), p. 306.
40. E. Kankeleit, in Proceedings of the Twelfth Summer School of Nuclear Physics, Mikolajki 1979, *Nukleonika* **25**, 253 (1980).
41. E. Kankeleit, R. Köhler, M. Kollatz, M. Krämer, R. Krieg, P. Senger, and H. Backe GSI-Sci. Rep. 1980 81-2 (1981), p. 195.
42. K. G. Kaun and S. A. Karamyan, JINR P 7–11420, Dubna (1978).
43. P. Kienle, in *Progress in Particle and Nuclear Physics*, Vol. 4, D. Wilkinson, ed., Plenum Press, New York (1980).
44. P. Kienle, in *Atomic Physics 7*, D. Kleppner and F. M. Pipkin, eds., Plenum Press, New York (1981), p. 1.
45. C. Kozhuharov, P. Kienle, D. H. Jakubassa, and M. Kleber, *Phys. Rev. Lett.* **39**, 540 (1977).
46. C. Kozhuharov, P. Kienle, E. Berdermann, H. Bokemeyer, J. S. Greenberg, Y. Nakayama, P. Vincent, H. Backe, L. Handschug, and E. Kankeleit, *Phys. Rev. Lett.* **42**, 376 (1979).
47. H. Krimm, Diplomarbeit, Universität Heidelberg (1978).
48. L. D. Landau and E. M. Lifschitz, *Lehrbuch der theoretischen Physik*, Vol. 11, Chap. 1, Akademie Verlag, Berlin (1973).
49. D. Liesen, P. Armbruster, H.-H. Behnke, F. Bosch, S. Hagmann, P. H. Mokler, H. Schmidt-Böcking, R. Schuch, in *Electronic and Atomic Collisions*, N. Oda and T. Takayanagi, eds., North-Holland, Amsterdam (1980), p. 337.
50. D. Liesen, P. Armbruster, F. Bosch, S. Hagmann, P. H. Mokler, J. Wollersheim, H.5Schmidt-Böcking, R. Schuch, and J. B. Wilhelmy, *Phys. Rev. Lett.* **44**, 983 (1980).
51. W. E. Meyerhoff, R. Anholt, Y. El Masri, D. Cline, F. S. Stephens, and R. Diamond, *Phys. Lett.* **B69**, 41 (1977).
52. W. E. Meyerhoff and K. Taulbjerg, *Ann. Rev. Nucl. Sci.* **27**, 279 (1977).
53. A. B. Migdal, *Qualitative Methods in Quantum Theory*, Don Mills, Ontario (1977), p. 115.
54. P. H. Mokler and F. Folkmann, in *Structure and Collisions of Ions and Atoms*, I. A. Sellin, ed., Springer-Verlag, Berlin (1978).

55. P. H. Mokler and D. Liesen, Chapter 8, this volume.
56. E. Moll and E. Kankeleit, *Nukleonik* **7**, 180 (1965).
57. B. Müller, G. Soff, W. Greiner, and V. Ceausescu, *Z. Phys.* **A285**, 27 (1978).
58. J. R. Nix, *Ann. Rev. Nucl. Sci.* **22**, 65 (1972).
59. I. Pomeranchuk and J. Smorodinsky, *J. Phys. USSR* **IX**, 97 (1945).
60. J. Rafelski, L. Fulcher, and A. Klein, *Phys. Rep.* **38C**, 227 (1978).
61. J. Rafelski, B. Müller, and W. Greiner, *Z. Phys.* **A285**, 49 (1978).
62. J. Reinhardt and W. Greiner, *Rep. Progr. Phys.* **40**, 219 (1977).
63. J. Reinhardt, U. Müller, B. Müller, and W. Greiner, *Z. Phys. A* **303**, 173 (1981).
64. J. Reinhardt, B. Müller, and W. Greiner, *Phys. Rev. A* **24**, 103 (1981).
65. F. Sauter, *Z. Phys.* **69**, 742 (1931).
66. F. Sauter, *Z. Phys.* **73**, 547 (1932).
67. L. I. Schiff, H. Snyder, and J. Weinberg, *Phys. Rev.* **57**, 315 (1940).
68. P. Schlüter, G. Soff, and W. Greiner, *Z. Phys.* **A286**, 149 (1978).
69. P. Schlüter and G. Soff, *ADANDT* **24**, 509 (1979).
70. P. Schlüter, G. Soff, and W. Greiner, *Phys. Rep.* **75(6)**, 329 (1981).
71. H. Skapa, Diplomarbeit, Universität Heidelberg (1980), unpublished.
72. G. Soff, W. Betz, B. Müller, W. Greiner, and E. Merzbacher, *Phys. Lett.* **65a**, 19. (1978).
73. G. Soff, J. Reinhardt, B. Müller, and W. Greiner, *Phys. Rev. Lett.* **43**, 43 (1979).
74. G. Soff, J. Reinhardt, B. Müller, and W. Greiner, *Z. Phys.* **A249**, 137 (1980).
75. G. Soff, private communication.
76. J. Soltani, Diplomarbeit, Universität Heidelberg (1980), unpublished.
77. H. Tsertos, Diplomarbeit, Technische Universität Munchen (1980), unpublished.
78. Ch. Urbanek, Diplomarbeit, Technische Hochschule Darmstadt (1979).
79. V. V. Voronkov and N. N. Kolesnikov, *Sov. Phys. JETP* **12**, 136 (1961).
80. F. G. Werner and J. H. Wheeler, *Phys. Rev.* **109**, 126 (1958).
81. G. Soff, B. Müller, and J. Rafelski, *Z. Naturforsch.* **29a**, 1267 (1974).

11
Impact Ionization by Fast Projectiles

RAINER HIPPLER

1. Introduction

The collision of two atomic particles is in general a rather complex phenomenon. The exchange of momentum and energy between the collision partners results in a wide variety of different processes involving elastic scattering dominating at low collision velocities, excitation processes to bound and continuum states (ionization), and electron exchange processes.

In this chapter we shall restrict ourselves to the discussion of ionization processes induced by structureless charged particle (electrons, protons) impact. They are of both fundamental and practical interest. Some arbitrarily selected examples of the latter will be given at the end of this chapter (Section 5). While we do not intend to give here a review of theoretical techniques, theoretical aspects of impact ionization by fast projectiles will be discussed briefly in Section 1. Total and differential cross sections will be discussed in Sections 2 and 3, respectively, while Section 4 is devoted to alignment studies reflecting the spatial distribution of atomic angular momenta. Excitation processes in electronic and atomic collisions have already been dealt with in Chapter 8 of Part A of this title.

Let us consider a sufficiently energetic projectile with charge $Z_1 \cdot e$ colliding with a target atom of nuclear charge $Z_2 \cdot e$. In the following we shall assume the projectile–target interaction to be weak and due to the Coulombic force between the projectile and target electrons and nucleus.

RAINER HIPPLER • Fakultät für Physik, Universität Bielefeld, Postfach 8640, D-4800 Bielefeld, Federal Republic of Germany.

Under the influence of this interaction the target atom may eventually undergo a transition in which an electron is brought into a continuum state and the atom remains ionized. This process is commonly referred to as Coulomb ionization.

1.1. Plane Wave Born Approximation (PWBA)

As a first step for a theoretical description of the ionization process one may start with a first-order perturbation theory. In the plane wave Born approximation (PWBA) the incident and the scattered projectile are approximated by plane waves. Since the interaction between projectile and target electrons is assumed to be weak, the wave functions of the initial (bound) and final (continuum) states may be regarded as undisturbed. No interaction between projectile and target nucleus is considered. Within these approximations the amplitude f_ε for a transition from an initial state $|i\rangle$ to a final state $|f\rangle$ may be expressed as[1,2]

$$f_\varepsilon = -\frac{Z_1 e^2}{2\pi^2 \hbar^2 q^2} F_\varepsilon(q) \tag{1}$$

where e is the electron charge, $q = |\mathbf{q}|$, $\hbar \cdot \mathbf{q}$ and ε are the momentum and energy transfer, respectively, $\hbar$ is Planck's constant, Ry is the Rydberg energy, and

$$F_\varepsilon(q) = \langle f| \exp(i\mathbf{q}\mathbf{r})|i\rangle \tag{2}$$

With v_0 the projectile velocity and $\mathbf{r}$ the target electron coordinate, the total ionization cross section σ is given by

$$\sigma = \frac{16\pi Z_1^2 e^4}{\hbar^2 v_0^2} \int d\varepsilon \int \frac{dq}{q^3} |F_\varepsilon(q)|^2 \tag{3}$$

The integration is carried out between the limits which follow from the conservation of energy and momentum during the collision. In principle, the atomic form factor $|F_\varepsilon(q)|^2$ can be calculated using properly antisymmetrized self-consistent Hartree–Fock wave functions for the bound and continuum state. While such a representation of the atomic electrons is feasible, calculations so far have been done with less sophisticated wave functions. Quite often, in the frame of the independent-electron model the wave function of the active electron is approximated by hydrogenic wave functions with an effective target nuclear charge. This effective charge Z_{eff} is chosen to be $Z_{\text{eff}} = Z_2 - 0.3$ and $Z_{\text{eff}} = Z_2 - 4.15$ for K and L shell, respectively.

PWBA calculations for K-shell ionization by electron impact using hydrogenic wave functions are given in Figure 1. The cross sections are presented in scaled form, i.e., $\sigma \cdot I^2$ versus E_0/I, as was suggested by McGuire.[3] E_0 is the incident and I the binding energy. The three PWBA calculations shown are for hydrogen ($I = 13.6$ eV), neon ($I = 867$ eV), and argon ($I = 3.20$ keV). All curves exhibit a maximum at about $E_0/I = 3$. Though the absolute cross sections for hydrogen and argon differ by almost five orders of magnitude, the scaled cross sections agree within about 5%–15%. This implies that this scaling procedure works rather well. The remaining difference is caused by the different $\theta = I/(Z_{\text{eff}}^2 \cdot \text{Ry})$, the ratio of the experimental binding energy I to the hydrogenic binding energy. In Figure 1 θ varies between $\theta = 1.0$ for hydrogen and $\theta = 0.68$ for neon. θ is a measure of the screening of inner shell electrons due to outer electrons. This outer-screening has the effect of lowering the inner-shell binding energy, as compared to the hydrogenic binding energy.

There are several restrictions imposed to the PWBA limiting the range of validity of PWBA calculations. The conditions for the PWBA to be valid may be written as[5]

$$Z_1 \cdot Z_2 \cdot e^2 \ll \hbar \cdot v_0 \tag{4}$$

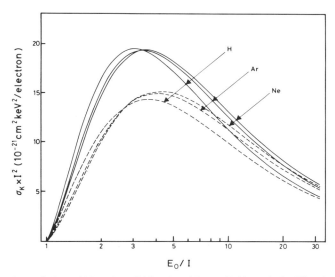

Figure 1. Scaled PWBA (solid lines) and Born–Ochkur (dashed lines) cross sections per electron for K-shell ionization of hydrogen ($\theta = 1.0$), neon ($\theta = 0.68$), and argon ($\theta = 0.75$) atoms by electron impact. (From Ref. 4.)

Hence, PWBA calculations should yield reliable results in very asymmetric collisions ($Z_1 \ll Z_2$) and/or high incident projectile velocities ($v_0 \gg v_e$, v_e the classical orbiting electron velocity).

PWBA calculations may be seriously in error at both low and very high incident velocities. At low velocities where the inequality (4) is no longer fulfilled the wave functions for the incident and scattered projectile may no longer be approximated by plane waves. Instead, Coulomb wave functions have to be used. Also, there is enough time for the bound electron wave function to adjust for the interaction, resulting in a polarization of the electron cloud for distant collisions and increased or decreased binding (depending on the sign of the projectile's charge) in close collisions.[6] Many of the problems associated herewith may be solved in a distorted wave Born approximation (DWBA), in which the wave functions for all particles are calculated as eigenfunctions of a single self-consistent Hartree–Fock potential.[7,8]

When the projectile is an electron exchange between incident and target electrons also has to be considered. Considering a one-electron atom we have two transition amplitudes f_ε and g_ε for "direct" and "exchange" ionization. For a hydrogenic atom, the cross section is proportional to[9,10]

$$\sigma \propto \tfrac{1}{4}|f_\varepsilon + g_\varepsilon|^2 + \tfrac{3}{4}|f_\varepsilon - g_\varepsilon|^2 \tag{5}$$

The exchange amplitude may be calculated in Born–Oppenheimer approximation.[9,10] Computational difficulties have led to an approximate form valid at not too low velocities, which is obtained by expanding the Born–Oppenheimer amplitude in powers of $1/k_0$, where $\hbar \cdot \mathbf{k}_0$ is the momentum of the incident electron, retaining only the first term.[9] This so-called Ochkur approximation connects the exchange amplitude to the direct amplitude by

$$g_\varepsilon = (q/k_0)^2 \cdot f_\varepsilon \tag{6}$$

Results of a Born–Ochkur calculation are given in Figure 1. The effect on the cross section is almost the same in all three cases presented: to lower the theoretical cross section by about 30% in the vicinity of and below the maximum. As is expected, the cross sections approach at higher velocities the PWBA cross sections.

At large velocities the exchange contribution becomes negligible. The high-velocity behavior of the ionization cross section is dominated by the direct amplitude, with the bulk of all ionizing events coming from small-angle, low-momentum transfer collisions. Then, the exponential in the form factor [Eq. (2)] may be expanded as

$$\exp(i\mathbf{qr}) = 1 + i\mathbf{qr} + \cdots \tag{7}$$

At large though still nonrelativistic velocities one may assume that $\mathbf{q} \perp \mathbf{k}_0$ (which is equivalent to the above assumption $q \ll k_0$). After integration over q and ε one arrives at the Bethe cross section[11,12]

$$\sigma = \frac{4\pi a_0^2}{T/\text{Ry}} M_i^2 \ln (4c_i T/\text{Ry}) \tag{8}$$

Here $T = (1/2)m \cdot v_0^2$, m the electron mass, a_0 is the Bohr radius, Ry the Rydberg energy, $M_i^2 = \int |\langle f|e\mathbf{r}|i\rangle|^2 d\varepsilon$ the square of the (dipole) matrix element integrated over all final states, and $1/c_i$ is a certain mean of the energy transfer and of the order of the experimental binding energy I. This shows that the ionization cross section in the nonrelativistic limit falls off as $\ln (T/\text{Ry})/T$. This high-energy behavior of the cross section is rather unrealistic, however. One reason is the neglect of terms of the order of $1/T$. The Bethe cross section [Eq. (8)] only includes contributions from optically allowed (dipole) transitions. A more realistic cross section additionally has to include monopole (and higher multipole) transitions. It may be expressed as[12]

$$\sigma = \frac{A}{T/I} \ln \{4T/I\} + \frac{B}{T/I} \tag{9}$$

Of course, if T is sufficiently large the logarithmic term in Eq. (9) still dominates. However, the domain for which this happens depends on the ratio B/A. If, for instance, ionization of a hydrogenic $2p$-state is considered, $B/A \cong 10$,[13] and the logarithmic term will be the larger of the two only for $T > 5000 \cdot Z_2^2 \cdot \text{Ry}$. Thus, even for the lightest atoms the cross sections of Eq. (8) are unrealistic, since such velocities are already relativistic.

At very high velocities $v_0 \lesssim c$, c the speed of light, a relativistic treatment of the interaction is necessary. The interaction between an incident and a target electron may be written as[12,14]

$$V = \frac{e^2}{|\mathbf{R} - \mathbf{r}|}(1 - \alpha^{(1)} \cdot \alpha^{(2)}) \cdot \exp\left(\frac{i}{\hbar \cdot c}|E_f - E_i| \cdot |\mathbf{R} - \mathbf{r}|\right) \tag{10}$$

here α symbolizes Dirac matrices, $\alpha^{(1)} \cdot \alpha^{(2)}$ is the magnetic interaction between the electron spins, $\mathbf{R}$ the projectile coordinate, E_i and E_f the energies of the initial and final states, respectively, and the exponent accounting for retardation. This relativistic interaction can be separated into two noninterfering contributions from longitudinal and transverse (with respect to the momentum transfer $\hbar \cdot \mathbf{q}$) fields. The longitudinal interaction results from the pure Coulomb field, whereas the transverse interaction stems from the probability for exchange of virtual photons.

Retardation effects are fully included in the transverse interaction.[15] The total cross section can be written as[16-18]

$$\sigma = \frac{16\pi Z_1^1 e^4}{\hbar^2 v_0^2} \int d\varepsilon \int q \, dq \cdot \left\{ \frac{|\langle f| \exp{(i\mathbf{qr})}|i\rangle|^2}{q^4} \right.$$

$$\left. + \frac{|\langle f|\boldsymbol{\alpha}\boldsymbol{\beta}_\perp \exp{(i\mathbf{qr})}|i\rangle|^2}{[q^2 - (\boldsymbol{\beta} \cdot q_{min})^2]^2} \right\}$$ (11)

with $q_{min} = (E_f - E_i)/\hbar v_0$, the minimum momentum transfer, $\boldsymbol{\beta} = \mathbf{v}_0/c$, and $\boldsymbol{\beta}_\perp = \boldsymbol{\beta} - (\boldsymbol{\beta} \cdot \hat{\mathbf{q}}) \cdot \hat{\mathbf{q}}$. The first term in Eq. (12) corresponds to the longitudinal, the second to the transverse contribution. The longitudinal interaction dominates the velocity dependence of the ionization cross section at low and medium velocities, predicting a constant cross section when the incident electron velocity approaches the speed of light. The transverse interaction is close to zero at low velocities; it is responsible for the logarithmic rise of the cross section at relativistic velocities (see Figure 2). This relativistic high-energy behavior of the cross section may be expressed in the relativistic Bethe theory as[11,12]

$$\sigma = \frac{8\pi a_0^2}{mv_0^2/\text{Ry}} M_i^2 \left[\ln\left(\frac{2c_i m v_0^2}{\text{Ry}}\right) - \ln\left(1 - \frac{v_0^2}{c^2}\right) - \frac{v_0^2}{c^2} \right]$$ (12)

An alternative method for K-shell ionization by relativistic electrons has been proposed by Kolbenstvedt.[19] Dividing the interactions into those arising from close and distant collisions, he calculated cross sections for the two regimes separately and in a different way. The contribution from

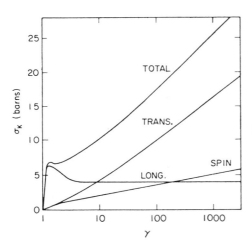

Figure 2. Theoretical cross sections for ionization of uranium by projectiles with kinetic energies $\gamma - 1$ [in units of $m_0 c^2$, $\gamma = (1 - \beta^2)^{-1/2}$]. The longitudinal, transverse (without spin-flip), and spin-flip contributions are shown separately as well as the total cross section (see text). (From Ref. 16.)

distant collisions was calculated by approximating the field of the incident electron by a spectrum of (virtual) photons. Photons with energy larger than the ionization energy will have a chance given by the photoelectric cross section to ionize a K-shell electron. The contribution from close collisions is calculated from the Møller interaction between two free electrons where the energy transfer must exceed the ionization energy (impulse approximation). The total cross section for K-shell ionization is obtained by summing the two contributions from close and distant collisions, and yields a relatively simple analytical expression

$$\sigma = \frac{0.275}{I_r} \frac{(E_r + 1)^2}{E_r(E_r + 2)} \left\{ \ln \left[\frac{1.19 E_r(E_r + 2)}{I_r} \right] - \frac{E_r(E_r + 2)}{(E_r + 1)^2} \right\}$$
$$+ \frac{0.99}{I_r} \frac{(E_r + 1)^2}{E_r(E_r + 2)} \left\{ 1 - \frac{I_r}{E_r} \left[1 - \frac{E_r^2}{2(E_r + 1)^2} + \frac{2E_r + 1}{(E_r + 1)^2} \ln \frac{E_r}{I_r} \right] \right\} \quad (13)$$

Here $I_r = I/(m \cdot c^2)$ and $E_r = E_0/(m \cdot c^2)$, with m the electron mass, and the cross section given in barns.

1.2. Semiclassical Approximation (SCA)

When heavy though still structureless projectiles (protons, deuterons, alpha-particles) are used, a full quantum-mechanical description of the ionization process seems to be unnecessary for a wide range of projectile velocities. In the semiclassical approximation (SCA) the motion of a projectile in the field of a nucleus is treated classically, whereas the interaction with the bound electron is calculated in a first-order time-dependent perturbation theory. The perturbing potential $V(t)$ is given by[20]

$$V(t) = -\frac{Z_1 \cdot e^2}{|\mathbf{R}(t) - \mathbf{r}|} \quad (14)$$

with $\mathbf{R}(t)$ the time-dependent position vector of the projectile. The transition amplitude f_ε takes the form

$$f_\varepsilon = i Z_1 e^2 \int_{-\infty}^{+\infty} dt \left\langle f \left| \frac{1}{|\mathbf{R}(t) - \mathbf{r}|} \right| i \right\rangle \cdot \exp(i\omega t) \quad (15)$$

with $\omega = (E_f - E_i)/\hbar$. The total cross section for ionization is obtained as

$$\sigma = \int_0^\infty P(b) b \, db = \int_0^\infty b \, db \int |f_\varepsilon|^2 \, d\varepsilon \quad (16)$$

with b the impact parameter. Though the SCA is conceptually simple, the computational difficulties with the application of the full concept are large. As a consequence, SCA calculations have been made in which the projectile's motion is assumed to be undisturbed and approximated by a straight line. This straight-line SCA is relatively easy to calculate. It is fully equivalent to the PWBA.[20]

As in the case of PWBA calculations the straight-line SCA gives unrealistic results at low projectile velocities where the deflection of the projectile in the field of the target nucleus drastically reduces the ionization cross section. Attempts have been made to calculate SCA cross sections using a hyperbolic path.[21,22] The hyperbolic SCA may be related to the straight-line SGA by[20]

$$\sigma^{\mathrm{Hyp}} = h(\xi_0)\sigma^{\mathrm{SL}} \tag{17}$$

with $\xi_0 = d \cdot q_0$, $d = Z_1 \cdot Z_2 e^2/(m_0 \cdot v_0^2)$ the half-distance of closest approach, $q_0 = I/(\hbar \cdot v_0)$. Several analytical approximations have been proposed for the so-called "Coulomb-deflection" factor $h(\xi_0)$. A simple approximate expression $h(\xi_0) = \exp(-\pi \cdot \xi_0)$ was found to underestimate the Coulomb deflection (and retardation) effect. A more reliable, though also more complicated analytical Coulomb deflection factor was obtained by Brandt and Lapicki.[23] Both corrections only account for such transitions, in which the initial and final electronic states have the same angular momentum (monopole transitions), whereas contributions resulting from higher multipole transitions and the recoil of the target nucleus were neglected. Recent theoretical calculations show that deflection of the projectile induces dipole transitions also at low incident velocities, which contribute significantly to the ionization cross section. In Figure 3 theoretical Coulomb deflection factors as calculated in SCA for K-shell ionization of nickel by proton and deuteron impact are given as a function of ξ_0.[22] The curves labeled $p(0)$ and $d(0)$ are the respective proton and deuteron Coulomb deflection factors for monopole transitions; those including dipole transitions are labeled $p(0 + 1)$ and $d(0 + 1)$, respectively. The complete Coulomb deflection factor also includes the so-called recoil contribution [$p(\mathrm{rec})$ and $d(\mathrm{rec})$, respectively]. It is seen that the inclusion of the dipole term enlarges the cross section, whereas the recoil contribution again decreases the cross section, so that the final cross section comes between the monopole and the monopole plus dipole cross section.

1.3. Perturbed Stationary States (PSS) Approximation

As has been already mentioned, the cross section is also influenced by a perturbation of the target electron wave function. At low velocities, this perturbation results in an increased binding of the target electron,

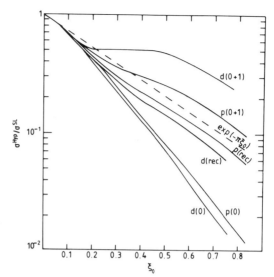

Figure 3. Theoretical Coulomb deflection factors $\sigma^{\text{Hyp}}/\sigma^{\text{SL}}$ for proton and deuteron impact on nickel vs. $\xi_0 = d \cdot q_0$ (see text). (From Ref. 22.)

which may be calculated in the perturbed stationary states approximation (PSS). It was shown by Basbas et al.[24] that increased binding may be incorporated in PWBA or SCA by replacing the experimental binding energy I by $I + \langle \Delta I \rangle$. In first-order perturbation theory, the change in binding energy, ΔI, is given by

$$\Delta I = \left\langle i \left| \frac{Z_1 \cdot e^2}{|\mathbf{R} - \mathbf{r}|} \right| i \right\rangle \tag{18}$$

where $|i\rangle$ is the unperturbed wave function of the initial state. In Figure 4

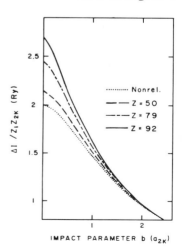

Figure 4. Increase in scaled binding energy as a function of impact parameter (in units of K-shell radius), for K-shell ionization by proton impact. (From Ref. 25.)

the relative change in K-shell binding energy for various atomic targets is calculated with relativistic hydrogenic wave functions and plotted vs. the impact parameter b in units of the K-shell radius. For comparison also the result with nonrelativistic hydrogenic wave functions is given. The average change in binding energy, $\langle \Delta I \rangle$, is obtained by integrating ΔI weighted by the differential cross section over the impact parameter. The total cross section is then calculated with the increased binding energy.

2. Total Cross Sections

2.1. Outer-Shell Ionization

2.1.1. Electron Impact

The study of ionization processes can be achieved in a rather straight-forward way by directing a beam of sufficiently energetic projectiles into a dilute gas target and measuring the yields of slow positive ions and/or free electrons produced in the collision with the help of two condensor plates. The measured currents of slow positive ions and electrons, i_+ and i_-, respectively, are readily related to corresponding cross sections, σ_+ and σ_-, by

$$\sigma_\pm = \frac{i_\pm}{L \cdot N \cdot i_0/(n \cdot e)} \qquad (19)$$

with i_0 the current of projectiles with charge $n \cdot e$, L the length of the condensor plates, and N the target gas density. For electron impact, the two cross sections σ_+ and σ_- are equal and related to a so-called total or gross ionization cross section

$$\sigma_+ = \sigma_- = \sum_n n \cdot \sigma_n = \sigma_g \qquad (20)$$

where σ_n is the cross section for producing slow positive ions of charge state n.

The condensor plate method is of little use, if cross sections of chemical unstable atoms are to be determined. In such cases crossed-beam techniques have been used, in which projectiles are cross-fired onto a (thermal) atomic beam. Figure 5 shows the apparatus used by Fite et al.[26] to study the production of slow positive hydrogen ions by proton impact on atomic hydrogen. In case of electron impact the ion source was replaced by an electron gun; the positive ions were detected by a mass-spectrometer. Due to low atomic beam densities, the atomic beam was modulated with the help of a chopper wheel.

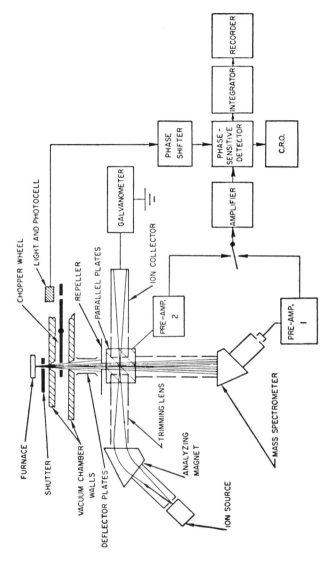

Figure 5. Experimental arrangement of Fite *et al.* (Ref. 26) to study production of slow positive ions for proton impact on atomic hydrogen.

As an example, the ionization cross section of atomic hydrogen by electron impact is given in Figure 6. This collision system has been studied experimentally from threshold (13.6 eV) up to incident electron energies of some hundred electron volts. In this range, the experimental data increase until they reach a maximum of the ionization cross section at about four times the threshold energy. From there the cross section decreases monototically towards higher energies. Despite some discrepancies between the experimental data, it is evident that theoretical data based on PWBA overestimate the experimental cross sections at low velocities and in the vicinity of the maximum. The PWBA cross section has a maximum at about 3 times the threshold energy, which differs from the experimental position. The theoretical cross section agrees well with experiment at the larger velocities. Some reasons for this failure of the Born calculation may be immediately seen. Firstly, the range of validity of PWBA is limited to regions where the perturbation of the target and projectile wave functions due to the interaction is weak. In the case considered here the inequality $Z_1 \cdot Z_2 \cdot e^2 \ll \hbar \cdot v_0$ is valid only at large impact velocities, and there the Born calculation is seen to work well. Secondly, important physical interactions have been neglected in the calculations. The inclusion of electron exchange in the calculations of Rudge and Seaton[30] and Prasad,[9] for instance, reduces the theoretical cross section at low velocities and brings it closer to the experimental data.

Though the agreement between experiment and theoretical calculations is considerably improved by the inclusion of exchange effects, there remain serious discrepancies at energies below the cross-section maximum.

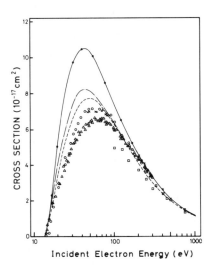

Figure 6. Experimental and theoretical cross sections for ionization of atomic hydrogen by electron impact: (——) PWBA calculation, (— · —), Born-exchange calculation[30]; (- - -), Born–Ochkur calculation.[4,9] Experiments: ○, ×, Fite and Brackmann[27]; △, Boyd and Boksenberg[28]; □, Roth et al.[29]

The origin of these can be studied in more detail by spin-polarized electron–atom experiments. Recently, such experiments have become feasible[31] and experimental results are available. In these experiments an asymmetry parameter A defined as[31]

$$A = \frac{\sigma^{(s)} - \sigma^{(t)}}{\sigma^{(s)} + 3\sigma^{(t)}} \tag{21}$$

is determined. $\sigma^{(s)}$ and $\sigma^{(t)}$ refer to the cross sections for singlet ($|f_\varepsilon + g_\varepsilon|^2$) and triplet ($|f_\varepsilon - g_\varepsilon|^2$) scattering, respectively. A determines the relative importance of the interference between the direct and exchange amplitude. In Figure 7 theoretical results in Born exchange[32] and Born–Ochkur[4] approximation are compared with experimental data of Alguard et al.[33] The good agreement with experiment for energies larger than 30 eV indicates that the exchange amplitude may be adequately described in Born–Ochkur approximation down to that energy. Reasonable agreement is also obtained with a Born-exchange calculation, which agrees well with experiment at even lower energies.

   (a) *Threshold Ionization.* A theoretical treatment of near-threshold ionization by electrons was given by Wannier.[34] Using phase-space arguments, the ionization cross section close to threshold is related to the volume of phase-space available for the two electrons to escape. If the two electrons move uncorrelated, the available phase-space and hence the ionization cross section increases linearly with the excess energy $E_0 - I$. Correlations between the two outgoing electrons reduce the available phase-space, and the resulting ionization cross section starts at threshold with an energy dependence given by[35]

$$\sigma \propto (E_0 - I)^n \tag{22}$$

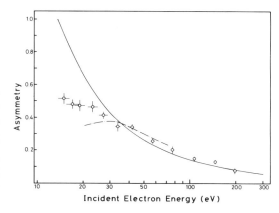

Figure 7. Asymmetry parameter A vs. incident energy for ionization of atomic hydrogen by electron impact: (——), Born–Ochkur calculation,[4] (–·–), Born-exchange calculation,[32] ○, experimental results of Alguard et al. (Ref. 33).

where

$$n = (1/4) \cdot \{(100 \cdot Z - 9)/(4 \cdot Z - 1)\}^{1/2} - 1/4 \qquad (23)$$

For single ionization $Z = 1$ and $n = 1.127$. Other predictions of the Wannier theory are, that the two outgoing electrons preferentially move out perpendicular to each other, and that the energy distribution of the outgoing electrons is uniform.

In contrast to the classical Wannier theory, early quantum-mechanical treatments by Rudge and Seaton[30] gave a linear energy dependence of the threshold ionization cross section. Recent quantum-mechanical calculations by Rau,[36] Peterkop,[37] and Klar and Schlecht[38] have confirmed the Wannier theory.

Early experimental results found values of n close to $n = 1.127$, but were not generally regarded as conclusive because of the finite energy spread in the electron beam.[35] Later, Cvejanovic and Read[39] have verified some of the predictions of the Wannier theory. Measuring the two outgoing electrons in coincidence, they studied the energy and angular distribution of the electrons. The experimental results agreed well with the Wannier theory. In a second experiment the rate of very slow electrons with kinetic energies ≤ 20 meV was studied as a function of the excess energy $E_0 - I$. This yield is proportional to the ionization cross section times the probability of finding the electrons in that energy interval. Since the energy distribution of the outgoing electrons was found to be uniform, in agreement with theory, the probability is proportional to $1/(E_0 - I)$ and the total yield of very slow electrons proportional to $(E_0 - I)^{n-1}$. The experimental results of Cvejanovic and Read[39] for helium are given in Figure 8. The solid line in Figure 8 is the predicted increase of slow electrons by $(E_0 - I)^{0.127}$. Above the ionization threshold the increase in the experimental yield of slow electrons is in excellent agreement with that prediction.

(b) *Ionization of Positive Ions.* A major theoretical problem in electron impact ionization studies is the interaction between the two final-state continuum electrons. A proper description of that interaction requires the correct representation of three free charged particles. Qualitatively, this final state interaction is the same for all electron–atom ionization events. Nevertheless, quantitative differences may arise for different collision systems. For instance, in electron–hydrogen scattering a significant interaction occurs between the two outgoing electrons due to the vanishing asymptotic Coulomb potential. In e–He^+ scattering the nuclear Coulomb potential dominates both the target and scattered electron orbitals, so that the coupling of target and scattered wave functions may be much weaker.[8] In Figure 9 experimental and theoretical data for e–He^+ ionization are given. The experimental data have been obtained from a crossed-beam

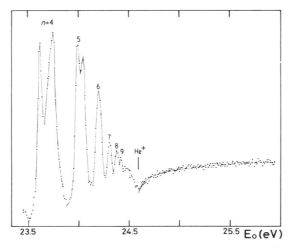

Figure 8. Yield of very low-energy electrons ($\leqslant 20$ meV) from electron–helium collisions. The solid curve through the data points above the ionization threshold is proportional to $(E_0 - I)^{0.127}$. (From Ref. 39.)

experiment; the theoretical calculation is a distorted wave exchange (DWEX) calculation.[8] As shown here for this collision system, the DWEX calculation gives a rather good agreement with experiment for simple systems with up to four initial target electrons, whereas for heavier systems such as Na^+ and Ne^+ the theoretical cross section overestimates the experimental data by more than 40% at low energies.

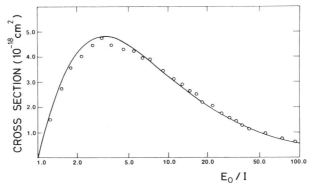

Figure 9. Electron-impact ionization of He^+. Solid line: DWEX calculation (Ref. 8); $\bigcirc$, experimental data of Peart *et al.* (Ref. 40). (From Ref. 8.)

A special aspect of electron impact ionization of positive ions is the contribution of other than direct ionization processes to the total cross section. Such a process is excitation to autoionizing states, which requires detailed knowledge of discrete atomic or ionic states embedded in and their coupling to the continuum. Though such processes also happen for neutral atoms, they are especially pronounced for positive ions with few valence electrons outside a many-electron closed-shell core. Available experimental data suggest that excitation autoionization may contribute to about 10%–20% of the total ionization cross section in lithiumlike ions, and may enhance the total cross section by about a factor of 2 in sodiumlike ions, and even more in some heavier isoelectronic sequences.

Figure 10 gives electron impact ionization data of Al^{2+} near threshold.[41] The solid curve is a DWEX calculation normalized to the experiment by multiplying with 0.65. Assuming that direct ionization and excitation autoionization are independent processes[42] theoretical excitation autoionization cross sections are added to the DWEX calculations. This results in a step-function-like increase of the theoretical cross section in the region of the autoionizing resonances, which does not show up to such extent in the experimental data. The discrepancy may be explained either by a deficiency of the independent processes model used here, or, more likely, by an inadequacy of the distorted wave model to properly account for the

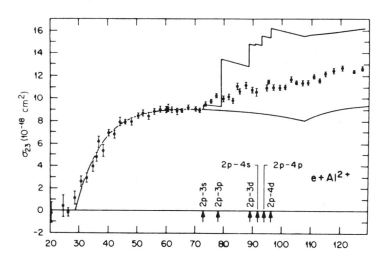

Figure 10. Electron-impact ionization of Al^{2+}. The solid curve is a DWEX calculation of Younger[8] normalized to experiment (normalization factor 0.65). Contributions from excitation autoionization calculated by Griffin *et al.* (Ref. 42) are added to Younger's data. (From Ref. 41.)

excitation to monopole-dominated $2p–np$ transitions.[42] Better agreement with experiment is obtained for excitation autoionization in, for instance, Ti^{3+} and Zr^{3+} ions, where $np–nd$ transitions are involved.[42]

2.1.2. Proton Impact

In Figure 11 cross sections for the ionization of helium by incident protons and electrons are given. In general, when comparing proton and electron ionization functions with each other, they display a rather similar behavior of the cross section. In fact, proton and electron impact cross sections become equal at large incident velocities. For a helium target this happens to occur at corresponding proton energies larger than about 1 MeV. The important differences occur at small incident velocities, in the vicinity and below the proton maximum. The position of maximum of the proton cross section is at about half the velocity compared to the electron

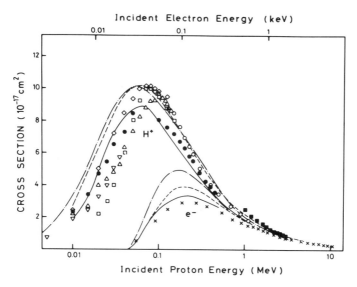

Figure 11. Cross section for ionization of helium by proton and electron impact on an equal velocity scale. (a) proton impact: PWBA calculations of (———), Peach (Ref. 10), (- - -), Bell and Kingston (Ref. 43), (– · –), McGuire (Ref. 44) are compared with experimental results of $\triangledown$, Keene (Ref. 45), $\diamond$, Solov'ev et al. (Ref. 46), $\square$, Fedorenko et al. (Ref. 46), $\triangle$, de Heer et al. (Ref. 47); ●, Gilbody et al. (Ref. 48); ○, Hooper et al. (Ref. 49); ■, Pivovar and Levchenko (Ref. 50); (– · –), PWBA (Ref. 10); (———), Born–Ochkur (Ref. 10); (- - -) Born-exchange (Ref. 10); ×, Schram et al. (Ref. 51); Adamczyk et al. (Ref. 51).

data. This essentially is caused by the different collision kinematics, imposing different upper and lower limits of the momentum transfer for proton and electron impact.

2.2. Outer s-Shell Ionization

If ionization processes in shells other than the outermost occur, the remaining ion after the collision is found in an excited state, from which it may decay into a lower or ground state via emission of characteristic radiation or Auger electrons. Such an experiment requires the detection of photons or energetic electrons.

Outer s-shell ionization has so far been studied over a wide range of projectile-target combinations and projectile energies (e.g., Refs. 52, 53). Almost exclusively rare gas target atoms having outer configurations $X(ns^2 \, np^6 \, {}^1S_0)$ were investigated. n is the main quantum number of the outer shell; $n = 2, 3, 4, 5$ for neon, argon, krypton, xenon, respectively. In case of proton impact the ionization process may be written as

$$H^+ + X(ns^2 \, np^6 \, {}^1S_0) \rightarrow H^+ + X^+(ns \, np^6 \, {}^2S_{1/2}) + e^- \tag{24}$$

The excited rare gas ion $X^+(ns \, np^6 \, {}^2S_{1/2})$ decays via emission of characteristic photons to the $X^+(ns^2 \, np^5 \, {}^2P_{3/2;1/2})$ ionic ground state. The two possible transitions have wavelengths lying in the spectral range of the vacuum ultraviolet (vuv) between 46 and 140 nm. The experimental setup for a study of outer s-shell ionization as used by Hippler and Schartner[52] is given in Figure 12. The main instrument besides the ion accelerator is a 1-m normal-incidence vuv monochromator suitable for wavelengths between about 35 and 600 nm. The observation angle was chosen to be 54.5° with respect to the incident particle's direction. At this "magic" angle the photon intensity becomes independent of the degree of linear polari-

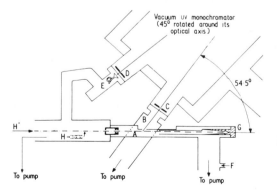

Figure 12. Experimental arrangement of Hippler and Schartner (Ref. 52) to study outer s-shell ionization by ion impact.

zation of the emitted light.[54] Moreover, the vuv monochromator is rotated 45° about its optical axis, in this way eliminating instrumental polarization. These precautions are not necessary in a study of photons originating from a $X^+(ns\,np^{6}\,{}^2S_{1/2})$ state, which are emitted isotropically. They are, however, necessary in the study of ionization processes leading to, for instance, ${}^1P_1$ states. Also, the observation angle of 54.5° with respect to the projectile's direction allows one to distinguish between projectile and target excitation in symmetric collision systems by means of Doppler shift. A typical vuv spectrum obtained from the impact of 500-keV He ions on neon is shown in Figure 13. The observed lines may be attributed to transitions in neutralized helium atoms and singly and multiply ionized neon ions. The most prominent lines around 46 and 49 nm originate from the single $2s$ and double $2s\,2p$ ionization of neon. The line at 52.8 nm results from (double) ionization of both neon $2s$ electrons.[53]

In Figure 14 cross sections for Ne-$2s$ ionization by proton and electron impact are presented on an equal velocity scale in a so-called Bethe–Fano plot. The proton data seem to approach the electron data at large velocities, though they are still about 30% larger even at a proton energy of 1 MeV. Here, PWBA calculations predict a much smaller relative difference. Also, the proton data seem to approach the electron data more slowly than predicted by theory. The high-velocity behavior of the experimental data suggests that the dipole matrix elements effective in the ionization are

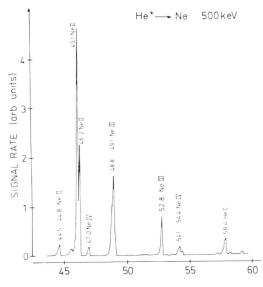

Figure 13. VUV spectrum obtained from 500 keV helium ion impact on neon (see text).

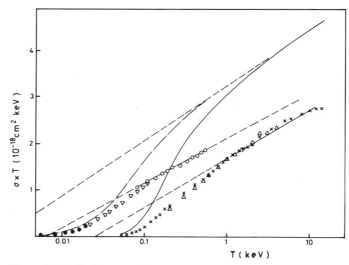

Figure 14. Bethe–Fano plot, i.e., $\sigma \cdot T$ vs. ln (T/Ry), for ionization of neon-2s-electrons by proton and electron impact. Electron impact: ×, Luyken *et al.* (Ref. 55) normalized to Beyer *et al.* (Ref. 53); △, Eckhardt (Ref. 56); (——), PWBA (Ref. 3). Proton impact: ○, Hippler and Schartner (Ref. 52) normalized to Beyer *et al.* (Ref. 53); ▽, Eckhardt (Ref. 56); ●, Van Eck *et al.* (Ref. 57); (– · –), PWBA (Ref. 44).

identical for proton and electron impact, though there seems to remain a constant absolute difference at large velocities. This might indicate either that the average energy transfer in the collision is different for proton and electron impact, and/or that nondipole transitions cannot be neglected even at large velocities.[12] PWBA calculations are in serious disagreement with the experimental data at large velocities. Since the slopes derived from the high-velocity behavior of the experimental and theoretical data widely differ, it is reasonable to assume that the wave functions used in the calculations are not appropriate for the case.

Outer s-shell ionization of neon is well suited for a study of multiple ionization events. By means of vuv spectroscopy it is possible to separate single ionization events resulting in $Ne^+(2s\,2p^6)$ from multiple ionization processes giving, for instance, $Ne^{++}(2s\,2p^5)$ or $Ne^{+++}(2s\,2p^4)$. In Figure 15 cross sections for single and double ionization of Ne-2s and 2p electrons by proton and electron impact are compared with each other. The largest cross sections is obtained for single 2p ionization; single 2s ionization is about one order of magnitude smaller. Double ionization events have a considerably smaller cross section compared to single ionization. Also, double ionization cross sections fall off more rapidly at large velocities than do cross sections for single ionization. The reason is that for large velocities single ionization is dominated by optical allowed (dipole) transitions, for which according to Bethe theory [Eq. (8)] a high-velocity behavior proportional to $\ln(T/\mathrm{Ry})/T$ is expected. For the same reasons optical forbidden double ionization events should decrease with $1/T$. However, in some cases a constant ratio of single to double ionization events is observed at large velocities. This observation indicates that double ionization at least partly may also proceed via dipole transitions. For instance, after ejection of a sufficiently fast δ electron during the collision the remaining ion may subsequently adjust to this sudden change in its electronic configuration by ejecting a second electron. This shake-off model is well known from photoionization studies[58]; it is also in satisfactory agreement with electron impact ionization studies of Ne KL vacancies. For multiple L-shell ionization the shake-off model is less satisfactory. In part this may be due to a deficiency of the model which requires a sudden change in electronic configuration; a condition which may no longer be fulfilled in case of outer shell ionization by particle impact, where a comparatively small amount of kinetic energy is transferred to ejected δ electrons.

2.3. Ionization of Inner Shells

To study ionization of inner atomic shells fast electron or ion beams from, for instance, electrostatic accelerators, are directed onto thin targets (foils or thin layers of material evaporated on carbon backings; only at low projectile velocities gas targets have to be used) inside a vacuum chamber. The production of inner-shell vacancies is detected by its subsequent decay

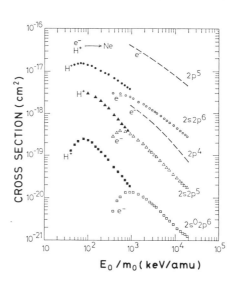

Figure 15. Total cross sections for ionization of neon by proton and electron impact. The final ionic states are indicated (see text). (From Ref. 56.)

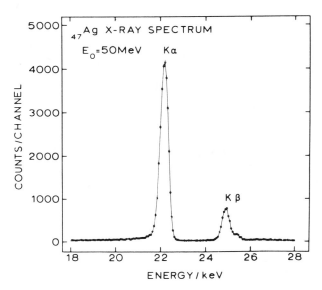

Figure 16. X-ray spectrum of silver following 40-MeV electron bombardment. (From Ref. 59.)

via emission of characteristic X rays. In most experiments only a moderate photon energy resolution is needed and solid-state detectors [Si(Li) or germanium detectors] or proportional counters are used. As an example an x-ray spectrum obtained from the impact of 40-MeV electrons on silver foils is shown in Figure 16.[59] The spectrum is taken with a Si(Li) solid state detector and displays the characteristic silver K_α and K_β lines.

2.3.1. Electron Impact

(a) *Low Energies.* In Figure 17 available experimental data for K-shell ionization of atomic targets with $Z = 6$ to 18, presented in scaled form are compared with PWBA and Born–Ochkur calculations for $\theta = 0.70$. The bulk of the experimental data closely follows the scaling law. As in the case of hydrogen, the PWBA cross sections agree well with experiment for $E_0/I \gtrsim 10$, whereas the Born–Ochkur approximation agrees well with experiment also at the cross section maximum. For not too heavy Z elements, the Born–Ochkur calculation thus agrees well with experiment from about twice the threshold energy up to incident electron energies where a relativistic treatment of the interaction becomes necessary. At lower energies the Born–Ochkur calculation increases the already existing

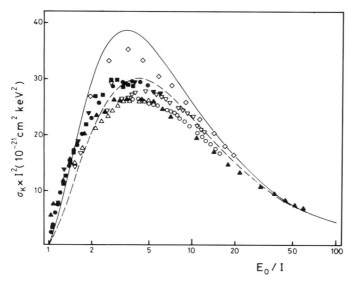

Figure 17. Comparison between experimental and theoretical cross sections for K-shell ionization of light atoms ($Z = 6$ to 18) by electron impact: (———), PWBA calculation for $\theta = 0.70$ (Ref. 4); (---), Born–Ochkur calculation for $\theta = 0.70$ (Ref. 4). Experimental data for $\bigcirc$, carbon; $\triangledown$, nitrogen; and $\triangle$, neon (Ref. 60); $\bullet$, neon (Ref. 61); $\diamond$, aluminum (Ref. 62); $\blacksquare$, argon (Ref. 63). The x-ray data of Tawara *et al.* (Ref. 64) for $\blacktriangle$, carbon, are normalized to Glupe and Mehlhorn's data (Ref. 60); those for $\blacktriangledown$, argon, have been converted to K-shell ionization cross sections using a fluorescence yield $\omega = 0.118$ (Ref. 65).

discrepancy between PWBA calculation and experiment. As in the case of hydrogen this indicates an inadequacy of both the PWBA and the Ochkur approximation at low velocities.

The threshold behavior of inner shell ionization was investigated by Hink *et al.*[61] They measured the Auger decay channel of K-shell ionized neon atoms and found a nonlinear dependence of the ionization cross section on the excess energy $E_0 - I$. As in the case of outer shell ionization (see above), the cross section was found to exhibit a threshold behavior proportional to $(E_0 - I)^n$, where the experimental n with $n = 1.13 \pm 0.02$ was in excellent agreement with the predictions of the Wannier theory ($n = 1.127$).

L-shell ionization by electron impact at incident electron energies a few times the ionization energy has been studied experimentally for the argon $2p$-subshell,[66] the $L2$ and $L3$ subshells of gold,[67,68] and the $L1$, $L2$, and $L3$ subshells of tungsten[69] and xenon.[70] In Figure 18 a typical

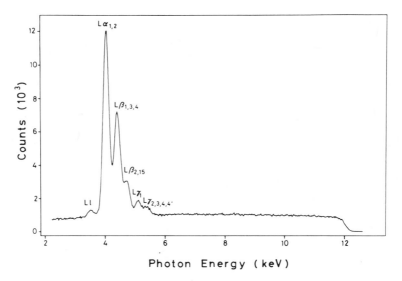

Figure 18. X-ray spectrum following 12-keV electron bombardment of xenon.

x-ray spectrum obtained from 12-keV electron bombardment of a xenon gas target is shown. The spectrum taken with a Si(Li) x-ray detector having an energy resolution of about 200 eV at 5.9 keV shows the characteristic xenon L-transitions superimposed on a continuous background. The characteristic lines are identified as

$$Ll(2P_{3/2} \rightarrow 3S_{1/2})$$

$$L\alpha_{1,2}(2P_{3/2} \rightarrow 3D_{5/2,3/2})$$

$$L\beta_1(2P_{1/2} \rightarrow 3D_{3/2}) \quad \text{and} \quad L\beta_{3,4}(2S_{1/2} \rightarrow 3P_{3/2,1/2}) \tag{25}$$

$$L\gamma_1(2P_{1/2} \rightarrow 4D_{3/2})$$

$$L\gamma_{2,3}(2S_{1/2} \rightarrow 4P_{1/2,3/2}) \quad \text{and} \quad L\gamma_{4,4'}(2S_{1/2} \rightarrow 5P_{1/2,3/2})$$

transitions. The continuous part of the spectrum has a high-energy cutoff at the incident electron energy of 12 keV. It is due to the slowing-down of electrons in the field of atomic nuclei (bremsstrahlung). The intensity of this atomic field bremsstrahlung can be calculated with fair accuracy[71]; the measured bremsstrahlung intensity in connection with theoretical bremsstrahlung data may be used for an *in situ* calibration of the x-ray detector.

Ionization cross sections for a given subshell i are obtained from the intensities of suitably chosen x-ray transition if the fluorescence yield ω_i, and the ratio of the partial (for the transition in question) to the total transition probability, Γ_{ik}/Γ_i, are known.[65] Cross sections for the $L2$ and $L3$ subshells have to be corrected for Coster–Kronig transitions, describing the nonradiative intrashell transfer of vacancies. In the present example, the $L\alpha_{1/2}$, $L\gamma_1$, and $L\gamma_{2,3,4,4'}$ transitions have been used in the evaluation of the cross-section data:

$$\sigma_1 = (\sigma_{\gamma_2}/\omega_1)(\Gamma_1/\Gamma_{\gamma_2})$$

$$\sigma_2 = (\sigma_{\gamma_1}/\omega_2)(\Gamma_2/\Gamma_{\gamma_1}) - f_{12}(\sigma_{\gamma_2}/\omega_1)(\Gamma_1/\Gamma_{\gamma_2}) \qquad (26)$$

$$\sigma_3 = (\sigma_\alpha/\omega_3)(\Gamma_3/\Gamma_\alpha) - f_{23}(\sigma_{\gamma_1}/\omega_2)(\Gamma_2/\Gamma_{\gamma_1}) - f_{13}(\sigma_{\gamma_2}/\omega_1)(\Gamma_1/\Gamma_{\gamma_2})$$

with σ_α, σ_{γ_1}, and σ_{γ_2} the measured x-ray production cross sections for the $L\alpha_{1,2}$, $L\gamma_1$, and $L\gamma_{2,3,4,4'}$ transitions, respectively, and f_{ij} the Coster–Kronig transition factors.

Cross section data for the ionization of xenon L-subshells are given in Figure 19 for incident electron energies of about 1.2 to 3 times the threshold energy I. The data are presented in scaled form, i.e., $\sigma \cdot I^2/n_i$ versus E_0/I. n_i is the number of electrons per subshell. The general tendency

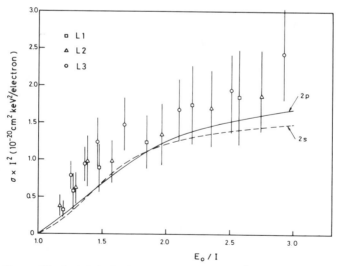

Figure 19. Scaled ionization cross sections $\sigma_i \cdot I^2/n_i$ of xenon L-subshells by electron impact vs. E_0/I. Experimental results for $\square$, $L1$, $\triangle$, $L2$; and $\bigcirc$, $L3$ subshells (Ref. 70) are compared with (——), (- - -), PWBA calculations (Ref. 3).

of all three subshell cross sections is to increase with increasing electron energy E_0. This shape of the cross section as well as the absolute magnitude is in agreement with a modified PWBA calculation of McGuire.[3] Only at the lower incident electron energies is a discrepancy between experiment and theoretical data larger than the experimental error bars observed. A similar observation has been reported for tungsten by Chang.[69] Most likely, this discrepancy is caused by the neglecting of exchange effects and the distortion of the incident electron wave due to the ionization process.

(b) *High Energies.* Nonrelativistic PWBA calculations become insufficient when the incident velocity becomes comparable to the speed of light. The projectile–target interaction may be divided into two noninterfering contributions from longitudinal and transverse fields. The longitudinal component is dominant at nonrelativistic velocities, giving a constant cross section at relativistic velocities. The transverse component becomes large at energies above about 1 MeV; it is responsible for the logarithmic rise of the ionization cross section.

In Figures 20 and 21 ionization cross sections for medium ($Z = 50$) and heavy ($Z = 79$) target atoms are given. In Figure 20 the ionization cross section for Sn ($Z = 50$) is given for incident electron energies up to 2 MeV, corresponding to $E_0/I \leqslant 70$. Relativistic and nonrelativistic PWBA calculations give almost the same cross section up to energies of about 0.1 MeV, with a maximum at about 4 times the threshold energy; they disagree at larger energies. In contrast to the nonrelativistic calculation, which decreases monotonically for energies larger than 0.2 MeV, the two

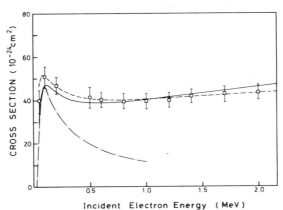

Figure 20. K-shell ionization cross section of tin ($Z = 50$) by electron impact. ○, Motz and Placious (Ref. 71); □, Rester and Dance (Ref. 72); (——) relativistic PWBA of Davidovic and Moiseiwitsch (Ref. 74); (- - -), Kolbenstvedt (Ref. 19); (- · -), non-relativistic PWBA (Ref. 4).

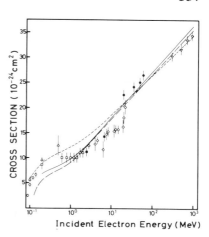

Figure 21. K-shell ionization cross section of gold ($Z = 79$) by electron impact. $\triangle$, Motz and Placious (Ref. 71); $\square$, Rester and Dance (Ref. 72); $\bigcirc$, Davis *et al.* (Ref. 68); $\blacksquare$, Berkner *et al.* (Ref. 76); $\triangledown$, Dangerfield and Spicer (Ref. 77); $\bullet$, Hoffmann *et al.* (Ref. 59); $\triangledown$, Middleman *et al.* (Ref. 78); relativistic PWBA of ($-\cdot-$), Davidovic and Moiseiwitsch (Ref. 74) and (———), Scofield (Ref. 79); (- - -), Kolbenstvedt (Ref. 19).

relativistic calculations predict an almost constant cross section in that range, with a small minimum at about 1 MeV and an increase for energies larger than 1 MeV. The experimental data agree well with the relativistic calculations. At still higher incident electron energies, the increase of the ionization cross section with increasing incident energy becomes pronounced. In Figure 21 K-shell ionization cross sections of gold are given for incident energies up to 1 GeV. The relativistic increase of the cross section at such large energies becomes evident. Here, the cross section is dominated by distant collisions. In a solid target as used in high-energy collisions, the far-reaching part of the projectile field interacting with a given target atom gets partially shielded by other atoms. This density effect is expected to cause a saturation of the cross section at incident energies larger than about 1 GeV. However, recent measurements at incident energies as large as 2 GeV could not find evidence for this density effect.[75]

(c) *Scaling Behavior.* At nonrelativistic velocities the ionization cross sections approximately obey a scaling behavior of the types $\sigma \cdot I^2 = f(E_0/I)$. At relativistic energies a scaling behavior of the type

$$\sigma \cdot I = f(E_0/I) \tag{27}$$

was discovered.[59] This scaling behavior holds for all the investigated K-, K-, and M-shell ionization processes (Figure 22). It shows that the ionization cross section at large velocities is insensitive to the individual atomic structure. This behavior is in agreement with numerical PWBA calculations for different atoms and shells, as well with the high-velocity form of Eq. (13), which predicts ($E_r \gg 4$, $I_r \ll 1$)

$$\sigma \cdot I_r \cong 0.55 \cdot \ln (E_r/I_r) + 0.275 \cdot \ln (I_r) + 0.765 \tag{28}$$

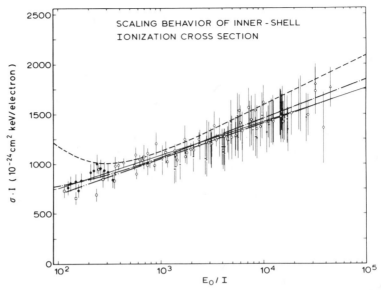

Figure 22. Scaling behavior of inner-shell ionization cross sections at relativistic energies. Experimental data for (a) K-shell ionization: □, Dangerfield and Spicer (Ref. 77); ●, Scholz *et al.* (Ref. 80); △, Middleman *et al.* (Ref. 78); ○, Hoffmann *et al.* (Ref. 59). (b) L-shell ionization: +, Hoffmann *et al.* (Ref. 59). (c) M-shell ionization: ×, Hoffmann *et al.* (Ref. 59). Relativistic PWBA calculations of Scofield (Ref. 79) for (- - -), K-shell ($Z = 79$); (– · –), K-shell ($Z = 18$); (– · · –), $L3$-shell ($Z = 79$). (After Ref. 59.)

2.3.2. Heavy-Particle Impact

We now consider the ionization of inner-shell electrons by sufficiently fast protons and heavier ions. Although ionization of inner-shell electrons by electron and proton impact is principally similar, there exist, mainly as a consequence of the different mass-to-charge ratios, characteristic differences. These differences are especially pronounced at low projectile velocities and reduce to about zero at larger velocities.

Though ionization of inner-shell electrons by proton impact is energetically allowed down to proton energies as small as the experimental binding energy, it was found experimentally that the cross section decreases extremely fast towards threshold, making it very difficult to observe inner-shell ionization at incident proton energies lower than about 5–10 times the threshold energy.

(a) *K-Shell Ionization.* In Figure 23 experimental and theoretical cross sections for K-shell ionization of a number of elements with atomic numbers ranging from $Z = 13$ (Al) to $Z = 30$ (Zn) are given. Since the cross sections for the different elements differ widely in the absolute magnitude, the data are displayed in scaled form. As has been pointed out by Madison and Merzbacher,[2] the PWBA suggests an approximate scaling behavior of the type

$$\sigma \cdot \theta \cdot Z_{\text{eff}}^4 / Z_1^2 = f(\eta / \theta^2) \tag{29}$$

with $\eta = E_0 / (Z_{\text{eff}}^2 \cdot \text{Ry})$, which holds at low incident velocities. The dashed and solid lines give the scaled PWBA calculations for Al ($Z = 13$, $\theta = 0.71$) and Ni ($Z = 28$, $\theta = 0.80$), respectively. At low velocities, the two calculation agree with each other. Here the theoretical cross section increases with the fourth power of the incident energy. At larger velocities the two PWBA calculations dissociate. Both calculations reach a broad maximum at about $\eta / \theta^2 = 1$ and fall off at higher energies. This maximum is sometimes referred to as the velocity-matching peak.[2] Over the range of relative incident energies plotted the scaled cross section varies by eight orders of magnitude. The experimental data for elements with atomic numbers $Z = 13$–30 are seen to follow the general trend given by the PWBA

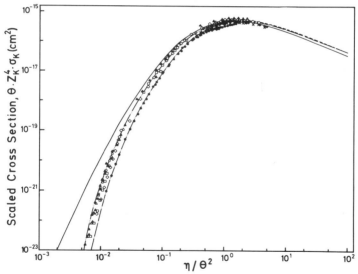

Figure 23. Scaled cross sections $\sigma_k \theta \cdot Z_k^4$ for K-shell ionization by proton impact vs. scaled proton energy η / θ^2 for atomic targets with $Z = 13$–30. Experimental data for ●, Al (Refs. 24 and 81); ◆, Si (Ref. 81); ◇, Ti (Refs. 82–84); ×, Ca (Ref. 84); □, Fe (Refs. 82, 83, and 85); ▲, Ni (Refs. 82–86); ○, Cu (Ref. 87) and △, Zn (Ref. 83) are compared with (- - -), (——) PWBA and (– · –), (– · · –) PWBA-BC calculations for Al and Ni, respectively. At low-scaled proton energies, the two PWBA calculations coincide.

calculations. It is obvious from the experimental data that they agree with the PWBA calculations at medium and large incident energies. At lower velocities the reduced cross sections of different target atoms at the same values of η/θ^2 scatter widely. The spread grows with decreasing η/θ^2 to nearly one order of magnitude on the ordinate scale.[24] Also, the experimental cross sections consistently fall below the PWBA values. The gap between theory and experiment grows with decreasing η/θ^2 and can exceed two orders of magnitude.

Basbas *et al.*[24] have pointed out that PWBA calculations are only in agreement with experiment for $v_0 > (Z_1/Z_2) \cdot v_e$. This, in fact, constitutes the condition that the amplitude of the wave scattered by the target electron is negligible compared to that of the incident wave. Below, PWBA calculations fail by factors of 2 or more. This discrepancy can be rectified somewhat by including the effect of Coulomb deflection of the impinging projectile in the field of a target nucleus. At low incident velocities the largest contribution to the ionization cross section stems from the high-momentum tail of the electronic wave function being located closest to the nucleus. Coulomb deflection obstructs deep particle penetration into the K-shell and reduces the likelihood of ionization since the impinging projectile is slower and farther from the center of the atom than it would be if it had followed a straight-line trajectory.

Another effect which has to be considered is the distortion of the target electron wave function caused by the interaction with the projectile. The lower the projectile's velocity, the larger is in general the influence on the electron wave function. At low velocities the projectile has to penetrate deep inside the K-shell to cause significant ionization. This results in an increased binding of the target electron to the target nucleus, reducing the already small cross section. At larger though still not fast velocities the dominant part of the ionization comes from impact parameters larger than the K-shell radius. The influence of the projectile is then to polarize the electron wave function, resulting in a larger cross section compared to the undisturbed case.

Increased binding and polarization effects can be treated in the perturbed stationary state (PSS) method.[24] Recently, also Coulomb (C) deflection and relativistic (R) wavefunction effects (see below) have been incorporated into the PSS theory.[23]

The influence of Coulomb deflection and of increased binding can be studied in detail using projectiles of different nuclear charges and charge-to-mass ratios. Of two isotopes with the same velocities, the heavier projectile penetrates deeper into the K-shell. Comparing, for instance, proton and deuteron impact with each other, any difference of the cross section at the same incident velocity may be attributed to Coulomb deflection which differs for these two projectiles. On the contrary, the binding energy increase only depends on the projectile's nuclear charge and is not influenced when deuterons instead of protons are used. On the other hand, projectiles with the same charge-to-mass ratio undergo equal Coulomb deflections in the field of a heavy target nucleus. Since, according to PWBA, the ionization cross section scales with Z_1^2, the square of the projectile's (nuclear) charge, any deviation from that scaling when using alpha particle instead of deuteron impact has to be attributed to increased binding. Figure 24 shows experimental results for the cross-section ratios mentioned, which clearly indicates the importance of these effects. The cross-section ratios deviate considerably from the predictions of the PWBA at the low velocities considered here. Good agreement is obtained with the CPSSR theory of Brandt and Lapicki.[23]

So far, effects have been considered where the wave function of the incident projectile or the target wave function gets altered by the projectile–target interaction. At the same time it was implicitly assumed that the

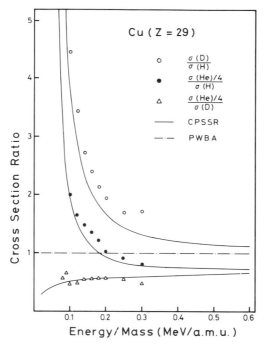

Figure 24. Cross section ratios for K-shell ionization of Cu by proton (H), deuteron (D), and helium-ion (He) impact. $(- \cdot -)$, PWBA; (———) CPSSR (Ref. 23); $\bigcirc$, experimental data of Rice *et al.* (Ref. 88).

(bound) target electron is adequately described by the target wave functions used. Especially for light and very heavy target atoms, however, the use of hydrogenic and/or nonrelativistic wave functions is no longer justified. For low Z elements, hydrogenic wave functions give a poor description and more sophisticated wave functions, for instance Hartree–Fock or Hartree–Slater wave functions, have to be used.[44,88] Such wave functions provide a more realistic approximation of the correct electron wave function, but do not, in general, allow one to predict a proper scaling behavior.

For heavy-Z elements relativistic wave functions have to be used. Since ionization processes at low incident velocities probe the high-momentum part of the wave function which is influenced the most by relativistic effects, relativistic wave-function effects are especially severe at low incident velocities. Contrary to the Coulomb deflection and binding increase effects, which decrease the total ionization cross section, the use of relativistic wave functions increases it. In Figure 25 the K-ionization cross sections of gold obtained for different plane wave Born calculations are compared. An excellent agreement with experiment is obtained for a PWBA calculation using relativistic Dirac wave functions (R) for the target electron, taking into account Coulomb deflection (C) and increased binding (B).[25]

(b) *L-Shell Ionization.* In general, L-shell ionization is similar to K-shell ionization. Differences are mainly due to wave-function effects. In

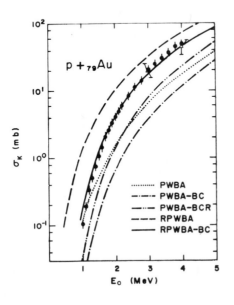

Figure 25. K-shell ionization cross sections of gold ($Z = 79$) by proton impact. Experimental data of ●, Kamiya *et al.* (Ref. 89) and △, Waltner *et al.* (Ref. 90) are compared with calculations in PWBA, taking into account increased binding and Coulomb deflection (PWBA-BC) and corrected for relativistic binding effects (PWBA-BCR), PWBA with relativistic wave functions (RPWBA) and corrections for increased binding and Coulomb deflection (RPWBA-BC). (From Ref. 25.)

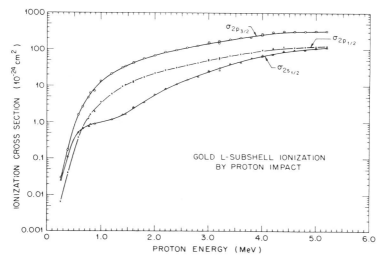

Figure 26. L-subshell cross sections of gold ($Z = 79$) vs. energy of incident protons. (From Ref. 91.)

Figure 26 cross sections for the $L1$, $L2$, and $L3$ subshells of gold ionized by proton impact are given.[91] In the region of interest here, all three cross sections increase with increasing proton energy, with the $L1$ cross section displaying a pronounced structure (knee) at about 1.5 MeV. The structure is more clearly seen when cross-section ratios σ_1/σ_2, σ_1/σ_3, and σ_2/σ_3 are compared. Whereas the cross-section ratio σ_2/σ_3 is a slowly varying function, the ratios σ_1/σ_2 and σ_1/σ_3 show a deep minimum at 1.5 MeV where the knee in the $L1$ cross section has been observed (Figure 27). The minimum is caused by the nodal structure of the $2s$ wave function (Figure 28). At low projectile velocities the ionization process probes the high-momentum tail of the wave function which is located close to the nucleus. At larger velocities also more distant parts of the wave function contribute to the cross section. When going from small to medium and large projectile velocities the total ionization cross section stems from increasing impact parameters, reaching an almost constant value (knee) when the minimum of the $2s$ wave function is reached. The $2p$ wave function does not display such a nodal structure, and structures in the cross section are therefore absent. There are numerous theoretical calculations based on PWBA, CPSSR, and SCA theory, to which the experimental cross section ratios are compared. The general trend of the σ_2/σ_3 ratio is correct described by all theories, though a quantitative agreement is still lacking. Better agreement is obtained for the σ_1/σ_2 and σ_1/σ_3 cross-section ratios over a wide range of proton energies. At large energies all theories predict almost

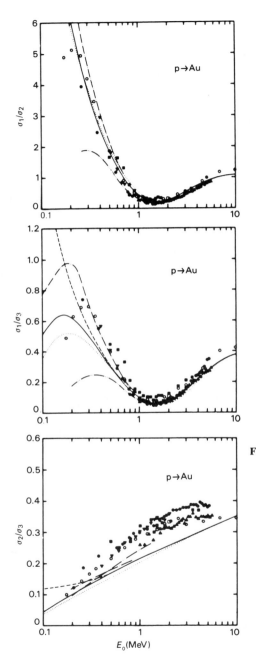

Figure 27. L-subshell cross section ratios σ_1/σ_2, σ_1/σ_3, and σ_2/σ_3 of gold vs. energy of incident protons. Experimental data of $\bigcirc$, Jitschin *et al.* (Ref. 92); $\blacksquare$, Barros Leite *et al.* (Ref. 93); $\blacktriangle$, Chang *et al.* (Ref. 94); $\blacktriangledown$, Chen (Ref. 95); $\bullet$, Datz *et al.* (Ref. 91); and $\square$, Sarkadi and Mukoyama (Ref. 96) are compared with calculations in (- - -) PWBA (Ref. 97); ($\cdots$), PWBA-BC allowing for increased binding (B) and Coulomb deflection (C) (Ref. 23); (——) CPSSR (Ref. 23); (—·—) RPWBA-BC (Ref. 98) and (—··—) RSCA (Ref. 21) with relativistic (R) wave functions. (From. Ref. 92.)

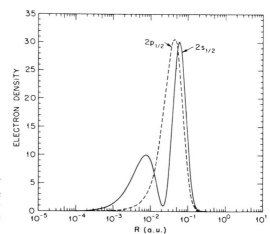

Figure 28. Radial distribution of electron density for $2s_{1/2}$ and $2p_{1/2}$ in gold calculated with relativistic Hartree–Fock–Slater wave functions. (From Ref. 91.)

identical cross-section ratios; they dissociate at low energies. None of the theories is here in full accord with the experimental data.

2.4. Electron Capture

Besides "direct" ionization in ion–atom collisions there exists a possibility of ionization due to rearrangement processes. One example is electron capture by the projectile ion, which leaves the target atom ionized without producing free electrons. At low velocities, the magnitude of the capture cross section may become comparable to or even larger than the direct ionization cross section.[99]

Cross sections for electron capture from individual shells of argon by incident protons are given in Figure 29. The experimental data are for total, L-, and K-shell capture. To separate L- and K-shell capture from other capture processes it was necessary to perform coincidence experiments. From about 10 keV to 10 MeV incident proton energy, the capture cross section decreases by almost eight orders of magnitude. At low velocities up to about 300 keV, the total capture cross section is dominated by M-shell capture. At proton energies between 400 keV and about 10 MeV the major part of the cross section comes from L-shell capture; above, K-shell capture dominates. The experimental data are compared with a theoretical calculation of Lin and Tunell,[100] which was performed within the atomic-expansion method. Starting from time-dependent Schrödinger equation, the electronic eigenfunctions are expanded in terms of traveling atomic eigenfunctions of the two collision partners. The numerical results of this two-state two-center method agree reasonably with experiments.

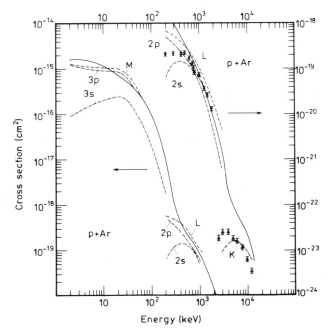

Figure 29. Total and subshell electron capture cross sections of argon by protons. The total experimental cross sections (solid lines) as compiled by Tawara and Russek (Ref. 101); ●, *L*-shell (Ref. 102); ▲, *K*-shell (Ref. 103) capture cross sections are compared with the theoretical cross sections of Lin and Tunnell (Ref. 100) for capture from indicated argon subshells. (From Ref. 100.)

3. Differential Cross Sections

So far, we have dealt with total cross sections which are a measure of a certain mean ionization probability averaged over projectile scattering angle, and angle and momentum of the ejected electron (δ electron). Finer details of the collision process are often obscured in such measurements; and specific ionization mechanisms may be overlooked if they do not show significant structure in the total cross-section measurements. Thus, one expects more insight into the collision process when differential cross-section measurements are performed and compared with theoretical predictions.

3.1. *Differential Cross Sections for δ-Electron Ejection*

As a consequence of the ionization process free electrons are produced. The angular and energy distribution of these δ electrons may be studied by conventional electron spectroscopy. Typically, the experimental arrangement consists of an electrostatic electron spectrometer inside a vacuum chamber, which may be rotated around a scattering center, being formed by the intersection of the projectile beam with a target. Since most of the interest concerns δ electrons of low energies in the eV or keV range, low-density targets have to be used to avoid absorption or scattering of electrons inside the target. Therefore, in most cases gas targets were used. From the measurements a double differential cross section (DDCS), differential in both δ-electron energy and ejection angle, is derived.

One of the best-studied collision systems is proton on argon. In Figure 30 experimental and theoretical DDCS for 350-keV protons on argon and δ-electron energies of (a) 46 eV, (b) 100 eV, and (c) 308 eV are plotted vs. electron ejection angle ϑ_e.[104,105] The two experimental curves labeled "total" and "L" refer to the total (i.e., δ-electrons from all argon shells)

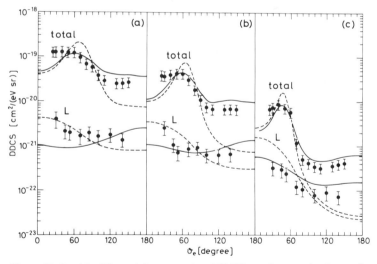

Figure 30. Double differential cross section (DDCS) vs. electron ejection angle ϑ_e for 350-keV proton–argon collisions and δ-electron energies of (a) 46 eV, (b) 100 eV, and (c) 308 eV. DDCS for total (M-, L-, and K-shell) and L-shell δ-electron ejection are shown. Experimental data of Gabler *et al.* (Ref. 104) and Sarkadi *et al.* (Ref. 105) are compared with PWBA calculations using (- - -) hydrogenic or (——) Hartree–Fock/Hartree–Slater wave functions. (From Ref. 105.)

and the L-shell DDCS, respectively. In effect, the total DDCS mainly reflects the M-shell DDCS since the L- and K-shell DDCS are comparatively small. To obtain L- or K-shell DDCS one has to perform a coincidence experiment between δ electrons and Auger electrons or characteristic X rays following the decay of specific inner-shell vacancies to effectively suppress detection of δ electrons from other shells. The experimental data are compared with PWBA calculations using Hartree–Fock/Hartree–Slater[106] or hydrogenic wave functions for the bound and continuum electronic states. Though the total cross sections of the two calculations agree reasonably with each other, the theoretical DDCS show pronounced differences especially in backward directions. As is expected, the experimental data are in better accord with the PWBA calculation using the more elaborate wave functions.

DDCS for the argon K shell following 6-MeV proton impact are shown in Figure 31. For two δ-electron ejection angles $\vartheta_e = 45°$ and $135°$ the K-shell DDCS is plotted vs. the δ-electron energy. Also given are the total δ-electron spectra at these two ejection angles. The experimental results[107] are compared with a PWBA calculation of Amundsen and Aashamar,[108] which gives a (single) energy differential cross section only. Hence, the comparison with experiment is rather limited and the good agreement for $\vartheta_e = 45°$ might be fortuitous. It should be noted here, that the calculations of Amundsen and Aashamar[108] predict a projectile energy-dependent structure in the energy differential cross section, which at the present projectile energy of 6 MeV is only weakly visible at an electron energy of about 250 eV. The other theoretical calculation was performed in SCA[109] and gives a reasonable though still incomplete agreement with experiment. In particular, the experimental K-shell DDCS shows a steeper energy dependence compared to the theoretical prediction. However, it cannot be ruled out completely that part of this discrepancy is caused by multiple ionization which affects the experimental data.

At forward ejection angles discrepancies between PWBA and experimental DDCS are observed (Figure 30). As is known from other collision systems, differences are largest when the electron velocity is comparable to the proton velocity. This observation invited speculations that electrons may get captured by the projectile into continuum states. Thus the electron moves along with the projectile until it is finally ejected with roughly the projectile's velocity and preferentially in forward directions.[110] This electron capture into continuum states (ECC) was later confirmed by experimental data of Crooks and Rudd,[111] who measured the energy distribution of electrons ejected in forward ($\vartheta_e = 0°$) directions for 300-keV H$^+$–He collisions. Figure 32 gives the measured DDCS, together with a theoretical calculation of Macek[112] allowing for ECC. In contrast, a PWBA without inclusion of ECC differs widely from both Macek's theory and experiment.

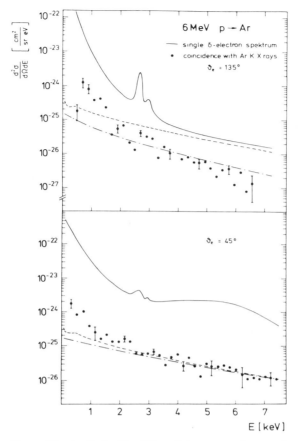

Figure 31. Ar-K double differential cross section vs. δ-electron energy for 6-MeV proton–argon collisions and two δ-electron ejection angles $\vartheta_e = 45°$ and $135°$. (- - -), PWBA (Ref. 108); (- · -), SCA (Ref. 109); (———), single δ-electron spectrum; ●, coincidence with Ar K X rays. (From Ref. 107.)

3.2. Impact Parameter Dependence

Complementary information may be obtained from measurements in which the scattering angle of the projectile is specified. To identify those scattering events which result in ionization of a given shell one has to either determine the energy loss of the projectile or perform a coincidence experiment between scattered projectile and emitted characteristic x-ray or Auger electron following the decay of inner-shell vacancies. Energy-loss

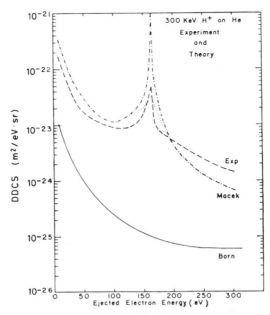

Figure 32. Energy distribution of electrons ejected at $\vartheta_e = 0°$ in proton–helium collisions. Experimental data of Crooks and Rudd (Ref. 111) are compared with theoretical calculations of Macek (Ref. 112) for $\vartheta_e = 0°$ and $\vartheta_e = 1.4°$ (dotted curve). (From Ref. 113.)

measurements usually require energy or momentum analysis by electrostatic or magnetic spectrometers with high resolution ($\Delta E_0/E_0 \cong 10^{-3}$), effectively resulting in a reduction of the solid angle of the particle detector. The detection of characteristic X rays or Auger electrons in coincidence with scattered ions has been employed by different groups (see, for instance, Lutz[114]). Though the method is not difficult, it is in general rather time consuming. Care has to be taken to avoid time structures in the ion beam which may falsify the results.[115]

First results of K-shell ionization studies in proton–copper and proton–silver collisions by the coincidence technique compared quite favorably with SCA calculations.[116] Excellent agreement with experiment is obtained when relativistic hydrogenic wave functions are used (Figure 33). For heavier target atoms the use of relativistic wave functions becomes essential.[117]

At low incident velocities the total ionization cross section is dominated by contributions from monopole transitions. Depending the size of the scattering angle, dipole transitions induced by the deflection of the projectile

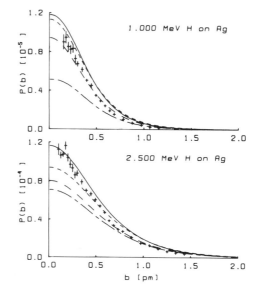

Figure 33. Impact parameter dependence of K-shell ionization probability in proton–silver collisions. +, experimental data of Andersen *et al.* (Ref. 117) are compared with (- - -) SCA calculations; (–––), SCA including binding; (– - –), SCA including Coulomb-repulsion and energy-loss corrections; (———), RSCA including corrections for Coulomb-repulsion, energy-loss, binding, and relativistic (wave function) effects. (From Ref. 117.)

contribute as well. Simultaneously, repulsion of the two nuclei transfers momentum and kinetic energy to the target nucleus. This acceleration of the target nucleus by itself may cause a significant distortion of electron wave functions and can contribute to the ionization amplitude. Best suited for a verification of the recoil effect should be collisions utilizing neutron bombardment. Mainly because of the large background from nuclear reactions, such experiments have not been successful so far. Evidence for the recoil effect has now been found in collisions of protons and deuterons with nickel targets. Figure 34 gives the differential probability for K-shell ionization in 1.2-MeV deuteron–nickel collisions vs. scattering angle.[118] The theoretical curves shown are SCA calculations of Rösel *et al.*[119] with and without inclusion of the recoil effect. From the two calculations one concludes that the recoil effect is most pronounced for backward directions where the recoil is largest. Since the dipole and recoil amplitudes have different sign, the inclusion of the recoil amplitude effectively reduces the ionization probability. The experiments are in reasonable agreement with the full calculation.

4. Alignment

So far, we have considered ionization of states characterized by atomic quantum numbers n, l or j, and s, referring to total, orbital or total angular

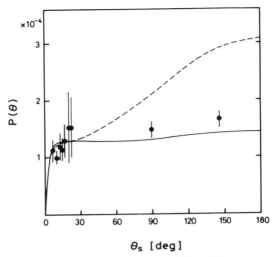

Figure 34. K-shell ionization probability $P(\theta)$ vs. projectile scattering angle for 1.2-MeV deuteron-nickel collisions. $\bigcirc$, experimental results of Schmidt-Böcking (Ref. 118) are compared with SCA calculations (Ref. 119) with (——) and without (- - -) inclusion of recoil.

momentum, and spin quantum number, respectively. For $l, j > 0$ such states split into a certain number of substates characterized by a magnetic quantum number m giving the projection of l or j to an arbitrarily selected quantization axis [in the following chosen to coincide with the incident particle's direction (z axis)]. Though without external field the magnetic substates are energetically degenerate, it turns out that ionization processes may preferentially ionize some of them.[120] In other words, the collision process may produce a nonisotropic atomic system, commonly referred to as being aligned or oriented.[121–123] However, the resulting state must not necessarily be a pure one. Instead, in most experiments averaging over unobserved initial (for instance, projectile's spin projection) and summation over unobserved final states (for instance, angular momentum of ejected electron) leads to a statistical mixture, which can be represented by a density matrix ρ, not by a wave function.

In the following we shall assume that the remaining ionic ensemble resembles an excited LS coupled state $(ls)j$, for which

Hyperfine splitting < level width < spin–orbit splitting

holds. Also, we consider only ionization processes for which the collision time (typically about 10^{-17} sec) is considerably shorter than the deexcitation time (lifetime of the autoionizing state, typically about 10^{-15} sec) (two-step process). If, as in the case of inner-shell ionization, the remaining ion is produced in an excited state from which it decays via emission of Auger electrons or x-ray photons, the angular distribution and polarization of these decay products contains the information about the alignment and orientation of the ionic state prior to the decay.

It is convenient to describe the anisotropy of the remaining ion by statistical tensors.[124] These tensors may be expressed as

$$\rho_{k\varkappa} = \sum_{m,m'} (-1)^{j-m'} \langle jmj - m' | k\varkappa \rangle \langle jm | \rho | jm' \rangle \qquad (30)$$

Here $\langle j_1 m_1 j_2 m_2 | jm \rangle$ is a Clebsch–Gordon coefficient. The density matrix is normalized so that its trace is equal to the (total or differential) ionization cross section

$$\text{Tr}\,(\rho) = \sigma = \sum_m \sigma_m \qquad (31)$$

where σm is the partial cross section for the magnetic substate $|jm\rangle$. The angular distribution and polarization can be completely described by a complete set of such tensors. Defining

$$A_{k\varkappa} = \rho_{k\varkappa}/\rho_{00} \qquad (32)$$

the angular distribution of emitted photons or Auger electrons may be expressed as[125]

$$W(\theta, \phi) = \frac{W_0}{4\pi} \left[1 + \sum_k \alpha_k \sum_\varkappa A_{k\varkappa} \left(\frac{4\pi}{2k+1} \right)^{1/2} Y_{k\varkappa}(\theta, \phi) \right] \qquad (33)$$

where θ is the emission angle with respect to the z axis, ϕ the azimuthal angle between scattering plane and emission plane (see Figure 35), W_0 the probability for photon or Auger electron emission, and $Y_{k\varkappa}(\theta, \phi)$ the sherical function. Equation (33) holds for both photons and Auger electrons; the difference between the two decay modes is contained in the α_k coefficients. Whereas for Auger decay the α_k coefficients in general depend on the reduced matrix elements for Auger decay, emission of photons being restricted by the selection rule $\Delta l = \pm 1$ to values of $k \leq 2$ results in α_2 coefficients independent of matrix elements and given by[125]

$$\alpha_2 = (-1)^{j+j'+1} [\tfrac{3}{2}(2j+1)]^{1/2} \begin{Bmatrix} 1 & j & j' \\ j & 1 & 2 \end{Bmatrix} \qquad (34)$$

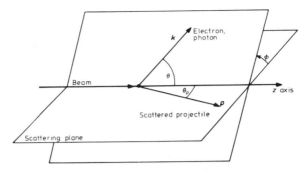

Figure 35. Scattering geometry. (From Ref. 122.)

where j, j' are the initial and final ionic angular momenta, and $\{^{abc}_{def}\}$ the Racah coefficients. The angular distribution of the decay products can be fully described by the complex parameters $A_{k\varkappa}$, which in turn are determined by the collision process. The number of independent parameters $A_{k\varkappa}$ is reduced by symmetry conditions. For instance, if both the projectile and the target atom are unpolarized and if the scattered particle and the ejected electron are not detected (noncoincidence experiment), the collision process is axially symmetric. Then only the alignment tensors A_{k0} with even k are nonzero. For photon emission (dipole radiation) only A_{20} may be different from zero, and the photon angular distribution determined from such a noncoincident experiment is given by

$$W_\gamma(\theta_\gamma) = \frac{W_{0\gamma}}{4\pi}[1 + \alpha_2 A_{20}P_2(\cos\theta_\gamma)] \tag{35}$$

The anisotropy $\alpha_2 A_{20}$ may be related to the polarization P of the emitted radiation by

$$P = 3 \cdot \alpha_2 \cdot A_{20}/(\alpha_2 \cdot A_{20} - 2) \tag{36}$$

The number of independent alignment parameters in a coincidence experiment defining a scattering plane is reduced by reflection symmetry. If both the collision partners are unpolarized, reflection invariance with respect to the scattering plane requires

$$A_{k\varkappa} = (-1)^{k+\varkappa}A_{k-\varkappa} \tag{37}$$

In case of photon emission, this reduces the number of independent parameters to four: A_{11}, A_{20}, A_{21}, A_{22}. They are fully equivalent to the alignment and orientation parameters as introduced by Fano and Macek[121] (see also Blum and Kleinpoppen[123]). It should be noted here that a further reduction to two parameters, for instance λ and χ as introduced by Eminyan *et al.*[126] for electron impact excitation of helium resonance levels, is in general only possible if additionally the ejected electron state is also specified. The (coincident) photon angular distribution is given by

$$W_\gamma(\theta_\gamma, \phi) = \frac{W_{0\gamma}}{4\pi}\{1 + \alpha_2[\tfrac{1}{2}A_{20}(3\cos^2\theta_\gamma - 1)$$
$$- \sqrt{\tfrac{3}{2}}A_{21}\sin 2\theta_\gamma \cos\phi + \sqrt{\tfrac{3}{2}}A_{22}\sin^2\theta_\gamma \cos 2\phi]\} \tag{38}$$

No information about tensors with $k > 2$ can be obtained from an experiment making use of the photon decay channel, but may be obtained via the Auger decay mode. A determination of the A_{11} tensor requires measurement of the photon circular polarization.

Experimentally, alignment studies have been achieved in two different ways, namely, (i) by investigating the angular distribution of emitted X rays or Auger electrons, and (ii) by measuring the degree of linear polarization of X rays. Both methods have been employed so far and comparable results of the two methods are in agreement with each other. Method (i) has the advantage that, as long as a solid-state [Si(Li)] detector can be used, it offers a large detection efficiency. It is handicapped by its comparatively low photon energy resolution of typically 200 eV for photon energies of about 6–10 keV. For Auger electron spectroscopy this is the only way to study alignment. Method (ii) offers a high photon energy resolution (typically about $E/\Delta E = 1000$), if a crystal spectrometer serving simultaneously as a linear polarizer is used. Its disadvantage is the extremely small detection efficiency (about 10^{-6} sr). A typical x-ray spectrum obtained with a crystal spectrometer for 2.6 MeV H^+–Ag collisions is given in Figure 36.

4.1. Electron Impact

First evidence for a nonisotropic emission of Auger electrons following the decay of argon $2p_{3/2}$ vacancies produced by electron impact was observed by Cleff and Mehlhorn.[128] In Figure 37 experimental and theoretical data for the argon and magnesium $L3(2p_{3/2})$ shells and the krypton $M4(3d_{3/2})$ and $M5(3d_{5/2})$ shells are presented. The relative alignment $A_{20}/A_{20}(\infty)$ is plotted versus E_0/I. $A_{20}(\infty)$ is the high-energy limit of A_{20} calculated in (nonrelativistic) PWBA with Hartree–Slater wave functions.

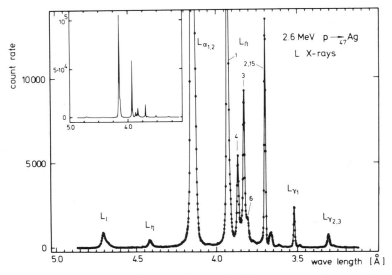

Figure 36. Ag-*L* x-ray spectrum excited by 2.6-MeV proton impact, taken with a curved crystal spectrometer. (From Ref. 127.)

In terms of cross sections for different magnetic substates m_j the $L3$-alignment A_{20} may be expressed as

$$A_{20}(2p_{3/2}) = \frac{\sigma\left(\frac{3}{2}\right) - \sigma\left(\frac{1}{2}\right)}{\sigma\left(\frac{3}{2}\right) + \sigma\left(\frac{1}{2}\right)} \tag{39}$$

If plotted in such a way, the high-velocity behavior of the alignment as predicted by theoretical (PWBA) calculations turns out to be insensitive to the atomic shell under consideration. This is caused by the small momentum transfer involved in such collisions, probing only the outermost and not very characteristic part of the bound electron's wave function. Of the two sets of calculations presented, the one with Hartree–Slater (HS) wave functions gives satisfactory agreement with experimental data at relative incident energies above $E_0/I > 5$. With hydrogenic (HL) wave functions, the shape of this alignment curve is similar to the one with HS wave functions, but appears to be shifted by about 50% on the energy scale towards lower energies. Above $E_0/I = 50$ there exists a serious discrepancy between the experimental and theoretical argon data, as well as between the experimental argon and krypton data. It has been attributed to relativistic effects which are expected to show up earlier than in the krypton data.

At lower energies the alignment becomes sensitive to the particular wave function and the alignment curves for different subshells dissociate.

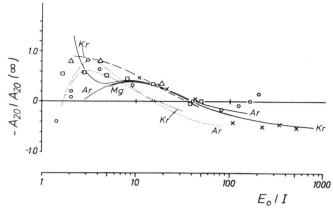

Figure 37. Relative $L3$-alignment of argon and magnesium and $M4$, $M5$ alignment of krypton by electron impact. Experimental data for △, argon (Ref. 128); ○, argon (Ref. 129); □, magnesium (Ref. 130); and ×, krypton (Ref. 131) are compared with PWBA calculations with (· · ·) hydrogenic (Ref. 125) and (——) (Ref. 125) and (- - -) (Ref. 132) Hartree–Slater wave functions. (From Ref. 131.)

Close to threshold one may expect a similar behavior as observed for excitation processes, effectively resulting in $\sigma(1/2) \gg \sigma(3/2)$ for $L3$ shells and $A_{20} = -1$. Whereas in excitation processes only s waves may effectively penetrate the centrifugal barrier, the long-range interaction between scattered electron and remaining ion also allows electrons with $l > 0$ to escape,[129] resulting in a smaller alignment. So far, no conclusive experimental determination of the threshold alignment has been reached. Measurements indicating a surprisingly large $L3$-shell alignment in xenon[133] have not been confirmed by argon measurements.[129] Recent measurements for magnesium[134] now give evidence that at threshold a large alignment ($|A_{20}| > 0.1$) may be observed.

4.2. Proton Impact

Figure 38 shows theoretical predictions for the alignment of $2p_{3/2}$, $3p_{3/2}$, $3d_{5/2}$, $4p_{3/2}$, and $4d_{5/2}$ subshells calculated in PWBA with unscreened hydrogenic wave functions.[135] For $V = v_0/v_e > 1$ the alignment for all five subshells looks similar. In this velocity regime collisions at large impact parameters and small momentum transfers dominate, probing as in the case of electron impact the outermost tail of the wave functions, which is not very different for all five subshells. At smaller velocities larger momentum transfers are required and the radial density distributions of the wave function show up in the alignment. Nodal structures in the radial part of

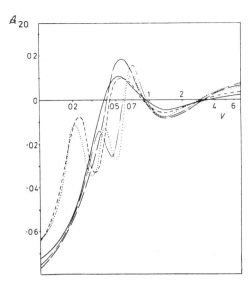

Figure 38. Theoretical alignment calculated in PWBA with hydrogenic wave functions for (——), $2p_{3/2}$; (- - -), $3p_{3/2}$; (— — —), $3d_{5/2}$; (· · ·), $4p_{3/2}$; and (— · —), $4d_{5/2}$ subshells by proton impact vs. scaled velocity $V = v_0/v_e$. (From Ref. 135.)

the wave function result in a strong variation of the alignment with impact velocity, as can be seen for the $3p_{3/2}$, $4p_{3/2}$, and $4d_{5/2}$ alignment. In contrast, the alignment of the $2p_{3/2}$ and $3d_{5/2}$ states decreases monotonically below $V < 0.5$.

Alignment data for different target atoms ($Z = 47$ to 79) are plotted vs. V in Figure 39. Within the quoted accuracy, the experimental data closely follow a universal curve. The experimental data agree well at the larger velocities with PWBA calculations. At low incident velocities, the PWBA calculations predict a large negative alignment. The experimental data follow that theoretical prediction down to only $V = 0.2$, below which the alignment increases again. This deviation from PWBA is explained by the Coulomb deflection of the incident projectile in the target nuclear field and the following rotation of the final target wave function towards the direction of the outgoing projectile. As was emphasized by Jitschin *et al.*[138] and Palinkas *et al.*,[137] the large alignment predicted by PWBA may be observed for a direction which is somehow also correlated to the outgoing rather than to the incoming projectile direction only. In a noncoincidence experiment the outgoing projectile direction is not specified; hence the observed total alignment is an integration over all azimuthal and polar scattered projectile's angles, resulting in a decrease of the observable alignment. Simple models, which account for Coulomb deflection, as well as theories in which Coulomb deflection is implicitly included, give in fact a reasonable agreement with the experimental observation (Figure 39).

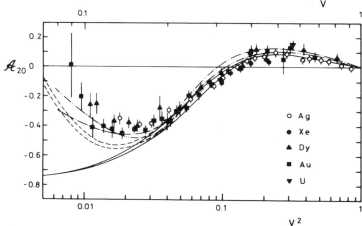

Figure 39. $L3$-alignment A_{20} for proton impact on different target atoms ($Z = 47$–92). Experimental data for $\bigcirc$, Ag (Ref. 127); $\bullet$, Xe (Ref. 136); $\blacktriangle$, Dy (Ref. 136); $\blacksquare$, Au (Refs. 136 and 137); and $\blacktriangledown$, U (Ref. 136) are compared with (——), PWBA calculations (Ref. 135); (- - -), including corrections for Coulomb-deflection (Ref. 138) for Ag (upper curves) and Au (lower curves), and RSCA calculations for (– · · –), Ag and (– · –) Au (Ref. 139). (From Ref. 136.)

Not included in Figure 39 are experimental data for $L3$ shell alignment of magnesium and argon by proton impact, which show an alignment different from the one observed for medium to heavy elements. The reason is that for these low-Z elements ionization of the $L3$ shell stems to a large extent from electron capture processes, whereas for heavy-Z elements electron capture is of little importance.[23,140] The importance of electron capture for these collision processes is demonstrated by calculations of Berezhko *et al.*[141] and Jakubassa-Amundsen.[142] Inclusion of electron capture in the calculation results in a better though still not complete agreement of theory with experiment (Figure 40).

To study M-shell alignment a comparatively high spectral resolution is needed. Accordingly, M-shell alignment studies by x-ray emission have been done with high-resolution crystal spectrometers. A typical x-ray spectrum obtained for 4 MeV H$^+$–Th collisions is given in Figure 41. The dominating lines are the $M_5 N_{6,7}$ (M_α) and $M_4 N_6$ (M_β) lines. Best suited for alignment studies of the $3p_{3/2}$, $3d_{3/2}$, and $3d_{5/2}$ states are the $M_3 N_1$, $M_4 N_2$, and $M_5 N_3$ lines, respectively.

Results for the alignment of the $3p_{3/2}$, $3d_{3/2}$, and $3d_{5/2}$ states are given in Figure 42. The experimental data are compared with PWBA calculations using hydrogenic (HL) or Hartree–Slater (HS) wave functions. In general,

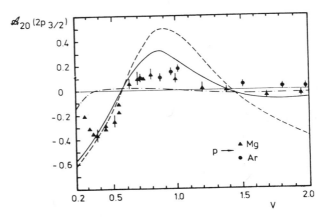

Figure 40. $L3$-alignment A_{20} of ●, argon, and ▲, magnesium by proton impact (Ref. 134). Theoretical curves (Ref. 141) are for (– · –) direct ionization, (- - -) electron capture, and (——) both direct ionization and electron capture. (From Ref. 143.)

the agreement between experiment and theory is good, though the $3p_{3/2}$ alignment seems to become positive at lower velocities, in contrast to the theoretical prediction. The minimum in the alignment caused by the nodal structure of the $3p_{3/2}$ wave function is clearly visible, whereas the $3d_{3/2}$ and $3d_{5/2}$ alignment does not show such a structure. Moreover, the predicted connection between the $3d_{3/2}$ and $3d_{5/2}$ alignment of

$$A_{20}(3d_{5/2}) = \sqrt{\tfrac{8}{7}}A_{20}(3d_{3/2}) \tag{40}$$

is in agreement with experiment.

4.3. Differential Alignment Measurements

Differential alignment measurements have been performed by Konrad et al.[146] They measured the angular distribution of Ll X rays in coincidence with scattered projectiles. From the measurements (differential) alignment parameters A_{20}, A_{21}, and A_{22} can be extracted. In Figure 43 the differential A_{20} for 16-MeV α particle impact on dysprosium is displayed. As a function of impact parameter, the A_{20} alignment increases from about -0.8 at $b \cong 150$ fm to about $+0.4$ at $b \cong 3500$ fm. For $b \geqslant 400$ fm, the experimental data agree well with SCA calculations using hydrogenic or relativistic wave functions. No explanation has been given yet for the discrepancy at small impact parameters.

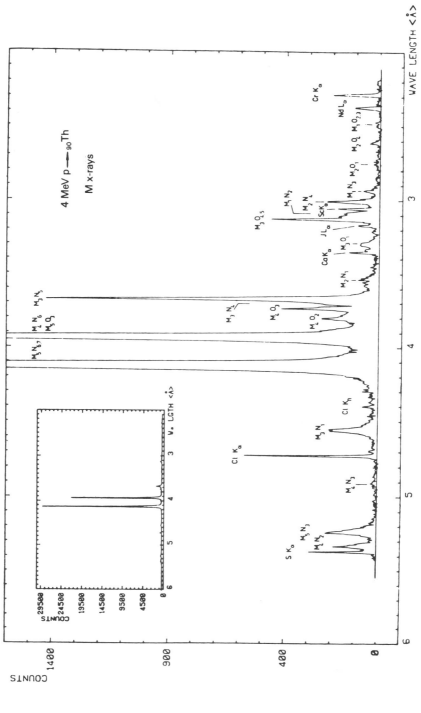

Figure 41. Th-M x-ray spectrum excited by 4-MeV proton impact. (From Ref. 144.)

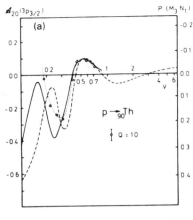

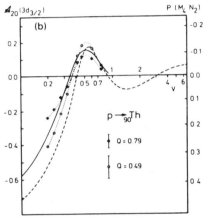

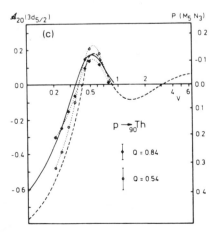

Figure 42. (a) $M3$, (b) $M4$, and (c) $M5$ subshell alignment A_{20} of thorium excited by proton impact. Experimental data of Wigger *et al.* (Ref. 144) taken with a curved crystal spectrometer and analyzed assuming, $\bigcirc$, a perfect Ge (111) crystal, and $\blacklozenge$, a mosaic crystal are compared with PWBA calculations (Ref. 135) using (- - -), hydrogenic and (——), Hartree–Slater wave functions. (From Ref. 144.)

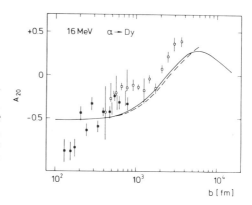

Figure 43. $L3$-Alignment A_{20} for 16 MeV α on Dy vs. impact parameter b. Experimental results of Konrad *et al.* (Ref. 146) are compared with SCA calculations using hydrogenic (- - -) (Ref. 147) or relativistic (——) (Ref. 139) wave functions. (From Ref. 146.)

5. Applications

There are numerous applications for which detailed knowledge of ionization processes are needed. Quite often the relevance for an understanding of astrophysical and fusion plasmas is mentioned. In fact, charge changing collisions play a dominant role in fusion plasmas. Also, electron impact ionization rates derived from total ionization cross sections are of critical importance for a understanding of high-temperature plasmas, both laboratory and astrophysical. There are other applications which are as important. It is not our intention to give here a full list of all possible applications rather than to present a few arbitrarily selected examples.

5.1. Auger Electron Spectroscopy (AES)

Auger electron spectroscopy may be used as an analytical tool for a quantitative elemental analysis of surfaces. Incident electrons with kinetic energies of a few keV have a mean free path of about 1 nm. Low-energy Auger electrons (<2000 eV) are emitted by all elements with the exception of hydrogen and helium; typical detection limits are about 0.1%. Auger transitions involving valence (V) band electrons may additionally be used for chemical state analysis.[148]

Much of the interest in surface physics has been devoted to *in situ* AES studies of chemical reactions on surfaces.[149] Interest is stimulated by catalytic properties of metallic surfaces like cobalt and platinum, but also by oxidation and corrosion processes. A major field whose economical importance is well known is corrosion of metals. The interaction of non-metals (C, O, S) with metallic surfaces may be studied by investigation of the elemental composition of surface layers. As an example of such an investigation KLL and MVV Auger spectra of 18/8 stainless steel during

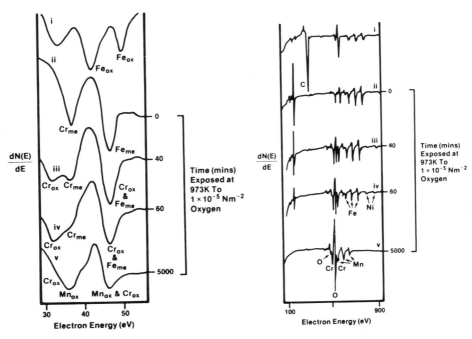

Figure 44. (a) Low-energy *MVV* and (b) high-energy *KLL* auger spectra of 18/8 stainless steel during oxidation (Ref. 150). (From Ref. 149.)

oxidation are given in Figure 44. To suppress background of continuous electrons, differentiated spectra are displayed. The metal surface was exposed at 973 K to 10^{-5} N/m² oxygen, while AES spectra were taken at intervals. The initial surface (i) showed the presence of iron oxide on the surface. After heating (ii), the iron oxide converted to metallic iron. During oxygen exposure iron and nickel disappeared from the surface, followed by the appearance of manganese. At the end (v), the top surface layers are composed by nearly equal mixtures of manganese and chromium oxide.[149]

5.2. Particle-Induced X-Ray Emission (Pixe)

X-ray emission following inner-shell ionization by impact processes may be used for analytical purposes. Johansson *et al.*[151] have demonstrated that the use of a sufficiently energetic proton beam and a Si(Li) detector to detect proton-induced characteristic X rays constitutes a powerful, multi-elemental analysis with high sensitivity. The high sensitivity is one of the main advantages of the Pixe method. It depends via the cross sections for inner-shell ionization on the energy of the incident projectile and the atomic number of the trace element. The limitation of the method largely results from background X rays induced, for instance, by bremsstrahlung (proton induced as well as secondary electron bremsstrahlung) processes. The minimum detectable concentration of an element is about 1 ppm (10^{-6}).[152] The use of the Pixe method for a rapid, multielemental determination of trace elements in biological or mineralogical materials has been explored by a great number of investigations.[153] A typical example of an analysis of human kidney cortex by the Pixe method is shown in Figure 45. Comparing the trace element contents of normal and malignant kidney it was found that cadmium is absent while the titanium concentration is significantly enhanced in malignant samples.[154]

Even lower trace element concentrations can be detected in water. Sub-ppb $(<10^{-9})$ water analysis is achieved by absorbing metals in the form of organic chelates onto activated carbon, which is pressed to a tablet and then analyzed with Pixe. In Figure 46 a sample made from distilled water is shown.[155] The observed metals are all in the lower ppb range.

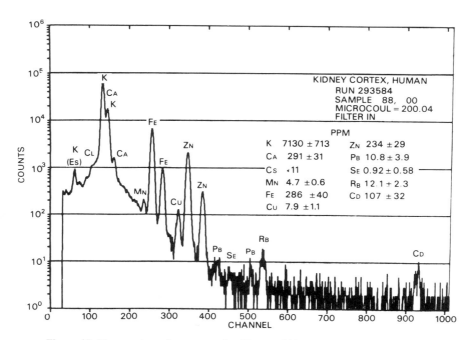

Figure 45. Pixe spectrum from a sample of human kidney cortex. (From Ref. 151.)

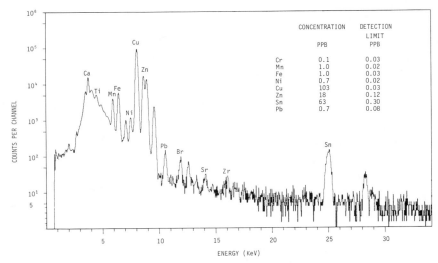

Figure 46. Pixe spectrum from a sample made from distilled water. The metals present are all in the lower ppb range, and the resulting detection limits are between 0.02 and 0.3 ppb. (From Ref. 155.)

5.3. Proton and Electron Microprobe

Proton and electron microprobes are useful devices for an analysis of microvolumes. The experimental setup consists of a well-focused and collimated particle beam. Typically, electron beams with energies of about 10–50 keV may be focused down to spot sizes of about 0.1 μm. Hence electron microprobe allows analysis of structures of multicomponent samples with a lateral resolution of 1 μm or less. Wheareas electron micro-probes are in frequent use and commercially available, proton microprobes using 2–4-MeV proton beams are more complicated, and comparatively few have been built so far. In connection with detection of secondary electrons, both microprobes allow imaging of irradiated surfaces. Figure 47 shows two micrographs obtained from proton and electron microprobes, respectively. The test specimen was a silver grid with 25 μm periodicity, supported by two copper grids. In contrast to the electron beam, which is completely stopped inside the copper bar of thickness 15 μm, the proton beam penetrates the copper bar and permits an imaging of the silver grid behind the copper bar with reasonable resolution.

Proton and electron microprobes may also be used in connection with detection of Auger electrons or X rays. The main difference in the perform-ance of the two microprobes using x-ray detection arises from the different slowing-down and scattering processes of protons and electrons. Electrons

3 MeV
Protons
SCANNING
PROTON
MICROPROBE

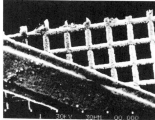

30 keV
Electrons
SCANNING
ELECTRON
MICROPROBE

Figure 47. Secondary electron images for proton and electron incident on silver grid (25 μm spacing) with 15-μm-thick copper strut overlay. (From Ref. 156.)

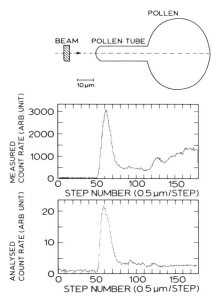

Figure 48. Schematic of lilium longiflorum pollen and measured and analyzed calcium gradient in the pollen. (From Ref. 157.)

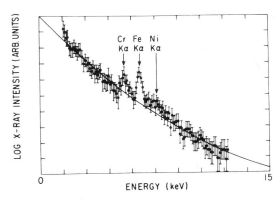

Figure 49. Typical x-ray spectra taken with a Si(Li) detector during Princeton large torus (PLT) discharge. (From Ref. 158.)

produce a large fraction of bremsstrahlung radiation, which reduces the detection limit to about 1%–10%. In contrast, protons produce comparatively little background radiation and the detection limit is as low as 1 ppm.

An investigation of the calcium concentration inside a *lilium longiflorum* pollen using micropixe has been performed by Chen *et al.*[157] Figure 48 shows the measured calcium profile and the calculated calcium concentration, together with a schematic view of the pollen. A pollen tube of the order of a few millimeters in length grows from the pollen grain. The measurements show that there is a strong increase of the calcium concentration at the growing tip of the pollen tube.

5.4. Determination of Charge State Distribution of Impurity Ions in Tokamaks

Many of the problems associated with fusion plasmas are caused by the relatively large concentration of impurity ions. Such impurities are a

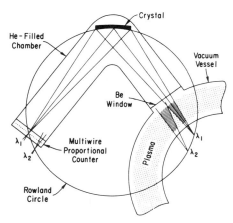

Figure 50. Schematic drawing of curved-crystal spectrometer used in the investigation of charge-state distribution of impurity ions in the Princeton large torus. X rays from the plasma pass through the Be window, are diffracted by the crystal, and are focused onto the position-sensitive, multiwire proportional counter. (From Ref. 158.)

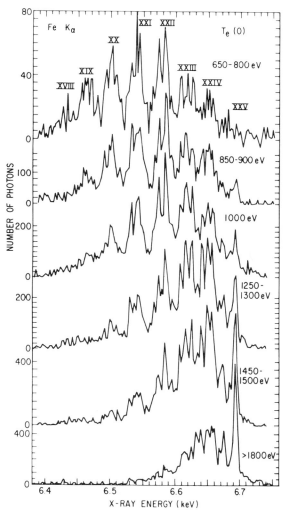

Figure 51. Fe $1s$-$2p$ x-ray spectra taken from PLT plasma discharges. The spectra were accumulated during the second time quadrant of the discharges, during which the central electron temperature was approximately constant. The corresponding electron temperatures are indicated. (From Ref. 158.)

result of the plasma-wall interaction and cause considerable cooling of the plasma due to the relatively high Z of the impurity material. A typical x-ray spectrum (Figure 49) taken with a Si(Li) detector during a single discharge at the Princeton Large Torus (PLT) shows as plasma impurities chromium, iron, and nickel which are the major components of the stainless-steel vacuum vessel.[158] X-ray spectra taken with high resolution reveal that the peaks of Figure 49 are composed of individual lines representing different ionic charge states. Figure 50 shows a schematic drawing of a curved-crystal spectrometer with which the high-resolution iron spectra of Figure 51 have been taken. The six spectra of Figure 51 are for different electron temperatures of PLT plasma discharges.[158] At $T_e = 650-850$ eV the charge states labeled Fe XX–Fe XXII (charge states 19+ to 21+) are the most abundant, while the 1800-eV discharge is dominated by Fe XXIII–Fe XXV (charge states 22+ to 24+) (Figure 51).

Acknowledgments

The author acknowledges helpful discussions with many colleagues in the field. Special thanks go to Professor Dr. H. O. Lutz (Bielefeld), Dr. W. Jitschin (Bielefeld), Professor Dr. K. Kleinpoppen (Stirling), Dr. L. Sarkadi (Debreceu), Dr. R. Scuch (Heidelberg), and Dr. I. McGregor (Stirling). Part of this work was supported by the Deutsche Forschungs-gemeinschaft (DFG).

References

1. N. F. Mott and H. S. W. Massey, *The Theory of Atomic Collisions*, 3rd ed., Oxford University Press, Oxford (1965).
2. D. H. Madison and E. Merzbacher, in *Atomic Inner Shell Processes*, Vol. I, B. Craseman, Ed., Academic Press, New York (1975), p. 1.
3. E. J. McGuire, *Phys. Rev. A* **3**, 267 (1971), and private communication; *Ibid.* **16**, 62, 73 (1977); **25**, 192 (1982).
4. R. Hippler and W. Jitschin, *Z. Phys. A* **307**, 287 (1982).
5. J. M. Hansteen, *Adv. Atom. Molec. Phys.* **11**, 299 (1975); B. Fricke, Chapter 5 of this book, Part A.
6. W. Brandt, R. Laubert, and I. Sellin, *Phys. Rev.* **151**, 56 (1966).
7. D. H. Madison and W. N. Shelton, *Phys. Rev. A* **7**, 499 (1973).
8. S. M. Younger, *Phys. Rev. A* **22**, 111 and 1425 (1980); *Comments At. Mol. Phys.* **11**, 193 (1982).
9. S. S. Prasad, *Proc. Phys. Soc.* **85**, 57 (1965).
10. G. Peach, *Proc. Phys. Soc.* **85**, 709 (1965); **87**, 375, 381 (1966).
11. M. Inokuti, *Rev. Mod. Phys.* **43**, 297 (1971).
12. S. C. McFarlane, *J. Phys. B* **5**, 1906 (1972); Ph.D. Thesis, University of Stirling (1972).
13. B. L. Schram and L. Vriens, *Physica* **31**, 1431 (1965).
14. C. Møller, *Ann. Phys.* (Leipzig) **14**, 531 (1932).

15. U. Fano, *Phys. Rev.* **102**, 385 (1956); *Ann. Rev. Nucl. Sci.* **13**, 1 (1963).
16. R. Anholt, *Phys. Rev. A* **19**, 1004 (1979).
17. P. A. Amundsen and K. Aashamar, *J. Phys. B* **14**, 4047 (1981).
18. D. M. Davidovic, B. L. Moiseiwitsch, and P. H. Norrington, *J. Phys. B* **11**, 847 (1978).
19. H. Kolbenstvedt, *J. Appl. Phys.* **38**, 4785 (1967).
20. J. Bang and J. M. Hansteen, *Mat. Fys. Medd. Dan. Vid. Selsk.* **31**, 13 (1959).
21. P. A. Amundsen, *J. Phys. B* **9**, 971 (1976); **10**, 1097, 2177 (1977).
22. R. Gundersen, L. Kocbach, and J. M. Hansteen, *J. Phys. B* **14**, L367 (1981); *Nucl. Instr. Methods* **192**, 63 (1982).
23. W. Brandt and G. Lapicki, *Phys. Rev. A* **10**, 474 (1974); **20**, 465 (1979).
24. G. Basbas, W. Brandt, and R. Laubert, *Phys. Rev. A* **7**, 983 (1973); **17**, 1655 (1978); G. Basbas, W. Brandt, and R. H. Ritchie, *Ibid.* **7**, 1971 (1973).
25. T. Mukoyama and L. Sarkadi, *Bull. Inst. Chem. Res.* (*Kyoto*) **57**, 33 (1979); **58**, 60, 95 (1980); *Phys. Rev. A* **23**, 375 (1981); *Nucl. Instr. Methods* **179**, 573 (1979).
26. W. L. Fite, R. T. Brackmann, and W. R. Snow, *Phys. Rev.* **112**, 1161 (1958).
27. W. L. Fite and R. T. Brackmann, *Phys. Rev.* **112**, 1141 (1958).
28. R. L. F. Boyd and A. Boksenberg, in *Proc. IVth Int. Conf. Ioniz. Phenom. Gases*, Vol. 1, North-Holland, Amsterdam (1960), p. 529.
29. E. W. Rothe, L. L. Marino, R. H. Neynaber, and S. M. Trujillo, *Phys. Rev.* **125**, 582 (1962).
30. M. R. H. Rudge and M. J. Seaton, *Proc. R. Soc. London Ser. A* **283**, 262 (1965).
31. M. S. Lubell, in *Coherence and Correlation in Atomic Collisions*, H. Kleinpoppen and J. F. Williams, Eds., Plenum Press, New York (1981), p. 663; D. Hils, K. Rubin, and H. Kleinpoppen, *loc. cit.*, p. 689; W. Raith, G. Baum, D. Caldwell, and E. Kisker, *loc. cit.*, p. 567.
32. M. R. H. Rudge, *J. Phys. B* **11**, L149 (1978).
33. M. J. Alguard, V. W. Hughes, M. S. Lubell, and P. F. Wainwright, *Phys. Rev. Lett.* **39**, 334 (1977).
34. G. H. Wannier, *Phys. Rev.* **90**, 817 (1953).
35. F. H. Read, in *Atomic Physics 7*, D. Kleppner and F. M. Pipin, Eds., Plenum Press, New York (1981), p. 429.
36. A. R. P. Rau, *Phys. Rev. A* **4**, 207 (1971).
37. R. Peterkop, *J. Phys. B* **4**, 513 (1971).
38. H. Klar and W. Schlecht, *J. Phys. B* **9**, 1699 (1976).
39. S. Cvejanovic and F. H. Read, *J. Phys. B* **7**, 1841 (1974).
40. B. Peart, D. S. Walton, and K. T. Dolder, *J. Phys. B* **2**, 1347 (1969).
41. D. H. Crandall, R. A. Phaneuf, R. A. Falk, D. S. Belic, and G. H. Dunn, *Phys. Rev. A* **25**, 143 (1982).
42. D. C. Griffin, C. Bottcher, and M. S. Pindzola, *Phys. Rev. A* **25**, 154, 1374 (1982).
43. K. L. Bell and A. E. Kingston, *J. Phys. B* **2**, 1125 (1969).
44. E. J. McGuire, *Phys. Rev. A* **22**, 868 (1980); and private communication.
45. J. P. Keene, *Phil. Mag.* **40**, 369 (1949).
46. N. V. Fedorenko, V. V. Afrosimov, R. N. Il'in, and E. S. Solov'ev, in *Proc. VIth Int. Conf. Ionization Phenomena in Gases*, North-Holland, Amsterdam (1960), p. 47; E. S. Solov'ev, R. N. Il'in, V. A. Oparin, and N. V. Fedorenko, *Sov. Phys.-JETP* **15**, 459 (1962).
47. F. J. De Heer, J. Schutten, and H. Moustafa, *Physica* **32**, 1766 (1966).
48. H. B. Gilbody and J. B. Hasted, *Proc. R. Soc. London Ser. A* **240**, 382 (1957); **274**, 365 (1963).
49. J. W. Hopper D. S. Harmer, D. W. Martin, and E. W. McDaniel, *Phys. Rev.* **121**, 1123 (1961); **125**, 2000 (1962).
50. L. I. Pivovar and Yu. Z. Levchenko, *Sov. Phys.-JETP* **25**, 27 (1967).

51. B. Adamczyk, A. J. H. Boerboom, B. L. Schram, and J. Kistemaker, *J. Chem. Phys.* **44**, 4640 (1966); B. L. Schram, F. J. de Heer, M. J. van der Wiel, and J. Kistemaker, *Physica* **31**, 94 (1965).

52. R. Hippler and K.-H. Schartner, *J. Phys. B* **7**, 618, 1167 (1974), **7**, 2528 (1975); *Z. Phys. A* **273**, 123 (1975).

53. H. F. Beyer, R. Hippler, and K.-H. Schartner, *Z. Phys. A* **289**, 239, 353 (1979); H. F. Beyer, M. Gros, R. Hippler, and K.-H. Schartner: *Phys. Lett.* **68A**, 215 (1978).

54. P. N. Clout and D. W. O. Heddle, *J. Opt. Soc. Am.* **59**, 715 (1969).

55. B. F. J. Luyken, F. J. de Heer, and R. Ch. Baas, *Physica* **61**, 200 (1972).

56. M. Eckhardt, Ph.D. Thesis, University of Giessen (1982).

57. J. Van Eck, F. J. de Heer, and J. Kistemaker, *Phys. Rev.* **130**, 656 (1963).

58. T. A. Carlson and C. W. Nestor, *Phys. Rev. A* **8**, 2887 (1973).

59. D. H. H. Hoffmann, C. Brendel, H. Genz, W. Löw, S. Müller, and A. Richter, *Z. Phys.* **A293**, 187 (1979); H. Genz, D. H. H. Hoffmann, W. Löw, and A. Richter, *Phys. Lett.* **73A**, 313 (1979).

60. G. Glupe and W. Mehlhorn, *Phys. Lett.* **25A**, 274 (1967).

61. W. Hink, L. Kees, H.-P. Schmitt, and A. Wolf, in *Inner-Shell and X-Ray Physics of Atoms and Solids*, D. Fabian, H. Kleinpoppen, L. M. Watson, Eds., Plenum Press, New York (1981), p. 327.

62. W. Hink and A. Ziegler. *Z. Phys.* **226**, 222 (1969).

63. R. Hippler, K. Saeed, I. McGregor, and H. Kleinpoppen, *Z. Phys. A* **307**, 83 (1982).

64. H. Tawara, K. G. Harrison, and F. J. de Heer, *Physica* **63**, 351 (1973).

65. M. O. Krause, *J. Phys. Chem. Ref. Data* **8**, 307 (1979); J. H. Scofield, *Phys. Rev. A* **10**, 1507 (1974); E. J. McGuire, *Ibid.* **5**, 1043 (1972); **9**, 1840 (1974).

66. H. J. Christofzik, Diploma thesis, Münster (1971); G. N. Ogurtsov, *Sov.-Phys. JETP* **37**, 584 (1973); A. Langenberg, F. J. de Heer, and J. van Eck, *J. Phys. B* **8**, 2079 (1975).

67. S. I. Salem and L. D. Moreland, *Phys. Lett.* **37A**, 161 (1971).

68. D. V. Davis, V. D. Mistry, and C. A. Quarles, *Phys. Lett.* **38A**, 169 (1972).

69. C.-N. Chang, *Phys. Rev. A* **19**, 1930 (1979).

70. R. Hippler, M. Aydinol, I. McGregor, and H. Kleinpoppen, *Phys. Rev. A* **23**, 1730 (1981).

71. R. H. Pratt, in *Inner-Shell and X-Ray Physics of Atoms and Solids*, D. Fabian, H. Kleinpoppen, and L. M. Watson, Eds., Plenum Press, New York (1981), p. 367.

72. J. W. Motz and R. C. Placious, *Phys. Rev.* **136**, 663 (1964).

73. D. H. Rester and W. E. Dance, *Phys. Rev.* **152**, 1 (1966).

74. D. M. Davidovic and B. L. Moiseiwitsch, *J. Phys. B* **8**, 947 (1975).

75. H. Genz, C. Brendel, P. Eschwey, U. Kuhn, W. Löw, A. Richter, P. Seserko, and R. Sauerwein, *Z. Phys. A* **305**, 9 (1982).

76. K. H. Berkner, S. N. Kaplan, and R. V. Pyle, *Bull. Am. Phys. Soc.* **15**, 786 (1970).

77. G. R. Dangerfield and B. M. Spicer, *J. Phys. B* **8**, 1744 (1975).

78. L. M. Middleman, R. L. Ford, and R. Hofstadter, *Phys. Rev. A* **2**, 1429 (1970).

79. J. H. Scofield, *Phys. Rev. A* **18**, 963 (1978).

80. W. Scholz, A. Li-Scholz, R. Colle, and I. L. Preiss, *Phys. Rev. Lett.* **29**, 761 (1972).

81. H. Tawara, Y. Hachiya, K. Ishii, and S. Morita, *Phys. Rev. A* **13**, 572 (1976).

82. E. Laegsgaard, J. U. Andersen, and F. Hogedal, *Nucl. Instr. Methods* **169**, 293 (1980).

83. A. R. Zander and M. C. Andrews III, *Phys. Rev. A* **20**, 1484 (1979).

84. G. A. Bissinger, J. M. Joyce, E. J. Ludwig, W. S. McEver, and S. M. Shafroth, *Phys. Rev. A* **1**, 841 (1970).

85. M. Milazzo and G. Riccobono, *Phys. Rev. A* **13**, 578 (1976).

86. G. Lapicki, R. Laubert, and W. Brandt, *Phys. Rev. A* **22**, 1889 (1980).

87. R. K. Rice, F. D. McDaniel, G. Basbas, and J. L. Duggan, *Phys. Rev. A* **24**, 758 (1981).

88. K. Aashamar and P. A. Amundsen, *J. Phys. B* **14**, 483 (1981).

89. M. Kamiya, K. Ishii, K. Sera, S. Morita, and H. Tawara, *Phys. Rev. A* **16**, 2295 (1978).
90. A. W. Waltner, D. M. Peterson, G. A. Bissinger, A. B. Baskin, C. E. Busch, P. A. Nettles, W. R. Scates, and S. M. Shafroth, in *Proc. Int. Conf. Inner Shell Ion. Phen. Fut. Appl.*, R. W. Fink, Ed., USAEC Technical Information Center, Oak Ridge (1972), p. 1080.
91. S. Datz, J. L. Duggan, L. C. Feldman, E. Laegsgaard, and J. U. Andersen, *Phys. Rev. A* **9**, 192 (1974).
92. W. Jitschin, A. Kaschuba, R. Hippler, and H. O. Lutz, *J. Phys. B* **15**, 763 (1982).
93. C. V. Barros Leite, N. V. de Castro Faria, and A. G. de Pinho, *Phys. Rev. A* **15**, 943 (1977); C. V. Barros Leite, A. G. de Pinho, and N. V. de Castro Faria, *Rev. Bras. Fis.* **7**, 311 (1977).
94. C.-N. Chang, J. F. Morgan, and S. L. Blatt, *Phys. Rev. A* **11**, 607 (1975).
95. J. R. Chen, *Phys. Rev. A* **15**, 487 (1977).
96. L. Sarkadi and T. Mukoyama, *J. Phys. B* **13**, 2255 (1980).
97. O. Benka and A. Kropf, *At. Data Nucl. Data Tables* **22**, 219 (1978).
98. T. Mukoyama and L. Sarkadi, *Phys. Rev. A* **23**, 375 (1981); **25**, 1411 (1982).
99. F. J. De Heer, in *Atomic and Molecular Processes in Controlled Thermonuclear Fusion*, M. R. C. McDowell, A. M. Ferendeci, Eds., Plenum Press, New York (1980), p. 351.
100. C. D. Lin and L. N. Tunell, *J. Phys. B* **12**, L485 (1979).
101. H. Tawara and A. Russek, *Rev. Mod. Phys.* **45**, 178 (1973).
102. M. Rødbro, E. Pedersen, C. L. Cocke, and J. R. Macdonald, *Phys. Rev. A* **19**, 1936 (1979).
103. J. R. Macdonald, C. L. Cocke, and W. W. Eidson, *Phys. Rev. Lett.* **32**, 648 (1974).
104. H. Gabler, N. Stolterfoht, and U. Leithäuser, from M. E. Rudd, L. H. Toburen, and N. Stolterfoht, *At. Data Nucl. Date Tables* **23**, 405 (1979).
105. L. Sarkadi, J. Bossler, R. Hippler, and H. O. Lutz, *J. Phys. B* **16**, 71 (1982).
106. D. H. Madison and S. T. Manson, *Phys. Rev. A* **20**, 825 (1979).
107. T. Weiter and R. Schuch, *Z. Phys. A* **305**, 91 (1982).
108. P. A. Amundsen and K. Aashamar, *J. Phys. B* **14**, L153 (1981).
109. D. Trautmann and F. Roösel, *Nucl. Instr. Methods* **169**, 259 (1980).
110. A. Salin, *J. Phys. B* **2**, 631 (1969).
111. G. B. Crooks and M. E. Rudd, *Phys. Rev.* **25**, 1599 (1970).
112. J. H. Macek, *Phys. Rev. A* **1**, 235 (1970).
113. M. E. Rudd, *Radiat. Res.* **64**, 153 (1975).
114. H. O. Lutz, in *Proc. 2nd Int. Conf. Inner-Shell Ioniz. Phenom.*, Freiburg (1976), W. Mehlhorn and R. Brenn, Eds., p. 104; and in *Inner-Shell and X-Ray Physics of Atoms and Solids*, D. J. Fabian, H. Kleinpoppen, and L. M. Watson, Eds., Plenum Press, New York (1981), p. 79.
115. J. U. Andersen, L. Kocbach, E. Laegsgaard, M. Lund, and C. D. Moak, *J. Phys. B* **9**, 3247 (1976).
116. J. U. Andersen, E. Laegsgaard, M. Lund, and C. D. Moak, *Nucl. Instr. Methods* **132**, 507 (1976).
117. J. U. Andersen, E. Laegsgaard, and M. Lund, *Nucl. Instr. Methods* **192**, 79 (1982).
118. H. Schmidt-Böcking, K. E. Stiebing, W. Schadt, N. Löchter, G. Gruber, S. Kelbsch, K. Bethge, R. Schuch, and I. Tserruya, *Nucl. Instr. Methods* **192**, 71 (1982).
119. F. Rösel, D. Trautmann, and G. Baur, *Nucl. Instr. Methods* **192**, 43 (1982).
120. W. Mehlhorn, *Phys. Lett.* **26A**, 166 (1968).
121. U. Fano and J. Macek, *Rev. Mod. Phys.* **45**, 553 (1973).
122. J. Eichler and W. Fritsch, *J. Phys. B* **9**, 1477 (1976).
123. K. Blum and H. Kleinpoppen, *Phys. Rep.* **52**, 203 (1979); K. Blum, Chapter 2 of this book, Part A.
124. U. Fano and G. Racah, *Irreducible Tensorial Sets*, Academic Press, New York (1959).

125. E. G. Berezhko, N. M. Kabachnik, and V. V. Sizov, *J. Phys. B* **11**, 1819 (1978); E. G. Berezhko and N. M. Kabachnik, *Ibid.* **10**, 2467 (1977); **12**, 2993 (1979).

126. M. Eminyan, K. B. MacAdam, J. Slevin, and H. Kleinpoppen, *J. Phys. B* **7**, 1519 (1974).

127. G. Richter, M. Brüssermann, S. Ost, J. Wigger, B. Cleff, and R. Santo, *Phys. Lett.* **82A**, 412 (1981).

128. B. Cleff and W. Mehlhorn, *J. Phys. B* **7**, 605 (1974); *Phys. Lett.* **37A**, 3 (1971).

129. W. Sandner and W. Schmitt, *J. Phys. B* **11**, 1833 (1978).

130. M. Rødbro, R. Dubois and V. Schmidt, *J. Phys. B* **11**, L551 (1978).

131. W. Sandner, M. Weber, and W. Mehlhorn, in *Coherence and Correlation in Atomic Collisions*, H. Kleinpoppen and J. F. Williams, Eds., Plenum Press, New York (1980), p. 215.

132. K. Omidvar, *J. Phys. B* **10**, L55 (1977).

133. M. Aydinol, R. Hippler, I. McGregor, and H. Kleinpoppen, *J. Phys. B* **13**, 989 (1980); in *Coherence and Correlation in Atomic Collisions*, H. Kleinpoppen and J. F. Williams, Eds., Plenum Press, New York (1981), p. 205.

134. R. DuBois and M. Rødbro, *J. Phys. B* **13**, 3739 (1980); R. Dubois, L. Mortensen, and M. Rødbro, *J. Phys. B* **14**, 1613 (1981).

135. V. V. Sizov and N. M. Kabachnik, *J. Phys. B* **13**, 1601 (1980).

136. W. Jitschin, A. Kaschuba, H. Kleinpoppen, and H. O. Lutz, *Z. Phys. A* **304**, 69 (1982); W. Jitschin, H. Kleinpoppen, R. Hippler, and H. O. Lutz, *J. Phys. B* **12**, 4077 (1979).

137. J. Palinkas, L. Sarkadi, and B. Schlenk, *J. Phys. B* **13**, 3829 (1980); J. Palinkas, B. Schlenk, and A. Valek, *J. Phys. B* **14**, 1157 (1981).

138. W. Jitschin, H. O. Lutz, and H. Kleinpoppen, in *Inner-Shell and X-ray Physics of Atoms and Solids*, D. J. Fabian, H. Kleinpoppen, L. M. Watson, Eds., Plenum Press, New York (1981), p. 89.

139. F. Rösel, D. Trautmann, and G. Baur, *Z. Phys. A* **304**, 75 (1982).

140. G. Lapicki and W. Losonsky, *Phys. Rev. A* **15**, 896 (1977).

141. E. G. Berezhko, N. M. Kabachnik, and V. V. Sizov, *Phys. Lett.* **77A**, 231 (1980); E. G. Berezhko, V. V. Sizov, and N. M. Kabachnik, *J. Phys. B* **14**, 2635 (1981).

142. D. H. Jakubassa-Amundsen, *J. Phys. B* **14**, 2647 (1981).

143. W. Mehlhorn, in *Proc. Int. Seminar High-Energy Ion-Atom Collision Processes*, Debrecen (1981), Akademiai Kiadó, Budapest (1983), p. 83.

144. J. Wigger, S. Ost, M. Brüssermann, G. Richter, and B. Cleff, *Phys. Lett.* **84A**, 110 (1981).

145. B. Cleff, *Acta Phys. Pol. A* **61**, 285 (1982).

146. J. Konrad, R. Hoffmann, H. Schmidt-Böcking, and R. Schuch, Arbeitsbericht EAS-3, J. Eichler, B. Fricke, R. Hippler, D. Kolb, H. O. Lutz, P. Mokler, Eds., p. 72 (1982).

147. L. Kocbach, in Ref. 146.

148. C. R. Brundle, and A. D. Baker, Eds., *Electron Spectroscopy: Theory, Techniques and Applications*, Academic Press, London (1981).

149. G. E. McGuire and P. H. Holloway, in *Electron Spectroscopy: Theory, Techniques and Applications*, C. R. Brundle, A. D. Baker, Eds., Vol. 4, Academic Press, London (1981), p. 1.

150. G. C. Allen and R. K. Wild, *J. Electron. Spectrosc.* **5**, 409 (1974).

151. T. B. Johansson, R. Akselsson, and S. A. E. Johansson, *Nucl. Instr. Methods* **84**, 141 (1970).

152. S. A. E. Johansson and T. B. Johansson, *Nucl. Instr. Methods* **137**, 473 (1976).

153. S. A. E. Johansson, Ed., Particle Induced X-Ray Emission and its Analytical Applications, *Nucl. Instr. Methods.* **181**, (1981).

154. J. A. Guffey, H. A. van Rinsvelt, R. M. Sarper, Z. Karcioglu, W. R. Adams, and R. W. Fink, *Nucl. Instr. Methods* **149**, 489 (1978).

155. E. M. Johansson and K. R. Akselsson, *Nucl. Instr. Methods* **181**, 221 (1981).

156. H. Kneis, B. Martin, R. Nobiling, B. Povh, and K. Traxel, *Nucl. Instr. Methods* **197**, 79 (1982).

157. J. R. Chen, H. Kneis, B. Martin, R. Nobiling, D. Pelte, B. Povh, and K. Traxel, *Nucl. Instr. Methods* **181**, 141 (1981).

158. K. W. Hill, S. von Goeler, M. Bitter, L. Campbell, R. D. Cowan, B. Fraenkel, A. Greenberger, R. Horton, J. Hovey, W. Roney, N. R. Sauthoff, and W. Stodiek, *Phys. Rev. A* **19**, 1770 (1979).

12

Amplitudes and State Parameters from Ion– and Atom–Atom Excitation Processes

T. ANDERSEN AND E. HORSDAL-PEDERSEN

1. Introduction

For many years, the study of inelastic processes in atomic collisions was directed towards the measurement or calculation of total cross sections. The interpretation of such measurements in terms of the interaction between the collision partners is, however, often very difficult except for some very simple systems.

In the period 1960–1972, this situation was radically changed with the introduction of new experimental techniques, which made it possible to measure differential scattering cross sections for each reaction channel as a function of energy and scattering angle. The first such differential measurements in ion–atom collisions were reported more than 20 years ago,[1] but an important stage in the development of this technique was not reached until 1965,[2] when careful analysis of the energy loss suffered by the scattered ion was made practical. Hence, detailed information on the impact-parameter dependence of the excitation probability for a given atomic state or, more generally, for a group of states, was obtainable. More recently, the time-of-flight technique has been introduced to differentiate the information obtained by the investigation of projectile-neutralization channels in ion–atom collisions[3] and to deal with neutral–neutral collisions.

T. ANDERSEN AND E. HORSDAL-PEDERSEN • Institute of Physics, University of Aarhus, DK-8000 Aarhus C, Denmark.

The two methods mentioned for investigation of inelastic processes are limited by their relatively poor energy resolution compared to other experimental techniques in atomic physics. However, a considerable improvement in resolution is possible if the identification of the scattering process comes from optical spectroscopic measurements, as is the case in the photon, scattered-particle coincidence technique.[4]

A new development in the study of ion–atom, atom–atom, or electron–atom collision started in the period 1973–1975,[5–7] when the photon, scattered-particle coincidence technique was combined with a polarization analysis of the emitted photons. By means of this technique, it has become possible to study binary atomic-collision processes in such detail that it is now possible to obtain information not only related to the size of the amplitudes for exciting individual magnetic substates (differential excitation cross sections) but also to the phase relations between these amplitudes. The experimental information obtainable can be expressed in various ways, which will be discussed in the following sections of this chapter, but one interesting possibility is the reconstruction of the angular part of the charge distribution excited in the particular collisions selected by the experimental arrangement. Such spatial distributions, however, do not reflect the full experimental information, which also includes expectation values of the electronic angular momentum of the excited state.

The information on the dynamics of atomic-excitation processes, which may be obtained by the polarized-photon, scattered-particle coincidence technique, may also be obtained by studying the inverse collision process. In this case,[8] the magnetic substates of the excited state in question are populated selectively by means of lasers prior to the collisional deexcitation process.

This chapter deals with single collisions between two atomic species, of which one is initially in a ^1S state. This means that there is only one initial spin channel. The collisions are further specified by a definite scattering plane and a definite orientation. The scattering plane is defined by the incoming and the outgoing wave vectors of the projectile, $\mathbf{k}$ and $\mathbf{k}'$, respectively, and the orientation by the vector product $\mathbf{k} \times \mathbf{k}'$. The coincidence experiments to be discussed do not resolve individual fine- or hyperfine-structure levels, and it is assumed that the forces causing the excitation of a given spectroscopic term are independent of spin such that different levels are populated according to the statistical weight. The excitation process itself is therefore assumed to populate pure eigenstates of the square of the electronic orbital angular momentum $\mathbf{L}^2$. These states subsequently evolve into the observed mixed states (spectroscopic terms) due to the fine and hyperfine interactions.

The following parts of the present chapter will include discussions of (i) the information to be obtained by studies of the angular correlation or the polarization of photons or electrons emitted from collisional-excited states, (ii) experimental methods currently used to measure the above angular correlations or polarizations, (iii) the results obtained so far for ion–atom and atom–atom collisions, and (iv) the prospects for future studies and for application of the polarized photon, scattered-particle coincidence technique in atomic spectroscopy.

Readers interested in having an insight into electron–atom or electron–ion collisions are referred to the recent review by Blum and Kleinpoppen[9] and to conference proceedings.[10]

2. Angular Correlation between Scattered Particles and Autoionization Electrons or Polarized Photons Emitted from States Excited in Atomic Collisions

Very detailed experimental information on atomic excitation processes may be obtained from the angular correlation of autoionization electrons or photons emitted from atomic states excited in individual collisions specified by scattering in a given direction. Even more details may be obtained from the precise state of polarization of the ensemble of photons emitted in a given direction. Alternatively, the same information is gained from experiments which study the deexcitation in specific collision of atoms selectively excited prior to the collision by absorption of photons in a given state of polarization.

The details of an observed excitation process can be expressed directly in terms of excitation amplitudes if a pure state of the projectile–target system is selected by the experiment or, in terms of more general parameters describing the anisotropy of the excited state if a mixed state is observed. The connection between such quantities and the above measurable angular-distribution or polarization functions is not trivial, but it has recently been worked out and discussed in detail.[11,12]

2.1. Photon Emission

The angular distribution and polarization of light excited in atomic collisions was treated by Macek and Jaecks[11] and later, in more general terms, by Fano and Macek.[12] The latter authors expressed their results in terms of the alignment and orientation of the radiating atomic system. The intensity of the electric-dipole radiation emitted in a given direction and in a given state of polarization is[12,13]

$$I = S(1 - \tfrac{1}{2}h^{(2)}A_0^{\text{det}} + \tfrac{3}{2}h^{(2)}A_{2+}^{\text{det}}\cos 2\beta + \tfrac{3}{2}h^{(1)}O_0^{\text{det}}\sin 2\beta) \tag{1}$$

where S is a constant of proportionality, and

$$A_0^{\text{det}} = A_0^{\text{col}}\tfrac{1}{2}(3\cos^2\theta - 1) + A_{1+}^{\text{col}}\tfrac{3}{2}\sin 2\theta \cos\phi + A_{2+}^{\text{col}}\tfrac{3}{2}\sin^2\theta \cos 2\phi$$

$$A_{2+}^{\text{det}} = A_0^{\text{col}}\tfrac{1}{2}\sin^2\theta \cos 2\psi$$
$$\qquad + A_{1+}^{\text{col}}(\sin\theta \sin\phi \sin 2\psi - \sin\theta \cos\theta \cos\phi \cos 2\psi) \qquad (2)$$
$$\qquad + A_{2+}^{\text{col}}[\tfrac{1}{2}(1 + \cos^2\theta)\cos 2\phi \cos 2\psi - \cos\theta \sin 2\phi \sin 2\psi]$$

$$O_0^{\text{det}} = O_{1-}^{\text{col}}\sin\theta \sin\phi$$

where, with the assumptions on the excitation process and the experimental resolution outlined in the introduction

$$A_0^{\text{col}} = a^{-1}\langle 3L_z^2 - L^2 \rangle$$
$$A_{1+}^{\text{col}} = a^{-1}\langle L_x L_z + L_z L_x \rangle$$
$$A_{2+}^{\text{col}} = a^{-1}\langle L_x^2 - L_y^2 \rangle \qquad (3)$$
$$O_{1-}^{\text{col}} = a^{-1}\langle L_y \rangle$$

The last parameters are the nonzero components of the alignment and orientation tensors characterizing the observed state expressed in terms of the components of the electronic orbital angular momentum $\mathbf{L}$ of this state, $h^{(1)}$ and $h^{(2)}$ are geometrical factors determined by the orbital angular-momentum quantum numbers of the upper and lower states of the observed emission; $\beta = 0$ corresponds to linearly polarized light; $\beta = +\pi/4$ corresponds to photons with helicity $\lambda = +1$ or left-hand circularly (LHC) polarized light, following the convention of classical optics; $\beta = -\pi/4$ corresponds to $\lambda = -1$ or right-hand circularly (RHC) polarized light. (θ, ϕ, ψ) are the Euler angles of the photon–detector coordinate system. The normalization factor a is $a = L(L + 1)\mathcal{P}$, where $\mathcal{P}$ is the total excitation probability.

Equation (1) expresses the polarization of the emitted light in terms of the alignment and orientation parameters A_0^{det}, A_{2+}^{det}, and O_0^{det}, which refer to the detector coordinate system. The alignment and orientation of the radiating atoms are, however, most naturally expressed in terms of alignment and orientation parameters, referring to a coordinate system fixed relative to the beam direction. These parameters are denoted A_0^{col}, A_{1+}^{col}, A_{2+}^{col}, and O_{1-}^{col}. The relation to the parameters of Eq. (1) is given in Eq. (2), and in Eq. (3), the components A_0^{col}, A_{1+}^{col}, and A_{2+}^{col} of the alignment tensor are expressed by expectation values of quadratic forms in the orbital angular momentum of the excited atomic state, but they could also be expressed, for example, in terms of the density matrix of the excited projectile–target system or the multipole moments of the excited charge cloud, from which the spatial angular distribution can be reconstructed.[8] The component O_{1-}^{col} of the orientation vector is proportional to the expectation value of the electronic angular momentum perpendicular to the collision plane.

The components of the alignment tensor or the orientation vector can be related to measurable quantities in various ways, to be outlined in the following.

2.1.1. Particle–Photon Angular Correlations

The angular correlation of scattered projectiles and photons is obtained from Eq. (1) by summing over two perpendicular directions of photon polarization

$$I = S(1 - \tfrac{1}{2} h^{(2)} A_0^{\text{det}}) \tag{4}$$

By inspection of Eq. (2), it is seen that the three alignment parameters, A_0^{col}, A_{1+}^{col}, and A_{2+}^{col}, are obtainable. They are determined most accurately from the angular distribution of photons emitted within the collision plane (i.e., for $\phi = 0°$ or $180°$, see Figure 1). Direct information on the orientation vector is not obtained. However, if one can prove or otherwise assume that the observed final state is a pure state, a relation between the above components of the alignment tensor and the orientation vector can be used to obtain the absolute value of O_{1-}^{col} (but not the sign).

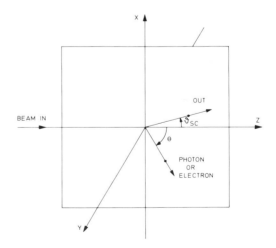

Figure 1. Determination of the angular distribution of photons/electrons within the x–z scattering plane by means of particle–photon and particle–electron correlation measurements.

2.1.2. Polarized-Photon, Scattered-Particle Correlations

The state of polarization of the photons emitted in a given direction is most sensitively related to the alignment and orientation parameters when the chosen direction is perpendicular to the scattering plane ($\theta = 90°$, $\phi = 90°$). The following refers to this situation (see Figure 2).

Linearly polarized photons are isolated by $\beta = 0$ in Eq. (1),

$$I = S(1 - \tfrac{1}{2}h^{(2)}A_0^{\text{det}} + \tfrac{3}{2}A_{2+}^{\text{det}}) \tag{5}$$

The direction of the linear polarization is selected by the angle ψ appearing in Eq. (2) for A_{2+}^{det}.

The relevant measurements are conveniently expressed in terms of the classical relative Stokes vector **P**, of which two components are defined by

$$P_1 \equiv [I(\psi = 0°) - I(\psi = 90°)]/[I(\psi = 0°) + I(\psi = 90°)]$$
$$P_2 \equiv [I(\psi = 45°) - I(\psi = 135°)]/[I(\psi = 45°) + I(\psi = 135°)] \tag{6}$$

or in terms of the alignment parameters,

$$P_1 = 3h^{(2)}(A_0^{\text{col}} - A_{2+}^{\text{col}})/I^Y \tag{7}$$

$$P_2 = 6h^{(2)}A_{1+}^{\text{col}}/I^Y \tag{8}$$

where

$$I^Y = 4 + h^{(2)}(A_0^{\text{col}} + 3A_{2+}^{\text{col}}) \tag{9}$$

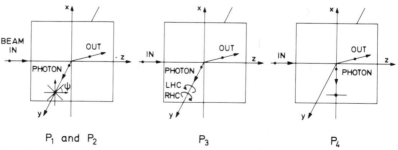

P_1 and P_2 P_3 P_4

Figure 2. The three types of measurements implied by the polarized-photon, scattered-particle coincidence technique. Left: Linearly polarized light is observed perpendicular to the x–z scattering plane (P_1 and P_2); middle: circularly polarized light is observed perpendicular to the x–z scattering plane; right: Linearly polarized light is observed in the x–z scattering plane along the x direction.

Circularly polarized photons are selected by $\beta = \pi/4$ (LHC polarized light) and $\beta = -\pi/4$ (RHC polarized light),

$$I(\text{RHC}) = S(1 - \tfrac{1}{2}h^{(2)}A_0^{\text{det}} - \tfrac{3}{2}h^{(1)}O_0^{\text{det}})$$
$$I(\text{LHC}) = S(1 - \tfrac{1}{2}h^{(2)}A_0^{\text{det}} + \tfrac{3}{2}h^{(1)}O_0^{\text{det}})$$
$$(10)$$

The third component of the Stokes vector $\mathbf{P}$ is defined by

$$P_3 \equiv [I(\text{RHC}) - I(\text{LHC})]/[I(\text{RHC}) + I(\text{LHC})] \qquad (11)$$

or, in terms of the orientation parameter,

$$P_3 = -6h^{(1)}O_{1-}^{\text{col}}/I^Y \qquad (12)$$

A measurement of the relative Stokes vector

$$\mathbf{P} = (P_1, P_2, P_3) \qquad (13)$$

in one given direction, however, does not uniquely determine the four alignment and orientation parameters A_0^{col}, A_{1+}^{col}, A_{2+}^{col}, and O_{1-}^{col} appearing in Eq. (2). One further measurement in the collision plane and perpendicular to the beam direction ($\theta = 90°$, $\phi = 0°$) of linear polarization parallel and perpendicular to the collision plane gives the extra independent piece of information needed,

$$P_4 = 3h^{(2)}(A_0^{\text{col}} + A_{2+}^{\text{col}})/I^X \qquad (14)$$

where

$$I^X = 4 + h^{(2)}(A_0^{\text{col}} - 3A_{2+}^{\text{col}}) \qquad (15)$$

It is thus possible to extract all alignment and orientation parameters from polarized-photon, scattered-particle correlations.

If a P → S transition from a pure state is observed, the emitted light is completely polarized, and consequently $|\mathbf{P}| = (P_1^2 + P_2^2 + P_3^2)^{1/2} = 1$. This relation reduces the number of independent alignment and orientation parameters to three and at the same time provides a possibility for verifying experimentally whether or not a pure state is observed.

Normally, the experiments do not select specific fine- or hyperfine-structure components. Consequently, there is often a loss of polarization due to fine and hyperfine interactions. This loss of polarization can normally be corrected for. The size of the corrected Stokes vector still expresses whether or not the selected excited state were a pure state before the system evolved into specific fine or hyperfine states. A mixed state may also be observed if both collision partners have internal degrees of freedom and only the excitation of one of the collision partners is observed while the state of the other remains undetermined.

2.2. Electron Emission

The angular distribution of autoionization electrons excited in atomic collisions was treated by Eichler and Fritsch.[14] The electron angular distributions are normally more complicated than those for photon emission. This higher degree of complexity originates from the different transition matrix elements for the two types of transitions. In practice, the photon emission has been limited to the electric-dipole term, whereas higher multipolarities are observable in electron emission, which is governed by a transition matrix element of the Coulomb interaction between a pair of electrons. This matrix element is characterized by selection rules less restrictive than the usual electric-dipole selection rules. The angular distribution of autoionization electrons may also depend on the relative size of the Coulomb matrix elements corresponding to the different multipolarities.

For a general discussion of the angular distribution of electrons emitted in the various types of autoionization transitions, we refer to the article by Eichler and Fritsch.[14] Here we limit ourselves to a particularly simple transition ($^1D \rightarrow {}^2S$), which has been studied experimentally.[15] The angular distribution is independent of transition matrix elements in this case and is given by

$$I = S(1 + A_0^{\text{det}})$$ (16)

where A_0^{det} is given in Eq. (2).

As is the case for photon emission, the alignment parameters in the collision frame are most sensitively determined from the angular distribution of electrons in the collision plane (Figure 1).

2.3. Selectively Excited Target Atoms

Details of inelastic collision amplitudes may also be obtained by studying the scattering of selectively excited target atoms.[8,16] The alignment parameters A_0^{col}, A_{1+}^{col}, and A_{2+}^{col} and the orientation parameter O_{1-}^{col} are determined from relative Stokes parameters as in Section 2.1.2, and the Stokes parameters are in turn given by intensity ratios of particles inelastically scattered off target atoms, which have been selectively excited to specific states.

The Stokes parameters P_1 and P_2 are given by

$$P_1 = (I_\parallel - I_\perp)/(I_\parallel + I_\perp) \qquad (17)$$

and

$$P_2 = (I_{45} - I_{135})/(I_{45} + I_{135}) \qquad (18)$$

where $I_\parallel$, $I_\perp$, I_{45}, and I_{135} are the intensities of projectiles scattered inelastically off target atoms excited by linearly polarized photons impinging on the target area in a direction perpendicular to the collision plane and polarized parallel and perpendicular to and at an angle of 45° and 135° to the beam direction, respectively (see Figure 3).

The Stokes parameter P_3 is given by

$$P_3 = [I(\text{LHC}) - I(\text{RHC})]/[I(\text{LHC}) + I(\text{RHC})] \qquad (19)$$

where $I(\text{LHC})$ and $I(\text{RHC})$ are the intensities of projectiles scattered inelastically off target atoms excited by left-hand and right-hand circularly-polarized photons, respectively, which impinge on the target area perpendicular to the collision plane.

A fourth Stokes parameter P_4 is related to projectiles scattered off target atoms excited by linearly polarized photons impinging on the target area in the collision plane and perpendicular to the beam direction, and

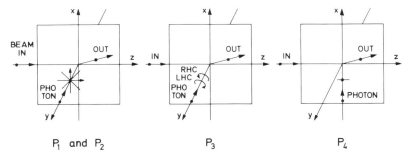

Figure 3. The three types of measurements implied by the laser-excitation technique. Left: Linearly polarized laser light falls perpendicularly onto the x–z scattering plane; middle: circularly polarized laser light falls perpendicularly onto the x–z scattering plane; right: linearly polarized laser light along the x direction, the linear polarization being parallel or perpendicular to the z direction.

further being polarized either parallel to, $I'_\parallel$, or perpendicular, $I'_\perp$, to the beam direction

$$P_4 = (I'_\parallel - I'_\perp)/(I'_\parallel + I'_\perp) \tag{20}$$

3. Experimental Methods for Obtaining Information on the Alignment and Orientation Parameters of Atoms or Ions Excited in Specific Collisions

So far, four experimental methods have been developed for studies of ion–atom or atom–atom collisions with the purpose of obtaining maximum information about collisional-induced magnetic-substate populations as expressed in terms of the alignment and orientation parameters. The four methods are as follows:

a. The particle–photon or the particle–electron angular-correlation technique.
b. The polarized-photon, scattered-particle coincidence technique.
c. The polarized-photon quasi-coincidence technique.
d. The selective laser-excitation technique.

(a) *The particle-photon angular-correlation technique* was first introduced by Kleinpoppen and co-workers[5] in 1973 to study the He($1s2p\ ^1P$) state formed in $e + $ He($1s^2$) collisions. Shortly after, Jaecks et al.[6] applied (b) *the polarized-photon, scattered-particle coincidence technique* to study the ($1s3p\ ^3P$) state formed in He$^+$–He collisions. Since then, these two methods have been applied to study a number of ion– and atom–atom collisions.

In a brief outline, a particle–photon angular-correlation experiment or a polarized-photon, scattered-particle coincidence experiment is carried out as follows: A well-collimated beam of particles (ions or atoms), which are energy- and mass-analyzed, are passed through a target under single-collision conditions. Inelastically scattered projectile particles, which have excited target atoms or have undergone collisional excitation, are detected in coincidence with photons emitted in the deexcitation process.

In the polarized-photon, scattered-particle coincidence method, polarized photons are usually detected in a direction perpendicular to the scattering plane, but to obtain complete information, it is necessary also to measure in a direction perpendicular to the beam direction within the collision plane. The polarized-photon, scattered-particle coincidence method is well suited at wavelengths above ~ 1800 Å, for which linear polarizers and $\lambda/4$ plates of acceptable area and transmission are commercially available.

In the particle–photon angular-correlation method, the angular distribution of the emitted photons (in coincidence with projectiles scattered a given angle) is normally measured by rotating the photon detector in the scattering plane. The particle–photon angular-correlation technique has mainly been used at photon wavelengths shorter than ~1800 Å. If the collisional-excited state decays through autoionization rather than by photon emission, which is the case for doubly excited states, e.g., the $2p^{2\ 1}D$ state created in He^+–He collisions,[15] the photon detector is replaced by an electron detector to obtain the angular distribution of the autoionizing electrons.

Coincidence experiments utilize the fixed time difference between the detection of particles or radiation from the same collision to unravel selected collisions from the totality of collisions. The time difference depends on the experimental setup, in the present context, mainly on the flight path from the collision cell to the scattered-particle detector. It is usually in the 0.1–10-μsec region, and typical time resolutions range from 10 to 100 nsec, the poorer resolutions being set by the flight time of projectiles through the target region.

Real coincidences are commonly identified by applying time-to-pulse-height converters. These convert differences in detection times to a spectrum of pulse heights, which is analyzed by conventional pulse-height analyzers. Figure 4 shows an example of a coincidence spectrum from a polarized-photon, scattered-particle coincidence experiment.

The counting rates of interest for an electron or photon, scattered-particle coincidence experiment are the following:

The singles rate for radiation (electrons or photons):

$$R_{\text{rad}} = I \left(\frac{d\sigma}{d\Omega}\right)_{\text{rad}} \varepsilon_{\text{rad}} \int N(z)\Omega_{\text{rad}}(z)\, dz \qquad (21)$$

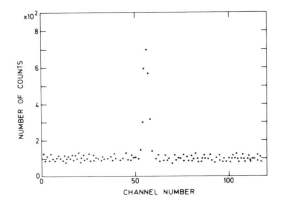

Figure 4. Polarized-photon, scattered-ion coincidence signal from the $3^2S \to 3^2P \to 3^2S$ excitation–deexcitation of $^{24}Mg^+$ ions in Mg^+–Ar collisions.

and for scattered particles:

$$R_{\text{part}} = I\left(\frac{d\sigma}{d\Omega}\right)_{\text{sc}} \Omega_{\text{part}}\varepsilon_{\text{part}} \int N(z)\, dz \tag{22}$$

where I is the beam intensity $(\sec^{-1})$, $N(z)$ is the target density along the beam direction, Ω and ε denote solid angles and efficiencies of the detectors, $(d\sigma/d\Omega)_{\text{rad}}$ is the total cross section per steradian for exciting the radiation being studied, and $(d\sigma/d\Omega)_{\text{sc}}$ is the total differential-scattering cross section.

The coincidence rate is

$$R_{\text{coinc}} = I\frac{d^2\sigma}{d\Omega_{\text{sc}}\, d\Omega_{\text{rad}}}\varepsilon_{\text{part}}\varepsilon_{\text{rad}}\Omega_{\text{part}} \int N(z)\Omega_{\text{rad}}(z)\, dz \tag{23}$$

where it is assumed that the target length seen by the radiation detector is part of the target length seen by the scattered-particle detector, and where $d^2\sigma/(d\Omega_{\text{sc}}\, d\Omega_{\text{rad}})$ is the double differential cross section for scattering *and* excitation of the radiation in question.

The random rate is

$$R_{\text{random}} = R_{\text{part}}R_{\text{rad}}\Delta\tau \tag{24}$$

where $\Delta\tau$ is an appropriate time interval. The total random rate is found if $\Delta\tau$ is the total time range scanned by the time-to-pulse-height converter. The random rate appropriate for an estimate of the real-to-random ratio (i.e., peak-to-background ratio) is found if $\Delta\tau$ is the time resolution of the experiment as given by, say, the FWHM of the coincidence peak.

One normally forms the ratio between R_{coinc} and one of the singles rates,

$$\frac{R_{\text{coinc}}}{R_{\text{rad}}} = \frac{d^2\sigma/(d\Omega_{\text{sc}}\, d\Omega_{\text{rad}})}{(d\sigma/d\Omega)_{\text{rad}}}\varepsilon_{\text{part}}\Omega_{\text{part}} \tag{25}$$

or

$$\frac{R_{\text{coinc}}}{R_{\text{part}}} = \frac{dP(\theta)}{d\Omega_{\text{rad}}}\varepsilon_{\text{rad}}\frac{\int N(z)\Omega_{\text{rad}}(z)\, dz}{\int N(z)\, dz} \tag{26}$$

The first form measures the double-differential scattering *and* excitation cross section. An absolute measurement requires knowledge of $(d\sigma/d\Omega)_{\text{rad}}$, $\varepsilon_{\text{part}}$, and Ω_{part}. The second form measures the differential excitation probability,

$$\frac{dP(\theta)}{d\Omega_{\text{rad}}} = \frac{d^2\sigma/(d\Omega_{\text{sc}}\, d\Omega_{\text{rad}})}{(d\sigma/d\Omega)_{\text{sc}}} \tag{27}$$

as a function of scattering angle θ. An absolute measurement of the excitation probability $P(\theta)$ requires knowledge of ε_{rad}, $\Omega_{rad}(z)$, $N(z)$, and the angular distribution of the radiation in question. The unknown efficiency and solid-angle factors are normally found by calibration procedures, using known probabilities or total cross sections. The angular distribution of the radiation from individual collisions is inferred from the measurements of the alignment parameters.

Coincidence measurements are normally time-consuming because each registered event has to be observed by two independent detectors, each covering only a finite solid angle. The type of measurement discussed here is slow in particular because both the scattering *direction* (i.e., the polar- as well as the azimuthal-scattering angle) and the direction of electron or photon emission must be defined experimentally. This naturally limits the solid-angle factors. The time needed to obtain, say, 10% statistics on real coincidences ranges from a few minutes for excitations occurring with large probabilities in the individual collisions to a few days for the least probable excitations studied so far.

To limit the time needed for the coincidence measurements, the scattered-particle detector can be replaced by a position-sensitive detector, which makes it possible to perform coincidence experiments at several scattering angles simultaneously.[17]

Apart from well-known sources of systematic error such as slit-scattering, departure from single-collision conditions, dead-time of the discriminators, and the time-to-pulse-height converter, there is one, which is particular to coincidence measurements with ion beams. This is time structure in the random coincidences, which may result from plasma oscillations of the discharge in the ion source.[18] If the intensity of the particle beam from the accelerator exhibits high-frequency time variations due to such plasma oscillations instead of being strictly a dc beam, the probability will be enhanced of obtaining random events at time delays corresponding to the difference in the flight time of the particles from the gas cell to the detectors plus or minus an integer number of oscillation periods. The beam behaves as if it were pulsed with a distribution of repetition frequencies. If this distribution is narrow, regular oscillations are observed in the random spectrum.[18] If, on the other hand, the distribution is broad, a flat spectrum results with only one peak of random events positioned at the same time correlation as the true coincidences. Figure 5 shows a very pronounced example of this effect. It is possible to test the operation of the ion-source plasma by measuring uncorrelated coincidences between two particle detectors placed at two different scattering angles[19] and if necessary to adjust the ion-source parameters to eliminate the effect. Time structure in the beam will lead to artificially high excitation probabilities.

(c) *The polarized-photon quasi-coincidence technique*, which was reported by Wittman and Andrä[20] in 1978, is limited to the study of excited

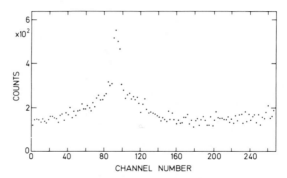

Figure 5. Coincidence signal perturbed by plasma oscillations in the ion source.

states of the projectile, but for such studies offer advantages with respect to the signal strength by several orders of magnitude. The technique is based upon the selection of a certain impact parameter by observing the light emitted from projectile ions (atoms) scattered in a given direction relative to the axis of the central beam, as shown schematically in Figure 6. This figure also indicates the drawback of the technique, the long flight path from the scattering cell to the detection region. Thus only excited states with rather long lifetimes can be studied by this technique and preferably at medium to high energies of the projectile.

(c) *The selective laser-excitation method* was first demonstrated by Hertel and Stoll.[8] It is based upon the use of selective magnetic-substate preparation of an atom $B^*(nl)$ colliding with another atom A. The selective magnetic-substate preparation can be obtained by means of optical pumping, using a single-mode tunable dye laser. Since the tunable dye lasers so

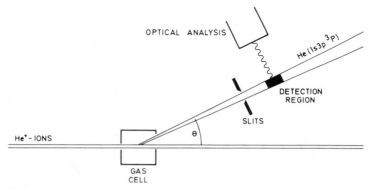

Figure 6. Schematic view of the experimental arrangement for polarized-photon, quasi coincidence measurements.

far cover only a limited wavelength region, the application of the method is somewhat limited. All the experiments reported until now have started by exciting the Na $3^2P_{3/2}$ ($F = 3$) state. The first experiments were concerned with electron and molecular scattering, but the technique could equally well be applied for atomic or ionic scattering.

The experimental setup is a conventional crossed-beam arrangement with access for laser excitation in the interaction zone. The scattering angle may be varied as well as the angle of incidence of the exciting photon beam. The connection between measurable quantities (Stokes parameters) and the primary parameters describing the dynamics of the inelastic process being studied (alignment and orientation parameters) was discussed in Section 2.3.

When applied to collisional deexcitation to the ground state, the laser-excitation method is equivalent to the polarized-photon, scattered-particle coincidence method for the inverse process in terms of the information that can be obtained. In practice, however, both methods have their specific advantages and shortcomings. The polarized-photon, scattered-particle coincidence method is a rather general method, limited only by the availability of suitable optical polarizers and sometimes by cascades from higher-excited states, but it is time-consuming because it requires coincidence technique. The laser-excitation method is selective and avoids the coincidence technique, but in practice it has been limited to atomic alkali–metal targets.

4. Results of Experiments and Numerical Calculations

The experimental results reported so far of particle–photon (or electron) angular correlations for binary atomic collisions are mainly concerned with the following two types of systems;

i. Quasi-one-electron systems, i.e., collison systems characterized by the presence of one valence electron outside two closed shells such as the alkali-atom and the alkaline-ion on rare-gas systems (e.g., Li–He, Be$^+$–Ne, etc.).

ii. The rare-gas ion on rare-gas systems, in particular the He$^+$–He system.

Prior to the particle-photon (or electron) angular-correlation studies of these selected systems, a relatively good knowledge concerning their excitation mechanisms had been obtained from measurements of total and/or differential cross sections and from total polarization data.

4.1. Quasi-One-Electron Systems

4.1.1. Introduction

Detailed experimental and theoretical studies of a large number of quasi-one-electron collision systems[21] have shown that these systems exhibit a rather simple behavior. The electrons of the alkali core and the rare-gas atom are much more tightly bound than the valence electron. At impact parameters large enough for the two cores not to interact significantly, the collision can to a good approximation be treated as a three-body system consisting of the active electron and the two cores. Besides ionization, the alkali $ns \rightarrow np$ resonance excitation is the dominating inelastic process in these distant collisions, which only to a minor degree excite higher-lying levels of the alkali atom. The core electrons are left unexcited. The $ns \rightarrow np$ excitation is induced by the direct interaction between the valence electron and the rare-gas atom and is qualitatively governed by the Massey criterion.

According to this, the cross section for an inelastic collision process goes through a maximum when the collision time a/v is at resonance with the transition frequency $\hbar/|\Delta E|$ for the given inelastic process, characterized by the energy defect ΔE and the interaction length a, which for the present system is of the order of the sum of the radii of the alkali atom and the noble-gas atom. The collision velocity is v. The Massey criterion can also be expressed by the relation $\Delta E a/\hbar v_m = \pi$, where v_m is the collision velocity, at which a maximum of the total cross section is expected. Theoretically, the excitation induced by the direct noble-gas-atom, valence-electron interaction has been treated in the quasi-one-electron model proposed by Nielsen and Dahler.[22] In this model, the scattering wave function is expanded on a limited set of atomic eigenstates of the quasi-one-electron projectile. The nuclear motion is treated classically and a straight-line, constant-velocity trajectory adopted. The model results in coupled equations for the time-dependent expansion coefficients (excitation amplitudes). The essential element of the theory is the noble-gas atom, valence-electron interaction, for which various model potentials have been used (frozen Hartree–Fock, Bottcher-type, or Bayliss–Gombas). For a given impact parameter, the model predicts oscillating excitation probabilities with maxima at $1/v_m = (\hbar/\Delta Ea)n\pi$, $n = 1, 2, \ldots$. The Massey criterion ($n = 1$) predicts the collision velocity, at which the most prominent or, in some cases, the only maximum of the total cross section occurs.

For impact parameters so small that the two cores interact significantly, resulting in large scattering angles, the systems may behave in a more complex way than outlined above since now also the core electrons are activated and may even be excited as a result of the collision. The so-called molecular orbital (MO) model adequately describes this many-electron problem.[23] The small impact-parameter collisions are important at low collision velocities [well below $v_m = |\Delta E| a / (\hbar \pi)$], whereas they play a less significant role at higher velocities. So far, only the specific $n^2S \to n^2P \to n^2S$ resonance ^2P state excitation–deexcitation process has been studied for this type of system.[19,24–27]

4.1.2. P → S Transitions

In a fast binary collision for which the spin-dependent forces can be neglected, as is the case for the lighter alkali atom or alkaline-ion, rare-gas systems (Li, Na, Be$^+$, Mg$^+$–He, Ne, Ar), a fully coherent excitation of the resonant p state can be described in terms of the orbital part of the wave function, expressed by the $|LM_L\rangle$ notation as

$$|\psi\rangle = a_0|10\rangle + a_1|11\rangle + a_{-1}|1-1\rangle \tag{28}$$

with $a_1 = -a_{-1}$ due to reflection symmetry in the collision plane. Together with the excitation probability $\mathscr{P} = |a_0|^2 + 2|a_1|^2$, the excited resonance state is commonly described by the two coherence parameters λ and χ, with $\lambda = |a_0|^2/\mathscr{P}$ and the phase angle χ defined by $a_0 = |a_0|$ and $a_1 = |a_1|e^{ix}$. Even though this parametrization, which is possible only for fully coherent states, has become a convention in the current literature for ion- and atom–atom collisions, it may not necessarily be a more convenient choice than the general alignment and orientation parameters used in Section 2.

The Stokes parameters for a $P \to S$ transition, for which the initial state is a pure state [Eq. (28)], can be derived from the formula given in Section 2

$$P_1 = 2\lambda - 1, \qquad P_2 = -2[\lambda(1-\lambda)]^{1/2}\cos\chi$$
$$P_3 = 2[\lambda(1-\lambda)]^{1/2}\sin\chi, \qquad P_4 = 1 \tag{29}$$

These formulas are modified if fine or hyperfine structure of the observed excited state is unresolved experimentally. In the present situation (an unresolved ^2P term), fine-structure effects will modify P_1, P_2, and P_4 to be

$$\tfrac{7}{3}P_1 = 2\lambda - 1, \qquad \tfrac{7}{3}P_2 = -2[\lambda(1-\lambda)]^{1/2}\cos\chi, \qquad \tfrac{7}{3}P_4 = 7\lambda/(4+3\lambda) \tag{30}$$

while P_3 is left unchanged. Hyperfine-structure effects may reduce the Stokes parameters still further.

The λ and χ parameters describing a pure p state are uniquely determined by P_1, P_2, and P_3. These three measurements also serve to check whether the observed excited state would have been a pure state in the absence of the spin–orbit coupling because, in this case, the degree of polarization P {given by $P = [(\frac{7}{3}P_1)^2 + (\frac{7}{3}P_2)^2 + P_3^2]^{1/2}$} is unity.

The observed excited state could be an incoherent, mixed state for other reasons than the spin–orbit coupling. Incoherence may also be introduced by cascades from higher-lying states or by contributions to the excitation from several unresolved inelastic channels leaving the other collision partner in some unspecified state of excitation. Such effects may in some situations prove interesting. Otherwise, they can in principle be eliminated experimentally by energy-loss analysis of the scattered particle, but this has not been done in practice. Disturbing cascade effects have been treated by means of isotropic cascade models which are based on the assumption that the cascade contributions from ^2S and ^2D levels act effectively as an isotropic, unpolarized background.[24]

4.1.3. Coherence in the Mg$^+$–He, Ne, Ar Systems

Figure 7 shows probabilities for excitation of the 3^2P state of Mg II in Mg$^+$ + rare-gas atom collisions[19] as a function of impact parameter. The degree of polarization of the 3^2P–3^2S optical emission is also given.

Although the experiment specifies the direction of motion of the scattered projectiles, the excitation of the Mg II 3^2P state and the direction

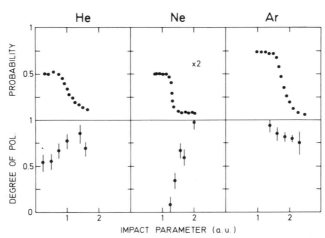

Figure 7. Excitation probability and degree of polarization for 15-keV Mg$^+$–He, Ne, Ar collisions as a function of impact parameter.

of the subsequent photon emission, it does not completely define the collision. The Mg II 3^2P state could be populated through cascades from the Mg II 4^2S, the Mg II 3^2D states, or from even higher states. It could also be excited simultaneously with the excitation of a target level. Since the amplitudes for these different channels, which all contribute to the observed photon emission, add incoherently, one would expect to observe a mixed state (i.e., a statistical mixture of the pure states corresponding to the different unresolved excitation channels) or in general, a rather low degree of polarization for the photon emission.

As seen in Figure 7, the observed degree of polarization turns out to be high for large impact-parameter collisions. One can conclude from this that essentially only one channel contributes to the excitation, i.e., the direct excitation of Mg II 3^2P with no target excitation. This is true also at small impact parameters for the asymmetric collision Mg$^+$ + Ar, but for the quasisymmetric system Mg$^+$ + Ne, the degree of polarization drops sharply when molecular excitations set in at impact parameters smaller than about 1.5 a.u.[19] Simultaneous target-excitation channels consequently contribute significantly to the Mg II 3^2P excitation for these particular collisions.

4.1.4. Direct Excitations. Comparisons with Theory: Be$^+$–Ne Collisions

By means of the quasi-one-electron model,[22] it has been possible quantitatively to reproduce the experimental total emission cross sections and polarizations for the lighter collision systems (Li–He, Be$^+$–He), and qualitatively the trends in the experimental data for the heavier system.[21] Measurements of the coherence parameters λ and χ for the projectile resonance state, however, constitute a much more sensitive test of the theoretical predictions of scattering amplitudes for the direct excitation mechanism (see Section 4.1.1) than do the total or the differential cross sections alone.

A comparison between experimental and theoretical coherence parameters has been made for the Be$^+$–He, Ne systems.[24,25] The theory predicts that at high energies and/or large impact parameters, λ will tend to zero and χ to $\pi/2$. For finite energies or impact parameters, the λ values should exhibit pronounced oscillations, reflecting the oscillations of the size of the excitation amplitudes (Section 4.1.1) for the magnetic substates, which, in general, are out of phase. The maxima ($\lambda = 1$) and minima ($\lambda = 0$) correlate with rapid changes (phase jumps $|\Delta\chi| \simeq \pi$) of the phase angle χ.

Figure 8 shows a comparison between experimental and calculated values for the total excitation probability $\mathscr{P}$ for the Be$^+(2s \to 2p)$ process together with the λ and χ parameters for the collisional-excited Be$^+$-resonance state formed in Be$^+$–Ne collisions. The general characteristics

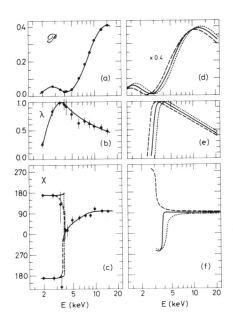

Figure 8. (a)–(c): Experimental emission probabilities $\mathscr{P}$ and coherence parameters λ and χ. (d)–(f): Theoretical $\mathscr{P}$, λ, and χ curves for $Be^+(2s \to 2p)$ excitation evaluated at $b = 1.5$ a.u. (dotted line), 1.6 a.u. (solid line), and 1.7 a.u. (dashed line) ($\mathscr{P}$ has been multiplied by 0.4).

of the experimental and theoretical curves are strikingly alike in most respects (same shape of $\mathscr{P}$ and λ, $|\Delta\chi| = \pi$ and $\chi \to \pi/2$ at large energies), but it should be noted that the theoretical and experimental phase angles jump from different values as the impact parameter is decreased.

The deviation between theory and experiment is seen more clearly in Figure 9, which shows spatial angular distributions of the excited $Be^+(2p)$ state at selected energies. The theoretical and experimental developments of the charge cloud are qualitatively different. Theory predicts an isotropic charge distribution at high energies, corresponding to full circular polarization of the emitted light. The charge distribution should then stretch along the beam direction as the energy is lowered and reach a maximum of polarization along this direction just prior to the rapid phase change. After that it should begin to depolarize while still being symmetric with respect to the beam axis and become isotropic once more when $\lambda = 0.5$, but the associated electronic angular momentum has changed sign. At still lower energies, the charge distribution stretches along an axis in the collision plane but perpendicular to the beam direction.

The experimental development of the charge cloud is different. The experimental charge cloud coincides with the theoretical one at high energies, but instead of stretching along a fixed axis, it stretches along a rotating axis as the energy is lowered, to coincide once more with the theoretical

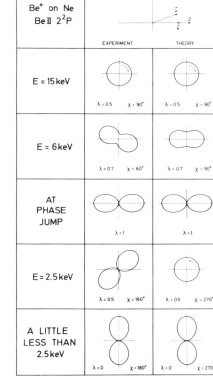

Figure 9. Polar diagrams of experimental and theoretical angular-charge distributions within the collision plane for Be$^+(2p)$ excited in Be$^+$–Ne collisions. The charge distributions are corrected for depolarization due to fine and hyperfine structure. The different collision energies E are chosen to represent typical situations from Figure 8. The arrows on the isotropic distributions indicate the sign of the associated electronic angular momentum. On top of the figure, the coordinate system used is indicated. The incoming and the outgoing wave vectors are denoted **k** and **k**′, respectively.

distribution just before the rapid phase change takes place. As the energy is further lowered, the experimental charge distribution keeps on rotating while still being strongly polarized until once more approaching the theoretical distribution as $\lambda \to 0$. The discrepancy between theory and experiment is most striking in the 2–3-keV energy region where the experimental charge distribution is strongly polarized (and the emitted light almost fully linearly polarized). It is also worthwhile noting that the rapid change of the phase angle near 4 keV is not reflected in the charge distribution, which in this region depends almost exclusively on λ.

Until now, the experiments[24] on Be$^+$–He have yielded no evidence of a similar dramatic energy and/or impact-parameter dependence of λ and χ. This may be due to a lack of overlap between the large impact parameters, at which these effects occur, and the experimentally accessible impact parameters.

4.1.5. Molecular Excitations at Localized Curve Crossings. Comparison with Theory: K–He and Mg$^+$–Ar Collisions

The molecular-excitation mechanism dominates the resonance-line excitation in most quasi-one-electron systems at small collision velocities and impact parameters. In *quasi-symmetric* systems (such as Mg$^+$ on Ne),[19] the molecular excitation of the one-electron resonance state is associated with a number of important noble-gas atom excitation channels. Experiments such as those carried out in the last few years, which do not determine the final state of the noble-gas atom, therefore fail to isolate the coherent parts of the excitation process. The observed mixed states turn out to be only weakly polarized (see Figure 7), and they depend in a complicated way on the various noble-gas atom excitations. The molecular-excitation mechanism is, on the other hand, highly selective in *asymmetric* systems (such as Mg$^+$ on Ar), for which essentially only the one channel leaving the noble-gas atom in the ground state contributes to the resonance-line excitation. For these systems, a pure state is therefore observed even in experiments which do not detect specifically the final state of the noble-gas atom.

To qualitatively understand the different behavior of the quasisymmetric and the asymmetric systems, it is useful to view a simplified molecular-state diagram (Figure 10), which in a schematic way illustrates the molecular-excitation mechanism. The excitation of the resonance state of the one-electron atom A*(np) takes place via intermediate states, which correlate with excited states B* of the noble-gas atom. For the *quasi-symmetric* systems, the relevant curve crossings C_1 and C_2 are highly diabatic (due to the quasisymmetry). Hence the probability for excitation at C_1 is small. In case of excitation at C_1, however, the probability for exciting the noble-gas atom B* simultaneously with the resonance state A*(np) is large since the crossings at C_2 are nearly diabatic. This qualitatively explains the relatively small excitation probabilities for the quasisymmetric systems and the strong simultaneous noble-gas-atom excitations.[21] In the asymmetric systems, the couplings at curve crossings C_1 and C_2 are

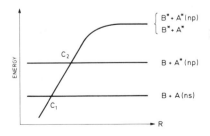

Figure 10. Simplified molecular-state diagram for quasi-one-electron systems. A(ns), A*(np), and A* denote the one-electron-ion and/or atom in its ground state, the excited-resonance state, and an arbitrary state of excitation, respectively. B and B* denote the rare-gas atom in ground state and some excited state, respectively.

stronger. This leads to a larger excitation probability at C_1 and to the strong selection of the $A^*(np) + B$ channel at the adiabatic crossings C_2.

The intermediate molecular states, which connect C_1 and C_2, involve the temporary excitation of electrons from both cores. The molecular excitation of even the $A^*(np) + B$ state (i.e., single-electron excitation of the quasi-one-electron ion) is therefore beyond the simple picture of the direct-excitation mechanism, in which only one electron is active and the cores are frozen or, at most, polarized during the collision.

(a) *The K–He System.* Differential cross sections for the excitation of the K I 4^2P resonance level in very slow collisions between accelerated, neutral potassium and noble-gas atoms was studied by Zehnle *et al.*[26] Apparent angular thresholds were observed for the heavy rare-gas targets. This points to excitation at localized curve crossings. For the helium target, the differential cross sections gave less definitive information.

In a subsequent study, the same authors[26] measured the alignment and orientation of the K I 4^2P state excited in the K + He collision at 93 eV center-of-mass energy. The results are shown in Figure 11. While the low collision velocity calls for a description of the excitation process in terms of molecular states, it is still not clear whether the K I 4^2P excitation

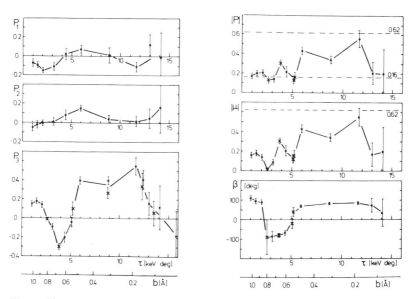

Figure 11. The Stokes parameters P_1, P_2, and P_3, the vector polarization P, the degree of coherence $|\mu| = (P_2 + iP_3)/(1 - P_1^2)^{1/2}$, and phase β of the complex correlation factor μ as a function of reduced scattering angle τ (keV degree) for K($4s \rightarrow 4p$) excitation in K–He collisions at 93 keV center-of-mass energy (Ref. 26).

in K + He collisions takes place at localized curve crossings or over an extended range of internuclear distances. The high degree of coherence of the observed excited states indicate that mainly one inelastic channel dominates the excitation process. The strong variation with impact parameter of the orientation (P_3) of the excited state, however, seems to exclude the simplest excitation mechanisms.

(b) *The Mg$^+$–Ar System.* Total excitation probabilities as well as the orientation and alignment of the Mg II 3^2P state formed in collisions between accelerated Mg$^+$ ions and argon atoms were recently measured[27] at energies and scattering angles, for which the molecular-excitation mechanism at localized curve crossings dominates. The study was in particular aimed at a mapping of the excitation process in the apparent angular threshold region.

The experimental results are shown in Figure 12, which in three columns show excitation probabilities, the components of the relative Stokes vector corrected for depolarization due to the fine structure of the ^2P term, the size of the relative Stokes vector, and finally the parameters λ and χ describing the relative amplitudes for populating the magnetic substates of the excited p state. These parameters were derived from the components of the relative Stokes vector $\mathbf{P}/|\mathbf{P}|$. The three columns illustrate the angular dependence (left column) and the energy dependence at constant scattering angle (middle column) and impact parameter (right column). The three cuts in the (E, θ) plane represented in Figure 12 go through the same point, $(E, \theta) = (3 \text{ keV}, \theta = 3.33°)$, corresponding to the impact parameter $b = 1.81$ a.u. The size of this impact parameter is close to the sum of the two core radii, i.e., 1.73 a.u., which is the internuclear distance at which the localized excitation is expected to set in.

The angular (or impact-parameter) dependence of the excitation probability does indeed exhibit a good correlation between the sharp rise of the excitation probability and this impact parameter (note that the vertical scale of the excitation probability is logarithmic and covers three orders of magnitude).

The degree of coherence of the excited states observed by the experiment, as measured by $P \equiv |\mathbf{P}|$, is seen to be large. The deviation of the P values from unity is to some extent due to the finite experimental definition of the collision plane, the finite solid angle of the photon counter, and the leaks of the optical polarizers. The large P values indicate that the observed excited p state to a good approximation is a pure state described by the relative size λ and the phase difference χ between the excitation amplitudes for the $m_l = 0$ and the $m_l = \pm 1$ magnetic substates.

Small variations of the λ and χ parameters are noted (the rise of λ in the deep angular-threshold region and the rise of χ at high energies), but the main result is a rather uniform and small value of λ corresponding

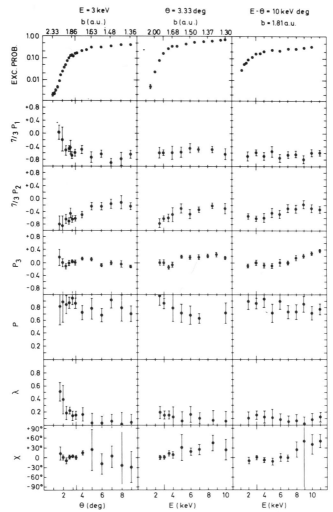

Figure 12. The parameters describing the excited $Mg^+(3p)$ state formed in Mg^+ on Ar collisions as functions of scattering angle θ and kinetic energy E. Primary experimental data: Excitation probabilities; components of the relative Stokes vector $\mathbf{P} = (\frac{7}{3}P_1, \frac{7}{3}P_2, P_3)$; the size of $\mathbf{P}$, $P = |\mathbf{P}|$. Data derived from $\mathbf{P}$: The parameters λ and χ describing the relative size and phase difference of the excitation amplitudes for the magnetic substates of the orbital angular momentum of the $Mg^+(3p)$ state. The points common to the three sets of data are indicated on the horizontal scales by enlarged vertical bars. The error bars show one standard deviation, calculated from the coincidence counting statistics alone.

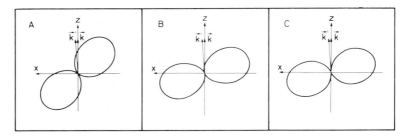

Figure 13. Polar diagrams showing the angular part, within the collision plane, of the observed excited clouds formed in three typical collisions (immediately after the collision or in the absence of spin–orbit coupling) A: $(\lambda, \theta) = (0.5, 0°)$; B: $(\lambda, \theta) = (0.1, 0°)$; C: $(\lambda, \theta) = (0.1, 30°)$. **k** and **k'** represent the incoming and the outgoing momenta of the Mg$^+$ ion respectively.

to the predominant population of the $m_l = \pm 1$ substates and a phase difference χ close to zero except at high energies, for which λ is very small, leaving the value of χ somewhat inconsequential. The overall result (see Figure 13) is therefore the excitation of a highly aligned state (i.e., a strongly polarized charge cloud) with the symmetry axis almost perpendicular to the beam direction within the collision plane. The symmetry axis turns a little as λ approaches 0.5, and at high energies, the excited state is also somewhat oriented (i.e., the expectation value of the electronic angular momentum is different from zero and points to the negative $\mathbf{k} \times \mathbf{k'}$ direction, where $\mathbf{k}$ and $\mathbf{k'}$ are the incoming and outgoing wave vectors of the projectile).

In terms of the simplified excitation mechanism of Figure 10, the data show that the crossing at C_2 is nearly adiabatic ($P \simeq 1$), and that the Σ states populated by the radial couplings at C_1 are strongly rotationally coupled to adjacent Π states to form a space-fixed state rather than a state which is fixed to the internuclear axis ($\lambda \simeq 0$).

Russek et al.[28] recently proposed a model for the population of the magnetic substates of atoms excited in atomic collisions at localized molecular curve crossings. According to this model, the excitation of the Mg II 3^2P state in Mg$^+$ + Ar collisions at impact parameters close to the crossing distance should result in a strong right-hand circular polarization of the subsequent photon emission. This effect is not seen experimentally at 3 keV (Figure 12). Earlier data[19] for the same system at 7 and 15 keV show high degrees of right-hand circular polarization at large impact parameters, but this agreement with the model is probably fortuitous because the direct-excitation mechanism may be important in this region.

4.2. He^+–He Collisions

4.2.1. Introduction

The interpretation of excitation or charge-exchange processes in collisions between rare-gas ions and atoms such as the He^+–He collision system at energies up to some keV is based upon a molecular description of the colliding-particle system. In most cases, the primary excitation mechanism involves an interaction between the ground state and excited molecular states, which occurs at a rather small internuclear distance. Many coherent scattering amplitudes, corresponding to several experimentally indistinguishable reaction channels, are in general produced at the same time by the collision. In spite of the fact that the primary excitation mechanisms now are rather well understood, it is still difficult to determine the excitation to specific atomic states since secondary interactions may blur the primary process. For the particular case of ^1P or ^3P states in helium generated either by excitation[7,29] or charge exchange[6,30] in He^+–He collisions, it is possible to gain more insight into the relative probabilities for population of a given P state via Σ or Π molecular states by measurements of the differential polarization of the coincident light or, alternatively, its angular distribution. In the studies described in the following, it is assumed that the observed excited states are coherent, but this is, in fact, not proved by measurements.

4.2.2. $He(3\,^3P)$ Formed by Charge Exchange in He^+–He Collisions

The pioneering work[6] performed with the polarized-photon, scattered-ion coincidence technique determined the magnetic substate population of the $He(3\,^3P)$ state formed in He^+–He charge-exchange collisions at 3 keV. The probabilities for exciting the $m_l = \pm 1$ and $m_l = 0$ substates as a function of scattering angle (1° to 2°) are shown in Figure 14 together with the phase angle χ. The $m_l = \pm 1$ population is nearly constant in the investigated scattering region, whereas $m_l = 0$ exhibits a maximum near 1.6°, and the phase angle χ appears to be nearly constant (90°).

The experimental data are used to distinguish between two excitation mechanisms within the electron-promotion model. The $He(3\,^3P)$ state can be formed by either the radial coupling at the Landau–Zener crossing between the $1s\sigma_g(2p\sigma_u)^2\,^2\Sigma_g$ and the $(1s\sigma_g)^24d\sigma_g\,^2\Sigma_g$ molecular-energy curves of He_2^+ at approximately 1.3 a.u. or by exciting the $1s\sigma_g2p\sigma_u2p\pi_u\,^2\Pi_g$ state by a rotational coupling at much smaller internuclear distances (Figure 15). To excite the $3\,^3P$ state in helium, the $^2\Pi_g$ state must couple at a large internuclear distance with states which dissociate to $1s3p\,^3P$ of helium.

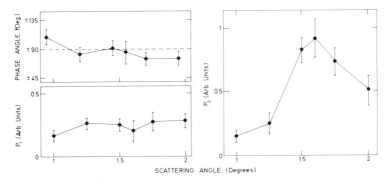

Figure 14. For the charge-exchange reaction $He^+ + He \rightarrow He(3\,^3P) + He^+$, the relative probabilities P_0 and P_1 for exciting the $m_l = 0$ and ± 1 magnetic substate are shown together with the relative phase angle between those two amplitudes.

The authors[6,30] propose a model for the former mechanism, which can account for the observed nearly constant χ value, whereas variations in χ are expected if other mechanisms contribute to the excitation. According to this model, the phase difference should be independent of any relation between impact parameter and scattering angle and any theoretical input concerning the detail of the potential energy, including the actual internuclear distance at which the crossing occurs. Conversely, the $|a_0|^2$ and $|a_1|^2$ values are very sensitive to these parameters, but a detailed comparison between experimental and theoretical data is still awaiting information about the requisite potential curves.

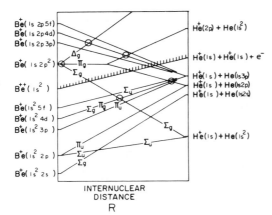

Figure 15. Correlation diagram for He^+–He.

4.2.3. Excitation of 2 ^1P and 3 ^3P of Helium by He^+–He Impact at 100–200 eV

Differential-scattering studies of low-energy He^+–He collisions leading to $2p$ ^1P excitation of the helium atom have shown[7] that at a few hundred eV, the primary excitation essentially occurs via radial Σ_g–Σ_g couplings and that molecular states with ungerade symmetry can be excluded from the excitation process. Angular-correlation measurements on the 2 ^1P–1 ^1S resonance line ($\lambda = 584$ Å)[29] have, however, indicated that this general picture is oversimplified. Figure 16 shows angular-correlation curves for an impact energy of 150 eV and a scattering angle of 13.5°. Within the experimental error, a maximum along the final molecular axis is found, and it is estimated that $|a_\Sigma|^2 \leq 10^{-1}|a_\Pi|^2$, i.e., the final molecular state is mainly a Π state. The experimental data for the He(2 ^1P) state indicate that the originally excited Σ_g-state potential curve is linked to the energetically very close-lying Π_g curve, also leading to the (2 ^1P), through a secondary Σ_g–Π_g coupling.

Contrary to the 2 ^1P state, the 3 ^3P state exhibits a clear preference for $m_l = 0$ population, meaning that the molecular state, which finally leads to the (3 ^3P) has Σ symmetry. This Σ symmetry makes the 3 ^3P excitation mechanism much more complex than for the equivalent singlet excitation. A supplementary primary coupling and many other secondary Σ_g–Σ_g radial couplings at large distances should be added to the model for the excitation of singlet states to account for the triplet excitation.[29] The present examples indicate that even collision systems, which could be expected to be rather simple, can exhibit very complex excitation mechanisms.

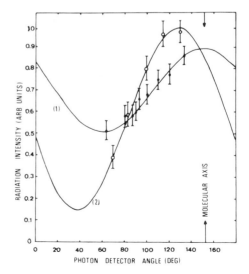

Figure 16. Ion–photon angular-correlation measurements for the He^+ + He → He^+ + He(2 ^1P) excitation at an incident energy of 150 eV and 13.5° scattering angle. The two data sets represent two different azimuthal angular spreads, filled circles 90°, open circles 30°. Curves (1) and (2) are least-squares fits to the data points. The incident beam direction is at 0°, the final molecular axis at 153°.

Table 1

State		$m = 0$	$m = 1$	$m = 2$
^1D	$\lvert a_m \rvert$	66	14.7	3.3
	χ_m / π	0	-0.5	0.05

4.2.4. He$(2p^2)$ ^1D Studied by Angular Correlation between Autoionizing Electrons and Scattered Ions in 2-keV He$^+$–He Collisions

Information about the spatial orientation of doubly excited states in helium, such as He$(2p^2)$ ^1D or He$(2s\,2p)$ ^1P must be extracted from electron scattered-ion coincidence measurements since the lifetimes of these states with respect to autoionization are too short to allow an application of the similar photon technique. It is natural to express the complex population amplitudes for the magnetic substates by using the symmetry axis of the electron distribution as a quantization axis. Studies[15] of the He$(2p^2)$ ^1D state created in 2-keV He$^+$–He collisions have shown that this axis, $\bar{z}$, nearly coincides with the direction of the internuclear line at the distance of closest approach. The size of the population amplitudes a_m and the phase parameters χ_m for the He$(2p^2)$ ^1D magnetic substates referring to the above quantization axis are given in Table 1.

According to the table, the angular distribution of the ejected electrons is similar to the d_{z^2} electron orbital. The authors[15] have described the excited $(2p^2)$ ^1D state as a coherent superposition of two pure $m_l = 0$ substates with quantization along two axes z_1 and z_2, which are oriented symmetrically with respect to the quantization axis $\bar{z}$ (Figure 17). In order to explain the observed relative phases within a two-step process such as that implied above, it must be assumed[15] that the excited-state wave functions do not rotate with the internuclear line but stay fixed in the laboratory frame.

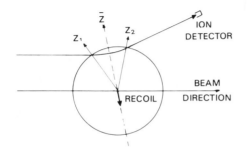

Figure 17. Trajectory diagram for the He$^+$ + He $\rightarrow$ He$(2p^2)$ ^1D + He$^+$ collision. The amplitudes of Table 1 support the assumption that an excitation of pure $m = 0$ sublevels occurs at orientations z_1 and z_2 of the internuclear line, which are symmetric to $\bar{z}$.

4.3. Other Collision Systems

4.3.1. H(2p) Formed by Charge-Exchange Collisions $H^+ + Ar \rightarrow$ $H(2p) + Ar^+$

The particle–photon, angular-correlation technique has also been applied to systems which exhibit a higher degree of complexity than the systems described above. Among these is the charge-exchange process $H^+ + Ar \rightarrow H(2p) + Ar^+$ at energies 0.7–2.5 keV,[31] for which it is not possible to express the alignment-tensor components simply by the three parameters $\mathcal{P}$, λ, and χ, referring to the H(2p) state. Assuming the residual Ar^+ to be dominantly populated in its ground state $Ar^+(^2P)$, the charge-exchange excitation process of H(2p) must be described by scattering amplitudes, depending on the magnetic substates of both H(2p) and $Ar^+(^2P)$. The alignment-tensor components can be expressed by means of five independent scattering amplitudes. They exhibit a dependence on the proton energy at fixed scattering angle, which so far has not been explained.

4.3.2. Ion- and Atom–Molecule Collisions

The experimental methods described in Section 3 have also been applied to gain more insight into the dynamics of triatomic collision systems such as He^+–H_2,[32] Li, Na–H_2, N_2.[33,34] Even though it is not possible to determine the final state of a collision system such as He^+–H_2 completely by a polarized-photon, scattered-particle coincidence measurement, it is still possible to gain new information on the triatomic collision process,

$$He^+ + H_2 \rightarrow He(3\ ^3P) + H_2^+ \rightarrow He(2\ ^3S) + H_2^+ + h\nu$$

from such a technique. The He^+ beam and the scattered $He(2\ ^3S)$ determine the scattering plane with the $He(3\ ^3P \rightarrow 2\ ^3S)$ photon being detected perpendicular to this plane. The degree of linear and circular polarization can be measured for specific scattering angles of $He(2\ ^3S)$. The polarization data exhibit no circular polarization, but at one scattering angle, a large linear polarization is seen. This alignment effect, which is even larger than seen in He^+–He,[6,30] is surprising in view of the many final states available for the recoiling H_2^+ ion. The high degree of alignment shows that referring to an axis of quantization parallel to the linear polarization, the excited atoms occupy almost exclusively the magnetic substates with $m_l = 0$.

Both linear and circular polarization have been observed in $Li(2p \rightarrow 2s)$ radiation generated in Li–H_2 collisions,[33] but the degree of polarization is low, as also seen for Na–N_2[34] collisions. The occurrence of circular polarization is difficult to account for. However, a complete analysis of the experimental results from these ion- and atom–molecule collisions awaits theoretical calculations, using reliable potential surfaces for the triatomic systems studied, such as the recently reported calculations for the $Na-N_2$ system.[35]

5. Future Aspects and Possible Applications of the Polarized-Photon, Scattered-Particle Coincidence Technique to Atomic Spectroscopy

5.1. Future Aspects

The application of coincidence techniques to the study of atomic collisions has been very successful during the last decade, and further expansion can be foreseen in the years to come. The addition of optical polarization analysis has opended up for more detailed investigations of selected collision systems as well as for critical tests of the validity of the theoretical models utilized to describe the binary collisions.

The experimental data available so far on ion–atom collisions relating directly to excitation amplitudes or to more general parameters reflecting the details of excited states formed in kinematically well specified collisions, touch on several different collision systems. In most cases, however, there are only very few systematic data available on the individual systems. The present theoretical understanding of the various collisions is based on model calculations and is only of fragmentary nature. For simple systems such as the quasi-one-electron systems, for which the gross features are well understood,[21] discrepancies exist, as described in Section 4, between theoretical predictions and experimental findings concerning the coherence parameters. No *ab initio* calculation capable of matching the details of the experimental data has yet been made.

It will be important in the future to concentrate the application of the various electron–photon, scattered-particle coincidence techniques to the study of relatively few ion–atom collision systems to obtain more systematic and complete information about a few representative collision systems. Examples would be the study of the rotational or the radial couplings taking place during slow collisions described within the MO model, or the direct coupling in fast collisions described within an atomic basis. It should be possible to select the collision systems such that essential features of these different mechanisms are isolated.

5.2. Possible Application of Polarized-Photon, Scattered-Particle Coincidence Technique to Atomic Spectroscopy

The polarized-photon, scattered-particle coincidence studies of systems such as the quasi-one-electron systems (Be^+, $Mg^+–He$, Ne, Ar)[19,24,27] have shown that the light emitted from the excited resonance P states can be highly linearly or circularly polarized ($\geq 50\%$). Collisional-induced alignment of an excited state, generated by collisions between fast beams and gases or foils, is commonly used to study lifetimes, fine or hyperfine structures by means of the zero-field (Hanle effect) and high-field level-crossing techniques, or by the quantum-beat technique. Collisional-induced orientation may also be used to study fine and hyperfine structures. Since states with $J = \frac{1}{2}$ can exhibit only orientation and not alignment, collisional-induced orientation may be of particular interest for hyperfine-structure studies of $^2P_{1/2}$ states. Oriented $^2P_{1/2}$ states can also be created by means of an asymmetric foil excitation (tilted foil) or by ion-beam, surface interaction. The many limitations associated with the tilted-foil technique (small orientation effects, short lifetimes of foils for heavier projectiles due to radiation damage, etc.) have so far prohibited the use of the tilted-foil technique for spectroscopic studies, whereas the ion-beam, surface-excitation technique is promising. For selected collision systems, the polarized-photon, scattered-particle coincidence technique may be a good alternative to the ion-beam, surface-excitation technique.

References and Notes

1. F. P. Ziemba and E. Everhart, *Phys. Rev. Lett.* **2**, 299 (1959).
2. D. C. Lorents and W. Aberth, *Phys. Rev. A* **139**, 1017 (1965). For a review, see M. Barat, *VIII Int. Conf. on Electronic and Atomic Collisions*, Beograd 1973, Invited Lectures and Progress Reports, B. Čobić and M. V. Kurepa, eds., p. 43 and references therein.
3. M. Barat, J. C. Brenot, and J. Pommier, *J. Phys. B* **6**, L105 (1973).
4. D. H. Jaecks, D. H. Crandall, and R. H. McKnight, *Phys. Rev. Lett.* **25**, 491 (1970). For a review, see D. H. Jaecks, *VIII Int. Conf. on Electronic and Atomic Collisions*, Beograd 1973, Invited Lectures and Progress Report, B. C. Čobić and M. V. Kurepa, eds., p. 137 and references therein.
5. M. Eminyan, K. B. MacAdam, J. Slevin, and H. Kleinpoppen, *Phys. Rev. Lett.* **31**, 576 (1973); *J. Phys. B* **7**, 1519 (1974).
6. W. de Rijk, F. J. Eriksen, and D. H. Jaecks, *Bull. Am. Phys. Soc.* **19**, 1230 (1974); D. H. Jaecks, F. J. Eriksen, W. de Rijk, and J. Macek, *Phys. Rev. Lett.* **35**, 723 (1975).
7. G. Vassilev, G. Rahmat, J. Slevin, and J. Baudon, *Phys. Rev. Lett.* **34**, 444 (1975).
8. I. V. Hertel and W. Stoll, *J. Phys. B* **7**, 570 (1974).
9. K. Blum and H. Kleinpoppen, *Phys. Rep.* **52**, 203 (1979).
10. *Coherence and Correlation in Atomic Collisions*, eds. H. Kleinpoppen and J. F. Williams, Plenum Press, New York (1980); *XI Int. Conf. on the Physics of Electronic and Atomic Collisions*, Kyoto 1979, Invited papers and progress reports, N. Oda and K. Takayanagi, eds., North-Holland, Amsterdam (1979), p. 55.
11. J. Macek and D. H. Jaecks, *Phys. Rev. A* **4**, 1288 (1971).
12. U. Fano and J. Macek, *Rev. Mod. Phys.* **45**, 553 (1973).
13. J. S. Briggs, J. H. Macek, and K. Taulbjerg, *J. Phys. B* **12**, 1457 (1979).
14. J. Eichler and W. Fritsch, *J. Phys. B* **9**, 1477 (1976).

15. Q. C. Kessel, R. Morgenstern, B. Müller, A. Niehaus, and U. Thielmann, *Phys. Rev. Lett.* **40**, 645 (1978); Q. C. Kessel, R. Morgenstern, B. Müller, and A. Niehaus, *Phys. Rev. A* **20**, 804 (1979).

16. I. V. Hertel and W. Stoll, *Adv. At. Mol. Phys.* **13**, 113 (1977); H. W. Hermann and I. V. Hertel, *Coherence and Correlation in Atomic Collisions*, eds. H. Kleinpoppen and J. F. Williams, Plenum Press, New York (1980), p. 625.

17. R. W. Wijnaendts van Resandt and J. Los, *XI Int. Conf. on the Physics of Electronic and Atomic Collisions*, Kyoto 1979, Invited papers and progress reports, N. Oda and K. Takayanagi, eds., North-Holland, Amsterdam (1979), p. 831; J. Hermann, B. Menner, E. Reisacher, L. Zehnle, and V. Kempter, *J. Phys. B* **13**, L165 (1980).

18. J. U. Andersen, L. Kocbach, E. Laegsgaard, M. Lund, and C. D. Moak, *J. Phys. B* **9**, 3247 (1976).

19. N. Andersen, T. Andersen, C. L. Cocke, and E. Horsdal-Pedersen, *J. Phys. B* **12**, 2541 (1979).

20. W. Wittman and H. J. Andrä, *Z. Phys. A* **288**, 335 (1978).

21. N. Andersen, *XI Int. Conf. on the Physics of Electronic and Atomic Collisions*, Kyoto 1979, Invited papers and progress reports, N. Oda and K. Takayanagi, eds., North-Holland, Amsterdam (1979), p. 301 and references therein.

22. S. E. Nielsen and J. S. Dahler, *J. Phys. B* **9**, 1383 (1976); *Phys. Rev. A* **16**, 563 (1977).

23. E. Horsdal-Pedersen *et al.*, *J. Phys. B* **11**, L317 (1978); C. Courbin-Gaussorgues, P. Wahnon, and M. Barat, *J. Phys. B* **12**, 3047 (1979).

24. N. Andersen, T. Andersen, J. Ø. Olsen, and E. Horsdal-Pedersen, *J. Phys. B* **13**, 2421 (1980); N. Andersen, T. Andersen, J. Ø. Olsen, E. Horsdal-Pedersen, S. E. Nielsen, and J. S. Dahler, *Phys. Rev. Lett.* **42**, 1134 (1979).

25. S. E. Nielsen and J. S. Dahler, *J. Phys. B* **13**, 2435 (1980).

26. L. Zehnle, E. Clemens, P. J. Martin, W. Schäuble, and V. Kempter, *J. Phys. B* **11**, 2133, 2865 (1978).

27. T. Andersen and E. Horsdal-Pedersen, *J. Phys. B* **14**, 3129 (1981).

28. A. Russek, D. B. Kimball, and M. J. Cavagnero, *Phys. Rev. A* **23**, 139 (1981).

29. G. Rahmat, G. Vassilev, and J. Baudon, *J. Phys. (Paris)* **41**, 319 (1980).

30. F. Eriksen, D. H. Jaecks, W. de Rijk, and J. Macek, *Phys. Rev. A* **14**, 119 (1976); D. H. Jaecks, F. Eriksen, and L. Fornari in, *Coherence and Correlations in Atomic Collisions*, eds. H. Kleinpoppen and J. F. Williams, Plenum Press, New York (1980), p. 473.

31. R. Hippler, H. Kleinpoppen, and H. O. Lutz, *XI Int. Conf. on the Physics of Electronic and Atomic Collisions*, Kyoto 1979, Invited papers and progress reports, N. Oda and K. Tagayanagi, eds., North-Holland, Amsterdam (1979), p. 611.

32. F. J. Eriksen and D. H. Jaecks, *Phys. Rev. Lett.* **36**, 1491 (1976).

33. K. H. Blattmann, B. Menner, W. Schäuble, B. Staudenmayer, L. Zehnle, and V. Kempter, *J. Phys. B* **13**, 3635 (1980).

34. I. V. Hertel, H. Hofmann, and K. A. Rost, *Phys. Rev. Lett.* **38**, 343 (1977).

35. P. Habitz, *Chem. Phys.* **54**, 131 (1980).

Index